ANNALS OF
THE NEW YORK ACADEMY
OF SCIENCES

Volume 923

EDITORIAL STAFF

Executive Editor
BARBARA M. GOLDMAN

Managing Editor
JUSTINE CULLINAN

Associate Editor
STEFAN MALMOLI

The New York Academy of Sciences
2 East 63rd Street
New York, New York 10021

THE NEW YORK ACADEMY OF SCIENCES
(Founded in 1817)

BOARD OF GOVERNORS, September 2000 – September 2001

BILL GREEN, *Chairman of the Board*
TORSTEN WIESEL, *Vice Chairman of the Board*
RODNEY W. NICHOLS, *President and CEO* [ex officio]

Honorary Life Governors
WILLIAM T. GOLDEN   JOSHUA LEDERBERG

JOHN T. MORGAN, *Treasurer*

Governors

ELEANOR BAUM	D. ALLAN BROMLEY	KAREN BURKE
LAWRENCE B. BUTTENWIESER		PRAVEEN CHAUDHARI
MICHAEL GOLDEN	JOHN H. GIBBONS	RONALD L. GRAHAM
ROBERT G. LAHITA	JACQUELINE LEO	WILLIAM J. McDONOUGH
JOHN F. NIBLACK	SANDRA PANEM	RICHARD RAVITCH
RICHARD A. RIFKIND	SARA LEE SCHUPF	JAMES H. SIMONS

HELENE L. KAPLAN, *Counsel* [ex officio]   PETER H. KOHN, *V.P. & Secretary* [ex officio]

THE UTEROGLOBIN/CLARA CELL PROTEIN FAMILY

ANNALS OF THE NEW YORK ACADEMY OF SCIENCES
Volume 923

THE UTEROGLOBIN/CLARA CELL PROTEIN FAMILY

Edited by Anil B. Mukherjee and Beverly S. Chilton

The New York Academy of Sciences
New York, New York
2000

Copyright © 2000 by the New York Academy of Sciences. All rights reserved. Under the provisions of the United States Copyright Act of 1976, individual readers of the Annals are permitted to make fair use of the material in them for teaching or research. Permission is granted to quote from the Annals provided that the customary acknowledgment is made of the source. Material in the Annals may be republished only by permission of the Academy. Address inquiries to the Permissions Department (editorial@nyas.org) at the New York Academy of Sciences.

Copying fees: *For each copy of an article made beyond the free copying permitted under Section 107 or 108 of the 1976 Copyright Act, a fee should be paid through the Copyright Clearance Center, Inc., 222 Rosewood Drive, Danvers, MA 01923 (www.copyright.com).*

♾ *The paper used in this publication meets the minimum requirements of the American National Standard for Information Sciences—Permanence of Paper for Printed Library Materials, ANSI Z39.48-1984.*

Library of Congress Cataloging-in-Publication Data

The uteroglobin/Clara cell protein family / editors, Anil B. Mukherjee, Beverly S. Chilton.
   p. cm. — (Annals of the New York Academy of Sciences ; v. 923)
   Includes bibliographical references and indexes.
   ISBN 1-57331-293-2 (cloth : alk. paper) — ISBN 1-57331-294-0 (pbk. : alk. paper)
   1. Uteroglobin—Congresses. I. Mukherjee, Anil B. II. Chilton, Beverly S. III. Series.

Q11.N5 vol. 923
[QP552.U86]
500 s—dc21
[572'.68]   00-051569

GYAT / B-M
Printed in the United States of America
ISBN 1-57331-293-2 (cloth)
ISBN 1-57331-294-0 (paper)
ISSN 0077-8923

ANNALS OF THE NEW YORK ACADEMY OF SCIENCES
Volume 923
December 2000

THE UTEROGLOBIN/CLARA CELL PROTEIN FAMILY

Editors
ANIL B. MUKHERJEE AND BEVERLY S. CHILTON

Conference Organizers
ANIL B. MUKHERJEE, BEVERLY S. CHILTON, FRANCESCO DEMAYO, AND GURMUKH SINGH

[This volume is the result of a conference entitled *Uteroglobin/Clara Cell 10kDa Family of Proteins* held by the New York Academy of Sciences, the National Institute of Child Health and Human Development (NIH), and the National Heart Lung and Blood Institute (NIH) on April 14–16, 2000, in Bethesda, Maryland.]

CONTENTS

Preface. *By* ANIL B. MUKHERJEE, BEVERLY S. CHILTON, GURMUKH SINGH, AND FRANCESCO DEMAYO ix

Keynote Addresses.
Blastokinin/Uteroglobin: Three Decades after Discovery

Discovery and Perspectives from the Blastokinin Era. *By* JOSEPH C. DANIEL .. 1

The Discovery of Uteroglobin and Its Significance for Reproductive Biology and Endocrinology. *By* HENNING M. BEIER 9

Part I. Uteroglobin/CC10 Family of Proteins

All Human Genes of the Uteroglobin Family Are Localized on Chromosome 11q12.2 and Form a Dense Cluster. *By* JIAN NI, MARTHA KALFF-SUSKE, REINER GENTZ, JEOFFREY SCHAGEMAN, MIGUEL BEATO, AND JÖRG KLUG 25

Clara Cell Proteins. *By* GURMUKH SINGH AND SIKANDAR L. KATYAL 43

Secretory Lipophilins: A Tale of Two Species. *By* ROBERT I. LEHRER, TUNG NGUYEN, CHENGQUAN ZHAO, CHEN XIAN HA, AND BEN J. GLASGOW .. 59

Clara Cell Secretory Protein (CC16): Features as a Peripheral Lung Biomarker. *By* F. BROECKAERT, A. CLIPPE, B. KNOOPS, C. HERMANS, AND A. BERNARD 68

Mammaglobin, a Breast-Specific Gene, and Its Utility as a Marker for Breast Cancer. *By* TIMOTHY P. FLEMING AND MARK A. WATSON 78

Part II. Structural Biology of Uteroglobin/CC10 Protein and UG-Derived Bioactive Peptides

The Uteroglobin Fold. *By* ISABELLE CALLEBAUT, ANNE POUPON, RÉNÉE BALLY, JEAN-PHILIPPE DEMARET, DOMINIQUE HOUSSET, JEAN DELETTRÉ, PAUL HOSSENLOPP, AND JEAN-PAUL MORNON 90

Crystal Structure Analysis of Recombinant Human Uteroglobin and Molecular Modeling of Ligand Binding. *By* N. PATTABIRAMAN, JOHN H. MATTHEWS, KEITH B. WARD, GIUDITTA MANTILE-SELVAGGI, LUCIO MIELE, AND ANIL B. MUKHERJEE 113

Antiflammins: Bioactive Peptides Derived from Uteroglobin. *By* LUCIO MIELE 128

Therapeutic Applications of Antiflammin Peptides in Experimental Ocular Inflammation. *By* CHI-CHAO CHAN, NADINE TUAILLON, QIAN LI, AND DE FEN SHEN 141

Antiflammin Peptides in the Regulation of Inflammatory Response. *By* JUAN J. MORENO 147

Part III. Regulation of the Uteroglobin Family of Genes by Hormones and Cytokines

Regulation of the Clara Cell Secretory Protein/Uteroglobin Promoter in Lung. *By* MAGNUS NORD, TOBIAS N. CASSEL, HARALD BRAUN, AND GUNTRAM SUSKE 154

Uteroglobin Gene Transcription: What's the RUSH? *By* BEVERLY S. CHILTON, AVELINE HEWETSON, JERRY DEVINE, ERICKA HENDRIX, AND MALINI MANSHARAMANI 166

Physiological Regulation of Uteroglobin/CCSP Expression. *By* ALBERT CHANG, PATRICIA RAMSAY, BIHONG ZHAO, MOON PARK, SUSAN MAGDALENO, MICHAEL J. REARDON, STEPHEN WELTY, AND FRANCESCO J. DEMAYO 181

Tumor Necrosis Factor Alpha Stimulation of Human Clara Cell Secretory Protein Production by Human Airway Epithelial Cells. *By* M. J. COWAN, X. HUANG, X. L. YAO, AND J. H. SHELHAMER 193

Part IV. Physiological Functions and Clinical Applications of Uteroglobin

Pulmonary Phenotype of CCSP/UG Deficient Mice: A Consequence of CCSP Deficiency or Altered Clara Cell Function? *By* BARRY R. STRIPP, SUSAN D. REYNOLDS, CHARLES G. PLOPPER, INGER-MARGRETHE BØE, AND JOHAN LUND 202

Insight into the Physiological Function(s) of Uteroglobin by Gene-Knockout and Antisense-Transgenic Approaches. *By* ZHONGJIAN ZHANG, GOPAL C. KUNDU, FENG ZHENG, C-J. YUAN, ERIC LEE, HEINER WESTPHAL, JERROLD WARD, FRANCESCO DEMAYO, AND ANIL B. MUKHERJEE 210

Uteroglobin Binding Proteins: Regulation of Cellular Motility and Invasion in Normal and Cancer Cells. *By* GOPAL C. KUNDU, ZHONGJIAN ZHANG, GIUDITTA MANTILE-SELVAGGI, ASIM MANDAL, CHIUN-JYE YUAN, AND ANIL B. MUKHERJEE 234

The Role of CC10 in Pulmonary Carcinogenesis: From a Marker to Tumor Suppression. *By* R. ILONA LINNOILA, EVA SZABO, FRANCESCO DEMAYO, HANSPETER WITSCHI, CAROL SABOURIN, AND AL MALKINSON   249

Development of an Enzyme-Linked Immunosorbent Assay for Clara Cell 10-kDa Protein: In Pursuit of Clinical Significance of Sera in Patients with Asthma and Sarcoidosis. *By* NORIHARU SHIJUBO, YOSHIHISA ITOH, TETSUJI YAMAGUCHI, AND SHOSAKU ABE   268

Rationale for the Development of Recombinant Human CC10 as a Therapeutic for Inflammatory and Fibrotic Disease. *By* APRILE L. PILON   280

Poster Papers

C/EBPα and TTF-1 Synergistically Transactivate the Clara Cell Secretory Protein Gene. *By* TOBIAS N. CASSEL, GUNTRAM SUSKE, AND MAGNUS NORD   300

Human Uteroglobin Gene Polymorphisms and Genetic Susceptibility to Asthma. *By* MOONSUK CHOI, ZHONGJIAN ZHANG, LEO P. TEN KATE, J. MARGRIET COLLEE, J. GERRITSEN, AND ANIL B. MUKHERJEE   303

Amino Acid Residues in α-Helix-3 of Human Uteroglobin Are Critical for Its Phospholipase A_2 Inhibitory Activity. *By* BHABADEB CHOWDHURY, GIUDITTA MANTILE-SELVAGGI, GOPAL C. KUNDU, LUCIO MIELE, ELEONORA CORDELLA-MIELE, ZHONGJIAN ZHANG, AND ANIL B. MUKHERJEE   307

Mammaglobin Complexes with BU101 in Breast Tissue. *By* T. L. COLPITTS, P. BILLING, E. GRANADOS, S. HODGES, N. MENHART, J. RUSSELL, AND S. STROUPE   312

Uteroglobin *In Situ* Hybridization: Novel Monitoring of Epithelial Differentiation in the Rabbit Endometrium. *By* C. A. KRUSCHE, A. HERRLER, AND H. M. BEIER   316

Prolactin Augments Progesterone-Dependent Expression of a Nuclear P-Type ATPase that Associates with the RING Domain of RUSH Transcription Factors in the Endometrium. *By* MALINI MANSHARAMANI AND BEVERLY S. CHILTON   321

A Novel *In Situ* Method of SV40 Transfection for the Establishment of Immortal Pulmonary Alveolar Type II Cell Lines. *By* KANAKO MOMEDA, ZHONGJIAN ZHANG, ANIL B. MUKHERJEE, AND R. DHANIREDDY   325

Uteroglobin Expression and Release in the Human Endometrium. *By* F. MÜLLER-SCHÖTTLE, I. CLASSEN-LINKE, K. BEIER-HELLWIG, K. STERZIK, AND H. M. BEIER   332

Expression of Inflammatory Cytokines in a Mouse Transformed Clara Cell Line by Tumor Necrosis Factor-α. *By* MOON S. PARK, BIHONG ZHAO, PATRICIA L. RAMSAY, ALBERT S. Y. CHANG, MICHAEL J. REARDON, AND FRANCESCO J. DEMAYO   336

Binding of rhCC10 to Fibronectin and Its Effect on Cellular Adhesion. *By* JEFFREY FARROW, JAMES MELBY, LAURA WIESE, JERRY LOHNAS, RICHARD WELCH, AND APRILE L. PILON   338

Mammaglobin as a Marker for the Detection of Tumor Cells in the Peripheral Blood of Breast Cancer Patients. *By* OTTO ZACH, HEDWIG KASPARU, HELGA WAGNER, OTTO KRIEGER, AND DIETER LUTZ   343

In Vivo and *In Vitro* Analysis of Hyperoxia-Induced Gene Expression in Mouse Lung and Mouse Transformed Clara Cells. *By* BIHONG ZHAO, PATRICIA L. RAMSAY, MOON S. PARK, STEPHEN E. WELTY, AND FRANCESCO J. DEMAYO 346

Nomenclature Session

Uteroglobin/Clara Cell 10-kDa Family of Proteins: Nomenclature Committee Report. *By* J. KLUG, H. M. BEIER, A. BERNARD, B. S. CHILTON, T. P. FLEMING, R. I. LEHRER, L. MIELE, N. PATTABIRAMAN, AND G. SINGH 348

Closing Remarks and Future Directions. *By* DOROTHY BERLIN GAIL 355

Index of Contributors 357

Financial assistance was received from:

Co-sponsors
- NATIONAL INSTITUTE OF CHILD HEALTH AND HUMAN DEVELOPMENT AND NATIONAL HEART LUNG AND BLOOD INSTITUTE, NIH

Supporter
- CLARAGEN, INCORPORATED

Contributor
- ASTRAZENECA R&D LUND

The New York Academy of Sciences believes it has a responsibility to provide an open forum for discussion of scientific questions. The positions taken by the participants in the reported conferences are their own and not necessarily those of the Academy. The Academy has no intent to influence legislation by providing such forums.

Preface

Research on the uteroglobin/Clara cell protein family continues to generate tremendous interest among researchers in diverse fields of biomedical sciences. It is surprising that even though the first publication on blastokinin and uteroglobin appeared in 1967 and 1968, respectively, to our knowledge no major conference was held on this subject until now. In more than three decades of research, uteroglobin has become one of the most thoroughly studied proteins from the standpoint of its structure, molecular biology, and biochemistry. What emerged as a result of three decades of work is the realization that uteroglobin is one member of a larger group of proteins with structural similarities that belong to what we now know as the uteroglobin/Clara cell protein family. Compelling evidence suggests that the genes encoding all members of this family of proteins, studied so far, are clustered at a single locus in human chromosome 11 (11q12-13). Because of these discoveries and the existence of numerous names currently in use to describe this family of proteins, the organizers decided to hold a nomenclature conference during this symposium. A Nomenclature Committee was formed and a report from this committee appears in the last section of this symposium volume.

This volume provides a permanent record of the historic symposium on the uteroglobin/Clara cell family of proteins. The symposium was held on April 14–16, 2000, at the Hyatt Regency Hotel, Bethesda, Maryland. This symposium was cosponsored by the New York Academy of Sciences, the National Institute of Child Health and Human Development (NICHD), and the National Heart, Lung, and Blood Institute (NHLBI) [the last two from the National Institutes of Health (NIH)]. Nearly four years ago, two of us (B. Chilton and A. Mukherjee) discussed the possibility of organizing a conference on uteroglobin and, for one reason or another, it failed to materialize. However, in the advent of a new millennium, everything came together. For this, we are especially grateful to Igor Dawid (Acting Scientific Director, NICHD), Dorothy Gail (Director of the Program on Lung Biology and Diseases, NHLBI), and Rashid Shaikh (Director of Science and Technology Meetings, New York Academy of Sciences) for their generous and enthusiastic support for holding this symposium. We are also very grateful to Sherryl Usmani (Meetings Coordinator of the New York Academy of Sciences) for her superb planning and organization skills that brought this conference to fruition.

Because of these kinds of support, we were able to invite 26 distinguished scientists from around the globe to present their work at this conference. We regret that we could not invite many more of our colleagues because, in a two-day conference, we could accommodate only so many presentations.

The purpose of this symposium was twofold: (i) to foster the exchange of ideas among scientists so that our knowledge base in this area of research is broadened and (ii) to facilitate discussions among participants regarding the future course of uteroglobin/Clara cell protein research. A third component was deemed essential, namely the consideration of a unified name for this family of proteins. We believe this symposium achieved most, if not all, of these objectives.

We began the conference with two keynote addresses, instead of one. This is because uteroglobin was discovered simultaneously by two distinguished scientists on two different continents: Joseph Daniel, Jr., in the United States and Henning Beier in Germany. We were very fortunate to have both of them at this symposium and to hear about their discoveries in their own words. A highlight of this symposium was when a plaque was presented to each of these scientists to recognize their seminal contributions to this area of research.

The remaining presentations in this symposium were divided into four sessions.

In session I, we have included presentations that described the proteins that make up the uteroglobin/Clara cell family. Between session I and II, we devoted an hour specially designated as the Nomenclature Conference. During this hour, we discussed various possibilities for a unified name for the uteroglobin/Clara cell family of proteins, but soon it was realized that there was not enough time to discuss all options and arrive at a consensus on this issue. Thus, a Nomenclature Committee consisting of nine members was formed. This committee was charged with finding a unified name for this family of proteins. We are happy to report that the committee has agreed on a unified family name (secretoglobin) for these proteins and a preliminary report on their deliberations appears in the last section of this volume.

In session II, we heard presentations that dealt with the structural aspects of this family of proteins and about the bioactive peptides derived from uteroglobin. The session was concluded with poster presentations and a reception.

In session III, we discussed uteroglobin/Clara cell protein gene regulation and, in session IV, we heard presentations on the physiological functions and possible clinical implications for this family of proteins.

The organizers would like to thank all the speakers, poster presenters, and participants for making this symposium a success.

The organizers are grateful to the New York Academy of Sciences for sponsoring the symposium and for publishing the presentations in the *Annals* series. We are especially grateful to Barbara Goldman, Executive Editor, for her approval for publication of this symposium volume and to Sheila Kane, Administrative Editor, for her persistence in obtaining the manuscripts from the presenters and superb organizational skills. We are also indebted to Stefan Malmoli, Associate Editor, for his fine editorial skills. Together, these individuals made the publication of this volume possible.

Last, but not the least, the organizers are grateful to Aprile Pilon and Mark Zimmer of Claragen, Inc., and Johan Lund and AstraZeneca R&D for their financial support for this symposium.

ANIL B. MUKHERJEE
BEVERLY S. CHILTON
GURMUKH SINGH
FRANCESCO DEMAYO

Discovery and Perspectives from the Blastokinin Era

JOSEPH C. DANIEL[a]

Department of Biological Sciences, Old Dominion University, Norfolk, Virginia 23529, USA

ABSTRACT: The events and rationalizations that led to one discovery of the rabbit uterine protein, then called blastokinin, are narrated in historical perspective and related to the independent discovery of the same protein, named uteroglobin. The period when the name "blastokinin" remained in partial use, roughly from the original publication in 1967 until the early 1980s, is considered here as the blastokinin era. Subsequent perspectives, originating from the era, are presented on the hormonal regulation of blastokinin, as well as its distribution, biological function, and potential relationships with other entities that exist coincidentally.

The discovery of blastokinin found its origin in two scientific quests that occupied the author in the early 1960s. One was focused on the problem of delayed implantation and the other on developing the technology for growing mammalian embryos *in vitro*. It quickly became apparent that both initiatives were addressing the same question: namely, what governed mammalian embryogenesis beyond blastulation? In the case of delayed implantation, the embryo enters a state of dormancy at the blastocyst stage, from which it does not emerge until its enveloping, heretofore quiescent uterus returns to a condition of active response. Similarly, many kinds of mammalian embryos that could proceed through cleavage, morulation, and early blastulation *in vitro* would not develop beyond the blastocyst stage.

From two possible causes for growth retardation of blastocysts, namely, the presence of inhibitors and the absence of promoters, the latter was chosen for further investigation. Endometrial inhibitors, however, could account for blastocyst dormancy *in utero*.[1]

The ensuing experiment sought to analyze the protein composition of the uterine fluids of a variety of mammals exhibiting delayed implantation in comparison with the rabbit, as a representative mammal that does not delay. For the comparison, fluids were collected from ranch mink, northern fur seal, black bear, nine-banded armadillo, roe deer, and ovariectomized rats during their dormant blastocyst periods and from rabbits at estrus and on days 3 through 9 *postcoitum* (*pc*). Quantitatively, the protein content of rabbit uterine fluids collected on day 5 *pc* was 10 to 100 times greater than that of the other animals. Qualitatively, when analyzed by polyacryl-

---

[a]Voice: 757-451-2920; fax: 757-683-5283.
mhd2jcd@aol.com

amide disk gel electrophoresis, a prominent postalbumin band appeared in the 4- to 9-day rabbit samples that was absent in estrous rabbit fluids and all the other samples.[2] This held promise of being a critical element for postblastocyst development and quickly became the subject of extensive investigation.

Initially, two tests were conducted to determine the effect of this special protein on embryonic growth *in vitro*—one with rabbit morulae and the other with dormant blastocysts of mink, fur seal, armadillo, and rat. In each case, the embryos were cultured in medium supplemented with the protein from rabbit uterine fluids, isolated by Sephadex gel filtration, and their growth *in vitro* measured over 24 hours.

Rabbit morulae, cultured in the protein-supplemented medium, underwent cavitation and blastocyst growth, for 1 day, significantly better than controls cultured in serum-supplemented medium. On this basis, the name "blastokinin" was conferred on the protein—the prefix to identify the embryonic stage concerned and kinin, taken from the Greek "kinein", meaning "to stimulate".[3] In the same study, the sequential occurrence of blastokinin in rabbit uterine fluids during early pregnancy was confirmed (i.e., appearance on day 3 *pc*, peak of concentration reached on day 5–6, with reduction thereafter through day 9), as was its presence in blastocoelic fluid and uterine fluids from pseudopregnant rabbits. Its absence or only trace presence in fluids from estrous animals, maternal serum, fetal serum, and amniotic fluid was also established. Some of its chemical characteristics were reported in a following study.[4]

Similarly, in the second test, the blastocysts of mink, fur seal, and armadillo showed measurable growth and improved mitotic index in culture medium supplemented with blastokinin.[5]

Coincidental with the blastokinin discovery, Henning Beier[6] reported his discovery of a rabbit uterine protein that he named "uteroglobin". Beier noted its immunoelectrophoretic similarity with β-U-globulin, a protein found by Schwick[7] from the rabbit uterus. Beier also reported the presence of his protein in pseudopregnant rabbits and in blastocoelic fluid and heightened concentrations when does were given specific timed injections of estradiol followed by progesterone. On the basis of their physicochemical characteristics and patterns of occurrence, blastokinin and uteroglobin were suspected to be identical proteins, a suspicion to be repeatedly confirmed thereafter by many structural, immunological, functional, and chronological studies.

There is reason to suspect that this protein may have been found even earlier. In the *Carnegie Institution of Washington Year Book* vol. 59, which was published in 1960 and included accounts of the scientific work done there between 1 July 1959 and 30 June 1960, there is a report from Abraham Kulangara about his studies of the thin fluid film that adheres to the mucosal surface of the rabbit uterus.[8] He records a "protein-polysaccharide complex that appears to be secreted by the uterus during the fourth to sixth days of gestation". Paper electrophoresis of washes from these uteri "showed two bands of protein with mobilities of fast γ- and between β1- and β2-globulins". In his report, Kulangara did not name these proteins.

These separate discoveries of the same substance caused the early literature to be burdened with two different names. "Blastokinin" persisted in the publications coming from this author's laboratory and from some other investigators for the latter years of the 1960s and through the following decade. On this basis, the "blastokinin era" is definable as being from discovery to 1980, or shortly thereafter.

For those investigators using the name "uteroglobin", there was no blastokinin era. Their numbers were greater and they had preferred and used "uteroglobin" from the time of its discovery. Both names were acknowledged for a while by the use of dual abbreviations—to wit, "BK/UG"—but gradually "uteroglobin" gained more favorable acceptance and so has now been in exclusive use for almost 20 years. For this report on the blastokinin era, the dual abbreviation is used.

The independent discoveries of blastokinin and uteroglobin (BK/UG) generated both excitement and skepticism and triggered an immediate rush to add new information about the protein, to confirm or to contest earlier findings, and even to explore and project its commercial potentials. Beier's initial demonstration[6] of the effect of ovarian steroid administration to pregnant does, followed by the report from Urzua et al.[9] showing that progesterone alone could induce BK/UG synthesis in the uteri of ovariectomized rabbits and that BK/UG binds progesterone, and that of Arthur and Daniel[10] showing that the uteri of such rabbits could support preimplantation embryonic growth and development, were especially effective in attracting interest. Investigators quickly saw the potential of the system for studying maternal-embryonic interchange and hormone–target tissue interaction, with the advantage of having a specific gene product resulting from that interaction. Extensive research was conducted by numerous laboratories throughout this period, defined here as the blastokinin era, on the steroid induction of BK/UG, on its chemical structure, and on questions of tissue and species specificity, molecular genetics, biochemistry, function, and embryo utilization.

To expect a review of all the work from the blastokinin era that emanated from these primary discoveries would be to misinterpret the intent of this paper and would be a redundant exercise at best. It has already been done, adequately and periodically to cover all aspects of the subject. The reader is recommended to references 11–16, all of the chapters in Beier and Karlson,[17] and ultimately the topical reviews that will proceed from this conference.

Discovery almost always requires perspective—the word defined as the capacity to view things in their true relations or relative importance. It implies the use of prophetic anticipation based on knowledge of the past, a type of hypothesis. As is true for every human endeavor, each success and each failure in the history of BK/UG was preceded by a perspective, sometimes correct, sometimes flawed, but all fundamental to the proper ordering of a research initiative. Literally, hundreds of perspectives marked the blastokinin era and, of course, continue to the present. Two have been selected here, not from a scale of importance, although they are, but rather as reflecting the author's involvement and highlighting the importance of recognizing relationships.

One of the most important advances in understanding the hormonal regulation of uterine secretory activity came with the perspective on the role of prolactin. Two lines of evidence contributed to this recognition. The first involved "castration atrophy", the condition where the tissues of the uterus are inactivated, shrink, and cease being secretory after ovariectomy. For 2 to 4 weeks thereafter (the short-term ovariectomy period typically used in most experiments), the rabbit uterus remains responsive to ovarian steroids, but with progressively longer periods, postcastration, the responsivity declines. It has reduced significantly after 6 weeks, is only detectable after 9, and has stopped by 12 weeks after ovariectomy.[18] With brain extract added to the steroid administration protocol, Daniel[18] showed some recovery of uterine response in long-term ovariectomized does and hypothesized the need for an "endometriotrophic factor" in the uterine-hormonal interplay.

The second line of evidence came again from studies on delayed implantation. In this case, the uterus is rendered nonproliferative and nonsecretory because of luteal quiescence so that its contained embryos become dormant for lack of a supportive, growth-promoting environment. Because the corpus luteum in these animals does not synthesize progesterone during this time and because the refractory uterus of short-term ovariectomized rabbits could be reactivated by exogenous progesterone, it was expected that progesterone administration to animals with obligative delayed implantation would shorten or eliminate the dormancy period. When tested with mink[19] and the northern fur seal,[20] that expectation did not materialize and the literature contains additional reports of similar failures with other species having a delayed implantation. These failures led to the general acceptance of the need for some additional factor(s). This perspective was specifically proposed by Daniel:[20] in his review of delayed implantation in the pinnipeds where, drawing from several lines of evidence, he again inferred the existence of an "endometriotrophic factor", separate from any of the ovarian steroids, but "which, acting synergistically with estrogen and progesterone, is needed for the active proliferation and secretion in the endometrium that is essential for dynamic embryogenesis."

The speculation about an additional essential "factor" was substantiated with the report of Murphy *et al.*[21] demonstrating that prolactin, administered to hypophysectomized mink, caused reactivation of the corpus luteum and, in turn, terminated the delay that precedes implantation in this species. Although its role in the mink was understood as being luteotrophic, prolactin nevertheless presented potential of being the heretofore evasive "endometriotrophic factor".

Although the perspective was conceived during the blastokinin era, studies on the role of prolactin as an endometriotrophic factor extended beyond that time and so exceed the scope of this paper. However, to complete the narration, in summary, it was shown that, synergized with progesterone in the treatment given to long-term ovariectomized rabbits, prolactin restored the uterine responses. Uteroglobin production, DNA synthesis, concentrations of estrogen and progesterone receptors, endometrial proliferation, and gland formation were reestablished comparable to those in 5-day pregnant rabbits.[22,23] Growth hormone had no effect.[24] Seasonal, animal strain, and dose variables were found in uteroglobin production, but in every case the responses to the prolactin-progesterone combination always exceeded those for progesterone alone and were absent for prolactin alone.[25] These variables did not influence steroid receptor concentrations, which were consistently and significantly increased by prolactin alone.[23] After Chilton *et al.*[26] hypothesized a servomechanism between these hormones for regulating uterine gene expression, Daniel and Juneja[27] provided confirming evidence by amplifying uteroglobin production through a series of alternating injections of prolactin and progesterone. Similarly, when endogenous secretion of prolactin was stimulated by perphenazine administration to progesterone-treated, ovariectomized does, BK/UG production increased two- to threefold over controls.[28] Conversely, when the prolactin inhibitor, bromocriptine, was administered, no effect was observed. Collectively, this mutual dependency of a steroid with a polypeptide hormone has added new perspective to the model for BK/UG synthesis. The possibility that prolactin interacts in a similar way with other inducers in other tissues merits further investigation.

Another perspective coming from the blastokinin era might be considered as one of coincidence. Believing that consistent natural associations are functionally significant and mutually dependent and that molecules might be defined by their associations, the author feels that the coincidental occurrence of other entities with BK/UG merits consideration.

This perspective was introduced by Daniel,[29] who displayed the chronological occurrence and uterine location of five other rabbit proteins in relation to blastokinin. The progression of prealbumin,[6,30] nonspecific esterases,[6,31] and gallium-binding protein,[32] in uterine fluids, parallels the occurrence of blastokinin, even to concentrating in implantation sites from days 6 through 9 pc.[30] Reverse transcriptase in the uterine lumen[33,34] also shows a gradual increase in activity through the first 3 or 4 days of pregnancy, but thereafter is less active at implantation sites than at nonsites. Crystalloid inclusion bodies (CIBs) (at least partially proteinaceous) appear in the rabbit endometrium on day 4, peak in concentration by day 7, and remain elevated through at least day 9, but with no demonstrable site preference.[35,36]

Calarco and Szollosi[37] noted similarities of crystalloid aggregates in mouse embryos to those in tissues infected with viruses. Their observation finds support in the fact that viruslike particles (VLPs) are regular components in cells of the same embryonic stages and progestational endometrium of the same species as the crystalloid inclusion bodies.[38]

For the rabbit, VLPs are found in blastocysts, but not cleavage-stage embryos,[39] in uterine flushings and luminal epithelium taken from does on days 4 to 6 pc, but not from days 1 or 2 or from estrous animals.[38] RNA-dependent DNA polymerase activity in uterine tissue parallels the occurrence of the VLPs through day 10, but is no longer detectable by the 25th day of pregnancy.[40] Any pattern of interaction between CIBs, VLPs, and polymerase remains obscure, but the situation has the elements for a system of communication between mother and embryo.

In the high-molecular-weight fractions (MW $\geq$ 100,000) of rabbit uterine fluids, Dunbar and Daniel[41] identified four antigens that increase from detectable at estrus to a proportion of as much as 30% of the total proteins by days 3 to 8 pc. Two of these were of uterine origin. Beier[13] listed nine different proteins in the endometrial secretion, two of which he classified as large molecules (i.e., 10^5–10^6).

Kirchner[42] illustrated a comparison of the relative and temporal occurrence of uteroglobin with nonuteroglobin proteins from the rabbit uterus through the first 8 days of pregnancy. In addition to prealbumin, which was portrayed similarly to that shown by Daniel,[29] he added albumin, postalbumin, transferrin, and a β-glycoprotein.

FIGURE 1 illustrates a composite representation of all of the components noted above relative to the ways they change during early pregnancy. These plus other reports of enzymatic activity and qualitative/quantitative changes in the seral proteins found in the uterus, carbohydrates, metabolites, and inorganic substances, all occurring in the rabbit coincident with BK/UG, plus similar findings in other species, provide ample evidence that the uterine milieu that presents to the mammalian blastocyst is a dynamic and highly specialized environment. Logic dictates that such an environment must be essential to support implantation (placentation?), embryonic growth, and the initiation of differentiation and/or to block immunological rejec-

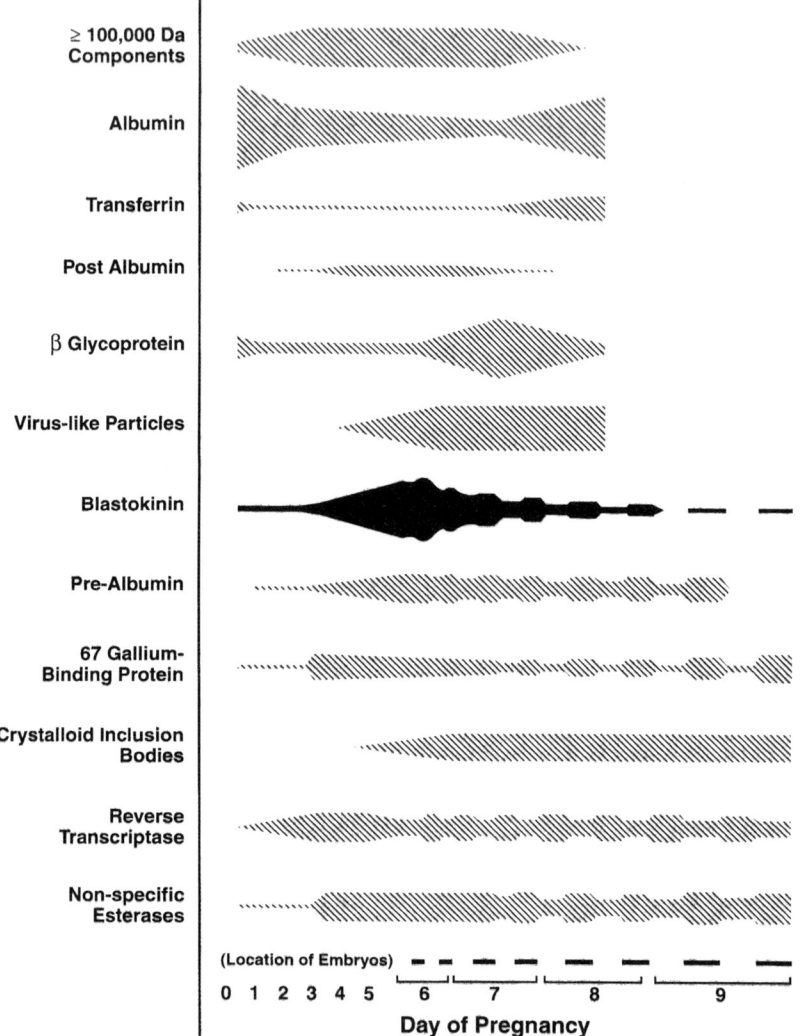

FIGURE 1. Temporal occurrence and relative concentration of rabbit uterine components prior to and during implantation.

tion. In its absence, embryogenesis stalls, implantation delays, and (in the extreme) reproduction fails.

The level of comprehension achieved by the blastokinin and uteroglobin research of the blastokinin era has been matched manyfold since that time. It did not, however, escape the attention of the early investigators that, although found originally in the uterine luminal fluids of the rabbit in a chronological pattern that implied a function in mammalian reproduction, BK/UG might have a broader distribution, both phylogenetically and organismally, and more varied physiological involvements. Subsequently, this perspective proved to be true as later studies established it in a number of secretory tissues, in a variety of mammals and other vertebrates, and having biological roles that included enzymatic, inhibitory, bonding, and immunoregulatory properties. Its discovery in the human was especially significant in stimulating research into its association with pathologies of secretory tissues and its applicability in contraception and fertility management. The depth of understanding extant today about all aspects of uteroglobin, in all of its manifestations, probably matches or exceeds that of most hormone-inducible, multifunctional proteins. It is the goal of this conference to expose the currency of that knowledge and to engage it as foundational for the challenges of the future.

REFERENCES

1. WEITLAUF, H.M. 1978. Factors in mouse uterine fluid that inhibit the incorporation of ^{3}H uridine by blastocysts *in vitro*. J. Reprod. Fertil. **52:** 321–325.
2. DANIEL, J.C. 1967. Comparison of electrophoretic patterns of uterine fluids from rabbits and mammals having delayed implantation. Comp. Biochem. Physiol. **24:** 297–300.
3. KRISHNAN, R.S. & J.C. DANIEL. 1967. "Blastokinin": inducer and regulator of blastocyst development in the rabbit uterus. Science **158:** 490–492.
4. KRISHNAN, R.S. & J.C. DANIEL. 1968. Composition of "blastokinin" from rabbit uterus. Biochim. Biophys. Acta **168:** 579–582.
5. DANIEL, J.C. & R.S. KRISHNAN. 1969. Studies on the relationship between uterine fluid components and the diapausing state of blastocysts from mammals having delayed implantation. J. Exp. Zool. **172:** 267–282.
6. BEIER, H.M. 1968. Uteroglobin: a hormone-sensitive endometrial protein involved in blastocyst development. Biochim. Biophys. Acta **160:** 289–291.
7. SCHWICK, H.G. 1965. Chemisch-entwicklungsphysiologische Beziehungen von Uterus zu Blastocyste des Kaninchens (*Oryctolagus cuniculus*). Wilhelm Roux' Arch. Entwicklungsmech. Org. **156:** 283–343.
8. KULANGARA, A.C. 1960. The macromolecular environment of the rabbit embryo before and during implantation. Carnegie Inst. Wash. Year Book **59:** 361–362.
9. URZUA, M.A., R. STAMBAUGH, G. FLICKINGER & L. MASTROIANNI. 1970. Uterine and oviduct fluid protein patterns in the rabbit before and after ovulation. Fertil. Steril. **21:** 860–865.
10. ARTHUR, A.J. & J.C. DANIEL. 1972. Progesterone regulation of blastokinin production and maintenance of rabbit blastocysts transferred into uteri of castrate recipients. Fertil. Steril. **23:** 115–122.
11. DANIEL, J.C. 1971. Uterine proteins and embryonic development. Adv. Biosci. **6:** 191–206.
12. GULYAS, B.J. & R.S. KRISHNAN. 1971. Current status of the chemistry and biology of "blastokinin". *In* The Biology of the Blastocyst, pp. 261–275. University of Chicago Press. Chicago.
13. BEIER, H.M. 1974. Oviducal and uterine fluids. J. Reprod. Fertil. **37:** 221–237.
14. DANIEL, J.C. 1976. Blastokinin and analogous proteins. J. Reprod. Fertil. Suppl. **25:** 71–83.
15. BEIER, H.M. & U. MOOTZ. 1979. Significance of maternal uterine proteins in the establishment of pregnancy. *In* Maternal Recognition of Pregnancy. Ciba Found. Symp. Volume 64-NS, pp. 111–140. Excerpta Medica. Amsterdam.
16. MIELE, L., E. CORDELLA-MIELE & A.B. MUKHERJEE. 1987. Uteroglobin: structure, molecular biology, and new perspectives on its function as a phospholipase A$_2$ inhibitor. Endocr. Rev. **8:** 474–490.
17. BEIER, H.M. & P. KARLSON. 1982. Proteins and Steroids in Early Pregnancy. Springer-Verlag. Berlin/New York.

18. DANIEL, J.C. 1980. Factors influencing uteroglobin synthesis. *In* Steroid Induced Uterine Proteins, pp. 87–103. Elsevier/North-Holland. Amsterdam/New York.
19. DANIEL, J.C. 1971. Termination of pregnancy in mink by repeated injections during the period preceding implantation. J. Anim. Sci. **33:** 659–661.
20. DANIEL, J.C. 1981. Delayed implantation in the northern fur seal (*Callorhinus ursinus*) and other pinnipeds. J. Reprod. Fertil. Suppl. **29:** 35–50.
21. MURPHY, B.D., P.W. CONCANNON, H.F. TRAVIS & W. HANSEL. 1981. Prolactin: the hypophyseal factor that terminates embryonic diapause in mink. Biol. Reprod. **25:** 487–491.
22. DANIEL, J.C., A.E. JETTON & B.S. CHILTON. 1984. Prolactin as an essential factor in the uterine response to progesterone in rabbits. J. Reprod. Fertil. **72:** 443–452.
23. CHILTON, B.S. & J.C. DANIEL. 1985. Influence of prolactin on DNA synthesis and glandular differentiation in rabbit uterine endometrium. *In* Prolactin: Basic and Clinical Correlates. FIDIA Research Series. Volume 1, pp. 351–359.
24. CHILTON, B.S., J.C. DANIEL & P.B. LONERGAN. 1986. Prolactin (PRL) and growth hormone (GH): differential response of rabbit uterine endometrium (abstract). Biol. Reprod. **34**(suppl. 1): 175a.
25. DANIEL, J.C., S.P. TAYLOR, S.C. JUNEJA *et al.* 1988. Variability in the response of the rabbit uterus to progesterone as influenced by prolactin. J. Reprod. Fertil. **84:** 13–21.
26. CHILTON, B.S., S.K. MANI & D.W. BULLOCK. 1988. Servomechanism of prolactin and progesterone in regulating uterine gene expression. Mol. Endocrinol. **2:** 1169–1175.
27. DANIEL, J.C. & S.C. JUNEJA. 1998. Amplification of uteroglobin secretion by alternating prolactin-progesterone administration. J. Endocrinol. **122:** R5–R6.
28. DANIEL, J.C., S.C. JUNEJA & B.S. CHILTON. Unpublished.
29. DANIEL, J.C. 1976. Temporal relationships between six uterine proteins associated with early pregnancy in the rabbit. *In* Protides of the Biological Fluids. Proc. XXIV Annual Colloquium, pp. 133–137.
30. DANIEL, J.C. 1972. Local production of protein during implantation in the rabbit. J. Reprod. Fertil. **31:** 303–306.
31. TYNDALL, R.L. & J.C. DANIEL. 1975. Alterations in uterine and serum esterases in pregnant mammals. Fertil. Steril. **86:** 1098–1104.
32. TYNDALL, R.L., S.J. CHASKES, J.E. CARLTON *et al.* 1976. Gallium–67 distribution in pregnant mammals. J. Exp. Zool. **195:** 417–424.
33. YANG, W.K., R.L. TYNDALL & J.C. DANIEL. 1976. DNA- and RNA-dependent DNA polymerases: progressive changes in rabbit endometrium during the preimplantation stage of pregnancy. Biol. Reprod. **15:** 604–613.
34. CHILTON, B.S. & J.C. DANIEL. 1978. Rabbit endometrial RNA- and DNA-dependent polymerase activity. Biol. Reprod. **18:** 371–378.
35. VAN BLERKOM, J., C. MANES & J.C. DANIEL. 1973. Development of preimplantation rabbit embryos *in vivo* and *in vitro*. I. An ultrastructural comparison. Dev. Biol. **35:** 262–282.
36. DANIEL, J.C. & J.R. KENNEDY. 1978. Crystalline inclusion bodies in rabbit embryos. J. Embryol. Exp. Morphol. **44:** 31–43.
37. CALARCO, P.G. & D. SZOLLOSI. 1973. Intercisternal A particles in ova and preimplantation stages of the mouse. Nat. New Biol. **243:** 91–93.
38. DANIEL, J.C. & B.S. CHILTON. 1978. Virus-like particles in embryos and female reproductive tract. *In* Development in Mammals. Volume 3, pp. 131–187. Elsevier/North-Holland. Amsterdam/New York.
39. MANES, C. 1974. Phasing of gene products during development. Cancer Res. **34:** 2044–2052.
40. BEGIDIAN, H.G., R.R. FOX & N. MEIER. 1976. Evidence for a particle-associated RNA-directed DNA polymerase in rabbit placental and uterine tissues. Cancer Res. **36:** 4687–4692.
41. DUNBAR, B.S. & J.C. DANIEL. 1979. High molecular weight components of rabbit uterine fluids. Biol. Reprod. **21:** 723–733.
42. KIRCHNER, C. 1980. Non-uteroglobin proteins in the rabbit. *In* Steroid Induced Uterine Proteins, pp. 69–86. Elsevier/North-Holland. Amsterdam/New York.

The Discovery of Uteroglobin and Its Significance for Reproductive Biology and Endocrinology

HENNING M. BEIER

Department of Anatomy and Reproductive Biology, School of Medicine, RWTH University of Aachen, 52057 Aachen, Germany

ABSTRACT: The discovery of uteroglobin resulted from investigations on the biochemical composition of oviductal and uterine secretions of the rabbit and other mammals. These determinations about physiological composition were urgently requested to prepare culture media for research on early mammalian development *in vitro*. Discovery of significant proteins during the sixties reflected the laboratory skills of that time. Protein characterization was achieved by isolation via Sephadex gels, electrophoresis on polyacrylamide gels, and finally immunoprecipitation using classical polyclonal antibodies. The molecular biology was not yet established. Uteroglobin could be found as the major protein component of rabbit uterine secretion. From the beginning, it was already identified as an unusually small, spheric uterine secretory molecule without any carbohydrates—hence its name. Uteroglobin was the first mammalian protein that turned out to be progesterone-regulated and, at the same time, released in mg amounts actually in one organ compartment. Moreover, uteroglobin and its gene proved to be a reliable model for the description of progesterone/progesterone receptor complex action at the DNA level. After its original observation in the uterus, however, uteroglobin was detected also in several other organs, for example, the epididymis, the seminal vesicle, and the lung. Initially, it could not be found in the blood, which challenged the hypothesis that uteroglobin specifically should operate by local activation rather than by a humoral or endocrine effect. Later, though, the human uteroglobin molecule, isolated from blood filtrate, was used for detailed structural analyses. The rabbit uteroglobin model certainly was beneficial for reproductive biological research. Experimental interference with steroid hormone regulation during preimplantation presented surprising effects, which led to the discovery of the transposition of the implantation window. The uterine secretion protein patterns, in particular the uteroglobin fraction and the β-glycoprotein fraction, served as decisive marker profiles to identify the biological stage of the intrauterine microenvironment during preimplantation. This diagnostic procedure, using only protein parameters, enabled us to precisely predict the receptive stage of the endometrium for donated blastocysts to achieve implantation successfully.

WHEN MAMMALIAN EMBRYOLOGY REQUIRED *IN VITRO* CULTURE

Fundamental understanding of the embryology of nonmammalian species was established early in the last century. Similar research in mammalian embryology was hampered by the lack of relevant *in vitro* models. Consequently, it was important to characterize oviductal and uterine secretions in order to determine the most suitable composition of culture media. Early efforts were successful for the culture of mouse embryos;[1,2] however, progress with other laboratory mammals was much slower. Embryologists such as Friedrich Seidel[3–5] (Marburg, Germany), Andrej Tarkowski[6] (Warsaw, Poland), Anne McLaren[7] (London, U.K.), Charles Thibault[8,9] (Paris, France), Cecilia Lutwak-Mann[10] (Cambridge, U.K.), and John Biggers[11,12] (Boston, MA) together with their students worked intensely on the analysis of oviductal and/or uterine secretions. This scenery helped to enlarge the scope of embryology quickly into the facets of numerous microenvironmental interactions of the embryo with the maternal compartments of the reproductive organs, and finally led to the concept of an early molecular embryo-maternal dialogue. More and more, the classical field of *Embryology* evolved into *Reproductive Biology* because researchers needed to understand the physiological and biochemical components of the oviductal and uterine compartments that facilitated embryonic development. During the 1960s, the discovery of important proteins in the female genital tract reflected the biochemical skills and insights of researchers at that time. Protein characterization was achieved by isolation procedures including Sephadex column chromatography, polyacrylamide gel electrophoresis (PAGE), and various methods of immunoprecipitation with polyclonal antibodies. At that time, *Molecular Biology* was not yet established.

INITIATION OF RESEARCH ON REPRODUCTIVE TRACT SECRETIONS

Friedrich Seidel and his Ph.D. students analyzed reproductive tract secretions, uterine histology, and endometrial cell biology of the rabbit.[13–20] For more than a decade, Seidel worked to culture rabbit cleavage stages during preimplantation development. Protein substitutions to culture media were considered a promising approach, and PAGE was an invaluable tool for analysis. Initially, this method was used as disk electrophoresis and did not involve protein denaturation. This technique capitalized on molecular sieving and brilliant staining of even the faintest of protein bands, and also used these polyacrylamide gels for photometric densitometry.

Proteins from rabbit uterine flushings that were dialyzed and concentrated by lyophilization represented the target of my research work. In rabbits, uterine protein patterns changed remarkably during the preimplantation phase of pregnancy and pseudopregnancy (FIG. 1). However, an equally intriguing research problem was whether the fluid of the blastocyst cavity during preimplantation time also contained these proteins of the uterine secretions, resembling the blastocyst's microenvironment. A predominant postalbumin protein band was detected by PAGE in rabbit uterine secretions, endometrial homogenates, and blastocyst fluid. A second intriguing question was whether or not this protein could be found in blood plasma or any other body fluids. The method of choice during the sixties was investigation by immunoreactions.

Unfortunately, the "magic" polyacrylamide gels did not allow direct exposition to antibodies because these immunoglobulins did not diffuse into the polyacrylamide gel due to their molecular size. Again, unfortunately, the Western blot technique had not yet been invented. Consequently, I was ready to solve this problem by developing a similar approach and successfully handled the localization of immunologically identical proteins under electrophoretic resolving conditions. I put agar layers on top

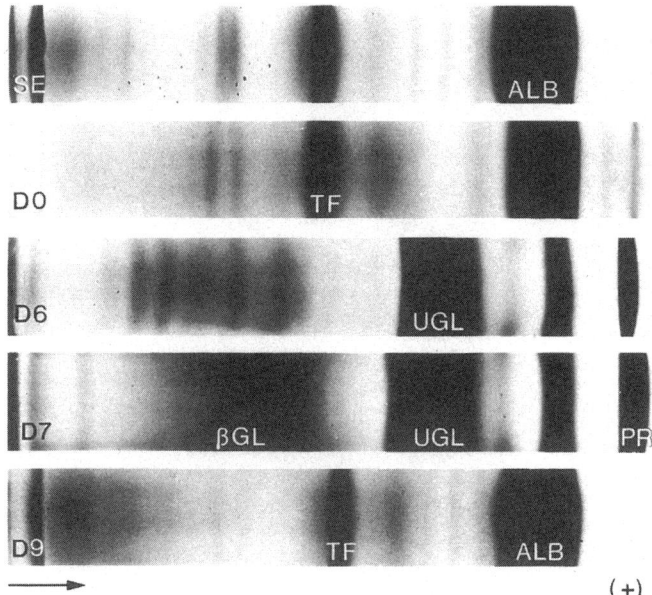

FIGURE 1. Protein patterns of uterine secretion of the rabbit. Polyacrylamide gel disk electrophoresis (PAGDE) of blood serum (SE) as standard reference compared to uterine flushings from an estrous rabbit (D0). The early pregnancy patterns characterize dynamic changes showing daily differences (D6 = day 6, D7 = day 7, D9 = day 9 postcoitum). The predominate fractions were prealbumin (PR), albumin (ALB), uteroglobin (UGL), transferrin (TF), and β-glycoprotein (β-GL). PAGDE was run in 7.5% polyacrylamide with trisglycine buffer, pH 9.0; protein staining by Amido Black 10B.

of the polyacrylamide gels for the application of antibodies to react with the resolved antigens adequately (FIG. 2). By direct antigen comparison of uterine secretion, endometrial homogenate, and blood serum, this technique enabled the discovery that the predominant protein of rabbit uterine fluid, resolved as a postalbumin band under nondenaturing conditions, was an endometrial protein. The protein was a small, diffusible molecule that appeared in considerable amounts in uterine secretions during early pregnancy (FIG. 3). Hence, I called this protein "uteroglobin".[21] When I presented this discovery to a distinguished scientific audience in November 1966, in Konstanz, Germany, these data were the focus of critical discussions by Max Delbrück and Peter Karlson. It was surprising to many colleagues that uteroglobin did not show any carbohydrate moieties and that it was an unusually small molecule with a sedimentation coefficient of 1.38 svedberg units (FIG. 4).[22]

A report on the description of uteroglobin was distributed from the meeting at Konstanz in 1966 by the German Research Council (Deutsche Forschungsgemeinschaft).[21] A second report appeared in the proceedings of the German Zoological Society from its annual meeting at Heidelberg in May 1967.[23] During the next month, my Ph.D. thesis[15] was available at the University of Marburg (Germany) and the journal version was published in 1968 by *Zoologische Jahrbücher*.[24] Ultimately, my work was published in English in *Biochimica Biophysica Acta* in February 1968.[22]

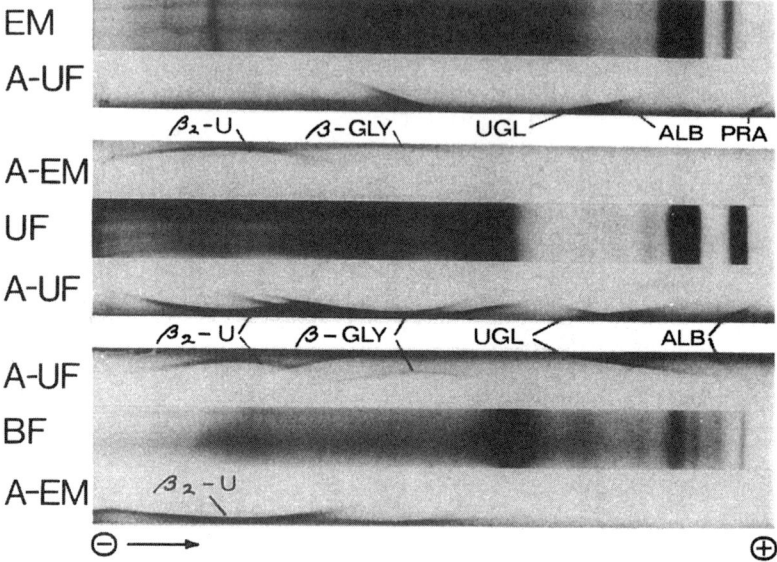

FIGURE 2. Agar-acrylamide immunoelectrophoresis. Uterine secretions (UF) were obtained by flushing the uterine cavity at day 6 postcoitum. Blastocyst fluid (BF) was withdrawn by puncture of flushed, unattached blastocysts. After flushing, the endometrial tissue was homogenized and the supernatant resolved (EM). Various polyclonal antisera were applied to precipitate proteins in the agar layer: anti-uterine secretion sera (A-UF) were obtained from immunized rats; (PRA) prealbumin, (ALB) albumin, (UGL) uteroglobin, (β-GLY) β-glycoprotein, ($β_2$-U) uterine immunoglobulin; PAGE in 5.5% polyacrylamide flat gel, tris-borate buffer, pH 8.7; agar gel 1.5%; protein and precipitate staining by Amido Black 10B.

In the August 1967 issue of *Science*, Krishnan and Daniel[25] published their independent observations on the same protein, which they called blastokinin. Scientific competition and constructive discussions on uteroglobin/blastokinin led to the establishment of the Gordon Research Conference Series on Genital Tract Secretions in 1973. Later, this conference was renamed Reproductive Tract Biology and it remains a lively forum for scientific exchange. The discovery of uteroglobin in epithelial cells of the lung emerged from those competitive interactions.[26–29] The next important question was whether uteroglobin was expressed in the human uterus. It took many years and the advent of molecular technology to describe convincingly the expression of uteroglobin in humans.[30–33] This protein, recently described in detail in its amino acid sequence and molecular structure,[33,34,37] belongs to the uteroglobin/CC10/CC16 gene family of mammalian proteins. It appears in the lung,[35,36] urine,[36] blood serum,[37–39] amniotic fluid,[38] and (in extremely low amounts) cerebrospinal fluid.[38] Finally, uteroglobin expression was demonstrated in the human reproductive organs, particularly in the endometrium.[40,41] Recently, hormone-dependent expression of human uteroglobin in the endometrium[41] and its release into uterine secretions during the menstrual cycle have been demonstrated.[41,42]

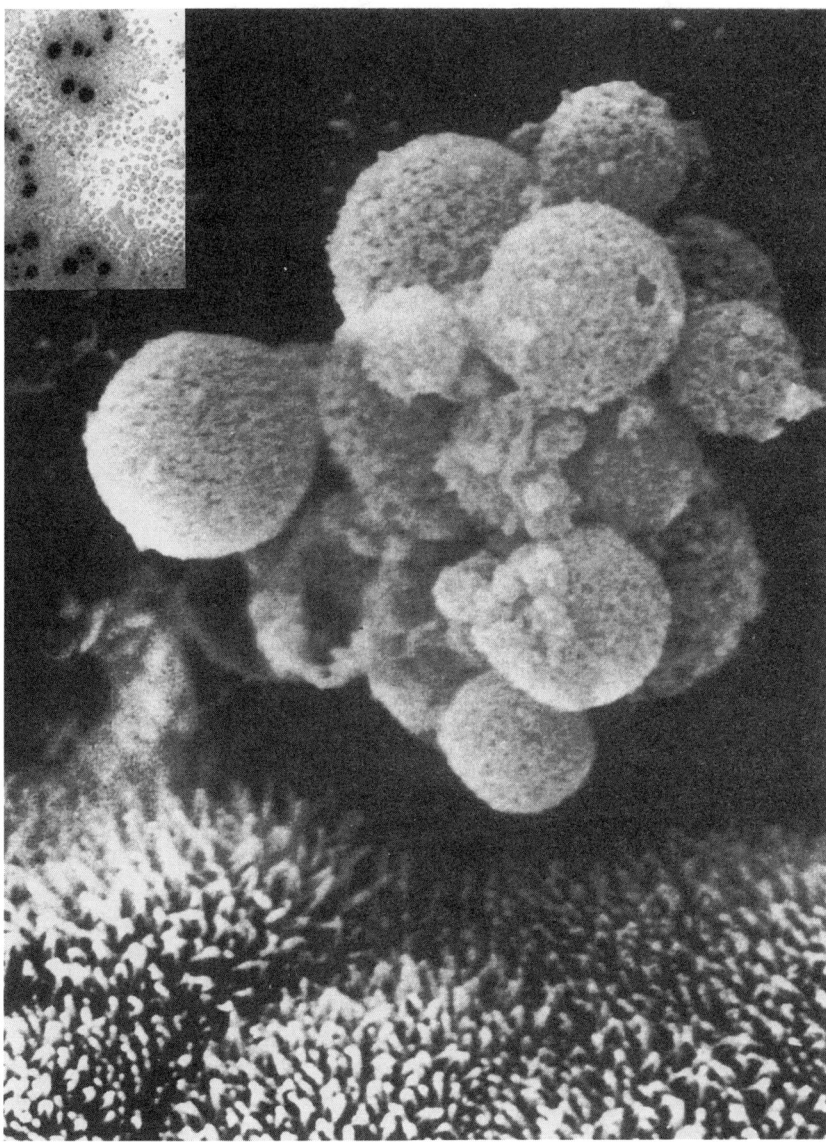

FIGURE 3. Scanning electron microscopy of the rabbit uterus. Globular secretion packets emerge from a gland opening and are released into the uterine cavity. Below are microvilli-covered surface epithelial cells at the gland opening. This secretory material contains uteroglobin, as demonstrated by peroxidase-antiperoxidase localization of uteroglobin clusters in transmission electron microscopy (upper left inset). SEM ×8000; TEM ×6500; courtesy of Ursula Mootz and Christa Hegele-Hartung. [Figure reduced to 90%.]

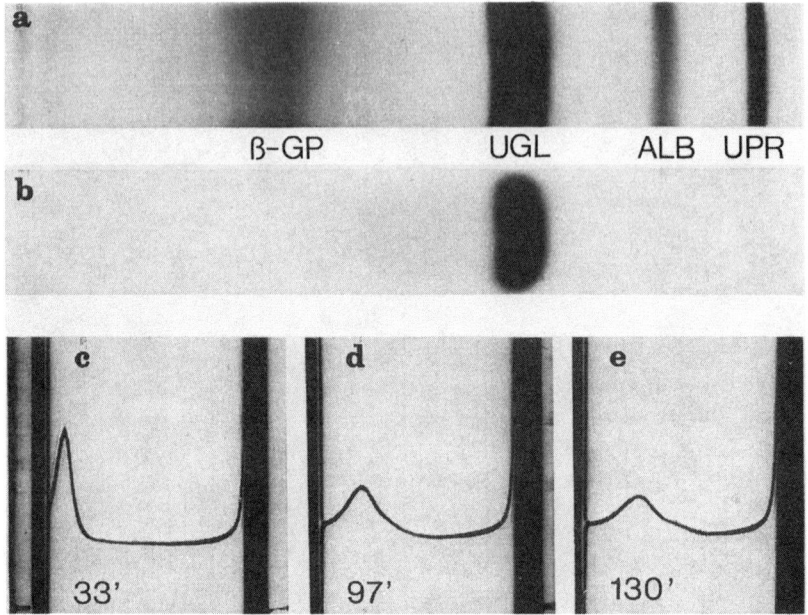

FIGURE 4. Analytical ultracentrifugation of isolated uteroglobin from pregnant rabbit uterine secretions (**a**). The uteroglobin fraction was obtained by cutting out the postalbumin zone for elution. The 90% purified sample (**b**) was analyzed in 10 mg/mL phosphate buffer, pH 6.8, at 68,000 rpm. Sedimentation diagrams are shown at (**c**) 33, (**d**) 97, and (**e**) 130 min, revealing S = 1.38 svedberg units. Terms: (ALB) albumin, (β-GP) β-glycoprotein, (UGL) uteroglobin, (UPR) uterine prealbumin.

UTEROGLOBIN AS THE PROGESTERONE-DEPENDENT PROTEIN

From the very beginning, it was evident that changes in the uteroglobin content of uterine secretions reflected the hormonally controlled proliferative and secretory transformation of the endometrium. Uteroglobin increased in its uterine luminal release in direct proportion to the increasing level of progesterone in peripheral blood after ovulation and corpus luteum establishment. However, the most interesting question was whether progesterone acted by stimulating either the synthesis or the release of uteroglobin, or both processes together. Evidence finally accumulated showing that progesterone and several synthetic progestins were capable of stimulating uteroglobin secretion. Uteroglobin synthesis is mainly activated by natural progesterone, although it is produced in small, yet immunohistochemically detectable amounts by the uterus of estrogen-treated ovariectomized rabbits. In response to estrogen alone, uteroglobin is present in the epithelial cells, but it is not released into the cavum uteri in amounts that could be detected radioimmunologically.[43] Progesterone-dependent uteroglobin secretion is dose-dependent and the most effective dosage range is 0.6–3.0 mg/day/animal. Increased doses have no additional

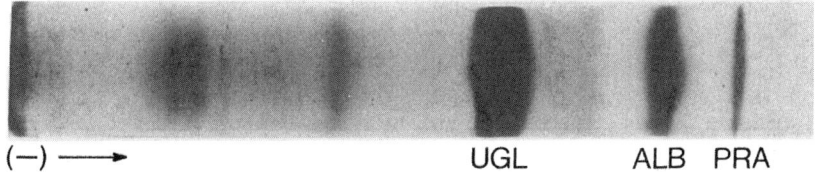

FIGURE 5. Protein pattern of uterine secretions from an ovariectomized rabbit. PAGDE after progesterone substitution with 3.0 mg/day/animal of Proluton® for 5 days. The protein pattern is completely restored and resembles the day-5 postcoitus pattern of pregnancy. Terms: (ALB) albumin, (PRA) prealbumin, (UGL) uteroglobin. PAGDE in 7.5% polyacrylamide, tris-glycine buffer, pH 9.0; protein staining by Amido Black 10B.

effect on uteroglobin concentrations in uterine secretions.[44,45] In 1969, I showed[46] that progesterone stimulated uteroglobin synthesis and release in ovariectomized rabbits by injecting 3.0 mg Proluton®/day/animal in oily solution (Schering AG, Berlin). Experimental results are shown in FIGURE 5.

Without a doubt, progesterone is required to establish the characteristic secretion pattern of uteroglobin during preimplantation *in vivo*. Beato and Arnemann[47] showed that secretion was unaffected when the uterus was removed from the mother, isolated, and perfused under defined experimental conditions. However, it was unclear what processes terminate uteroglobin secretion. Since the protein pattern observed during pseudopregnancy suggests that implantation itself terminates uteroglobin release or synthesis, there may be a specific message delivered from the blastocyst or from the decidual tissue as proposed by Johnson.[48] Perfusion experiments on isolated uteri and uteroglobin production[47] indicate that the off switch *in vivo* is a termination of synthesis rather than secretion. These studies showed there was no long-term accumulation of uteroglobin within the cytoplasm of the endometrial cells. Recent investigations by Krusche[49] suggest that progesterone receptor downregulation plays a key role in this process.

THE UTEROGLOBIN MODEL: A MAJOR ADVANTAGE TO MOLECULAR ENDOCRINOLOGY

Production of uteroglobin as the major secretory protein required characterization of its actual synthesis. The endometrium of progesterone-stimulated rabbits was used as the source of uteroglobin mRNA. The translation of uteroglobin message was demonstrated in different systems.[50–52] Identification of uteroglobin, newly synthesized *in vitro*, was proven by means of monospecific antibodies. These results showed that uteroglobin is synthesized *de novo* in the uterus and provided molecular proof that progesterone stimulated this synthesis.[22,46] Bullock and coworkers[53–55] demonstrated that the specific mRNA for uteroglobin accounts for an increasing proportion of endometrial mRNA during the preimplantation period, reaching a maximum at day 4 postcoitum in normal pregnancy. The pattern of change in uteroglobin mRNA precedes the change in the pattern of secretion during this time.

This animal model in which mg quantities of uteroglobin could be isolated from uterine flushings after progesterone treatment proved to be very useful in the resolution of questions about steroid hormone action at the level of receptor interaction and gene transcription. The rabbit uteroglobin gene was cloned[56,57] and its structure analyzed.[58–63] The gene is encoded by three exons, and progesterone regulation is accomplished by binding of the progesterone receptor complexes to response elements located approximately 3 kb upstream of the transcription start site. In addition, progesterone-stimulated proteins bind to sequences more proximal to the promoter and function as *trans*-activating factors. Binding of these proteins constitutes a putative mechanism by which the tissue specificity of uteroglobin protein expression is maintained.[64–70]

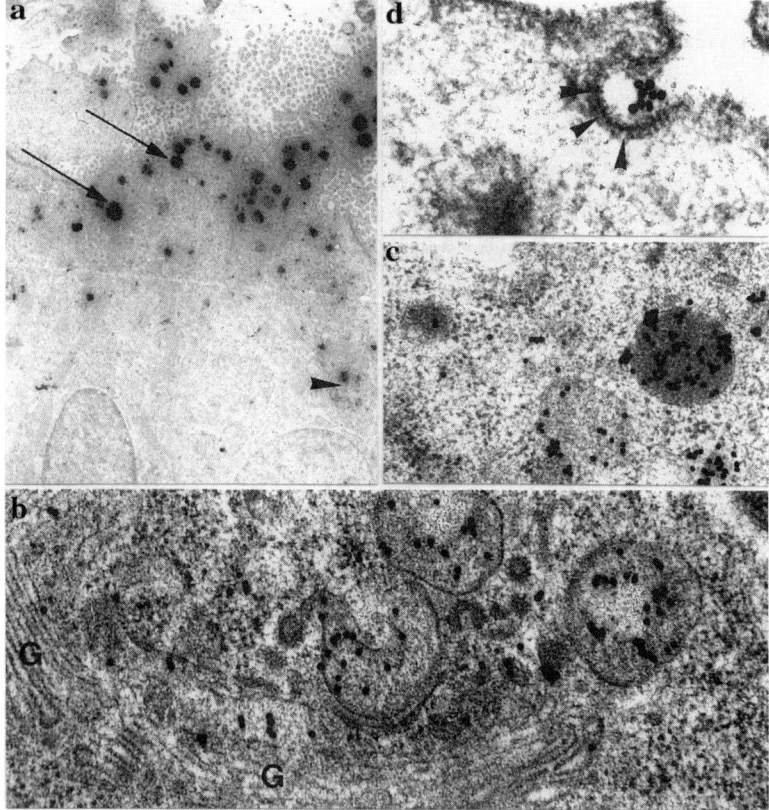

FIGURE 6. Immunocytochemical localization of uteroglobin in epithelial cells of the rabbit endometrium on day 5 of pseudopregnancy. (**a**) Uteroglobin is found in apical cluster-like accumulations (arrows), and some clusters are found at the Golgi region (arrowhead) (×6500). (**b**) Uteroglobin is seen (black round particles) at the *trans*-face of the Golgi complex (×48,000). (**c**) After processing, it is well demonstrated in apical secretory vesicles (×48,000). (**d**) Extrusion may appear, like when clathrin-coated vesicles are released (×120,000). Transmission electron microscopy, peroxidase-antiperoxidase method (a), and protein A–gold method (b, c, d). Courtesy of Christa Hegele-Hartung. [Figure reduced to 67%.]

Testosterone regulation of uteroglobin gene expression in male reproductive organs was reported by my research group in collaboration with colleagues from Schering Research Laboratories at Berlin. The appearance of uteroglobin in seminal plasma and in secretions of the seminal vesicle of the rabbit was described in 1975.[71] Its localization in epithelial cells of the epididymis, in ampulla of the ductus deferens, and on the surface of spermatozoa was published later.[72] Finally, testosterone control of epididymal uteroglobin expression was shown in 1984 by my group[73] and confirmed in 1988 by López de Haro et al.[74]

The process of rabbit uteroglobin synthesis and release was investigated at the subcellular level. *In situ* hybridization demonstrated the presence of uteroglobin message in endometrial epithelial cells.[75,76] Furthermore, the dynamic differentiation of epithelial cells was monitored by nonradioactive *in situ* hybridization.[77] Uteroglobin was localized to secretory vesicles of epithelial gland cells by electron microscopic immunogold-labeling[78] (FIG. 6).

THE UTEROGLOBIN MODEL: CERTAINLY A BENEFIT TO REPRODUCTIVE BIOLOGY AND MEDICINE

Experimental interference from steroid hormone regulation in the preimplantation period led to the discovery of the transposition of the implantation window.[43,79–83] The pattern of secreted uterine proteins, in particular the uteroglobin fraction and the β-glycoprotein fraction, served to identify the biological stage of the intrauterine microenvironment during preimplantation. These protein profiles allowed us to precisely predict the receptive stage of the endometrium so that donated transferred blastocysts achieved successful implantation. Such assessment is not possible by using the histology of the endometrium alone.

Injections of estradiol benzoate at 6 h and 20 h after mating revealed a significant delay in the secretory pattern of the rabbit endometrium (FIG. 7). Depending on the stage when estradiol was given, a delay of 2–5 days resulted. This feature is true not only for the protein patterns, but also for the histology of the endometrium and for the enzyme histochemistry of the endometrial epithelia. I described this phenomenon as *Delayed Secretion* and showed that asynchronous protein patterns correlated with a uterine environment that is unreceptive for blastocyst development. In these experiments, complete implantation failure of native blastocysts was demonstrated.[43,47,84] Subsequent embryo transfer experiments showed that normal blastocysts require a well-timed uterine environment to accomplish implantation and further development.[85,86] For proof of concept, normal day-4 blastocysts were transferred into the uteri of day-8 pseudopregnant recipients. From these transferred blastocysts, 74% implanted normally around day 12 (recipient's reproductive stage) and produced 40% normal fetuses. Several of these fetuses were allowed to develop into viable offspring.[42]

In conclusion, we showed that the synchronized uterine environment, particularly its protein secretion pattern, is required for successful mammalian embryogenesis. The model of Delayed Secretion was not studied just to demonstrate a questionable growth-inhibition effect on the native blastocysts. Clearly, the asynchronous egg transfer into delayed secretory uteri provides evidence for the necessity of a proper uterine environment to support blastocyst development and implantation. The essen-

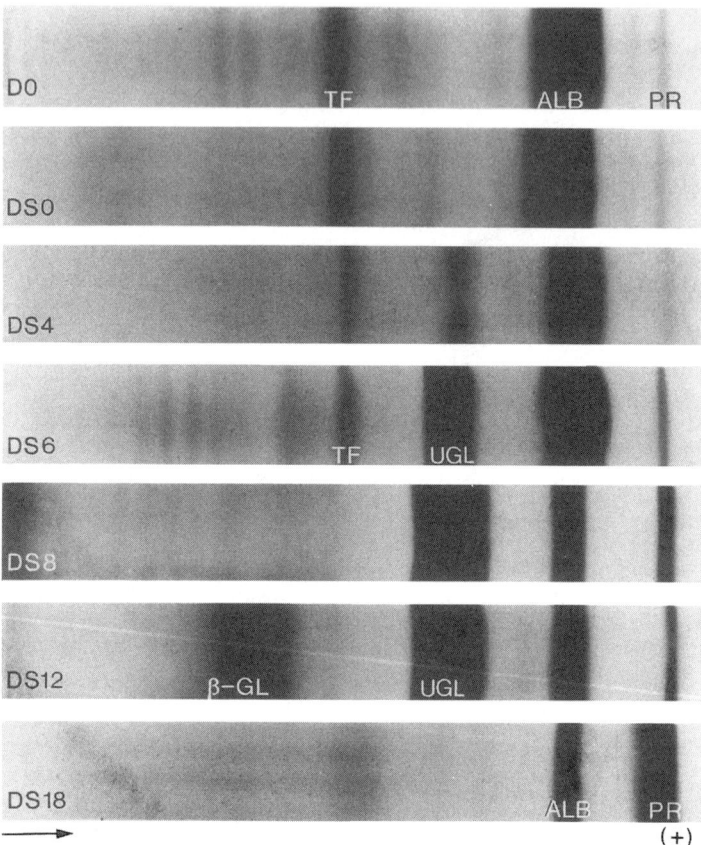

FIGURE 7. Protein patterns of uterine secretion of the rabbit. PAGDE of flushed uterine secretions throughout Delayed Secretion (DS), which was experimentally induced by postcoital 17β-estradiol injections (100 μg at 6 h postcoitum; 150 μg at 30 h postcoitum). Note the delayed development of characteristic protein patterns of early pregnancy: for example, DS8 resembles day 4 postcoitus and DS12 resembles day 7 postcoitus. The predominating fractions are prealbumin (PR), albumin (ALB), uteroglobin (UGL), transferrin (TF), and β-glycoprotein (β-GL). PAGDE was run in 7.5% polyacrylamide with tris-glycine buffer, pH 9.0; protein staining by Amido Black 10B.

tial protein environment for normal blastocyst development is composed of a considerable number of macromolecules. Within this context, uteroglobin is a specific marker molecule for ovarian hormonal status and for the reproductive stage of the endometrium, that is, its differentiation towards receptivity for implantation.

The endocrinological explanation for this phenomenon may be found after more detailed investigations on the expression or downregulation of estrogen and progesterone receptors in endometrial tissue compartments. Earlier studies in our laboratory demonstrated that this research avenue should help to resolve this problem.[87]

Experimentally induced delay of uterine secretion is not the only way to desynchronize maternal and embryonic systems. Prefertilization treatment of estrous rabbits with progesterone (up to 2 mg/day/animal) for 8 days (days −6 to +1) and induction of ovulation (by HCG injection on day 0), with subsequent artificial insemination, resulted in normal egg development during oviductal passage. However, degenerate blastocysts appeared within about 2 days of exposure to the uterine environment.[88] With this treatment, the intrauterine milieu is altered such that protein patterns are advanced compared to the norm. The exogenous progesterone induces uteroglobin synthesis and release before ovulation and fertilization. There is maximum uteroglobin secretion from day −1 until day +2—hence the designation *Advanced Secretion*. This phenomenon leads to implantation failure identical to Delayed Secretion.

In the early eighties, we investigated the endocrine regulation of preimplantation by progesterone antagonists.[89] The uteroglobin model was as good a model to investigate progesterone antagonist effects as it had been to study progesterone effects and actions.[62,90–92] Treatment of pregnant or pseudopregnant rabbits with progesterone antagonists after ovulation and during various preimplantation times resulted in inhibition of progesterone-dependent endometrial transformation and luteal phase differentiation and, perhaps most importantly, in inhibition of implantation. However, endometrial differentiation and protein secretion start again when progesterone antagonist treatment is complete. This leads to a new, but clearly chronologically shifted stage of uterine receptivity. When normal blastocysts from untreated donor rabbits are transferred at the appropriate stage of development, they implant successfully. Here again, protein patterns of uterine secretion enabled us to predict the appropriate stage of receptivity of the rabbit endometrium, as previously described for Delayed Secretion. In the rabbit, uteroglobin is the key marker molecule within the protein pattern of receptivity.[82,83] The transfer of successful achievements in reproductive biology from the rabbit model to human reproductive biology and medicine will lead to beneficial applications in assisted reproduction.[92,93]

From the very beginning, uteroglobin was implicated in blastocyst formation, expansion, and/or attachment because it was detected in blastocyst fluid *in vivo*[21] and *in vitro*[94] as well as in the uterine milieu. However, uteroglobin was lost from growing and expanding blastocysts *in vitro*.[94,95] The initial observations on blastocyst formation *in vitro* by Krishnan and Daniel[25] could not be confirmed by Maurer and Beier[96] because unfractionated uterine secretions supported significantly better blastocyst development *in vitro* than the isolated uteroglobin fraction. Clearly, uteroglobin is not solely responsible for supporting rabbit blastocyst development. By now, we have evidence that paracrine factors in uterine secretions such as insulin and IGF-I promote blastocyst formation and development *in vitro*.[97,98] Experimental evidence shows that this is accomplished by an increase in cell proliferation and prevention of apoptosis.[98] Modern molecular research will undoubtedly lead to a better understanding of early blastocyst development and early events of implantation than we could achieve 30 years ago, when we had discovered an exciting new uterine protein.

REFERENCES

1. BRINSTER, R.L. 1964. Studies on the development of mouse embryos *in vitro*. Ph.D. thesis, University of Pennsylvania, Philadelphia.
2. BRINSTER, R.L. 1970. *In vitro* cultivation of mammalian ova. Adv. Biosci. **4**: 199–232.

3. SEIDEL, F. 1952. Die Entwicklungspotenzen einer isolierten Blastomere des Zweizellenstadiums im Säugetierei. Naturwissenschaften **39**: 355–356.
4. SEIDEL, F. 1954. Das entwicklungsphysiologische Verhalten des Säugerkeims beim Beginn seiner Uteruswanderung. Verh. Dtsch. Zool. Ges. Tübingen **1954**: 371–380.
5. SEIDEL, F. 1960. Die Entwicklungsfähigkeiten isolierter Furchungszellen aus dem Ei des Kaninchens, Oryctolagus cuniculus. Wilhelm Roux' Arch. Entwicklungsmech. Org. **152**: 43–130.
6. TARKOWSKI, A.K. 1962. Inter-specific transfer of eggs between rat and mouse. J. Embryol. Exp. Morphol. **10**: 467–495.
7. MCLAREN, A. & D. MICHIE. 1956. Studies on the transfer of fertilized mouse eggs to uterine foster-mothers. I. Factors affecting the implantation and survival of native and transferred eggs. J. Exp. Biol. **33**: 394–416.
8. THIBAULT, C. 1966. La culture in vitro de l'œf de vache. Ann. Biol. Anim. Biochim. Biophys. **6**: 158–164.
9. WINTENBERGER-TORRES, S., L. DAUZIER & C. THIBAULT. 1953. Le développement in vitro de l'œuf de la brebis et de celui de la chèvre. C. R. Soc. Biol. **147**: 1971–1973.
10. LUTWAK-MANN, C. 1971. The rabbit blastocyst and its environment: physiological and biochemical aspects. In The Biology of the Blastocyst, pp. 243–260. University of Chicago Press. Chicago.
11. BIGGERS, J.D., W.K. WHITTEN & D.G. WHITTINGHAM. 1971. The culture of mouse embryos in vitro. In Methods in Mammalian Embryology, pp. 86–116. Freeman. San Francisco.
12. BORLAND, R.M., S. HAZRA, J.D. BIGGERS & C.P. LECHENE. 1977. The elemental composition of the environments of the gametes and preimplantation embryo during the initiation of pregnancy. Biol. Reprod. **16**: 147–157.
13. SCHWICK, H.G. 1963. Untersuchungen über die Zusammensetzung der Uterus- und Blastocystenflüssigkeit des Kaninchens (Oryctolagus cuniculus). Dissertation, Philosophische Fakultät, Philipps-Universität Marburg/Lahn.
14. SCHWICK, H.G. 1965. Chemisch-entwicklungsphysiologische Beziehungen von Uterus zu Blastocyste des Kaninchens (Oryctolagus cuniculus). Wilhelm Roux' Arch. Entwicklungsmech. Org. **156**: 283–343.
15. BEIER, H.M. 1967. Biochemisch-entwicklungsphysiologische Untersuchungen am Proteinmilieu für die Blastocystenentwicklung des Kaninchens (Oryctolagus cuniculus). Dissertation, Naturwissenschaftliche Fakultät, Philipps-Universität Marburg/Lahn.
16. KIRCHNER, C. 1969. Untersuchungen an uterusspezifischen Glycoproteinen während der frühen Gravidität des Kaninchens (Oryctolagus cuniculus). Wilhelm Roux' Arch. Entwicklungsmech. Org. **164**: 97–133.
17. DENKER, H.W. 1970. Topochemie hochmolekularer Kohlenhydratsubstanzen in Frühentwicklung und Implantation des Kaninchens. Zool. Jb. Allg. Zool. Physiol. **75**: 141–245; 246–308.
18. PETZOLDT, U. 1971. Untersuchung über das anorganische Milieu in Uterus und Blastocyste des Kaninchens. Zool. Jb. Allg. Zool. Physiol. **75**: 547–593.
19. PETZOLDT, U. 1972. Protein patterns of the rabbit blastocyst tissues. Cytobiologie **6**: 473–475.
20. DIES, R. 1972. Hormonale Beeinflussung des Sekretproteinmusters und der Keimesentwicklung im Genitaltrakt von Oryctolagus cuniculus. Dissertation, Naturwissenschaftliche Fakultät, Philipps-Universität Marburg/Lahn.
21. BEIER, H.M. 1966. Das Proteinmilieu in Serum, Uterus, und Blastocysten des Kaninchens vor der Nidation. In Coll. Biochemie d. Morphogenese, Konstanz 1966, pp. 1–10. Deutsche Forschungsgemeinschaft. Bonn.
22. BEIER, H.M. 1968. Uteroglobin: a hormone-sensitive endometrial protein involved in blastocyst development. Biochim. Biophys. Acta **160**: 289–291.
23. BEIER, H.M. 1967. Veränderungen am Proteinmuster des Uterus bei dessen Ernährungsfunktion für die Blastocyste des Kaninchens. Verh. Dtsch. Zool. Ges. Heidelberg **31**: 139–148.
24. BEIER, H.M. 1968. Biochemisch-entwicklungsphysiologische Untersuchungen am Proteinmilieu für die Blastocystenentwicklung des Kaninchens (Oryctolagus cuniculus). Zool. Jb. Anat. **85**: 72–190.

25. KRISHNAN, R.S. & J.C. DANIEL. 1967. "Blastokinin": inducer and regulator of blastocyst development in the rabbit uterus. Science **158:** 490–492.
26. NOSKE, I.G. & M. FEIGELSON. 1976. Immunological evidence of uteroglobin (blastokinin) in the male reproductive tract and in non-reproductive ductal tissues and their secretions. Biol. Reprod. **15:** 704–713.
27. BEIER, H.M., C. KIRCHNER & M. MOOTZ. 1978. Uteroglobin-like antigen in the pulmonary epithelium and secretion of the lung. Cell Tissue Res. **190:** 15–25.
28. BEIER, H.M. 1978. Physiology of Uteroglobin. In Novel Aspects of Reproductive Physiology, pp. 219–248. Halsted Press. New York.
29. DANIEL, J.C., JR. & J.T. MILAZZO. 1976. Continuity of a rabbit antigen between generations. Cancer Res. **36:** 3409–3411.
30. SHIRAI, E., R. IIZUKA & Y. NOTAKE. 1972. Analysis of human uterine fluid protein. Fertil. Steril. **23:** 522–528.
31. BEIER, H.M. 1978. Control of implantation by interference with uteroglobin synthesis, release, and utilization. In Human Fertilization, pp. 191–200. Thieme. Stuttgart.
32. COWAN, R.D., D.H. NORTH, N.S. WHITWORTH, R. FUJITA, E.K. SHUMACHER & A.B. MUKHERJEE. 1986. Identification of a uteroglobin-like antigen in human uterine washings. Fertil. Steril. **45:** 820–823.
33. MANTILE, G., L. MIELE, E. CORDELLA-MIELE, G. SINGH, S.L. KATYAL & A.B. MUKHERJEE. 1993. Human Clara cell 10-kDa protein is the counterpart of rabbit uteroglobin. J. Biol. Chem. **268:** 20343–20351.
34. UMLAND, T.C., S. SWAMINATHAN, G. SINGH, V. WARTY, W. FUREY, J. PLETCHER & M. SAX. 1994. Structure of a human Clara cell phospholipid-binding protein-ligand complex at 1.9 Å resolution. Nat. Struct. Biol. **1:** 538–545.
35. SINGH, G., J. SINGH, S.L. KATYAL, W.E. BROWN, J.A. KRAMPS, I.L. PARADIS, J.H. DAUBER, T.A. MACPHERSON & N. SQUEGLIA. 1988. Identification, cellular localization, isolation, and characterization of human Clara cell–specific 10 kD protein. J. Histochem. Cytochem. **36:** 73–80.
36. BERNARD, A., H. ROELS, R. LAUWERY, R. WITTERS, C. GIELENS, A. SOUMILLION, J. VAN DAMME & M. DE LEY. 1992. Protein 1 is a secretory protein of the respiratory and urogenital tracts identical to the Clara cell protein. Clin. Chem. **38:** 434–435.
37. AOKI, A., H.A. PASOLLI, M. RAIDA, M. MEYER, P. SCHULZ-KNAPPE, H. MOSTAFAVI, A.G. SCHEPKY, R. ZNOTTKA, J. ELIA, D. HOCK, H.M. BEIER & W.G. FORSSMANN. 1996. Isolation of human uteroglobin from blood filtrate. Mol. Hum. Reprod. **2:** 489–497.
38. DEJONGH, R., J. VRANKEN, G. KENIS, E. BOSMANS, M. MAES, G. STANS, M. DE LEY & R. HEYLEN. 1998. Clara cell protein: concentration in cerebrospinal fluid, serum, and amniotic fluid. Cytokine **6:** 441–444.
39. KIKUKAWA, T. & A.B. MUKHERJEE. 1989. Detection of a uteroglobin-like phospholipase A2 inhibitory protein in the circulation of rabbits. Mol. Cell. Endocrinol. **62:** 177–187.
40. PERI, A., B.D. COWAN, D. BHARTIYA, L. MIELE, L.K. NIEMAN, I.O. NWAEZE & A.B. MUKHERJEE. 1994. Expression of Clara cell 10 kD gene in the human endometrium and its relationship to ovarian menstrual cycle. DNA Cell Biol. **13:** 495–503.
41. MÜLLER-SCHÖTTLE, F., I. CLASSEN-LINKE, J. ALFER, C. KRUSCHE, K. BEIER-HELLWIG, K. STERZIK & H.M. BEIER. 1999. Expression of uteroglobin in the human endometrium. Mol. Hum. Reprod. **12:** 1155–1161.
42. MÜLLER-SCHÖTTLE, F., I. CLASSEN-LINKE, K. BEIER-HELLWIG, K. STERZIK & H.M. BEIER. 2000. Uteroglobin expression and release in the human endometrium. This volume.
43. BEIER, H.M. 1974. Oviducal and uterine fluids. J. Reprod. Fertil. **37:** 221–237.
44. ARTHUR, A.T. & J.C. DANIEL, JR. 1972. Progesterone regulation of blastokinin production and maintenance of rabbit blastocyst transferred into uteri of castrated recipients. Fertil. Steril. **23:** 115–122.
45. NISHINO, Y. 1982. Uteroglobin as a sensitive indicator for the biological activity of progestogens. In Proteins and Steroids in Early Pregnancy, pp. 83–87. Springer-Verlag. Berlin/New York.
46. BEIER, H.M., G. PETRY & W. KÜHNEL. 1970. Endometrial secretion and early mammalian development. In Mammalian Reproduction. Colloquium der Gesellschaft für Biologische Chemie in Mosbach, vol. 21, pp. 264–285. Springer-Verlag. Berlin/New York.

47. BEATO, M. & J. ARNEMANN. 1975. Hormone-dependent synthesis and secretion of uteroglobin in isolated rabbit uterus. FEBS Lett. **58:** 126–129.
48. JOHNSON, M.H. 1974. Studies using antibodies to the macromolecular secretions of the early pregnant uterus. *In* Immunology in Obstetrics and Gynecology, pp. 123–133. Excerpta Medica. Amsterdam.
49. KRUSCHE, C.A. 1998. Molekulare und zellbiologische Untersuchungen zur Differenzierung des Endometriumepithels: Eine Studie mit Progesteron-Antagonisten. Dissertation, Mathematisch-Naturwissenschaftliche Fakultät, RWTH Aachen.
50. BEATO, M. & D. RUNGGER. 1975. Translation of the messenger RNA for rabbit uteroglobin in *Xenopus* oocytes. FEBS Lett. **59:** 305–309.
51. BULLOCK, D.W., S.L.C. WOO & B.W. O'MALLEY. 1976. Uteroglobin messenger RNA: translation *in vitro*. Biol. Reprod. **15:** 435–443.
52. ATGER, M. & E. MILGROM. 1977. Progesterone-induced messenger RNA: translation, purification, and preliminary characterization of uteroglobin mRNA. J. Biol. Chem. **252:** 5412–5418.
53. BULLOCK, D.W. 1977. Progesterone induction of messenger RNA and protein synthesis in rabbit uterus. Ann. N.Y. Acad. Sci. **286:** 260–272.
54. KUMAR, N.M. & D.W. BULLOCK. 1982. Hybridization analysis of steady-state levels of uteroglobin mRNA in rabbit uterus and lung during early pregnancy. J. Endocrinol. **94:** 407–414.
55. SHEN, X.Z., M.J. TSAI, D.W. BULLOCK & S.L. WOO. 1983. Hormonal regulation of rabbit uteroglobin gene transcription. Endocrinology **112:** 871–876.
56. ATGER, M., M. PERRICAUDET, P. TIOLLAIS & E. MILGROM. 1980. Bacterial cloning of the rabbit uteroglobin structural gene. Biochem. Biophys. Res. Commun. **93:** 1082–1088.
57. ATGER, M., J.F. SAVOURET & E. MILGROM. 1980. Synthesis, purification, and characterization of a DNA complementary to uteroglobin messenger RNA. J. Steroid Biochem. **13:** 1157–1162.
58. MENNE, C., G. SUSKE, J. ARNEMANN, M. WENZ, A.C. CATO & M. BEATO. 1982. Isolation and structure of the gene for the progesterone-inducible protein uteroglobin. Proc. Natl. Acad. Sci. U.S.A. **79:** 4853–4857.
59. SUSKE, G., M. WENZ, A.C. CATO & M. BEATO. 1983. The uteroglobin gene region: hormonal regulation, repetitive elements, and complete nucleotide sequence of the gene. Nucleic Acids Res. **11:** 2257–2271.
60. BAILLY, A., M. ATGER, P. ATGER, M.A. CERBON, M. ALIZON, M.T. VU HAI, F. LOGEAT & E. MILGROM. 1983. The rabbit uteroglobin gene: structure and interaction with the progesterone receptor. J. Biol. Chem. **258:** 10384–10389.
61. SAVOURET, J.F. & E. MILGROM. 1983. Uteroglobin: a model for the study of progesterone action in mammals. DNA **2:** 99–104.
62. RAUCH, M., H. LOOSFELT, D. PHILIBERT & E. MILGROM. 1985. Mechanism of action of an antiprogesterone, RU486, in the rabbit endometrium: effects of RU486 on the progesterone receptor and on the expression of the uteroglobin gene. Eur. J. Biochem. **148:** 213–218.
63. CATO, A.C. & M. BEATO. 1985. The hormonal regulation of uteroglobin gene expression. Anticancer Res. **5:** 65–72.
64. SAVOURET, J.F., H. LOOSFELT, M. ATGER & E. MILGROM. 1980. Differential hormonal control of a messenger RNA in two tissues: uteroglobin mRNA in the lung and the endometrium. J. Biol. Chem. **255:** 4131–4136.
65. BEATO, M. 1989. Gene regulation by steroid hormones. Cell **56:** 335–344.
66. JANTZEN, K., H.P. FRITTON, K.T. IGO, E. ESPEL, S. JANICH, A.C. CATO, K. MUGELE & M. BEATO. 1987. Partial overlapping of binding sequences for steroid hormone receptors and DNase I hypersensitive sites in the rabbit uteroglobin gene region. Nucleic Acids Res. **15:** 4535–4552.
67. RIDER, V. & D.W. BULLOCK. 1988. Progesterone-dependent binding of a *trans*-acting factor to the uteroglobin promoter. Biochem. Biophys. Res. Commun. **156:** 1368–1375.
68. RIDER, V. & C.J. PETERSON. 1991. Activation of uteroglobin gene expression by progesterone is modulated by uterine-specific promoter-binding proteins. Mol. Endocrinol. **5:** 911–920.

69. KLEIS-SANFRANCISCO, S., A. HEWETSON & B.S. CHILTON. 1993. Prolactin augments progesterone-dependent uteroglobin gene expression by modulating promoter-binding proteins. Mol. Endocrinol. **7:** 214–223.
70. GRAHAM, J.D. & CH. L. CLARKE. 1997. Physiological action of progesterone in target tissues. Endocr. Rev. **18:** 502–519.
71. BEIER, H.M., H. BOHN & W. MÜLLER. 1975. Uteroglobin-like antigen in the male genital tract secretions. Cell Tissue Res. **165:** 1–11.
72. EL-ETREBY, M.F., H.M. BEIER, W. ELGER, A-T. MAHROUS & M. TÖPERT. 1983. Immunocytochemical localization of uteroglobin in the genital tract of male rabbits. Cell Tissue Res. **229:** 61–73.
73. BEIER, H.M., W. ELGER, M. TÖPERT, M.F. EL-ETREBY & A. BÄUMER. 1984. Uteroglobin im männlichen Genitaltrakt und im Seminalplasma des Kaninchens. Verh. Anat. Ges. **78:** 337–339.
74. LÓPEZ DE HARO, M.S., L. ALVAREZ & A. NIETO. 1988. Testosterone induces the expression of the uteroglobin gene in rabbit epididymis. Biochem. J. **250:** 647–651.
75. WAREMBOURG, M., O. TRANCHANT, M. ATGER & E. MILGROM. 1986. Uteroglobin messenger ribonucleic acid: localization in rabbit uterus and lung by *in situ* hybridization. Endocrinology **119:** 1632–1640.
76. KRUSCHE, C.A. & H.M. BEIER. 1994. Localization of uteroglobin mRNA by nonradioactive *in situ* hybridization in the pregnant rabbit endometrium. Ann. Anat. **176:** 23–31.
77. KRUSCHE, C.A., A. HERRLER & H.M. BEIER. 2000. Uteroglobin *in situ* hybridization: novel monitoring of epithelial differentiation in the rabbit endometrium. This volume.
78. HEGELE-HARTUNG, CH. & H.M. BEIER. 1985. Immunocytochemical localization of uteroglobin in the rabbit endometrium. Anat. Embryol. **172:** 295–301.
79. BEIER, H.M. 1976. Uteroglobin and related biochemical changes in the reproductive tract during early pregnancy in the rabbit. J. Reprod. Fertil. (Suppl.) **25:** 53–69.
80. BEIER, H.M. 1976. Biochemical approach to ovum implantation. *In* Implantation of the Ovum, pp. 81–101. Harvard University Press. Cambridge, Massachusetts.
81. BEIER, H.M. & K. BEIER-HELLWIG. 1973. Specific secretory protein of the female genital tract. Acta Endocrinol. (Copenhagen) **74**(suppl. 180)**:** 404–425.
82. HEGELE-HARTUNG, CH., U. MOOTZ & H.M. BEIER. 1992. Luteal control of endometrial receptivity and its modification by progesterone antagonists. Endocrinology **131:** 2446–2460.
83. BEIER, H.M., CH. HEGELE-HARTUNG, U. MOOTZ & K. BEIER-HELLWIG. 1994. Modification of endometrial cell biology using progesterone antagonists to manipulate the implantation window. Hum. Reprod. **9**(suppl. 1)**:** 98–115.
84. BEIER, H.M., W. KÜHNEL & G. PETRY. 1971. Uterine secretion proteins as extrinsic factors in preimplantation development. Adv. Biosci. **6:** 165–189.
85. BEIER, H.M., U. MOOTZ & W. KÜHNEL. 1972. Asynchrone Eitransplantationen während der verzögerten Uterussekretion beim Kaninchen. Proc. Seventh Int. Congr. Anim. Reprod. Artif. Insem. Munich **3:** 1891–1896.
86. ADAMS, C.E. 1974. Asynchronous egg transfer in the rabbit. J. Reprod. Fertil. **3:** 613–614.
87. NEULEN, J., M. BEATO & H.M. BEIER. 1982. Cytosol and nuclear progesterone-receptor concentrations in the rabbit endometrium during early pseudopregnancy under different treatments with estradiol and progesterone. Mol. Cell. Endocrinol. **25:** 183–191.
88. DEVISSER, J. 1982. Prefertilization progesterone treatment: effect on endometrium, uteroglobin secretion, and embryonic development. *In* Proteins and Steroids in Early Pregnancy, pp. 89–98. Springer-Verlag. Berlin/New York.
89. BEIER, H.M. & I. SPITZ. 1994. Progesterone antagonists in reproductive medicine and oncology. Hum. Reprod. **9:** suppl. 1.
90. HEGELE-HARTUNG, CH. & H.M. BEIER. 1986. Distribution of uteroglobin in the rabbit endometrium after treatment with an anti-progesterone (ZK 98.734): an immunocytochemical study. Hum. Reprod. **1:** 497–505.
91. BEIER, H.M., CH. HEGELE-HARTUNG, B. BONN & W. ELGER. 1986. Effect of an antigestagen (ZK 98.734) on uteroglobin synthesis and release in the rabbit endometrium. Acta Endocrinol. (Copenhagen) **111**(suppl. 274)**:** 147–148.

92. BEIER-HELLWIG, K., K. STERZIK, B. BONN, U. HILMES, M. BYGDEMAN, K. GEMZELL-DANIELSSON & H.M. BEIER. 1994. Hormone regulation and hormone antagonist effects on protein patterns of human endometrial secretion during receptivity. Ann. N.Y. Acad. Sci. **734:** 143–156.
93. BEIER, H.M. & K. BEIER-HELLWIG. 1998. Molecular and cellular aspects of endometrial receptivity. Hum. Reprod. Update **5:** 448–458.
94. BEIER, H.M. & R.R. MAURER. 1975. Uteroglobin and other proteins in rabbit blastocyst fluid after development *in vivo* and *in vitro*. Cell Tissue Res. **159:** 1–10.
95. HEGELE-HARTUNG, CH., U. DREINER & H.M. BEIER. 1991. Effect of *in vitro* culture on the dynamics of uteroglobin distribution in rabbit blastocysts. Anat. Embryol. **183:** 119–128.
96. MAURER, R.R. & H.M. BEIER. 1976. Uterine proteins and development *in vitro* of rabbit preimplantation embryos. J. Reprod. Fertil. **48:** 33–41.
97. HERRLER, A. & H.M. BEIER. 1994. Influence of IGF-I on blastocyst formation in the rabbit. J. Reprod. Fertil. Abstr. Ser. **13:** 1.
98. HERRLER, A., C.A. KRUSCHE & H.M. BEIER. 1998. Insulin and insulin-like growth factor-I promote rabbit blastocyst development and prevent apoptosis. Biol. Reprod. **59:** 1302–1310.

All Human Genes of the Uteroglobin Family Are Localized on Chromosome 11q12.2 and Form a Dense Cluster

JIAN NI,[a] MARTHA KALFF-SUSKE,[b] REINER GENTZ,[a] JEOFFREY SCHAGEMAN,[c] MIGUEL BEATO,[d] AND JÖRG KLUG[d,e]

[a]*Human Genome Sciences, Incorporated, Rockville, Maryland 20850-3338, USA*

[b]*Medizinisches Zentrum für Humangenetik, Philipps-Universität Marburg, D-35033 Marburg, Germany*

[c]*GESTEC, University of Texas Southwestern Medical Center, Dallas, Texas 75235-8591, USA*

[d]*Institut für Molekularbiologie und Tumorforschung (IMT), Philipps-Universität Marburg, D-35033 Marburg, Germany*

ABSTRACT: Rabbit uteroglobin is the founder member of a family of mammalian proteins that has expanded to more than 20 members within the last few years. All members are small, secretory, rarely glycosylated dimeric proteins with unclear physiological functions and are mainly expressed in mucosal tissues. A phylogenetic analysis shows that the family can be grouped into five subfamilies, A to E. Subfamily A contains rabbit uteroglobin and its orthologues from various species; most of these have been described to form antiparallel homodimers via two intermolecular disulfide bonds. All other subfamily members contain a third conserved cysteine and, from existing biochemical data, it can be predicted that a member of subfamily B or C will likely form heterodimers with a partner from subfamily E or D, respectively. Besides the mentioned cysteines, only one central lysine is conserved in all family members. In the known uteroglobin structures, this lysine forms an exposed salt bridge with an aspartate side chain, which is conserved in almost all sequences. Using radiation hybrid mapping and P1 clone analysis and utilizing data from the human genome project, we show that all known five human family members (Clara cell 10-kDa protein, lipophilins A and B, lacryglobin, mammaglobin) and a new member, we call lymphoglobin, are localized on chromosome 11q12.2 in a dense cluster spanning not more than approximately 400 kbp.

INTRODUCTION

Uteroglobin,[1] the founder member of a family of related proteins, was first identified in rabbit uterus secretions some 30 years ago and was originally called blastokinin.[2] It is a small 8.5-kDa α-helical protein forming homodimers via two disulfide bonds.[3] From X-ray diffraction studies, its structure is well known.[4–6] The two

---

[e]Author for correspondence: J. Klug, Institut für Molekularbiologie und Tumorforschung (IMT), Philipps-Universität Marburg, Emil-Mannkopff-Strasse 2, D-35033 Marburg, Germany. Voice: +49 6421 28 66545; fax: +49 6421 28 65398.
Klug@IMT.Uni-Marburg.de

monomers are oriented in an antiparallel fashion, forming an internal hydrophobic cavity accessible through a channel that is closed by the two disulfide bonds in the oxidized form. *In vitro*, it was shown that progesterone, which is also required for inducing transcription of the uteroglobin gene in the uterus, is bound by reduced uteroglobin with moderate affinity. Because progesterone binding could not be demonstrated *in vivo* and is not well conserved in its human orthologue, the physiological function of steroid hormone binding remains elusive.

In rat and mouse, it was observed that certain polychlorinated biphenyl (PCB) metabolites accumulate in lung Clara cells that are lining the bronchiolar epithelium.[7,8] Accumulation of PCBs occurs through high-affinity binding to a protein initially referred to as PCB–binding protein in the rat[9] and in humans.[10] During the investigation of proteins found in rat lung lavage fluid, Clara cell secretory protein C was identified,[11] which was later also called rat and human Clara cell 10-kDa protein (CC10).[12,13] The name Clara cell secretory protein (CCSP), originally coined for the rabbit protein,[14] is nowadays also in common use for the mouse and rat protein. Moreover, in human urine, a protein called protein 1 or urinary protein 1 (UP-1) was characterized,[15] which was later shown to be identical to CC10/CCSP. The rat protein was also called Clara cell 17-kDa protein (CC17) in order to account for the correct molecular weight,[16] and the human protein was also named Clara cell phospholipid-binding protein (HCCPBP) in order to account for its phospholipid-binding ability.[17] Orthologous proteins have been identified in the mammalian orders Rodentia (hamster),[18] Lagomorpha (hare, pica),[19] Primates (macaque),[20] and Artiodactyla (pig).[21] Because all these proteins are clearly orthologues,[22] the name uteroglobin with a species prefix will be used in order to simplify nomenclature.

A number of paralogous proteins have been described in increasing number over the past two decades. Among them, the rat prostatic binding protein (PBP),[23] also called prostatein[24] and prostate alpha-protein,[25] is known best. Since it was found to bind steroids, the name prostatic steroid binding protein (PSBP) is also in common use.[26] Prostatein is known to bind the estrogen carbamate estramustine and, hence, was initially also called estramustine binding protein (EBP).[27] Estramustine ("estrogen mustard", Estracyt), which inhibits the assembly of microtubules, has long been used in the treatment of metastatic hormone-refractory prostate cancer. Prostatein is an oligomer with the subunit composition (C1/C3)(C2/C3).[28] Subunit C3 forms heterodimers with subunits C1 and C2 that are covalently linked by disulfide bonds similar to the uteroglobin homodimer. Although all uteroglobin family members are classical secretory proteins, glycosylation is not a common modification. Therefore, it is noteworthy that prostatein C3 is known to be moderately glycosylated at N17,[29] whereas C1 and C2 are not glycosylated. Two C3 isoforms, C3(1) and C3(2), which are encoded by distinct genes and differ in only 6 amino acid residues, have been described.[30] In the rat prostate, the ratio of mRNA transcripts derived from the two genes is C3(1):C3(2) = 18:1, and both transcripts are coordinately induced by androgens.

A number of proteins, homologous to the rat prostatein subunits, are known in various species. In humans, mammaglobin[31] and lacryglobin[32] are highly homologous to rat prostatein C3(1), showing 41% and 34% identity, respectively. Mammaglobin and lacryglobin differ in only 39 out of 93 amino acid residues (58% identity). Although the protein sequence of lacryglobin is not of full length, over its reported length it is identical to mammaglobin B[32] and lipophilin C.[33] Lipophilins A and B have been described as homologues of the rat prostatein subunits C1 and C2[33,34] and, similarly, lipophilin AC heterodimers have been shown to occur in human tears where lipophilin B is absent. The lipophilins A, B, and C should not be confused with the lipophilins described as a family of hydrophobic integral membrane proteins in myelin.[35] In the female harderian gland of the Syrian hamster (a specific dimorphic endocrine/exocrine organ common in rodents), another protein, homologous to the rat prostatein subunits C1 and C2, has been identified.[36] FHG22 (female harderian gland clone 22) has highest homology to lipophilins A and B, with 43% and 41% identity, respectively. The major allergen of the domestic cat, named Fel dI according to allergen nomenclature and known to elicit rhinitis, conjunctivitis, and even potentially life-threatening asthmatic episodes in humans, was identified as a heterodimeric member of the uteroglobin family.[37] Fel dI chain 2 (Fel dI C2) is most closely related to lipophilin B (29% identity), whereas Fel dI chain 1 (Fel dI C1) and the α subunit of mouse salivary androgen-binding protein (ABPa)[38,39] form a new subfamily with closest homology to the uteroglobins. Like rat prostatein subunit C3, Fel dI C2 is N-glycosylated, most likely at N33, which is part of a consensus sequence for N-glycosylation (N33-A-T). Fel dI C1 can be equipped with two different signal peptides, designated A and B, encoded by different exons.[40] Chain 2 also exists in two forms, a long form of 92 amino acid residues (C2L) and a short one with 90 residues (C2S), differing by 5 contiguous residues and a 2-residue deletion/insertion. Most likely, alternative splicing is responsible for the generation of these Fel dI C1 and C2 isoforms.

Because a generally accepted physiological function of none of the uteroglobin family members is known, the expression pattern of each member is a decisive characteristic. Although each member has a peculiar expression pattern, the expression spectrum of all family members is limited to a characteristic set of epithelial tissues separating the body interior from the exterior world. Apart from the uterus, rabbit uteroglobin is expressed in the lung,[41,42] esophagus, male genital organs,[41,43] and oviduct.[44] Strong expression in the lung is observed for all orthologous proteins in hamsters,[18] humans,[45] mice,[46] monkeys,[20] hares,[19] pigs,[21] and rats,[47] but strong expression in the uterus seems to be a peculiarity of the Lagomorpha.[19] Expression in the salivary glands has been reported for rabbit uteroglobin,[48] cat Fel dI,[40] rat prostatein C3,[49] FHG22,[36] mouse salivary ABPa,[38] and lacryglobin/mammaglobin B/lipophilin C.[50]

Human uteroglobin has been mapped to chromosome 11q11-qter by *in situ* hybridization to a human-mouse somatic cell hybrid panel.[22] This assignment was refined by meiotic break-point analysis to 11p12-q13 between markers D11S16 and D11S97.[51] The human mammaglobin and lacryglobin/mammaglobin B/lipophilin C genes are closely linked and were mapped by fluorescent *in situ* hybridization to chromosome 11q13.[50,52] In another study using the same technique, the genes for lacryglobin/mammaglobin B/lipophilin C, lipophilin B, and lipophilin A were mapped to chromosome 11q12-q13.1, chromosome 10q23, and chromosome 15q12-q13, respectively.[34]

Contrary to the latter assignments, we present evidence that all human members of the uteroglobin family, including the ones mentioned above and one new member reported here, we call lymphoglobin (YGB), are localized in a dense cluster on the long arm of chromosome 11. Furthermore, we present a comprehensive multiple alignment of all known protein sequences and a phylogenetic tree that allows grouping of the proteins into five subfamilies. Both analyses form a basis for nomenclature considerations, dimerization potential of individual members, and the prediction of glycosylation sites.

MATERIALS AND METHODS

Radiation Hybrid Mapping

The Genebridge 4 radiation hybrid panel (Research Genetics, Huntsville, Alabama) consisting of 93 radiation hybrid clones was used. The PCR amplifications performed on the panel samples and controls were analyzed on 2.0% agarose gels. A second independent amplification of panel samples and controls was only performed in some cases.

The primer pairs listed in TABLE 1 were used in standard polymerase chain reactions with an annealing temperature of 55°C to amplify short sequences from exon 2 or 3. The primers were selected from regions with low homology between cDNAs in order to achieve gene specificity. PCR reactions using these primer pairs with human genomic DNA yielded only one specific fragment of the expected size. When PCR reactions were performed with each primer pair on highly homologous cDNA templates (like MGB primer pair on an LGB cDNA template), no amplified product was obtained.

For testing for the presence of the lipophilin A gene in a P1 clone, another PCR primer pair was used that is not listed in TABLE 1—upper primer: CTTATTAGCTG-GAAAACCTG; lower primer: CCCAATGTTTTTGTAATTAGC. This 144-bp amplicon is from exon 2.

P1 Library Screening and Preparation of P1 DNA

The human P1 library used was made from digests of DNA from the male cell line Bristol 8 ligated into pAd10SacBII.[53] Approximately 47,000 P1 clone DNAs were spotted in a 4 × 4 array in duplicates on three 21 × 21–cm nylon membranes supplied by the German Genome Project RZPD (http://www.rzpd.de). Probes were ^{32}P-labeled using random hexanucleotide priming (Rediprime DNA labeling system, Amersham Pharmacia Biotech). Library filters were prehybridized at 65°C for 1 h in hybridization buffer containing 0.5 M sodium phosphate (pH 7.2), 7% SDS, and 1 mM EDTA.[54] Hybridization was performed at 65°C in the same solution with denatured probe added overnight. Filters were washed in 40 mM sodium phosphate/ 0.1% SDS twice at room temperature and then twice at 65°C for 20 min each. Exposure to X-ray film (Kodak Biomax MR) was for 1 day at −80°C with intensifying screen.

Positive P1 clones in the host NS3145 were requested from the RZPD and then DNA was prepared by a modified version of the alkaline lysis procedure.[55] Briefly, a 25-mL expanded overnight culture is grown in 800 mL 2×YT plus kanamycin for 1 h at 37°C. Cells were harvested after induction with 1 mM IPTG for 4 h. Using 10 mL, 30 mL, and 22.5 mL of solutions I, II, and III, respectively, the cells were lysed and nucleic acids were precipitated with 45 mL of isopropanol. After centrifugation, the dried DNA pellet was resuspended in TE8 and purified through a CsCl gradient.[55]

Identification of cDNAs Encoding Proteins with Homology to the Rat Prostatein Subunits

A database containing more than one million ESTs obtained from >750 different cDNA libraries has been generated by Human Genome Sciences, Incorporated, and The Institute for Genomic Research (TIGR) using high-throughput automated DNA sequence analysis of randomly selected human cDNA clones.[56,57] Sequence homology comparisons of each EST were performed against the GenBank database using the BLAST and BLASTN algorithms.[58] ESTs having homology to previously identified sequences were collected in a database. A specific homology and motif search using the known amino acid sequences of rat prostatic binding protein against this database revealed several ESTs with >30% homology. Three different clones encoding intact N-terminal signal peptides were identified in an endometrial tumor library and selected for further investigation. The complete cDNA sequences of both strands of these clones were determined, and their homology to the rat prostatein subunit cDNAs was confirmed.

Northern Blot Analysis

cDNA probes were ^{32}P-labeled using random hexanucleotide priming (Rediprime DNA labeling system, Amersham Pharmacia Biotech). Unincorporated nucleotides were removed using CHROMA SPIN-100 columns (Clontech). Two human Multiple Tissue Northern (MTN) blots containing approximately 2 µg of poly (A)+ RNA per lane from various human tissues were obtained from Clontech. Hybridizations were performed with Express Hybridization Solution (PT1190-1, Clontech) according to the manufacturer's manual.

RESULTS

Radiation Hybrid Mapping

In a database containing more than one million human ESTs, three initially new sequences with homology to the rat prostatein subunit cDNAs had been identified (see MATERIALS AND METHODS). The full-length cDNAs were cloned from an endometrial tumor library and are known as lacryglobin/mammaglobin B/lipophilin C and lipophilin B today. The third sequence, most closely related to lipophilin B, we report here and refer to the encoded protein as lymphoglobin (see below). The chromosomal location of the lymphoglobin gene was mapped relative to lacryglobin/mammaglobin B/lipophilin C, mammaglobin, and human uteroglobin using radiation hybrid mapping. Gene-specific amplification of clone DNAs from the Genebridge 4 radiation hybrid panel[59,60] was performed with primer pairs YGB (lymphoglobin), LGB (lacryglobin/mammaglobin B/lipophilin C), MGB (mammaglobin), and UGB (human CC10/CCSP). The amplicons had the expected product sizes. The data vectors (TABLE 1) were sent to the WICGR mapping service (http://

TABLE 1. Results of radiation hybrid mapping experiments

Primer name	Primer sequence	Amplicon (bp)	Data vector	Flanking markers
YGB	GCTGCGGTAAACCTCCAAG CAATGAGAGTCGTTTCTTAAA	114	10010010011001010110010 01011100001100001110111 00111000000000000101100 11002101010100002011	D11S1965 WI-1409
LGB	CAGTGTACGACAGCATTTGG AACACAAGAAGAAAGTGGTTT	128	10010000011001011011010 01011100021100000210111 00111000000000000001100 01001100010010000001	WI-8652 CHLC.GATA46A12
MGB	CAGCAGTCTTTGTGATTTATT ATAAGAAAGAGAAGGTGTGG	114	10010010011001010110010 01011100001100001110111 00111000000000000101100 11001101010100000011	D11S1965 WI-1409
UGB	CCCAAAGCTCACTGTGTAATT TGGGCGTGGACTCAAAGCAT	90	10010010011001010110010 01011100001100001110111 00111000000000000101100 21001202010010000011	D11S1965 WI-1409

NOTE: The upper primer of each pair is the upstream primer; the lower primer is the downstream primer. The amplicon from YGB is located in exon 2, whereas all other amplicons are in exon 3. The lod score threshold for initial two-point analysis was set to 15.0. The digits in a data vector denote the following: 0 = no PCR signal; 1 = positive PCR signal; 2 = uncertain data.

www.genome.wi.mit.edu), and the linkage of all four genes to their respective centromeric and telomeric genetic framework markers resulting from the two-point maximum-likelihood analysis of the data vectors was received. As shown in TABLE 1, the genes for lymphoglobin, mammaglobin, and human uteroglobin map together and cannot be separated by the radiation hybrid panel used. The gene for lacryglobin/mammaglobin B/lipophilin C is located a few cR apart into the direction of the centromere. These results lead us to conclude that all human genes of the uteroglobin family might map to this region on chromosome 11.

Identification of a Large Genomic Fragment Encompassing the Genes for Lacryglobin, Mammaglobin, and Lipophilins A and B

In an effort to clone the genes for mammaglobin and lacryglobin/mammaglobin B/lipophilin C and to further characterize the uteroglobin gene cluster, high density array filters carrying the ICRF human P1 library no. 700[53,61] were hybridized with a number of cDNA probes. Clone ICRFP700J1347Q6 (http://www.rzpd.de) gave a positive hybridization signal with lacryglobin, mammaglobin, lymphoglobin, and lipophilin B cDNA probes. No positive clone could be identified with a human uteroglobin cDNA probe. Because lacryglobin and mammaglobin as well as lymphoglobin and lipophilins A and B easily cross-hybridize, gene-specific primer pairs (see MATERIALS AND METHODS) were employed for PCR analyses (FIG. 1). The results show that the P1 clone contains the genes for lipophilins A and B as well as lacryglobin, whereas the primers for lymphoglobin and human uteroglobin did not yield an amplification product. The PCR results for mammaglobin were not fully convincing because in some cases only a faint amplicon band was observed.

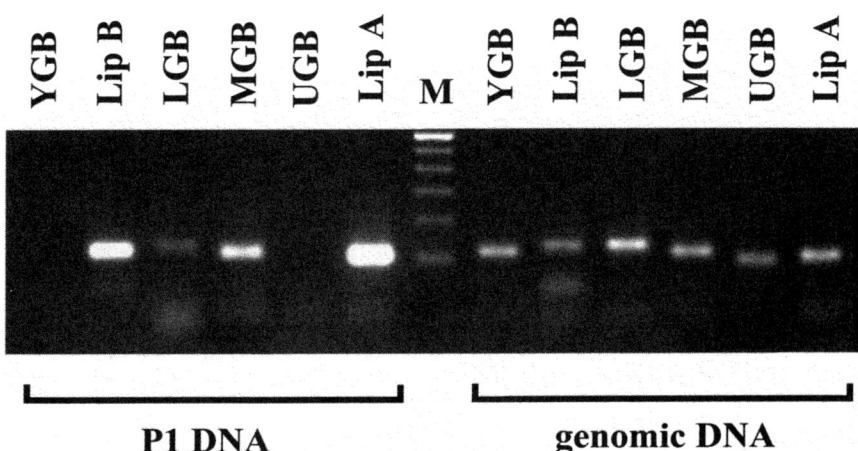

FIGURE 1. Identification of uteroglobin family genes in the P1 clone ICRFP700J1347Q6 (http://www.rzpd.de). PCR reactions were set up with specific primer pairs for the genes indicated and either P1 DNA (left) or genomic DNA (right) as template. As molecular weight marker, a 100-bp DNA ladder (Gibco) was employed (M).

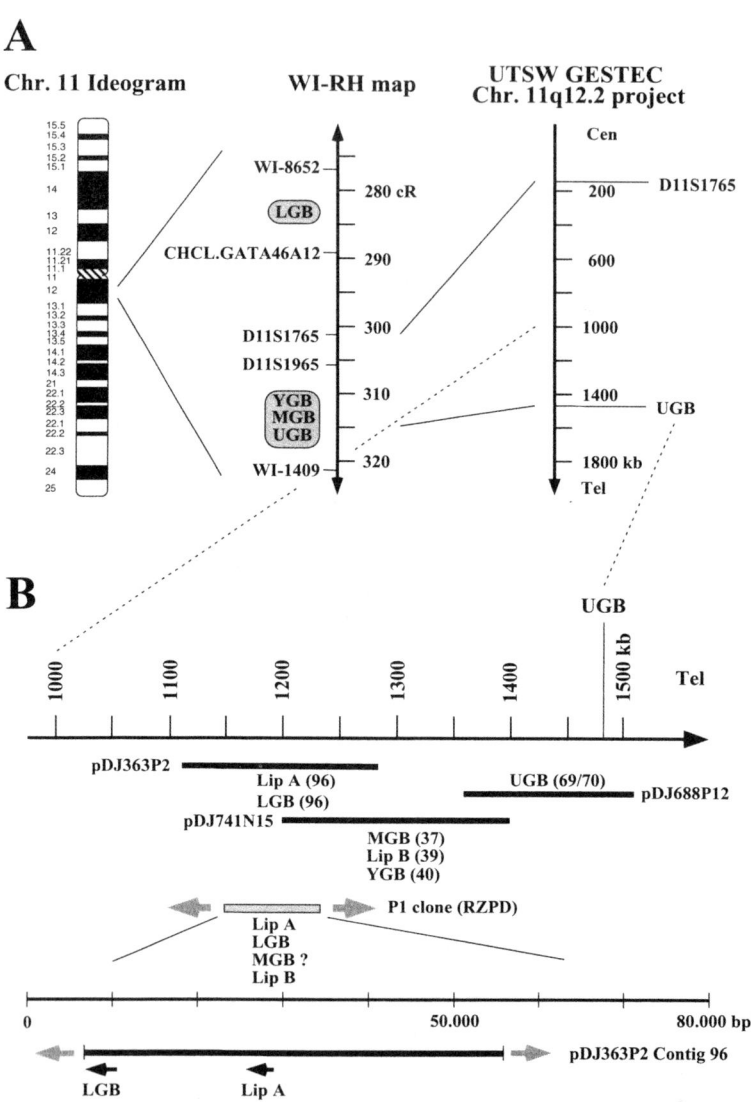

FIGURE 2. Overview of the uteroglobin family gene cluster. **(A)** A chromosome 11 ideogram is shown on the left. Chromosome band 11q12 explodes into the radiation hybrid map of this region established by the Whitehead Institute (http://www.genome.wi.mit.edu) with relevant genetic framework markers indicated. The distance from the top of the chromosome 11 linkage group is given in centirad (cR). The genes for lacryglobin (LGB), lymphoglobin (YGB), mammaglobin (MGB), and uteroglobin (UGB) are placed on this map according to the results of their data vectors. The region between D11S1765 and UGB explodes into a physical map of the chromosome 11q12.2 region established by the University of Texas Southwestern Medical Center (UTSW GESTEC). The unit of distance is kb, and Cen and Tel indicate the direction of the centromere and telomere, respectively. **(B)** A

At the time we had identified this P1 clone, mass sequencing in the chromosome 11q12.2 Best's disease region, which maps between STS D11S461 and EST AHNAK, was in progress.[62–64] This region spans over 1.5 Mb, and a high coverage PAC (P1 artificial chromosome) contig map is available. In three overlapping PACs spanning 400 kb, all six human genes of the uteroglobin family known so far were identified (see FIG. 2). Four of them (Lip A and B; LGB; and MGB) are located on one P1 clone spanning approximately 80 kb at the overlap of clones pDJ363P2 and pDJ741N15. The orientation of the P1 clone as well as the precise distance and relative orientation of all genes are not known yet, except for LGB and Lip A, which are located on a single 50-kb contig. They are only 20 kb apart and transcribed into one direction (see FIG. 2). All genes encompass three exons in each case and several kb of promoter and downstream sequence. Only in the LGB gene, which is located at the very end of the contig, exon 3 is missing. The contiguous DNA sequence of this region will be available through the various genome projects in the near future.

Northern Analyses

Because the expression pattern of each uteroglobin family member is characteristic and informative and has implications for its dimerization capabilities, two MTN blots were hybridized with ^{32}P-labeled cDNA probes for YGB, Lip B, LGB, and MGB (FIG. 3). Interestingly, YGB is expressed only in spleen and peripheral blood lymphocytes and is thus the only uteroglobin family gene that is specifically expressed in lymphocytes. Because a generally accepted physiological function of none of the uteroglobin family members is known, the "guide tissue", where strong expression is observed, is a good basis for naming a new family member. Therefore, the name lymphoglobin, which cannot be found in the literature yet, was chosen. Lipophilin B is found to be moderately expressed in heart, skeletal muscle, kidney, and pancreas, whereas lacryglobin is strongly expressed in pancreas, prostate, testis, and ovary, but only weakly in thymus. Prostate is the only tissue in which weak expression of mammaglobin could be detected. Because the Northern blots were hybridized with homologous cDNA probes, care was taken to choose stringent washing conditions. Moreover, the observation that in most cases the expression patterns are not overlapping is a good proof of specificity. Just for the weak mammaglobin signal in the prostate, it cannot be ruled out that this band is due to cross-hybridization with lacryglobin mRNA.

500-kb region explodes into a map of three P1 artificial chromosome (PAC) clones, where large parts are sequenced The sequenced contigs are available through Genbank: pDJ363P2 (acc. no. AC003023), pDJ741N15 (acc. no. AC004127), and pDJ688P12 (acc. no. AC0060789). All known human uteroglobin family genes could be identified within these three PACs. The contig in which each gene was found is listed in parentheses (unofficial numbering). Based on the PCR typing results (see text), the P1 clone ICRFP700J1347Q6 (http://www.rzpd.de) could be positioned at the overlap of pDJ363P2 and pDJ741N15. A large part of the P1 clone is represented by the almost 50-kb-long contig 96 of pDJ363P2, which contains the genes for lacryglobin (LGB) and lipophilin A (Lip A). Both genes are separated by approximately 20 kb and are transcribed in the same direction. The orientation of contig 96 and the P1 clone is not known (indicated by gray arrows). All genes are complete, except for LGB, where the third exon is missing.

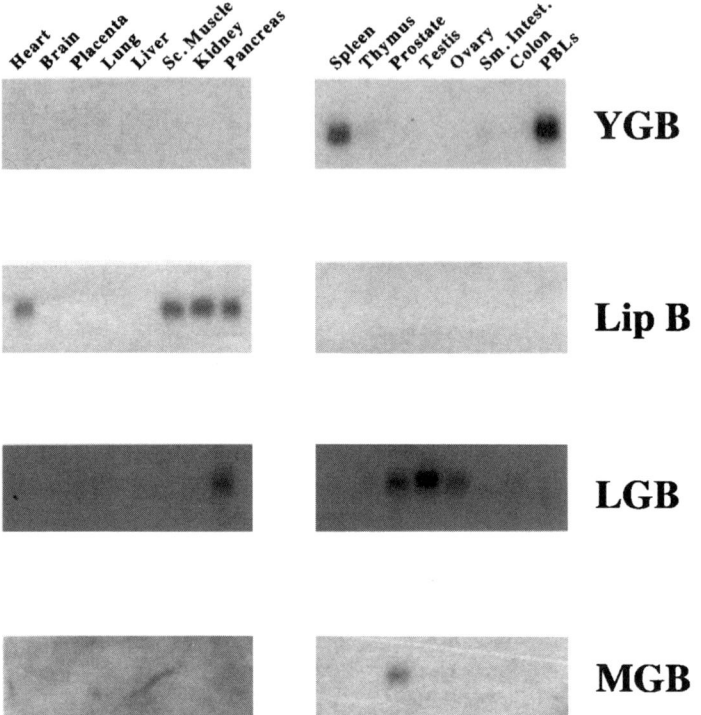

FIGURE 3. Northern blot analyses of the expression of lymphoglobin (YGB), lipophilin B (Lip B), lacryglobin (LGB), and mammaglobin (MGB) in multiple human tissues. Two human tissue Northern blots (Clontech) containing 2 μg of poly (A)+ RNA of the indicated tissue per lane were consecutively hybridized with the indicated ^{32}P-labeled cDNAs.

Amino Acid Sequence Comparison of All Uteroglobin Family Members

All 23 known uteroglobin family protein sequences were aligned using the CLUSTAL W program (FIG. 4). Three sequences are not entered in Genbank (nos. 4, 5, and 22; see legend to FIG. 4), whereas mouse lacrimal gland protein can be found in Genbank, but is not published yet. Remarkably, only 3 amino acid positions are generally conserved. These are the 2 N- and C-terminal cysteines, required for homo- and heterodimerization, and a central lysine, which is lysine 42 in helix 3 of Oc UGB (gray bars in FIG. 4). In 3 further positions, hydrophobic residues are invariably used (Leu 14, Leu 25, and Ile 63 in Oc UGB). All other positions are varied extensively when the whole set of 23 sequences is considered, with one exception. When the alignment is viewed with a program like CLUSTAL X[65] that provides a sequence coloring scheme, it can be noted that sequence nos. 9–23, the "non-uteroglobins", all contain a third conserved cysteine at 2 positions C-terminal to the conserved lysine. In the most distantly related sequence, Fc Fel dI C2, this cysteine is located at 3 amino acid positions further to the C-terminus. When the 2 cysteines

Acc.	Name	Signal Peptide		Helix 1	Helix 2	Helix 3	Helix 4	Length (aa)
				HHHHHHHHHHH	HHHHHHHHHHH	HHHHHHHHHHHHHH	HHHHHHHHHHHHH	
1. X01423	Oc UGB	MKLTAITFLALVTLALLCCSPASA	-21	GICPRFAHVIENLLLGTPS---SYETSLKEFEPDDTMKDAGMQMKKVLD-SLPQTTRENIMKL TEKIVKSPLC M------				70
2. M25609	Lc UGB	MKLTITLALVTLALLCCSPASA	-21	GICPGFAHVIENLLLGTPS---SYETSLKEFQPDDAMKDAGMQMKKVLD-TLPQTTRENIIKL TEKIVKSPLC-------				69
3. X13197	Hs UGB	MKLAVTLTVTLALLCCSS GSA	-21	EICPSFQRVIETLLMDTPS---SYEAAMELFSPDQDMRRAGAQLKLVD-TLPQKPRESIIKL MEKIAQSSLCN------				70
4. see 1)	Mf UGB		+1	EICPTFLRVIESLFLDTPS---SFEAAMGFFSPDQMSEAGAQLKKLVD-TLPAKARDSIIKLMEKIDKSLICN------				70 (no cDNA)
5. see 2)	Ss UGB	MKIAITFTLVALIFFCSPASA	-21	EVCPSFVGVIQNLPKGTL---ASYQASVEPFSPNEDMKKAGAQLKTLVD-TLSPEAKDSVLKLQEKIIKSPLCA-----				91 (prepro)
6. Y13765	Ma UGB	MKIAITMAVVMLSVCCSSA	-19	SSDTCPGFPOVLEFLFMGSES---SYEAALKFYNPGSDLQDSGTQLKKLVD-TLPQKTRMNIMKLSEIILTSPLCNQDLSV				78
7. J05536	Rn UGB	MKIAITMVVMLSICCSSA	-19	SSDICPGFLQVLEALLLGSES---SYEAALKFPNPASDLQNAGTQLKRLVD-TLPQETRINIVKLTEKILTSPLCEQDLRV				78
8. L04503	Mm UGB	MKIAITITVVMLSICCSSA	-19	SSDICPGFLQVLEALLMESES---GYVASLKFPNPGSDLQNAGTQLKRLVD-TLPQETRINIMKL TEKILTSPLCKQDLRF				78
9. AF008595	Mm LGP	MKLTGALVILGATLLLLTSSGDC	+1	GICPAIKEDVHLFLFGTPE---EVNYVEKYKDDPETLENTEKLKICVDRTLTKENKEHAAAFIEKIESSPLC--------	?			93 (prepro)
10. S68136	Mm ABP a			DCGLGPALQRKVDLFLNGTTG---EVQYLKQFNENRVDLDNAANIKKGSDRTLTEEDKAQATSL INKITASRTC-------	?			72
11. M74952	Fc Fel dl C1A	MKGARVLVLLMAALLLIWGG-NC	-22	EICPAVKRDVDLFLTGTPD---EVVEQVAQYNALPVVLENARILKNCVDAKMTEEDKENALSVLQKIYTSPLC-------				70
12. M74953	Fc Fel dl C1B	MLDAALPCCPFVAATA----DC	-18	EICPAVKRDVDLFLTGTPD---EVVEQVAQYNALPVVLENARILKNCVDAKMTEEDKENALSVLQKIYTSPLC-------				70
13. J00777	Rn PBP C3(1)	MKLVFLFLLVFIPICCY-A	-18	SGSG-CSILDEVIRGTINSTVTLHDYMKLVKPYVQDHFTEKAVKQFKQCFL-DQTDETLSNVEVMQFAIFNSESCQOPS--				77
14. J00777	Rn PBP C3(2)	MKLVFLFLLVFIPICCY-A	-18	SGSG-CSILDEVIRGTINSTVTLHDYMKLVKPYVHDHFTANAVKQFKQCFL-DQTNKTVENVGWMIEAIFNSESCQOPS--				77
15. U33147	Hs MGB	MKLLMVLMLAALSQHCY-A	+1	GPC-PLLENVISKTINPQVSKVEYKELLQEFIDDNATTNAIDELKEFCL-NQTDETLSNVEVPMQLIYDSSLC-DI-F	?			93 (prepro)
16. AJ224173	Hs LGB	MKLLMVLMLAALLLHCY-A	+1	DSG-CKLLEDMVEKTINSDISIPEYKELLQEFIDSDAAAEAMGKFQCFL-NQSHRTLRNFGLAMHFTVTDSIKCNMKSN-	?			95 (prepro)
17. J00774	Rn PBP C1	MSSTIELSLCLLIMLAVCCTEANA	-23	SQICELVAHETISFLMKSEE--ELKRELEMYNAPPAAVEAKLEVRKCVDQ-MSNGDRLVVAETL--VYIFLEC-G-----				69 (trunc.)
18. J00776	Rn PBP C2	M---RLSLCLLTILVVCCTEANG	-20	QTLAGVQAQDVTITFLLNPEE--ELKRELREFDAPPEAVEANLKVRCINK-IMYGDRLSMGTSL--VFTMLKCD------				72 (trunc.)
19. Z66540	Ma FHG 22	MKLSLCLLVILAVHCYEAN	+1	AANVCPAVLSVSKSFLFDKVE--KPEAYLQTFNAPPEAVEAKVEVKKCIDSTMYLERKEMGKIL--AEVVGYCKGTEN--				95 (prepro)
20. AJ224171	Hs Lip A	MRLSVCLLLLTIALCCYRANA	+1	VVOALGSEITGFLLAGKP--TMAFEKPLEVAAKMEVKKCVD-TMAYEKRVLITKTLGKIAEK--CDR-----	?			90 (prepro)
21. AJ224172	Hs Lip B	MKLSVCLLLVTLALCCYQANA	+1	EFCPALVSELLDFFISEP--LFKLSLAKFDAPPEAVAAKLGVERCTDQ-MSLQKRSLIAEVLVKILKK--CSV-----	?			90 (prepro)
22. here	Hs YGB	MRLSVCLLMVSLALCCYQAHA	+1	LVCPAVASEITVFLFLSDA--AVNLQVAKLNPPPEALAAKLEVKHCTDQ--ISFKKRSLEKVIVEIVK--CGV------	?			90 (prepro)
23. M77341	Fc Fel dl C2	MRGAL-LVLALLVTQAILG	-17	VKMAETCPIFYDV--FFAVANGNELLLDLS-LTKVNATEPERTAMKIQDEYVENGLISRVLDGLVMIAINEYCMGE----				76 (trunc.)
		*		*	*		*	

FIGURE 4. Multiple sequence alignment of uteroglobin family proteins. The alignment was performed using CLUSTAL W 1.8[75] with standard parameter settings (http://www2.ebi.ac.uk/clustalw). Genbank accession numbers are given for the cognate cDNA entries. Sequence nos. 4, 5, and 22 are not listed in Genbank. (1) The monkey uteroglobin sequence (Mf UGB) can be found in reference 20; (2) pig uteroglobin (Ss UGB) is reported in reference 21; whereas the lymphoglobin (Hs YGB) sequence is reported here. The cDNA sequence of Mf UGB has not been reported yet. The name is composed of the species initials. Oc, *Oryctolagus cuniculus*; Lc, *Lepus capensis*; Hs, *Homo sapiens*; Mf, *Macaca fuscata*; Ss, *Sus scrofa*; Ma, *Mesocricetus auratus*; Rn, *Rattus norvegicus*; Mm, *Mus musculus*; Fc, *Felis catus*) and the abbreviation for the common name found in the literature. The proteins most closely related to rabbit uteroglobin and with only two cysteines in the mature polypeptide are all called uteroglobins (UGB). Sequence no. 9 (Mm LGP = lacrimal gland protein) is not published; it is only listed in Genbank. If the signal peptide is known, +1 is set to the first amino acid of the mature polypeptide; if it is not known, the most likely point of proteolysis is indicated (?) and numbering starts with the initiating methionine. In the headline, the positions of the four α-helices found in the rabbit UGB X-ray structure are marked (HHH…). The length refers to the number of amino acids (aa) in the mature protein or the preprotein (prepro). Sequence no. 17 (111 aa), 18 (98 aa), and 23 (109 aa) were truncated at the C-terminus. A bold underline indicates the start of an intron either within the codon for the underlined amino acid or directly afterwards. The two known N-glycosylation sites within sequence nos. 13 and 23 are highlighted in underlined bold type. The three amino acids conserved in all 23 sequences are marked by asterisks and gray columns. The third cysteine present in sequence nos. 9–23 is also emphasized in gray.

flanking the mature proteins are used as anchoring residues, the alignment is very simple and only few gaps have to be introduced. Most mature proteins have a length of 69–78 amino acid residues, with only rat PBP C1 and C2 and cat Fel dI C2 being significantly longer (98–111 residues).

In those cases where the cognate genes are known, the position of the two introns has been marked. The first intron is always located in one of the last codons of the signal peptide or in the first two codons of the mature protein. The second intron always interrupts the sequence coding for helix 4. Thus, the position of unknown introns can be estimated with an accuracy of 3 bp.

Although all uteroglobin family proteins are classical secretory proteins, only a few proteins are known to be glycosylated. The precise N-glycosylation site within exon 2 is known only for rat PBP C3(1) and cat Fel dI C2 (FIG. 4).

Based on the multiple alignment shown in FIGURE 4, an unrooted phylogenetic tree connecting all family members has been computed with CLUSTAL W (FIG. 5). Using the lastmost internal branch as criterion, five subfamilies A–E can be defined. Subfamily A contains the "true uteroglobins", which are well known to homodimerize. Subfamily B contains four proteins that are known to be expressed in salivary or lacrimal glands. Cat Fel dI C1A and C1B are known to heterodimerize with the only representative of subfamily E, cat Fel dI C2. Also, mouse androgen-binding protein is known to heterodimerize with other ABP subunits (beta and gamma), which have not been fully characterized yet.[38] Subfamily C contains the two rat PBP C3 isoforms as well as human mammaglobin and lacryglobin. PBP C3(1) has been shown to form heterodimers with PBP C1 and C2, two members of subfamily D. Likewise, LGB has been reported to form heterodimers with another subfamily D member, lipophilin A. In addition to these mentioned dimerization partners, subfamily D contains three more homologous proteins: lipophilin B, lymphoglobin, and the hamster FHG22 protein.

DISCUSSION

Based on three independent lines of evidence—radiation hybrid mapping, P1 clone analysis, and human genome sequence data—we have shown that all six human members of the uteroglobin gene family are located on chromosome 11q12.2, forming a very homogeneous paralogous group within an approximately 400-kb sequence. This chromosome 11 region is already notorious for its increased gene density.[66] A gene search program like FGENES (http://genomic.sanger.ac.uk/gf/gf.shtml) does not find more genes belonging to the uteroglobin family in the known nucleotide sequences of this chromosomal region. Human UGB,[22,51] MGB, and LGB[50,52] had been mapped to the same chromosome region before, but lipophilins B and A were assigned to chromosome 10q23 and 15q12-q13, respectively.[34] The latter assignments should be reinvestigated in the light of our results. Moreover, the rat prostatic binding protein genes C1, C2, and C3 have been mapped to rat chromosome 5q21-31 by *in situ* hybridization,[67] and the genes for mouse salivary ABP have been shown to be located on chromosome 7 near the glucose phosphate isomerase-1 locus.[68] Regions of mouse chromosome 7 and human chromosome 11 have been shown already to be syntenic (http://www.informatics.jax.org). Based on our results,

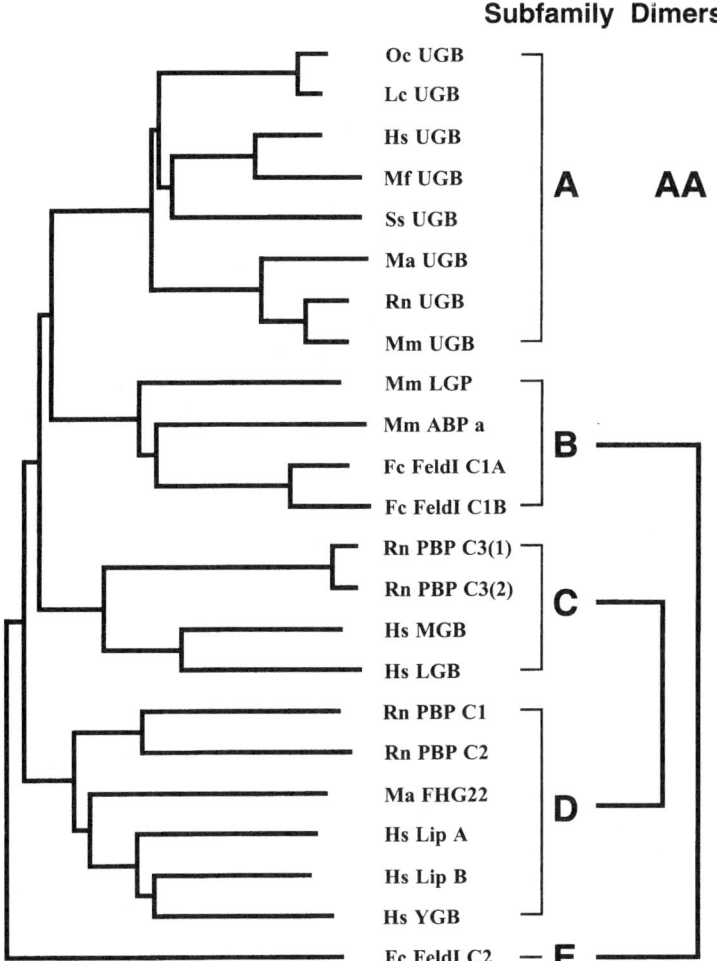

FIGURE 5. An unrooted phylogenetic tree of the uteroglobin family made with CLUSTAL W (see FIG. 4) and njplot (a program by Manolo Gouy, University of Lyon, France). The branch lengths are drawn to scale. Subfamilies A–E are defined as the lastmost internal branches. Members of subfamily A are all known to form homodimers, whereas at least one member of subfamilies B and C is known to form heterodimers with at least one subfamily E or D member, respectively.

we suggest that human chromosome 11q12.2 should also be considered syntenic to rat chromosome 5q21-31.

A PROSITE search against a pattern that describes the conserved characteristics of all known uteroglobins does not identify more homologous proteins in the data bank than already known. Therefore, the uteroglobin family must be considered as a recent invention of evolution restricted to mammalian species that has expanded into a paralogous group by gene duplication events. This is also reflected in a very similar gene structure. All genes are composed of three exons with intron positions strongly conserved. The encoded proteins are also very similar in length and contain two conserved cysteines, each close to the two polypeptide termini, and one central lysine. Apart from these generally conserved residues, the amino acid sequences are very divergent. The orthologous uteroglobins of subfamily A (FIG. 5), for example, are much more divergent than the histones of the same species. Clearly, the physiologic function(s) of the uteroglobins does not require longer patches of conserved amino acid sequences. This is also confirmed by the X-ray structures available for rabbit,[4–6] human,[17] and rat UGB,[16] which show that the 3-D fold is conserved despite considerable sequence divergence. For most of the better-characterized proteins, a quaternary structure with a dimer as the fundamental unit has been shown. For subfamily A, only homodimerization via two disulfide bonds formed by the conserved cysteines has been described. Whereas in the wild type the N-terminal cysteine of one monomer forms a disulfide bond with the C-terminal cysteine of the other antiparallel monomer, in mutants with one of the cysteines mutated to serine[69] parallel homodimers are still formed in which the N-terminal cysteines or the C-terminal cysteines form a disulfide bond. Therefore, a certain degree of promiscuity in dimerization seems to be possible. In subfamilies B–D, a heterodimer, which is also formed by the two terminal cysteines, is the fundamental unit of the quaternary structure. Moreover, in case of the rat prostatic binding proteins, the holoprotein is well characterized as a heterodimer of heterodimers [(PBP C1/C3)(PBP C2/C3)].[28] Interestingly, all subfamily B–D members contain a third conserved cysteine that could be responsible for dimerization of the heterodimers. This implies that, in humans, heterotetramers composed of MGB, LGB, Lip A, Lip B, and/or YGB could be observed. The X-ray or solution structure of prostatein or its putative human homologue would be much welcome in this respect.

The third generally conserved residue is a central lysine (FIG. 4) that is positioned in a region described as a calcium binding site of the phospholipase A2 type for human UGB.[70] By site-directed mutagenesis, it was shown that Asp 46 in Hs UGB acts as a cap residue that traps the calcium ion in its binding site. In the unliganded state, the same Asp 46 forms a salt bridge with Lys 42.[71] Because an aspartate at 4 or 5 residues apart from the conserved lysine is present in 20 of the 23 proteins, this salt bridge also seems to be a common feature of the uteroglobins. Because exposed salt bridges often, and in the UGB in particular, do not contribute to the stability of the protein,[71] it is well conceivable that calcium binding after dissolution of the salt bridge is another general characteristic of the whole family. Only the observation that an Oc UGB K42I mutant could not be expressed in *E. coli*,[72] indicating that the salt bridge is important for the stability of the whole protein, does not fit into this picture.

Only two of the proteins, rat PBP C3(1) and cat Fel dI C, have been shown to be glycosylated at defined asparagine residues (FIG. 4). Glycosylation of MGB has been observed,[73] but the glycosylated residue is unknown. Because in MGB the same sequence NAT is present that has been shown to be the glycosylation site of Fel dI C2, it can be predicted that this also applies for MGB.

On the basis of the phylogenetic tree presented in FIGURE 5, the uteroglobin family can be split into five subfamilies, A–E. In an attempt to introduce a unified nomenclature system into this confusing field, the phylogenetic tree could be used to name each member according to its subfamily affiliation with an appended arbitrary Arabic number (Oc UGB = uteroglobin A1). This official name for a family member could be complemented by a trivial name that is based on its expression pattern as the only noncontroversial protein characteristic and that should be unique in the scientific literature. A similar system is already in use for a couple of large protein families like the nuclear receptors.[74]

The expression profiles presented here for YGB, Lip B, LGB, and MGB overlap with data already reported[31,34] and add some more tissues where expression was unknown so far. For the first time, we could show that a uteroglobin family member is expressed in lymphocytes and which we named lymphoglobin accordingly. Whereas up to now most, if not all, tissues in which uteroglobin expression was described were epithelial tissues separating the body interior from the exterior world, this notion can no longer be maintained in the light of proteins like lipophilin B and lacryglobin, which are expressed in tissues like pancreas, kidney, skeletal muscle, and heart. Therefore, an attractive hypothesis saying that uteroglobins might be involved in microbial defense has lost some ground.

Although the homologies between uteroglobin family genes also extend into regulatory regions (data not shown), the expression pattern of each gene is remarkably distinct. Therefore, it will be interesting to explore which mechanisms govern tissue-specific expression and how these mechanisms are varied between different genes.

ACKNOWLEDGMENTS

We thank Timothy Fleming for providing mammaglobin cDNA and the students, Vincent Beuger, Harald Junge, Catrin Pracht, and Markus Uhrig, for their help. The Resource Center of the German Human Genome Project at the Max-Planck-Institut for Molecular Genetics is acknowledged for providing array filters and P1 clones. This work was supported by a grant from the Kempkes Foundation.

REFERENCES

1. BEIER, H.M. 1968. Uteroglobin: a hormone-sensitive endometrial protein involved in blastocyst development. Biochim. Biophys. Acta **160:** 289–291.
2. KRISHNAN, R.S. & J.C. DANIEL, JR. 1967. "Blastokinin": inducer and regulator of blastocyst development in the rabbit uterus. Science **158:** 490–492.
3. NIETO, A., H. PONSTINGL & M. BEATO. 1977. Purification and quaternary structure of the hormonally induced protein uteroglobin. Arch. Biochem. Biophys. **180:** 82–92.
4. MORNON, J.P., F. FRIDLANSKY, R. BALLY et al. 1980. X-ray crystallographic analysis of a progesterone-binding protein: the C2221 crystal form of oxidized uteroglobin at 2.2 Å resolution. J. Mol. Biol. **137:** 415–429.
5. MORIZE, I., E. SURCOUF, M.C. VANEY et al. 1987. Refinement of the $C222_1$ crystal form of oxidized uteroglobin at 1.34 Å resolution. J. Mol. Biol. **194:** 725–739.
6. BALLY, R. & J. DELETTRE. 1989. Structure and refinement of the oxidized P21 form of uteroglobin at 1.64 Å resolution. J. Mol. Biol. **206:** 153–170.
7. BRANDT, I., J. LUND, A. BERGMAN et al. 1985. Target cells for the polychlorinated biphenyl metabolite 4,4′-bis(methylsulfonyl)-2,2′,5,5′-tetrachlorobiphenyl in lung and kidney. Drug Metab. Dispos. **13:** 490–496.
8. LUND, J., I. BRANDT, L. POELLINGER et al. 1985. Target cells for the polychlorinated biphenyl metabolite 4,4′-bis(methylsulfonyl)-2,2′,5,5′-tetrachlorobiphenyl: charac-

terization of high affinity binding in rat and mouse lung cytosol. Mol. Pharmacol. **27:** 314–323.
9. NORDLUND-MÖLLER, L., O. ANDERSSON, R. AHLGREN *et al.* 1990. Cloning, structure, and expression of a rat binding protein for polychlorinated biphenyls: homology to the hormonally regulated progesterone-binding protein uteroglobin. J. Biol. Chem. **265:** 12690–12693.
10. ANDERSSON, O., L. NORDLUND-MÖLLER, M. BRONNEGARD *et al.* 1991. Purification and level of expression in bronchoalveolar lavage of a human polychlorinated biphenyl (PCB)–binding protein: evidence for a structural and functional kinship to the multihormonally regulated protein uteroglobin. Am. J. Respir. Cell Mol. Biol. **5:** 6–12.
11. SINGH, G., S.L. KATYAL & S.A. GOTTRON. 1985. Antigenic, molecular, and functional heterogeneity of Clara cell secretory proteins in the rat. Biochim. Biophys. Acta **829:** 156–163.
12. SINGH, G., S. SINGAL, S.L. KATYAL *et al.* 1987. Isolation and amino acid composition of the isotypes of a rat Clara cell specific protein. Exp. Lung Res. **13:** 299–309.
13. SINGH, G., S.L. KATYAL, W.E. BROWN *et al.* 1988. Amino-acid and cDNA nucleotide sequences of human Clara cell 10 kDa protein. Biochim. Biophys. Acta **950:** 329–337.
14. GUPTA, R.P., S.E. PATTON, A.M. JETTEN *et al.* 1987. Purification, characterization, and proteinase-inhibitory activity of a Clara-cell secretory protein from the pulmonary extracellular lining of rabbits. Biochem. J. **248:** 337–344.
15. BERNARD, A., H. ROELS, R. LAUWERYS *et al.* 1992. Human urinary protein 1: evidence for identity with the Clara cell protein and occurrence in respiratory tract and urogenital secretions. Clin. Chim. Acta **207:** 239–249.
16. UMLAND, T.C., S. SWAMINATHAN, W. FUREY *et al.* 1992. Refined structure of rat Clara cell 17 kDa protein at 3.0 Å resolution. J. Mol. Biol. **224:** 441–448.
17. UMLAND, T.C., S. SWAMINATHAN, G. SINGH *et al.* 1994. Structure of a human Clara cell phospholipid-binding protein-ligand complex at 1.9 Å resolution. Nat. Struct. Biol. **1:** 538–545.
18. SAGAL, R.G. & A. NIETO. 1998. Molecular cloning of the cDNA and the promoter of the hamster uteroglobin/Clara cell 10-kDa gene (ug/cc10): tissue-specific and hormonal regulation. Arch. Biochem. Biophys. **350:** 214–222.
19. NIETO, A. & M. LOMBARDERO. 1982. Uteroglobin-like antigens in species of Lagomorpha. Comp. Biochem. Physiol. **B71:** 511–514.
20. HASHIMOTO, S., K. NAKAGAWA & K. SUEISHI. 1996. Monkey Clara cell 10 kDa protein (CC10): a characterization of the amino acid sequence with an evolutional comparison with humans, rabbits, rats, and mice. Am. J. Respir. Cell Mol. Biol. **15:** 361–366.
21. SAGAL, R.G. & A. NIETO. 1998. Cloning and sequencing of the cDNA coding for pig pre-uteroglobin/Clara cell 10 kDa protein. Biochem. Mol. Biol. Int. **45:** 205–213.
22. WOLF, M., J. KLUG, R. HACKENBERG *et al.* 1992. Human CC10, the homologue of rabbit uteroglobin: genomic cloning, chromosomal localization, and expression in endometrial cells. Hum. Mol. Genet. **1:** 371–378.
23. HEYNS, W. & P. DE MOOR. 1977. Prostatic binding protein: a steroid-binding protein secreted by rat prostate. Eur. J. Biochem. **78:** 221–230.
24. LEA, O.A., P. PETRUSZ & F.S. FRENCH. 1979. Prostatein: a major secretory protein of the rat ventral prostate. J. Biol. Chem. **254:** 6196–6202.
25. CHEN, C., K. SCHILLING, R.A. HIIPAKKA *et al.* 1982. Prostate alpha-protein: isolation and characterization of the polypeptide components and cholesterol binding. J. Biol. Chem. **257:** 116–121.
26. PARKER, M., M. NEEDHAM & R. WHITE. 1982. Prostatic steroid binding protein: gene duplication and steroid binding. Nature **298:** 92–94.
27. FORSGREN, B., P. BJORK, K. CARLSTROM *et al.* 1979. Purification and distribution of a major protein in rat prostate that binds estramustine, a nitrogen mustard derivative of estradiol-17β. Proc. Natl. Acad. Sci. U.S.A. **76:** 3149–3153.
28. HEYNS, W., B. PEETERS, J. MOUS *et al.* 1978. Purification and characterisation of prostatic binding protein and its subunits. Eur. J. Biochem. **89:** 181–186.
29. PEETERS, B., W. ROMBAUTS, J. MOUS *et al.* 1981. Structural studies on rat prostatic binding protein: the primary structure of its glycosylated component C3. Eur. J. Biochem. **115:** 115–121.

30. HURST, H.C. & M.G. PARKER. 1983. Rat prostatic steroid binding protein: DNA sequence and transcript maps of the two C3 genes. EMBO J. **2:** 769–774.
31. WATSON, M.A. & T.P. FLEMING. 1996. Mammaglobin, a mammary-specific member of the uteroglobin gene family, is overexpressed in human breast cancer. Cancer Res. **56:** 860–865.
32. MOLLOY, M.P., S. BOLIS, B.R. HERBERT et al. 1997. Establishment of the human reflex tear two-dimensional polyacrylamide gel electrophoresis reference map: new proteins of potential diagnostic value. Electrophoresis **18:** 2811–2815.
33. LEHRER, R.I., G. XU, A. ABDURAGIMOV et al. 1998. Lipophilin, a novel heterodimeric protein of human tears. FEBS Lett. **432:** 163–167.
34. ZHAO, C., T. NGUYEN, T. YUSIFOV et al. 1999. Lipophilins: human peptides homologous to rat prostatein. Biochem. Biophys. Res. Commun. **256:** 147–155.
35. BOGGS, J.M., W.J. VAIL & M.A. MOSCARELLO. 1976. Preparation and properties of vesicles of a purified myelin hydrophobic protein and phospholipid: a spin label study. Biochim. Biophys. Acta **448:** 517–530.
36. DOMINGUEZ, P. 1995. Cloning of a Syrian hamster cDNA related to sexual dimorphism: establishment of a new family of proteins. FEBS Lett. **376:** 257–261.
37. MORGENSTERN, J.P., I.J. GRIFFITH, A.W. BRAUER et al. 1991. Amino acid sequence of Fel dI, the major allergen of the domestic cat: protein sequence analysis and cDNA cloning. Proc. Natl. Acad. Sci. U.S.A. **88:** 9690–9694.
38. KARN, R.C. & R. RUSSELL. 1993. The amino acid sequence of the alpha subunit of mouse salivary androgen-binding protein (ABP), with a comparison to the partial sequence of the beta subunit and to other ligand-binding proteins. Biochem. Genet. **31:** 307–319.
39. KARN, R.C. 1994. The mouse salivary androgen-binding protein (ABP) alpha subunit closely resembles chain 1 of the cat allergen Fel dI. Biochem. Genet. **32:** 271–277.
40. GRIFFITH, I.J., S. CRAIG, J. POLLOCK et al. 1992. Expression and genomic structure of the genes encoding FdI, the major allergen from the domestic cat. Gene **113:** 263–268.
41. NOSKE, I.G. & M. FEIGELSON. 1976. Immunological evidence of uteroglobin (blastokinin) in the male reproductive tract and in nonreproductive ductal tissues and their secretions. Biol. Reprod. **15:** 704–713.
42. SAVOURET, J.F., H. LOOSFELT, M. ATGER et al. 1980. Differential hormonal control of a messenger RNA in two tissues: uteroglobin mRNA in the lung and the endometrium. J. Biol. Chem. **255:** 4131–4136.
43. BEIER, H.M., H. BOHN & W. MÜLLER. 1975. Uteroglobin-like antigen in the male genital tract secretions. Cell Tissue Res. **165:** 1–11.
44. KIRCHNER, C. 1976. Uteroglobin in the rabbit. I. Intracellular localization in the oviduct, uterus, and preimplantation blastocyst. Cell Tissue Res. **170:** 415–424.
45. SINGH, G., J. SINGH, S.L. KATYAL et al. 1988. Identification, cellular localization, isolation, and characterization of human Clara cell–specific 10kD protein. J. Histochem. Cytochem. **36:** 73–80.
46. SANDMÖLLER, A., A.K. VOSS, J. HAHN et al. 1991. Cell-specific, developmentally and hormonally regulated expression of the rabbit uteroglobin transgene and the endogenous mouse uteroglobin gene in transgenic mice. Mech. Dev. **34:** 57–67.
47. HAGEN, G., M. WOLF, S.L. KATYAL et al. 1990. Tissue-specific expression, hormonal regulation, and 5′-flanking gene region of the rat Clara cell 10 kDa protein: comparison to rabbit uteroglobin. Nucleic Acids Res. **18:** 2939–2946.
48. SANDMÖLLER, A., R. HALTER, E. GOMEZ-LAHOZ et al. 1994. The uteroglobin promoter targets expression of the SV40 T antigen to a variety of secretory epithelial cells in transgenic mice. Oncogene **9:** 2805–2815.
49. AUMÜLLER, G., E.A. ARCE, W. HEYNS et al. 1995. Immunocytochemical localization of seminal proteins in salivary and lacrimal glands of the rat. Cell Tissue Res. **280:** 171–181.
50. BECKER, R.M., C. DARROW, D.B. ZIMONJIC et al. 1998. Identification of mammaglobin B, a novel member of the uteroglobin gene family. Genomics **54:** 70–78.
51. HAY, J.G., C. DANEL, C.S. CHU et al. 1995. Human CC10 gene expression in airway epithelium and subchromosomal locus suggests linkage to airway disease. Am. J. Physiol. **268:** L565–L575.

52. WATSON, M.A., C. DARROW, D.B. ZIMONJIC et al. 1998. Structure and transcriptional regulation of the human mammaglobin gene, a breast cancer associated member of the uteroglobin gene family localized to chromosome 11q13. Oncogene **16:** 817–824.
53. FRANCIS, F., G. ZEHETNER, M. HÖGLUND et al. 1994. Construction and preliminary analysis of the ICRF human P1 library. Genet. Anal. Tech. Appl. **11:** 148–157.
54. CHURCH, G.M. & W. GILBERT. 1984. Genomic sequencing. Proc. Natl. Acad. Sci. U.S.A. **81:** 1991–1995.
55. MANIATIS, T., E.F. FRITSCH & J. SAMBROOK. 1989. Molecular Cloning: A Laboratory Manual. Second edition. Cold Spring Harbor Laboratory Press. Cold Spring Harbor, New York.
56. ADAMS, M.D., A.R. KERLAVAGE, R.D. FLEISCHMANN et al. 1995. Initial assessment of human gene diversity and expression patterns based upon 83 million nucleotides of cDNA sequence. Nature **377**(suppl.): 3–20.
57. FENG, G.S., Y.B. OUYANG, D.P. HU et al. 1996. Grap is a novel SH3-SH2-SH3 adaptor protein that couples tyrosine kinases to the Ras pathway. J. Biol. Chem. **271:** 12129–12132.
58. ALTSCHUL, S.F., W. GISH, W. MILLER et al. 1990. Basic local alignment search tool. J. Mol. Biol. **215:** 403–410.
59. GYAPAY, G., K. SCHMITT, C. FIZAMES et al. 1996. A radiation hybrid map of the human genome. Hum. Mol. Genet. **5:** 339–346.
60. BRAY-WARD, P., J. MENNINGER, J. LIEMAN et al. 1996. Integration of the cytogenetic, genetic, and physical maps of the human genome by FISH mapping of CEPH YAC clones. Genomics **32:** 1–14.
61. ZEHETNER, G. & H. LEHRACH. 1994. The Reference Library System—sharing biological material and experimental data. Nature **367:** 489–491.
62. COOPER, P.R., N.J. NOWAK, M.J. HIGGINS et al. 1997. A sequence-ready high-resolution physical map of the best macular dystrophy gene region in 11q12-q13. Genomics **41:** 185–192.
63. STOHR, H., A. MARQUARDT, A. RIVERA et al. 1998. A gene map of the Best's vitelliform macular dystrophy region in chromosome 11q12-q13.1. Genome Res. **8:** 48–56.
64. COOPER, P.R., N.J. NOWAK, M.J. HIGGINS et al. 1998. Transcript mapping of the human chromosome 11q12-q13.1 gene-rich region identifies several newly described conserved genes. Genomics **49:** 419–429.
65. THOMPSON, J.D., T.J. GIBSON, F. PLEWNIAK et al. 1997. The CLUSTAL X windows interface: flexible strategies for multiple sequence alignment aided by quality analysis tools. Nucleic Acids Res. **25:** 4876–4882.
66. JAMES, M.R., C.R. RICHARD, J.J. SCHOTT et al. 1994. A radiation hybrid map of 506 STS markers spanning human chromosome 11. Nat. Genet. **8:** 70–76.
67. ZHANG, J., L. DIRCKX, P. MARYNEN et al. 1988. Mapping of rat prostatic binding protein genes C1, C2, and C3 to rat chromosome 5 by in situ hybridization. Cytogenet. Cell Genet. **48:** 121–123.
68. DLOUHY, S.R., B.A. TAYLOR & R.C. KARN. 1987. The genes for mouse salivary androgen-binding protein (ABP) subunits alpha and gamma are located on chromosome 7. Genetics **115:** 535–543.
69. PETER, W., R. DUNKEL, P.F. STOUTEN et al. 1992. Interchain cysteine bridges control entry of progesterone to the central cavity of the uteroglobin dimer. Protein Eng. **5:** 351–359.
70. BARNES, H.J., L. NORDLUND-MÖLLER, M. NORD et al. 1996. Structural basis for calcium binding by uteroglobins. J. Mol. Biol. **256:** 392–404.
71. HENDSCH, Z.S. & B. TIDOR. 1994. Do salt bridges stabilize proteins? A continuum electrostatic analysis. Protein Sci. **3:** 211–226.
72. PETER, W. 1990. Identifizierung und Charakterisierung des Progesteron-Bindungsortes von Uteroglobin durch spezifische Mutagenese und Expression in *E. coli*. Dissertation (Marburg), pp. 63–68.
73. WATSON, M.A., S. DINTZIS, C.M. DARROW et al. 1999. Mammaglobin expression in primary, metastatic, and occult breast cancer. Cancer Res. **59:** 3028–3031.
74. NUCLEAR RECEPTOR NOMENCLATURE COMMITTEE. 1999. A unified nomenclature system for the nuclear receptor superfamily. Cell **97:** 161–163.
75. HIGGINS, D.G., J.D. THOMPSON & T.J. GIBSON. 1996. Using CLUSTAL for multiple sequence alignments. Methods Enzymol. **266:** 383–402.

Clara Cell Proteins

GURMUKH SINGH[a] AND SIKANDAR L. KATYAL

Department of Veterans Affairs and Department of Pathology, University of Pittsburgh School of Medicine, Pittsburgh, Pennsylvania 15240, USA

> ABSTRACT: Clara cells are nonciliated, nonmucous, secretory cells of the pulmonary airways. These cells are known to secrete a variety of proteins, including Clara cell 10-kDa protein/uteroglobin. This protein consists of a homodimer of 70–77 amino acid polypeptides arranged in antiparallel fashion. *In vitro* testing suggests that the protein suppresses inflammation. The physiologic role of the protein remains to be determined.

Gas exchange between the environment and the circulating blood is carried out by transporting air through the tracheobronchial tree to the alveoli. The architectural and functional integrity of the airway and alveolar epithelia is essential for maintaining the gas transport function of the lung.[1–3]

During embryogenesis, the lung bud develops from the foregut epithelium and grows into the surrounding mesenchyme. Repetitive branching and growth of the lung bud lead to the formation of the tracheobronchial tree. The branching of the tracheobronchial tree itself is controlled by differential regulation of the growth of the epithelium.[3,4]

PULMONARY EPITHELIAL CELLS

The adult lung consists of some 40 different cell types derived from the foregut epithelium and the surrounding mesenchyme. The major types of epithelial cells lining the parenchyma are ciliated cells, mucous cells, serous cells, basal cells, neuroepithelial/Kulchitsky cells, Clara cells, type II alveolar pneumocytes, and type I pneumocytes.[4] In addition to these major cell types, there are other less-recognized cell types that may be subtypes of the major cells: for example, the Clara cells at the bifurcation of the bronchial tree act as progenitor cells for the bronchial epithelium and facilitate regeneration of the epithelium following injury. Similarly, in human lungs, there appears to be a cuboidal cell population at the junction of the bronchial tree and the alveoli that has characteristics of both Clara cells and type II pneumocytes.[5–7]

The pleural-lining mesothelial cells appear to be of different embryologic origin than the parenchymal epithelial cells. Morphologically, the mesothelial cells are

---

[a]Address for correspondence: Gurmukh Singh, VA Medical Center, University Drive, Pittsburgh, PA 15240. Voice: 412-688-6745; fax: 412-699-6135.
gurmukh@pol.net

similar to epithelial cells and perform similar functions. The mesothelial cells even share surface protein markers with ciliated cells of the tracheobronchial tree. It should be noted that the antigen shared by the mesothelial cells and the ciliated cells is not associated with cilia per se, but is present on the luminal surface of these cell types.[8,9]

Clara cells are nonmucous, nonserous, columnar to cuboidal secretory cells lining the bronchial tree. A smaller number of cells with similar characteristics may also be present in the trachea in some species. Clara cells were first described based on the morphology and histochemistry by Kolliker in 1881–82.[10] However, it was the description of the same cell types by Max Clara in 1937 that earned these cells a recognized place in human anatomy and histology.[11] Clara's description of the cells that now bear his name was based on the study of human lungs. According to his characterization, Clara cells are nonciliated, nonmucous, and nonserous secretory cells present in the segmental bronchi. Subsequent morphologic studies of these cells in different species have revealed considerable structural and morphologic variation among the species.[12–15] However, their secretory nature and the lack of mucus are consistent features among the various species examined. In general, Clara cells are club-shaped, with a dome-shaped luminal surface projecting above the surrounding cells. The presence of secretory granules recognized in the initial description by Kolliker and by all subsequent morphologic studies allowed morphologic identification of the cells and also led to studies on the regulation of their secretion through morphometric quantitation of secretory granules before the nature of Clara cell secretory proteins was known.[16,17] Based on the digestion of histologic sections with proteolytic enzymes, it was concluded that Clara cell granules contain proteins. The major secretory protein of Clara cells was later identified as a 10- to 16-kDa protein, which is abundant in the bronchoalveolar lavage fluid.[18–24]

Clara cells entered the lexicon of most biologists following a description by Niden stating that these cells secrete surfactant lipids, which are taken up by type II pneumocytes. Even though this notion was later shown to be erroneous, the paper is credited with raising the status of Clara cells.[25]

Despite exquisite and detailed studies of the structure of Clara cells, the physiologic role of these cells in pulmonary biology has not been entirely settled. The major roles ascribed to these cells include (a) secretion of the hypophase layer of bronchoalveolar fluid, (b) function as reparative cells in the bronchial epithelium, and (c) metabolism of xenobiotics through their abundant P450 cytochrome-dependent mixed-function oxygenases.[6,12,25–31]

CLARA CELL PROTEINS

Clara cells obviously share a large number of the housekeeping and structural proteins with other cells in the body. However, they also have many proteins that are selectively expressed in such cells. A number of the Clara cell–selective/specific proteins are those secreted by these cells. Of the nonsecretory Clara cell proteins, those associated with the P450 cytochrome-dependent mixed-function oxygenases are perhaps the most important. This enzymatic system is geared toward metabolism of xenobiotics. However, in the process of metabolizing some molecules, Clara cells turn these entities into toxic metabolites as in the case of ipomeanol and naphthalene.[26–31] Activation of these compounds by the P450 cytochrome oxygenases accounts for selective toxicity of these compounds to Clara cells. However, the promise of selective toxicity to Clara cells has not translated into any clinically useful pharmaceuticals. The following proteins are known to be secreted by Clara cells: Clara cell 10-kDa protein (CC10); Clara cell 55-kDa protein; surfactant proteins A, B, and D; Clara cell tryptase; β-galactoside-binding lectin; and perhaps a specific phospholipase.[24–32]

Two major approaches were taken to study Clara secretory proteins: (a) analysis of bronchoalveolar lavage fluid proteins and immunohistochemical localization of these proteins in various lung cell types; (b) examination of proteins released into the culture medium by cultures of isolated Clara cells. CC10 protein/uteroglobin was identified to be the dominant secretory protein of Clara cells by both of these approaches.[21–24,33]

CC10 PROTEIN

A low-molecular-weight protein present in bronchoalveolar lavage fluid has been conclusively localized to secretory granules of Clara cells.[21–24] The structure of this protein is similar to that of a previously described protein secreted by rabbit endometrium called uteroglobin.[34–40] The protein has been referred to by various names in the literature. In the initial description of the protein, related to its appearance in the rabbit reproductive tract around the time of implantation, it was called blastokinin. The protein is also referred to in the literature by the following names: Clara cell secretory protein (CCSP), Clara cell 16-kDa protein, Clara cell 10-kDa protein, human protein 1, urine protein 1, and polychlorinated biphenyl–binding protein. In this communication, the protein will be referred to as Clara cell 10-kDa protein (CC10).[24]

The human CC10 protein was isolated from bronchoalveolar lavage fluid and shown to be present in the secretory granules of nonserous, nonmucous, nonciliated columnar cells lining the bronchial tree. The protein is present in developing human lung and is detectable by immunohistochemistry at about 21 weeks of gestation. The protein is also detectable in the amniotic fluid. Examination of human lung tumors revealed that 10% to 30% of pulmonary adenocarcinomas exhibit focal staining for CC10. The staining in malignant cells is usually much less intense than that seen in normal Clara cells.[23] Among the experimentally induced tumors in hamsters, staining for CC10 is exhibited only in early phases of tumor development. These tumors on further growth often differentiated into squamous cell carcinomas and did not express CC10. Most other animal models of tumor induction, for example, nitrosourea-induced tumors in rat and mouse, do not exhibit CC10 expression. In fact, almost all of these tumors stain for surfactant proteins, suggesting their origin from or differentiation into type II pneumocytes.[41–47]

CC10 protein appears to be secreted constitutively at a fairly high level. Secretion of the protein may be stimulated by adrenergic agonists. Massaro *et al.* demonstrated that the cholinergic secretagogue activity seen in intact animals was attributed to an indirect sympathetic stimulation.[17] Later studies by Plopper *et al.* suggest that the bulk of the Clara cell secretory activity is probably constitutive with minimal effects of secretagogues.[16,17,48]

STRUCTURE OF CC10 PROTEIN/UTEROGLOBIN

CC10 protein/uteroglobin consists of two identical polypeptides arranged in an antiparallel fashion. The molecular weight of the protein based on the amino acid sequence and resonance studies appears to be about 16,000 Da. The protein has an apparently high mobility in SDS polyacrylamide gels and has an apparent molecular weight of 10 to 12 kDa. Hence, in the initial descriptions, the protein was described as CC10 protein. The polypeptide chain of CC10 consists of 70 to 77 amino acids. In primates, including humans, and Lagomorpha, the peptide consists of 70 amino acids, whereas the rodent protein (rat, mouse, and hamster) has 2 additional amino acid residues at the N-terminal and 5 additional amino acids at the C-terminal for a total of 77 amino acids. The sequence of the CC10 monomer from various species is shown in TABLE 1. The numbering of residues was based on the sequence of human protein with the N-terminal glutamic acid labeled as residue 1. The two N-terminal serines in rodent protein are labeled as −1 and −2. With this numbering system, all of the proteins have cysteines at positions 3 and 69. In the dimeric, mature secreted protein, the cysteine at position 3 of one residue is linked to the cysteine at position 69 of the second chain. The nascent protein has a 19- to 21-amino-acid-long signal peptide, typical of most secretory proteins.[49–61]

Each of the CC10 monomers has 4 alpha helices separated by beta bends. The 8 alpha helices of the dimer enclose a hydrophobic pocket. The presence of a hydrophobic pocket is one of the consistent findings in the proteins from various species. The residues that line the hydrophobic pocket are highly conserved and the substitutions almost invariably are conservative.

Even though uteroglobin was initially thought to exert its action by binding progesterone, progesterone has not been demonstrated to be present in association with native uteroglobin. However, X-ray diffraction crystallography of the purified human CC10 protein showed a lipid occupying the hydrophobic cavity. Extraction of the isolated protein and analysis of the lipid extract by gas chromatography revealed that the protein binds phosphatidylcholine and phosphatidylinositol in nearly equal amounts.[61]

The amino acid sequence of CC10 is about 50% identical among the various species. If conservative substitutions of amino acids are discounted, the match of residues among the species increases to about 70%. Comparisons of the amino acid sequences among the species as well as of the proteins generated by site-directed mutagenesis suggest that phenylalanine 6, tyrosine 21, and threonine 60 are important for the binding of progesterone by CC10. Because progesterone binding requires prior chemical reduction of the protein, it is doubtful that this activity is physiological.[56,57]

FUNCTION OF CC10 PROTEIN

Despite elucidation of the complete amino acid sequence and the detailed X-ray diffraction crystallographic structure of the protein, the primary physiologic function of CC10 remains unknown. A number of properties/functions have been ascribed to CC10/uteroglobin. A listing of these functions/properties is given below with brief comments.

TABLE 1. Amino acid sequence of CC10 protein in different species

	-2	-1	1	2	3	4	5	6	7	8	9	10	11	12	13	14	15	16
Human			Glu	Ile	Cys	Pro	Ser	Phe	Gln	Arg	Val	Ile	Glu	Thr	Leu	Leu	Met	Asp
Monkey			Glu	Ile	Cys	Pro	Thr	Phe	Leu	Arg	Val	Ile	Glu	Ser	Leu	Phe	Leu	Asp
Rat	Ser	Ser	Asp	Ile	Cys	Pro	Gly	Phe	Leu	Gln	Val	Leu	Glu	Ala	Leu	Leu	Leu	Gly
Mouse	Ser	Ser	Asp	Ile	Cys	Pro	Gly	Phe	Leu	Gln	Val	Leu	Glu	Ala	Leu	Leu	Met	Glu
Hamster	Ser	Ser	Asp	Thr	Cys	Pro	Gly	Phe	Phe	Gln	Val	Ile	Glu	Phe	Leu	Phe	Met	Gly
Rabbit			Gly	Ile	Cys	Pro	Arg	Phe	Ala	His	Val	Leu	Glu	Asn	Leu	Leu	Leu	Gly
Hare			Gly	Ile	Cys	Pro	Gly	Phe	Ala	His	Val	Leu	Lys	Glu	Phe	Gln	Pro	Asp
Pig			Glu	Val	Cys	Pro	Ser	Phe	Val	Glu	Val	Ile	Gln	Asn	Leu	Phe	Lys	Gly

	17	18	19	20	21	22	23	24	25	26	27	28	29	30	31	32
Human	Thr	Pro	Ser	Ser	Tyr	Glu	Ala	Ala	Met	Glu	Leu	Phe	Ser	Pro	Asp	Gln
Monkey	Thr	Pro	Ser	Ser	Phe	Glu	Ala	Ala	Met	Gly	Phe	Phe	Ser	Pro	Asp	Gln
Rat	Ser	Glu	Ser	Asn	Tyr	Glu	Ala	Ala	Leu	Lys	Pro	Phe	Asn	Pro	Ala	Ser
Mouse	Ser	Glu	Ser	Gly	Tyr	Val	Ala	Ser	Leu	Lys	Pro	Phe	Asn	Pro	Gly	Ser
Hamster	Ser	Glu	Ser	Ser	Tyr	Glu	Ala	Ala	Leu	Lys	Pro	Tyr	Asn	Pro	Gly	Ser
Rabbit	Thr	Pro	Ser	Ser	Tyr	Gly	Thr	Ser	Leu	Lys	Glu	Phe	Glu	Pro	Asp	Asp
Hare	Thr	Pro	Ser	Ser	Tyr	Gly	Thr	Ser	Leu	Lys	Glu	Phe	Gln	Pro	Asp	Asp
Pig	Thr	Leu	Ala	Ser	Try	Glu	Ala	Ser	Val	Glu	Pro	Phe	Ser	Pro	Asn	Glu

TABLE 1. Continued

	33	34	35	36	37	38	39	40	41	42	43	44	45	46	47	48
Human	Asp	Met	Arg	Glu	Ala	Gly	Ala	Gln	Leu	Lys	Lys	Leu	Val	Asp	Thr	Leu
Monkey	Asp	Met	Ser	Glu	Ala	Gly	Ala	Gln	Leu	Lys	Lys	Leu	Val	Asp	Thr	Leu
Rat	Asp	Leu	Gln	Asn	Ala	Gly	Thr	Gln	Leu	Lys	Arg	Leu	Val	Asp	Thr	Leu
Mouse	Asp	Leu	Gln	Asn	Ala	Gly	Thr	Gln	Leu	Lys	Arg	Leu	Val	Asp	Thr	Leu
Hamster	Asp	Leu	Gln	Asp	Ser	Gly	Thr	Gln	Leu	Lys	Lys	Leu	Val	Asp	Thr	Leu
Rabbit	Thr	Met	Lys	Asp	Ala	Gly	Met	Gln	Met	Lys	Lys	Val	Leu	Asp	Ser	Leu
Hare	Ala	Met	Lys	Asp	Ala	Gly	Met	Gln	Met	Lys	Lys	Val	Leu	Asp	Thr	Leu
Pig	Asp	Met	Lys	Lys	Ala	Gly	Ala	Gln	Leu	Lys	Thr	Leu	Val	Asp	Thr	Leu

	49	50	51	52	53	54	55	56	57	58	59	60	61	62	63	64
Human	Pro	Gln	Lys	Pro	Arg	Glu	Ser	Ile	Ile	Lys	Leu	Met	Glu	Lys	Ile	Ala
Monkey	Pro	Ala	Lys	Ala	Arg	Asp	Ser	Ile	Ile	Lys	Leu	Met	Glu	Lys	Ile	Asp
Rat	Pro	Gln	Glu	Thr	Arg	Ile	Asn	Ile	Val	Lys	Leu	Thr	Glu	Lys	Ile	Leu
Mouse	Pro	Gln	Glu	Thr	Arg	Ile	Asn	Ile	Met	Lys	Leu	Thr	Glu	Lys	Ile	Leu
Hamster	Pro	Gln	Lys	Thr	Arg	Met	Asn	Ile	Met	Lys	Leu	Ser	Glu	Ile	Ile	Leu
Rabbit	Pro	Gln	Thr	Thr	Arg	Glu	Asn	Ile	Met	Lys	Leu	Thr	Glu	Lys	Ile	Val
Hare	Pro	Gln	Thr	Thr	Arg	Glu	Asn	Ile	Ile	Lys	Leu	Thr	Glu	Lys	Ile	Val
Pig	Ser	Pro	Glu	Ala	Lys	Asp	Ser	Val	Leu	Lys	Leu	Gln	Glu	Lys	Ile	Ile

TABLE 1. Continued

	65	66	67	68	69	70	71	72	73	74	75
Human	Gln	Ser	Ser	Leu	Cys	Asn					
Monkey	Lys	Ser	Leu	Leu	Cys						
Rat	Thr	Ser	Pro	Leu	Cys	Glu	Gln	Asp	Leu	Arg	Val
Mouse	Thr	Ser	Pro	Leu	Cys	Lys	Gln	Asp	Leu	Arg	Phe
Hamster	Thr	Ser	Pro	Leu	Cys	Asn	Gln	Asp	Leu	Ser	Val
Rabbit	Lys	Ser	Pro	Leu	Cys	Met					
Hare	Lys	Ser	Pro	Leu	Cys	Met					
Pig	Lys	Ser	Pro	Leu	Cys	Ala					

Progesterone Binding

The binding of progesterone by reduced rabbit uteroglobin was the first function ascribed to the protein. It was considered to be important in local sequestration and concentration of progesterone to facilitate the development of endometrium in early pregnancy. However, there is wide variation in the progesterone-binding activity of the protein among species, even among the Lagomorpha.[51,53,56] There is no evidence that progesterone is associated with the native protein. Transgenic animals deficient in CC10/uteroglobin continue to reproduce normally, further calling into question the physiologic importance of progesterone binding in reproductive functions.[52]

A Substrate for Transglutaminase

By serving as a substrate for transglutaminase, uteroglobin was thought to facilitate the immunomodulation required for carrying a histo-incompatible fetus. As mentioned earlier, transgenic animals deficient in uteroglobin/CC10 continue to reproduce normally, calling into question the physiologic importance of this property.[61–65]

Protease Inhibition

Rabbit uteroglobin was shown to be an inhibitor of papain even though the inhibition was not particularly strong. As in the case of progesterone binding, there is marked variation in the protease inhibitory activity of CC10 from various species, including actual enhancement of such activity in the case of human CC10.[56]

Inhibition of Phospholipase A2

By inhibiting phospholipase A2, CC10 is thought to downregulate inflammatory activity. Inhibition of phospholipase A2 is a consistent finding among CC10 proteins of all species examined.[56,66–70] Inhibition of phospholipase A2 may be a physiologically important function. It is worth noting that the lung is exposed to various environmental agents in the inspired air, many of which reach the alveoli. Either through their primary irritant property or because of immune response to the inhaled molecules, the lung could be a site of constant inflammatory activity designed to counteract and eliminate the inhaled molecules. Presence of an anti-inflammatory agent in the lung would appear logical. The human CC10 gene has been mapped to the chromosome 11q12.2-13.1 region. The chromosomal region also encodes genes for other proteins associated with anti-inflammatory functions. However, at a basal level, the lungs in transgenic animals deficient in CC10 do not exhibit unchecked inflammatory activity.[62,63,71,72]

Polychlorinated Biphenyl (PCB) Binding

CC10 has been shown to bind PCBs *in vitro*. In animals exposed to PCBs, the chemical is also found in the secretory granules of Clara cells, suggesting that the binding takes place *in vivo* as well. However, PCBs are man-made molecules and are otherwise not present in nature. Therefore, it is unlikely that CC10 was created to bind PCBs.[69,70]

Calcium Binding

CC10 has been shown to bind and sequester calcium. It is possible that some of the functions of CC10 may be mediated through its calcium-chelating activity.[73] The spatial disposition of atoms that could form ligands with calcium in the calcium-binding site appears similar to that of the primary calcium-binding site of secretory phospholipase A2 enzymes.[74]

Binding to Phosphatidylcholine and Phosphatidylinositol

As mentioned earlier, isolated human native CC10 contains phosphatidylcholine and phosphatidylinositol in nearly equal amounts. The binding to phosphatidylcholine may be relevant to its anti-inflammatory activity by sequestering the substrate for phospholipase A2. However, it is more likely that the phosphatidylcholine and phosphatidylinositol binding by CC10 is associated with transport of surfactant lipids and probably of other hydrophobic molecules in the lung.[61]

Retinoid Binding

Uteroglobin was shown to bind retinoic acid and retinol in a nonsaturable manner and the binding increased by about 10-fold by chemical reduction of uteroglobin. The physiological role of this property, like others, has not been elucidated. Requirement for prior chemical reduction of the protein further casts doubt on the physiologic importance of this property.[75]

Binding to Microsomes and Plasma Membranes

Uteroglobin has been shown to bind microsomes and plasma membranes isolated from rat liver. The binding is saturable and depends on prior chemical reduction of uteroglobin. It is possible that the binding is related to the presence of specific uteroglobin-binding proteins on the surface of such membranes. Once again, the need for a reduced protein makes a physiologic role for this property less likely.[76]

Anti-inflammatory Activity

As mentioned earlier, the primary physiological role of CC10 may be as an inhibitor of inflammation. Inhibition of phospholipase A2 is the major reason for suspecting this to be the main role for the protein. Attempts to test the anti-inflammatory activity of the protein and its peptides have produced varying results in the hands of different investigators.[56,61–64,71,72]

Immunomodulatory Activity

By serving as a substrate for transglutaminase, uteroglobin was thought to function in dampening the immune response to a histo-incompatible fetus.[65,77–79] More recently, CC10 has been shown to have anticytokine activity in that it inhibits the production of IFN-gamma, IL-1, and TNF-alpha. The protein has also been demonstrated to exhibit suppression of antiviral activity and augmentation of phagocytosis induced by INF-gamma. In addition, CC10/uteroglobin has been shown to inhibit leukocyte chemotaxis.[72,80]

Tumor Inhibition

While CC10 is present in normal epithelial cells, prostate cancer cells lose CC10, suggesting a role for the protein in inhibiting malignant growth. *In vitro* studies of CC10 reveal an inhibition of penetration of the matrix by tumor cells. It is possible that the protein may be an inhibitor of malignancy.[81–83]

Prevention of Renal Disease

The development of transgenic animals in whom the CC10 gene had been inactivated was expected to provide the definitive answer for the physiological role for CC10/uteroglobin. However, the two models, one developed by Mukherjee *et al.* and the other by Stripp *et al.*, demonstrate very different phenotypes. An explanation for these marked differences has not been forthcoming. In Mukherjee's model, the animals develop excessive deposition of fibronectin in the kidney and develop renal failure. Based on this model, CC10 may also be important in prevention of IgA nephropathy in humans.[62,63,84–86]

Prevention of Oxygen Toxicity

The transgenic CC10 null mouse model developed by Stripp *et al.* does not exhibit any renal disease or any other abnormalities under baseline conditions. Stripp *et al.* showed that the CC10-deficient mice are more susceptible to oxygen toxicity. However, this increase in toxicity was not considered to be due to enhancement of the inflammatory response.[24,62]

Protection against Asthma

Genetic variation in the noncoding region of the human CC10 gene affects the level of CC10 in plasma. Asthmatic children had lower levels of plasma CC10/CC16. The 38A gene sequence was associated with lower levels of CC10/CC16 and individuals with lower levels had a higher incidence of asthma. It is possible that CC10 protects against asthma through its anti-inflammatory activity.[71,72,79,87]

OTHER CLARA CELL PROTEINS

Clara Cell 55-kDa Protein

While analyzing bronchoalveolar lavage fluid, a 55-kDa protein was detected in rat lung lavage that was localizable to Clara cells. The protein is resistant to digestion with trypsin. However, further studies on the structure and function of the protein remain to be performed.[22,32]

Leukocyte Protease Inhibitor

An acid-stable 11.7-kDa protein that inhibits neutrophil elastase and cathepsin G has been detected in airway secretions and in other organs secreting mucus. This protein has been localized to the Clara cells as well as mucus cells of the lung, salivary glands, uterine cervical glands, etc.[88–91]

β-Galactosidase-Binding Protein

Analysis of rat lung lavage fluid has revealed two additional proteins (of 14 and 29 kDa) that react with an antiserum to β-galactosidase-binding lectin. The antiserum specifically stains Clara cells, suggesting that Clara cells secrete β-galactosidase-binding lectins.[92]

Phospholipase

Rat lungs stained positively by cytochemistry for phospholipase A, lysophospholipase, and lipase. Because the activity was localizable to the secretory granules, it is assumed that the enzymes are secreted and may be important in the catabolism of surfactant lipids.[32,93]

Tryptase Clara

A 30-kDa protein has been detected in the lung lavage fluid. The protein is localizable to Clara cells. It has protease activity similar to that of trypsin and appears to be important in the activation of influenza virus and its pathogenicity to the lung.[94,95]

Surfactant-Associated Proteins

Surfactant-associated proteins A, B, and D have been localized to the epithelial cells in the distal-most respiratory tree. These proteins and the associated surfactant lipids may be important in maintaining the patency of distal airways as well as in the antibacterial activity of surfactant proteins A and D.[96–101]

ACKNOWLEDGMENTS

We are grateful to the Department of Pathology and the Department of Veterans Affairs, University of Pittsburgh School of Medicine, for support.

REFERENCES

1. PLOPPER, C.G. & K.E. PINKERTON. 1991. Overview of diversity in the respiratory system of mammals. *In* Comparative Biology of the Normal Lung. Volume I, chapter 1, pp. 3–5. CRC Press. Boca Raton, FL.
2. TYLER, W.S. & M.D. JULIAN. 1991. Gross and subgross anatomy of lungs, pleura, connective tissue septa, distal airways, and structural units. *In* Comparative Biology of the Normal Lung. Volume I, chapter 4, pp. 37–48. CRC Press. Boca Raton, FL.
3. SPENCER, H. 1985. The anatomy of the lung. *In* Pathology of the Lung, pp. 70–78. Pergamon. Elmsford, New York.
4. GAIL, D.B. & C.J.M. LENFANT. 1983. Cells of the lung: biology and clinical implications. Am. Rev. Respir. Dis. **127**: 366–387.
5. BASSET, F., J. POIRIER, M. LE CROM & J. TURIAF. 1971. Etude ultrastructurale de l'epithelium bronchiolaire humain. Z. Zellforsch. **116**: 425–442.
6. KHOOR, A., M.E. GRAY, G. SINGH & M.T. STAHLMAN. 1996. Ontogeny of Clara cell–specific protein and its mRNA: their association with neuroepithelial bodies in human fetal lung and in bronchopulmonary dysplasia. J. Histochem. Cytochem. **44**: 1429–1438.

7. STRIPP, B., K. MAXSON, R. MERA & G. SINGH. 1995. Repair of mouse lung epithelium from naphthalene injury involves distinctly regulated temporal and spatial expression of markers for cellular differentiation and proliferation. Am. J. Physiol. (Lung Cell Mol. Physiol.) **13:** L791–L799.
8. HERMANS, C. & A. BERNARD. 1999. Lung epithelium-specific proteins. Am. J. Respir. Crit. Care Med. **159:** 645–677.
9. SINGH, G., J. SINGH, N.G. ORDONEZ *et al.* 1995. Expression of a 130 kDa mesothelial and ciliated cell antigen (MCp130) in normal and developing human and rat lung, and its role as a diagnostic marker for mesotheliomas and tumors of the female reproductive system. Lab. Invest. **73:** 48–58.
10. KOLLIKER, A. 1881. Zurkeniniss des Baues der Lunge des Menschen. Verh. Phys. Med. Ges. Wurzburg **16:** 1–24.
11. CLARA, M. 1937. Zur Histobiologie des Bronchiaepithels. Z. Mikrosk. Anat. Forsch. **41:** 321–347.
12. PLOPPER, C.G. 1997. Clara cells. *In* Lung Biology in Health and Disease. Volume 100. Dekker. New York.
13. PLOPPER, C.G., A.P. MARIASSY & L.H. HILL. 1980. Ultrastructure of the non-ciliated bronchiolar epithelial (Clara) cell of mammalian lung: I. A comparison of rabbit, guinea pig, rat, hamster, and mouse. Exp. Lung Res. **1:** 139–154.
14. PLOPPER, C.G., A.P. MARIASSY & L.H. HILL. 1980. Ultrastructure of the non-ciliated bronchiolar epithelial (Clara) cell of mammalian lung: II. A comparison of horse, steer, sheep, dog, and cat. Exp. Lung Res. **1:** 155–169.
15. PLOPPER, C.G., A.P. MARIASSY & L.H. HILL. 1980. Ultrastructure of the non-ciliated bronchiolar epithelial (Clara) cell of mammalian lung: III. A study of man with comparison of 50 mammalian species. Exp. Lung Res. **1:** 171–180.
16. YONEDA, K. 1977. Pilocarpine stimulation of the bronchiolar Clara cell secretion. Lab. Invest. **37:** 447–452.
17. MASSARO, G.D., C.M. FISCHMAN, M-J. CHIANG *et al.* 1981. Regulation of secretion in Clara cells. J. Clin. Invest. **67:** 345–351.
18. CUTZ, E. & P.E. CONEN. 1971. Ultrastructure and cytochemistry of Clara cells. Am. J. Pathol. **62:** 127–134.
19. KUHN, C., III, L.A. CALLAWAY & F.B. ASKIN. 1974. The formation of granules in the bronchiolar Clara cells of the rat. J. Ultrastruct. Res. **49:** 387–400.
20. KUHN, C., III & L.A. CALLAWAY. 1975. The formation of granules in the bronchiolar Clara cells of the rat. J. Ultrastruct. Res. **53:** 66–76.
21. SINGH, G. & S.L. KATYAL. 1984. An immunologic study of the secretory products of rat Clara cells. J. Histochem. Cytochem. **32:** 49–54.
22. SINGH, G. & S.L. KATYAL. 1985. Antigenic, molecular, and functional heterogeneity of Clara cell secretory proteins in the rat. Biochim. Biophys. Acta **829:** 156–163.
23. SINGH, G., J. SINGH, S.L. KATYAL *et al.* 1988. Identification, cellular localization, isolation, and characterization of human Clara cell specific 10 kD protein. J. Histochem. Cytochem. **36:** 73–80.
24. SINGH, G. & S.L. KATYAL. 1997. Clara cells and Clara cell 10 kD protein (CC10). Am. J. Respir. Cell Mol. Biol. **17:** 141–143.
25. NIDEN, A.H. 1967. Bronchiolar and large alveolar cell in pulmonary phospholipid metabolism. Science **158:** 1323–1324.
26. BRODY, A.R., G. HOOK, G.S. CAMERON *et al.* 1987. The differentiation capacity of Clara cells isolated from the lungs of rabbits. Lab. Invest. **57:** 219–229.
27. BOYD, M.R. 1977. Evidence for the Clara cell as a site of cytochrome P450–dependent mixed-function oxidase activity in lung. Nature **269:** 713–715.
28. SERABJIT-SINGH, C.J., C.R. WOLF & R.M. PHILPOT. 1980. Cytochrome P-450: localization in rabbit lung. Science **207:** 1469–1470.
29. JI, C-M., W.V. CARDOSO, A. GEBREMICHAEL *et al.* 1995. Pulmonary cytochrome P-450 monooxygenase system and Clara cell differentiation in rats. Am. J. Physiol. (Lung Cell Mol. Physiol.) **13:** L394–L402.
30. CHICHESTER, C.H., R.M. PHILPOT, A.J. WEIR *et al.* 1991. Characterization of the cytochrome P-450 monooxygenase system in nonciliated bronchiolar epithelial (Clara) cells isolated from mouse lung. Am. J. Respir. Cell Mol. Biol. **4:** 179–186.

31. DOMIN, B.A., T.R. DEVEREUX & R.M. PHILPOT. 1986. The cytochrome P-450 monooxygenase system of rabbit lung: enzyme components, activities, and induction in the nonciliated bronchiolar epithelial (Clara) cell, alveolar type II cell, and alveolar macrophage. Mol. Pharmacol. **30:** 296–303.
32. SINGH, G. & S.L. KATYAL. 1991. Secretory proteins of Clara cells and type II cells. *In* Treatise on Pulmonary Toxicology. Volume I, pp. 93–103. CRC Press. Boca Raton, FL.
33. PATTON, S.E., L.B. GILMORE, A.M. JETTEN *et al.* 1986. Biosynthesis and release of proteins by isolated pulmonary Clara cells. Exp. Lung Res. **11:** 277–294.
34. KRISHNAN, R.S. & J.C. DANIEL, JR. 1967. "Blastokinin": inducer and regulator of blastocyst development in the rabbit uterus. Science **148:** 490–492.
35. BEATO, M. & H.M. BEIER. 1978. Characteristics of the purified uteroglobin-like protein from rabbit lung. J. Reprod. Fertil. **53:** 305–314.
36. AUMULLER, G., J. SEITZ, W. HEYNS & C. KIRCHNER. 1985. Ultrastructural localization of uteroglobin immunoactivity in rabbit lung and endometrium, and rat ventral prostate. Histochemistry **83:** 413–417.
37. HACKENBERG, R. & H.M. BEIER. 1982. Proteinase identification in the uteroglobin fraction of rabbit uterine secretions. Arch. Gynecol. **231:** 289–297.
38. HEGEL-HARTUNG, C. & H.M. BEIER. 1985. Immunocytochemical localization of uteroglobin in the rabbit endometrium. Anat. Embryol. **172:** 295–301.
39. NIETO, A., H. PONSTINGL & M. BEATO. 1977. Purification and quaternary structure of the hormonally induced protein uteroglobin. Arch. Biochem. Biophys. **180:** 82–92.
40. NIETO, A. & M. LOMBARDEO. 1982. Uteroglobin-like antigens in species of Lagomorpha. Comp. Biochem. Physiol. **71:** 511–514.
41. REHM, S., M. TAKAHASHI, J. WARD *et al.* 1989. Immunohistochemical demonstration of Clara cell antigen in lung tumors of bronchiolar origin induced by N-nitrosodiethylamine in Syrian Golden hamster. Am. J. Pathol. **134:** 79–87.
42. REHM, S., J.M. WARD, A.A.W. TEN HAVE–OPBROEK *et al.* 1988. Mouse papillary lung tumors transplacentally induced by N-nitrosoethylurea originate from alveolar type II cells: a light microscopic, ultrastructural, and immunohistochemical study. Cancer Res. **48:** 148–160.
43. REHM, S., W. LIJINSKY, G. SINGH & S.L. KATYAL. 1991. Mouse bronchiolar cell carcinogenesis: histologic characterization and expression of Clara cell antigen in lesions induced by N-nitrosobis-(2-chloroethyl) ureas. Am. J. Pathol. **139:** 413–422.
44. WARD, J.M., G. SINGH, S.L. KATYAL *et al.* 1985. Immunocytochemical localization of the surfactant apoprotein and Clara cell antigen in chemically induced and naturally occurring pulmonary neoplasms of mice. Am. J. Pathol. **118:** 493–499.
45. BROERS, J.L.V., S.M. JENSEN, W.D. TRAVIS *et al.* 1992. Expression of SP-A and Clara cell 10 kDa mRNA in neoplastic and non-neoplastic human lung tissue as detected by *in situ* hybridization. Lab. Invest. **66:** 337–346.
46. HERBERT, R.A., N.A. GILLETT, A.H. REBAR *et al.* 1994. Plutonium-induced proliferative lesions and pulmonary neoplasms in the rat: immunohistochemical and ultrastructural evidence for their origin from type II pneumocytes. Vet. Pathol. **31:** 366–374.
47. OHSHIMA, M., J.M. WARD, G. SINGH & S.L. KATYAL. 1985. Immunocytochemical and morphological evidence for the origin of N-nitrosomethylurea-induced and naturally occurring primary lung tumors in F344/CNr rats. Cancer Res. **45:** 2785–2792.
48. DODGE, D.E., R.B. RUCKER, G. SINGH & C.G. PLOPPER. 1993. Quantitative comparison of intracellular concentration and volume of Clara cell 10 kDa protein in rat bronchi and bronchioles based on laser scanning confocal microscopy. J. Histochem. Cytochem. **41:** 1171–1183.
49. LOPEZ DE HARO, M.S. & A. NIETO. 1983. Isolation and characterization of uteroglobin from the lung of the hare (*Lepus capensis*). Arch. Biochem. Biophys. **225:** 539–547.
50. LOPEZ DE HARO, M.S. & A. NIETO. 1985. Primary structure of rabbit lung uteroglobin as deduced from the nucleotide sequence of a cDNA. FEBS Lett. **193:** 247–249.
51. MORNON, J.P., F. FRIDLANSKY, R. BALLY & E. MILGROM. 1980. X-ray crystallographic analysis of a progesterone-binding protein: the C2221 crystal form of oxidized uteroglobin at 2.2 Å resolution. J. Mol. Biol. **137:** 415–429.
52. PONSTINGL, H., A. NIETO & M. BEATO. 1978. Amino acid sequence of progesterone-induced rabbit uteroglobin. Biochemistry **17:** 3908–3912.

53. POPP, R.A., K.R. FORESMAN, L.D. WISE & J.C. DANIEL, JR. 1978. Amino acid sequence of a progesterone-binding protein. Proc. Natl. Acad. Sci. U.S.A. **75:** 5516–5519.
54. SINGH, G., S.L. KATYAL, W.E. BROWN et al. 1988. Amino acid and cDNA nucleotide sequences of human Clara cell 10 kDa protein. Biochim. Biophys. Acta **950:** 329–337.
55. KATYAL, S.L., G. SINGH, W.E. BROWN et al. 1990. Clara cell secretory (10 kDa) protein: amino acid and cDNA nucleotide sequences and developmental expression. Prog. Respir. Res. **25:** 29–35.
56. SINGH, G., S.L. KATYAL, W.E. BROWN et al. 1990. Clara cell 10 kDa protein (CC10): comparison of structure and function to uteroglobin. Biochim. Biophys. Acta **1039:** 348–355.
57. SINGH, G., S.L. KATYAL, W.E. BROWN & A.L. KENNEDY. 1993. Mouse Clara cell 10 kDa (CC10) protein: cDNA nucleotide sequence and molecular basis for the variation in the progesterone binding of CC10 from different species. Exp. Lung Res. **19:** 67–75.
58. RUBEN, G.S. & A. NIETO. 1998. Molecular cloning of the cDNA and the promoter of the hamster uteroglobin/Clara cell 10-kDa gene (ug/cc10): tissue-specific and hormonal regulation. Arch. Biochem. Biophys. **350:** 214–222.
59. GUTIERREZ, S.R. & A. NIETO. 1998. Cloning and sequencing of the cDNA coding for pig pre-uteroglobin/Clara cell 10 kDa protein. Biochem. Mol. Biol. Int. **45:** 205–213.
60. UMLAND, T.C., S. SWAMINATHAN, W. FUREY et al. 1992. The refined structure of rat Clara cell 17 kDa protein at 3.0 Å resolution. J. Mol. Biol. **224:** 441–448.
61. UMLAND, T., S. SWAMINATHAN, G. SINGH et al. 1994. The structure of a novel protein-phospholipid complex formed *in vivo*: human Clara cell phospholipid binding protein–(phosphatidylcholine/phosphatidylinositol) complex at 1.9 Å resolution. Nat. Struct. Biol. **1:** 538–545.
62. JOHNSTON, C.J., G.W. MANGO, J.N. FINKELSTEIN & B.R. STRIPP. 1997. Altered pulmonary response to hyperoxia in Clara cell secretory protein deficient mice. Am. J. Respir. Cell Mol. Biol. **17:** 147–155.
63. ZHANG, Z., G.C. KUNDU, C.J. YUAN et al. 1997. Severe fibronectin-deposit renal glomerular disease in mice lacking uteroglobin. Science **276:** 1408–1412.
64. MIELE, L., E. CORDELLA-MIELE, A. FACCHIANO & A.B. MUKHERJEE. 1988. Novel antiinflammatory peptides from the region of highest similarity between uteroglobin and lipocortin I. Nature **335:** 726–730.
65. MUKHERJEE, D.C., A.K. AGRAWAL, R. MANJUNATH & A. MUKHERJEE. 1983. Suppression of epididymal sperm antigenicity in the rabbit by uteroglobin and transglutaminase *in vitro*. Science **219:** 989–991.
66. BAILLY, A., M. ATGER, P. ATGER et al. 1983. The rabbit uteroglobin gene. J. Biol. Chem. **258:** 10384–10389.
67. LEVIN, S.W., J. BUTLER, U.K. SCHUMACHER et al. 1986. Uteroglobin inhibits phospholipase A_2 activity. Life Sci. **38:** 1813–1819.
68. MANTILE, G., L. MIELE, E. CORDELLA-MIELE et al. 1993. Human Clara cell 10-kDa protein is the counterpart of rabbit uteroglobin. J. Biol. Chem. **268:** 20343–20351.
69. ANDERSON, O., L. NORDLUND-MOLLER, H.J. BARNES & J. LUND. 1994. Heterologous expression of human uteroglobin/polychlorinated biphenyl–binding protein: determination of ligand binding parameters and mechanism of phospholipase A2 inhibition *in vitro*. J. Biol. Chem. **269:** 19081–19087.
70. ANDERSON, O., L. NORDLUND-MOLLER, M. BRONNEGARD et al. 1991. Purification and level of expression in bronchoalveolar lavage of a human polychlorinated biphenyl (PCB)–binding protein: evidence for a structural and functional kinship to the multihormonally regulated protein uteroglobin. Am. J. Respir. Cell Mol. Biol. **5:** 6–12.
71. HAY, J.G., C. DANIEL, D.S. CHU & R.G. CRYSTAL. 1995. Human CC10 gene expression in airway epithelium and subchromosomal locus suggest linkage to airway disease. Am. J. Physiol. **268:** L565–L575.
72. MUKHERJEE, A.B., G.C. KUNDU, G. MANTILE-SELVAGGI et al. 1999. Uteroglobin: a novel cytokine? Cell. Mol. Life Sci. **55:** 771–787.
73. NORD, M., J.A. GUSTAFSSON & J. LUND. 1995. Calcium-dependent binding of uteroglobin (PCB-BP/CCSP) to negatively charged phospholipid liposomes. FEBS Lett. **374:** 403–406.

74. BARNES, H.J., L. NORDLUND-MOLLER, M. NORD et al. 1996. Structural basis for calcium binding by uteroglobins. J. Mol. Biol. **256:** 392–404.
75. LOPEZ DE HARO, M.S., M.M. PEREZ, C. GARCIA & A. NIETO. 1994. Binding of retinoids to uteroglobin. FEBS Lett. **349:** 249–251.
76. GONZALEZ, K.D. & A. NIETO. 1995. Binding of uteroglobin to microsomes and plasmatic membranes. FEBS Lett. **361:** 255–258.
77. MUKHERJEE, A.B., R.E. ULANE & A.K. AGRAWAL. 1982. Role of uteroglobin and transglutaminase in masking the antigenicity of implanting rabbit embryo. Am. J. Reprod. Immunol. **2:** 135–141.
78. DIERYNCK, I., A. BERNARD, H. ROELS & M. DE LEY. 1996. The human Clara cell protein: biochemical and biological characterization of a natural immunosuppressor. Mult. Scler. **1:** 385–387.
79. LAING, I.A., C. HERMANS, A. BERNARD et al. 2000. Association between plasma CC16 levels, the A38G polymorphism, and asthma. Am. J. Respir. Crit. Care Med. **161:** 124–127.
80. ZHANG, Z., D.B. ZIMONJIC, N.C. POPESCU et al. 1997. Human uteroglobin gene: structure, subchromosomal localization, and polymorphism. DNA Cell Biol. **16:** 73–83.
81. LEYTON, J., M.J. MANYAK, A.B. MUKHERJEE et al. 1994. Recombinant human uteroglobin inhibits the *in vitro* invasiveness of human metastatic prostate tumor cells and the release of arachidonic acid stimulated by fibroblast-conditioned medium. Cancer Res. **54:** 3696–3699.
82. KUNDU, G.C., G. MANTILE, L. MIELE et al. 1996. Recombinant human uteroglobin suppresses cellular invasiveness via a novel class of high-affinity cell surface binding site. Proc. Natl. Acad. Sci. U.S.A. **93:** 2915–2919.
83. WEERARATNA, A.T., J.A. CAJIGAS, A. SCHWARTZ et al. 1997. Loss of uteroglobin expression in prostate cancer: relationship to advancing grade. Clin. Cancer Res. **3:** 2295–2300.
84. REYNOLDS, S.D., G.W. MANGO, R. GELEIN et al. 1999. Normal function and lack of fibronectin accumulation in kidneys of Clara cell secretory protein/uteroglobin deficient mice. Am. J. Kidney Dis. **33:** 541–551.
85. VOLLMER, M., R. KRAPF & F. HILDEBRANDT. 1998. Exclusion of the uteroglobin gene as a candidate for fibronectin glomerulopathy (GFND) (letter). Nephrol. Dial. Transplant. **13:** 2417–2418.
86. ZHENG, F., G.C. KUNDU, Z. ZHANG et al. 1999. Uteroglobin is essential in preventing immunoglobulin A nephropathy in mice. Nat. Med. **5:** 1018–1025.
87. SHIJUBO, N., Y. ITOH, T. YUAMAGUCHI et al. 1999. Clara cell protein-positive epithelial cells are reduced in small airways of asthmatics. Am. J. Respir. Crit. Care Med. **160:** 930–933.
88. DEWATER, R., L.N.A. WILLEMS, G.N.P. VAN MUIJEN et al. 1986. Ultrastructural localization of bronchial antileukoprotease in central and peripheral airways by a gold-labelling technique using monoclonal antibodies. Am. Rev. Respir. Dis. **11:** 882–890.
89. FRANKEN, C., C.J.L.M. MEIJER & J.H. DIJKMAN. 1989. Tissue distribution of antileukoprotease and lyzozyme in humans. J. Histochem. Cytochem. **37:** 493–498.
90. HEINZEL, R., H. APPELHANS, G. GASSEN et al. 1986. Molecular cloning and expression of cDNA for human antileukoprotease from cervix uterus. Eur. J. Biochem. **160:** 61–67.
91. THOMPSON, R.C. & K. OHLSSON. 1986. Isolation, properties, and complete amino acid sequence of human secretory leukocyte protease inhibitor, a potent inhibitor of leukocyte elastase. Proc. Natl. Acad. Sci. U.S.A. **83:** 6692–6696.
92. WASANO, K. & T. YAMAMOTO. 1989. Rat lung 29 kD beta-galactoside-binding lectin is secreted by bronchiolar Clara cells into airways. Histochemistry **90:** 447–451.
93. YONEDA, K. 1978. Ultrastructural localization of phospholipases in the Clara cell of the rat bronchiole. Am. J. Pathol. **93:** 745–752.
94. KIDO, H., Y. YOKOGOSHI, K. SAKAI et al. 1992. Isolation and characterization of a novel trypsin-like protease found in rat bronchiolar epithelial Clara cells. J. Biol. Chem. **267:** 13573–13579.
95. KENTARO, S., Y. KAWAGUCHI, Y. KISHINO & H. KIDO. 1993. Electron immunohistochemical localization in rat bronchiolar epithelial cells of tryptase Clara, which

determines the pneumotropism and pathogenicity of Sendai virus and influenza virus. J. Histochem. Cytochem. **41:** 89–93.
96. PERSSON, A., D. CHANG, K. RUST *et al.* 1989. Purification and biochemical characterization of CP4 (SP-D), a collagenous surfactant-associated protein. Biochemistry **28:** 6361–6367.
97. WALKER, S.R., M.D. WILLIAMS & B. BENSON. 1986. Immunocytochemical localization of the major surfactant proteins in type II cells, Clara cells, and alveolar macrophages of rat lung. J. Histochem. Cytochem. **34:** 1137–1148.
98. WILLIAMS, M.C., S. HAWGOOD, D.B. SCHENK *et al.* 1988. Monoclonal antibodies to surfactant proteins SP 28–36 labelled type II and non-ciliated bronchiolar cells by immunofluorescence. Am. Rev. Respir. Dis. **137:** 399–405.
99. BALIS, J.U., J.F. PATERSON, J.E. PACIGA *et al.* 1985. Distribution and subcellular localization of surfactant-associated glycoproteins in human lung. Lab. Invest. **52:** 657–669.
100. CROUCH, E., D. PARGHI, S. KUAN & A. PERSSON. 1992. Surfactant protein D: subcellular localization in nonciliated bronchiolar epithelial cells. Am. J. Physiol. **263:** L60–L66.
101. VOORHOUT, W.F., T. VEENENDAAL, Y. KUROKI *et al.* 1992. Immunocytochemical localization of surfactant protein D (SP-D) in type II cells, Clara cells, and alveolar macrophages of rat lung. J. Histochem. Cytochem. **40:** 1589–1597.

Secretory Lipophilins: A Tale of Two Species

ROBERT I. LEHRER,[a,b,c] TUNG NGUYEN,[b] CHENGQUAN ZHAO,[b] CHEN XIAN HA,[b] AND BEN J. GLASGOW[d,e]

[b]*Department of Medicine,* [c]*Molecular Biology Institute,* [d]*Department of Pathology, and* [e]*Jules Stein Eye Institute, UCLA School of Medicine, Los Angeles, California 90095, USA*

ABSTRACT: Secretory lipophilins are "lipid-loving" proteins that are major constituents of several mammalian secretions, including the prostatic fluid of rats and the tears of humans and rabbits. These proteins form covalent heterodimers that are stabilized by three intramolecular cystine disulfide bonds. The heterodimers, some of which are glycosylated, may undergo additional non-covalent assembly to form tetramers. The peptide components found in secretory lipophilins are from two subfamilies: lipophilins A/B and lipophilin C. The C subfamily members described in this report are three rabbit and one human lipophilin, plus human mammaglobin and the C3 subunit of rat prostatein. Human A/B and C lipophilins are expressed by many tissues and are especially prominent in endocrine-responsive organs. The gene for human lipophilin B resides at chromosome 10q22-23. This region harbors the PTEN/MMAC1 gene and is believed to contain additional tumor suppressor genes. Although the functions of secretory lipophilins are imperfectly understood, their abundance in glandular secretions and in hormone-responsive tissues suggests that they deserve considerably more attention than they have received to date.

INTRODUCTION

Secretory lipophilin is a nonglycosylated, heterodimeric protein that occurs in human tears at concentrations above 100 µg/mL.[1,2] It has two subunits, components A and C, that are connected by three intermolecular cystine disulfide bonds. We discovered it by accident while mapping the peptides of human tears during a search for antimicrobial molecules. Although secretory lipophilins had unimpressive antimicrobial properties, we obtained N-terminal sequence data to identify this unfamiliar molecule. When it proved to be a novel homologue of rat prostatein, the major secretory product of the ventral prostate, it caught our eye and has remained there since.

Rat prostatein has received several alternative names, including "prostatic binding protein", "prostatic α-protein", "estramustine binding protein" (estramustine is an antineoplastic agent that contains estradiol linked to nornitrogen mustard), "prostatic secretion protein", and "prostatic steroid-binding protein".[3-6] For simplicity,

---

[a]Address for correspondence: Robert I. Lehrer, Department of Medicine, CHS 37-062, UCLA Center for the Health Sciences, 10833 LeConte Avenue, Los Angeles, CA 90095-1690. Voice: 310-825-5340; fax: 310-206-8766.

rlehrer@mednet.ucla.edu

we will call it prostatein in this discussion. The subunit structure of rat prostatein was defined at the peptide level almost two decades ago.[4] Fully assembled prostatein is a tetramer that contains two molecules of component 3 ("C3") and one molecule each of components 1 and 2 ("C1", "C2"). Rat C1 (88 residues) and C2 (92 residues) are homologous to each other, but neither shows homology to C3 (77 residues).[5,6]

The two C3 components of the tetramer associate noncovalently with each other, and each C3 subunit is covalently attached via three intermolecular cystine disulfide bonds to either a C1 or a C2 molecule. Thus, prostatein consists of two heterodimers that are stabilized by covalent and noncovalent bonds. The C3 component of prostatein is heavily N-glycosylated,[7] making its mass ($\approx$14 kDa) larger than its 77 amino acid residues would suggest. Aware that two homologous peptides, C1 and C2, existed in rat prostatein, we looked carefully (but without avail) for the analogous situation in human tears. Only later did our cloning experiments discover the missing component B and show its expression by nonlacrimal tissues.

Rat lacrimal glands also express a secretory lipophilin. Vercaeren and associates detected abundant C3-like mRNA in rat lacrimal glands, without finding mRNAs for C1 and C2. Expression of C3 was developmentally regulated and androgen-responsive.[8,9] From rat lacrimal gland homogenates, they recovered a glycosylated protein containing the prostatic C3 component. When reduced by dithiothreitol and run on SDS gels, the protein resolved into glycosylated and nonglycosylated subunits. The investigators concluded that "the C3 component of rat prostatein was present in the lacrimal gland not as a component of prostatein per se, but as a component of a strikingly similar, multimeric protein." Aumuller et al.[10] showed that the lacrimal gland of male Wistar rats was strongly immunoreactive for the C3 epitope of rat prostatein.

The conserved cysteines of all three prostatein components are crucial for assembly of the tetrameric prostatein molecule.[11] When prostatein tetramers are boiled in SDS or in 6 M guanidine, they dissociate into C1/C3 and C2/C3 dimers, which can then be reduced and separated into their component monomers.[4] The dissimilar structures of rat prostatein components C1 and C2 have functional consequences related to steroid binding. When tetrameric prostatein is dissociated, the C2-C3 heterodimers bind steroids and estramustine, but their C1-C3 counterparts do not.[4,12] The pregnenolone binding site has an affinity of $1.2 \times 10^6 \, M^{-1}$. Besides binding steroids, prostatein binds various proline-rich peptides found in rat prostatic secretions, suggesting that it may be a general delivery vehicle for hydrophobic molecules.[13–15] Although proline-rich peptides were recently described in human lacrimal secretions, little is known about their function.[16,17]

We recently began to study rabbit tears, a species that is widely used for ophthalmology research. To our surprise, rabbit and human tears showed many differences when analyzed by gel electrophoresis. Not the least of these was that a secretory lipophilin was present at concentrations exceeding 3 mg/mL, far higher than any other tear proteins or peptides. Thus far, by a combination of cloning and peptide-level chemistry, we have identified six new members of the secretory lipophilin family in rabbits. They include three different homologues of lipophilin C and three homologues of lipophilins A/B. One of the lipophilin C peptides (C^L) was cloned from a rabbit lacrimal gland cDNA library. The others, C^S and C^P, were PCR-cloned directly from rabbit salivary gland and prostate gland cDNA, respectively.

METHODS

Because the methods used in our study are described in detail elsewhere,[1,2] this section will be short. Briefly, human tears were obtained from healthy adult volunteers after brief exposure to the vapors of freshly minced onions, and stored at −20°C until used. Rabbit tears were collected without restraining or stimulating the animals. Pooled tears were subjected to RP-HPLC on a C-18 column (Vydac, The Separations Group, Hesperia, CA), using linear gradients of acetonitrile in 0.1% trifluoroacetic acid (TFA). Lipophilin heterodimers were identified by comparing SDS-PAGE gels run under reducing (dithiothreitol) and nonreducing conditions. Once identified, they were further purified by RP-HPLC using various linear gradients of acetonitrile in 0.1% TFA or 0.13% heptafluorobutyric acid.

RESULTS

We first became acquainted with secretory lipophilin as an abundant protein that we had transferred to a PVDF membrane for N-terminal sequencing. Its analysis returned two residues in most of the initial 25 cycles, but then gave a single sequence (AKFKAPLXAVAAKMEVK) during cycles 26 to 42. A BLAST search performed on a "best guess" composite sequence showed substantial homology to the C1 component of rat prostatein. At that point, we were hooked.

Lipophilins were surprisingly easy to purify because they eluted from the C-18 columns well after most other peptides had emerged.[1] We obtained over 400 μg of the purified dimer from 9 mL of pooled human tears and estimated its concentration to be in excess of 100 μg/mL. We found similar lipophilin contents in tears from males and females. The human molecule's mass was 16,424.01 ± 1.17 by ESI-MS. When we reduced the molecule with dithiothreitol (DTT) before performing RP-HPLC, we recovered its monomeric components, as component A emerged before component C. Reduced lipophilins A and C had masses of 7574.7 and 8854.94, respectively. When they were alkylated with iodoacetamide, these masses increased by ≈171.75 amu, indicating that each contained three cysteines. The masses and sequences of component A (88 residues) and component C (77 residues) agreed perfectly.

FIGURE 1 shows an RP-HPLC profile of rabbit tears. The very large peak that emerged after 48–51 minutes was composed almost entirely of secretory lipophilin dimer. The inset shows an SDS-PAGE of a similar lipophilin peak, taken from another preparatory run. Because the samples had been reduced with DTT, only the ≈7-kDa monomer is evident. Unlike the lipophilin C from human tears, the rabbit C molecule is heavily glycosylated and stains poorly with silver or Coomassie blue. Thus, although it is also present in these fractions, it is not visible on the gel. The faint bands at 68 kDa are human albumin, which was added to all of the fractions to prevent absorptive loss of the less abundant tear proteins and peptides.

FIGURE 2 shows the primary amino acid sequences of six members of the lipophilin A/B group, two each from rabbit, human, and rat. Residues identical to rabbit lipophilin A are shown in bold type and the conserved residues are underlined. The midportion of the molecule, from residues 30–46, is highly conserved and presum-

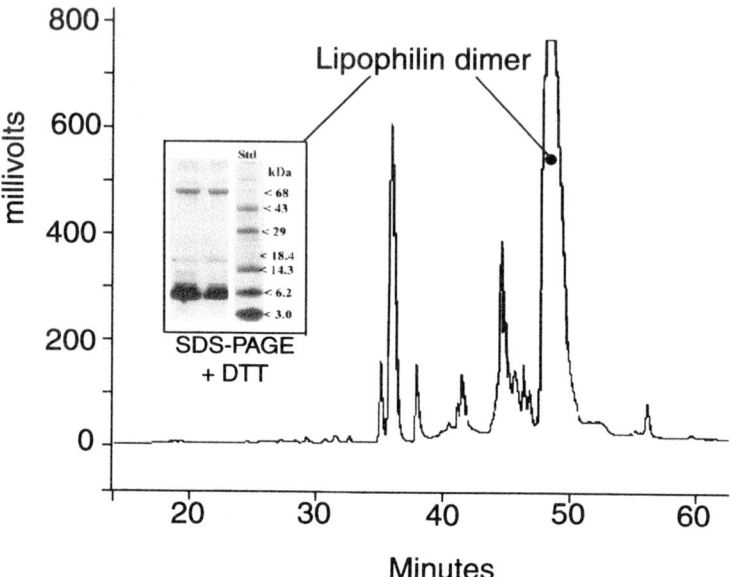

FIGURE 1. RP-HPLC profile of rabbit tears. The sample was run on a Vydac C-18 column using a linear gradient of acetonitrile in 0.1% TFA. Elution was monitored at 225 nm. The large late peak contains almost pure lipophilin dimer. An analytical SDS-PAGE gel, run with samples that had been reduced with DTT, is shown in the inset.

Rabbit Lipophilin A	1. **AACPAFVL**DS**VGFLF**D**PKPV**YR**QKLAKYD**	29
Human Lipophilin A	1. VVCQALGSE**ITGFLLAGKPV**FKF**QLAK**FK	29
Rabbit Lipophilin B	1. **AACP**TFIAE**S**TANLLASE**S**VFRASLSK**Y**G	29
Human Lipophilin B	1 EFCPALVSELLDFFF**IS**EPL**F**KLS**LAK**FD	29
Rat prostatein C1	1 QICELVAHET**IS**FLMKSEEEL**K**KELEMY**N**	29
Rat Prostatein C2	1 GVCQALQDVT**I**TFLLNP**E**EELKRELEEFD	29
	* * *	

Rabbit Lipophilin A	30. **APPEAVEAKLQVK**E**CTD**EI**DKGKRVLIAAVLTKIVK**E-**CAL**
Human Lipophilin A	30. AP**L**EAVAAKM**E**VK**K**CVDTMAYEKRVLITKTLGKIA**EK**-CDR
Rabbit Lipophilin B	30. **APPEAVEAKLQVK**RCTD**KMSLGKRVLFGKVLGEIVL**R**TCTL**
Human Lipophilin B	30. APP**A**AVEAKLE**VK**RCVD**QMSNG**DRLVVAETLVYIFL**E**-CGVKQWVE
Rat Prostatein C2	30. **APPEAVEA**NL**KVK**RCI**NK**IMYGDRLSMGTSLVFTMLK-CDVKYGYK
	** ** * ** * * * *

FIGURE 2. Primary amino acid sequences of six members of the lipophilin A/B group. The peptides are aligned with rabbit lipophilin A. Identical residues are in bold font and conservative substitutions are underlined. The asterisks denote residues that are identical in all six peptides.

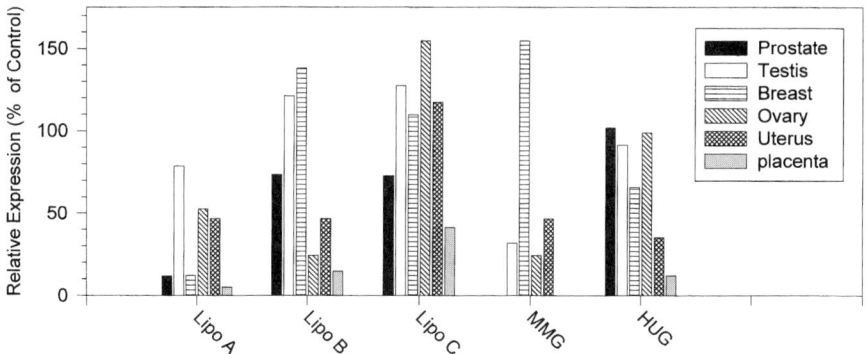

FIGURE 3. Relative expression of human lipophilins and uteroglobin. Results in these endocrine-responsive tissues were obtained by quantitative PCR and are normalized to actin expression. Additional technical information can be found in reference 2.

ably constitutes a domain that is especially important for lipophilin structure and/or function.

Sequences of several C group lipophilins have been published to date, including those of human lipophilin C,[1,2] mammaglobin,[18] and rat prostatein C.[7] Human lipophilin C was also cloned independently by investigators who named it mammaglobin B.[19] To date, information about mammaglobin exists only at the nucleotide level, and it is not known whether it forms lipophilin and prostatein-like heterodimers or higher order structures (as it almost certainly must).

FIGURE 3 shows the expression of human lipophilins A, B, and C; mammaglobin; and uteroglobin by six endocrine-responsive tissues: prostate, testis, breast, ovary, uterus, and placenta. The experiments were performed by quantitative PCR, using highly specific probes, and the data have been normalized with respect to β-actin expression by the same tissues. Whereas ovary and uterus expressed mRNA for secretory lipophilins A and B to a similar extent, lipophilin B mRNA was over 10 times more abundant than lipophilin A in the mammary gland. The mammary gland expressed lipophilin C and mammaglobin at about the same level as lipophilin B. This suggests that lipophilin C and/or mammaglobin may form a covalent heterodimer with lipophilin B. Despite its name, we found greater expression of uteroglobin in the prostate, testis, breast, and ovary than in the uterus. Mammaglobin expression was most prominent in the breast, but also occurred to a significant degree in several other tissues (FIGS. 3 and 4).

FIGURE 4 shows the expression of these peptides in six additional human tissues. None was significantly expressed by bone marrow. Prominent expression of lipophilin A was noted in thymus and high levels of lipophilin B were observed in skeletal muscle. Considerable mammaglobin expression was noted in the salivary gland. The cells within these tissues that are responsible for this expression pattern remain to be identified.

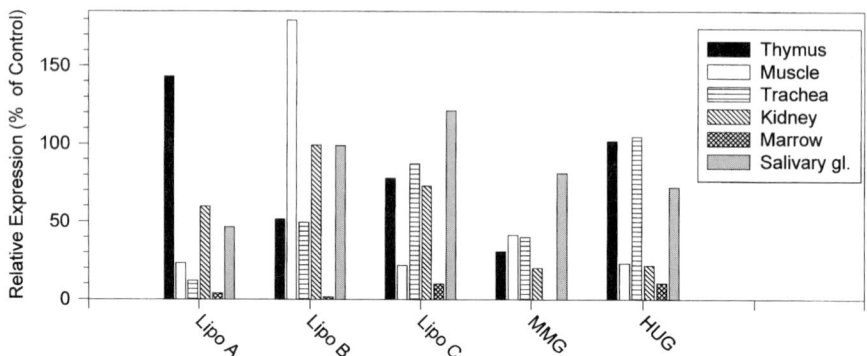

FIGURE 4. Relative expression of human lipophilins and uteroglobin. Results in these additional tissues were obtained by quantitative PCR and are normalized to actin expression. Additional technical information can be found in reference 2.

How are secretory lipophilins related to uteroglobins? The genes for human secretory lipophilin C,[2] mammaglobin,[20] and uteroglobin[21] are all located on or near chromosome 11q12, a region that is frequently amplified in breast cancers.[18,22] This region also contains a major atopy locus as well as a locus for insulin-dependent diabetes mellitus.[21] Both mammaglobin and lipophilin C (mammaglobin B) show considerable promise as markers for detecting micrometastatic cancer lesions in the lymph nodes and bone marrow.[23,24] The chromosomal proximity of the lipophilin C, mammaglobin, and uteroglobin genes strongly suggests that all three genes evolved by repeated duplications of a common precursor.

However, FIGURE 5 suggests that lipophilins and uteroglobins parted company very long ago, and that the lipophilin gene family diverged into two main branches in the interim. One of these branches contains the lipophilin C and its related peptides, rat prostatein C3 and human mammaglobin. The other branch contains human and rabbit lipophilins A and B as well as rat prostatein components C1 and C2. The human genes for lipophilins A and B have evidently separated during evolution. Lipophilin B resides on chromosome 23, in a region of considerable interest. Not only does chromosome 10q22-23 harbor an anti-oncogene strongly implicated in prostate cancer,[24] it has also been implicated in Cowden syndrome, an autosomal dominant hamartoma syndrome associated with a high risk of thyroid and breast cancer.[25,26] Although interest has justifiably centered on PTEN/MMAC1, evidence suggests that additional prostatic cancer tumor suppressor genes may be situated close to the PTEN/MMAC1 locus on 10q23.[27,28] It will be important to determine if lipophilin B plays such a role. The lipophilin A gene maps to chromosome 15q12-q13, a region amplified in some breast cancers[29] and also implicated in Angelman syndrome,[30] whose symptoms include microcephaly, severe mental retardation, ataxic gait, and seizures.

The functions of secretory lipophilins in human or rabbit tears are not yet understood. The presence of high concentrations of these peptides in tears suggests they play an important functional role in these secretions. Since secretory lipophilin does not exert significant antimicrobial activity, it is unlikely to play a direct role in host

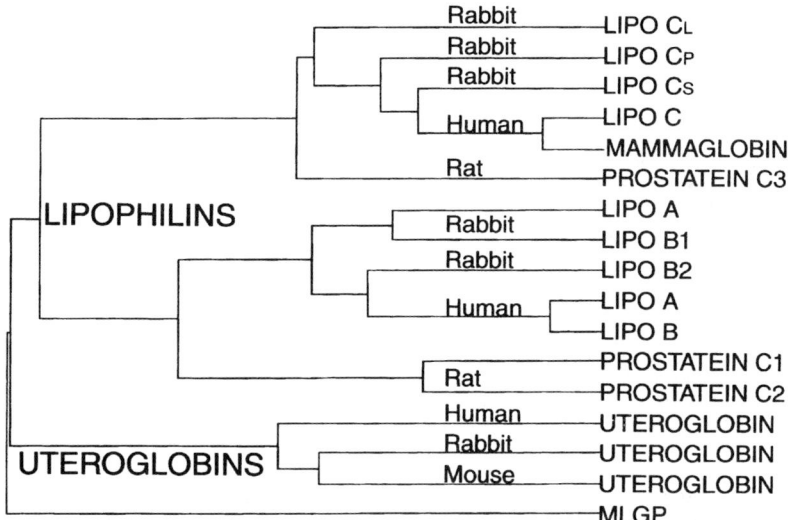

FIGURE 5. Relationship of lipophilins and uteroglobin. This dendrogram is based on the nucleotide sequences of the precursor peptides and was constructed with the CLUSTAL program.

defense. Its ability to bind apolar molecules remains to be studied in depth, but could give it a role in transporting essential phospholipids, retinoids, or steroids to the cornea and conjunctiva and in removing lysolipids. Although secretory lipophilin does not inhibit type II phospholipase A2 (pLA2), a potently antimicrobial enzyme found at high concentrations in human and rabbit tears, by binding lysolipids it could prevent pLA2 from causing collateral damage. Since the tear film drains into the nasal cavity via the nasolacrimal ducts, lipophilin could have olfactory functions, for example, by capturing pheromones or other lipophilic molecules from the atmosphere and conveying them to receptors for further sensory processing. Whatever their roles, the presence of multiple lipophilin isoforms in humans (A, B, C, and mammaglobin) and in rabbits (A, B-1, B-2, C^L, C^S, and C^P) suggests that they are likely to subserve important functions in many different tissues and organs.

ACKNOWLEDGMENTS

Our studies of lipophilins have been supported by NIH Grant No. EY 12080.

REFERENCES

1. LEHRER, R.I., G. XU, A. ABDURAGIMOV et al. 1998. Lipophilin, a novel heterodimeric protein of human tears. FEBS Lett. **432:** 163–167.
2. ZHAO, C., T. NGUYEN, T. YUSIFOV et al. 1999. Lipophilins: human peptides homologous to rat prostatein. BBRC **256:** 147–155.

3. HEYNS, W. & P. DEMOOR. 1977. Prostatic binding protein: a steroid binding protein secreted by rat prostate. Eur. J. Biochem. **78:** 221–230.
4. HEYNS, W., B. PEETERS, J. MOUS et al. 1978. Purification and characterisation of prostatic binding protein and its subunits. Eur. J. Biochem. **89:** 181–186.
5. PARKER, M., M. NEEDHAM & R. WHITE. 1982. Prostatic steroid binding protein: gene duplication and steroid binding. Nature **298:** 92–94.
6. PARKER, M., M. NEEDHAM, R. WHITE et al. 1982. Prostatic steroid binding protein: organization of C1 and C2 genes. Nucleic Acids Res. **10:** 5121–5131.
7. PEETERS, B., W. ROMBAUTS, J. MOUS et al. 1981. Structural studies on rat prostatic binding protein: the primary structure of its glycosylated component C3. Eur. J. Biochem. **115:** 115–121.
8. VERCAEREN, I., H. VANAKEN, J. VAN DORPE et al. 1998. Expression of cystatin-related protein and of the C3-component of prostatic-binding protein during postnatal development in the rat ventral prostate and lacrimal gland. Cell Tissue Res. **292:** 115–128.
9. VERCAEREN I., H. VANAKEN, A. DEVOS et al. 1996. Androgens transcriptionally regulate the expression of cystatin-related protein and the C3 component of prostatic binding protein in rat ventral prostate and lacrimal gland. Endocrinology **137:** 4713–4720.
10. AUMULLER, G., E.A. ARCE, W. HEYNS et al. 1995. Immunocytochemical localization of seminal proteins in salivary and lacrimal glands of the rat. Cell Tissue Res. **280:** 171–181.
11. PEETERS, B., W. HEYNS, J. MOUS et al. 1983. Structural studies on rat prostatic binding protein: the primary structure of component C2 from subunit S. Eur. J. Biochem. **132:** 669–679.
12. HEYNS, W. & D. BOSSYNS. 1983. A comparative study of estramustine and pregnenolone binding to prostatic binding protein: evidence for subunit cooperativity. J. Steroid Biochem. **19:** 1689–1694.
13. HEYNS, W., D. BOSSYNS, B. PEETERS et al. 1982. Study of a proline-rich polypeptide bound to the prostatic binding protein of rat ventral prostate. J. Biol. Chem. **257:** 7407–7413.
14. HEYNS, W., B. PEETERS & D. BOSSYNS. 1983. Multiple forms of the proline-rich polypeptide (PRP) bound to rat prostatic steroid binding protein. BBRC **111:** 172–179.
15. HEMSCHOOTE, K., B. PEETERS, L. DIRCKX et al. 1988. A single 12.5-kilobase androgen-regulated mRNA encoding multiple proline-rich polypeptides in the ventral prostate of the rat. J. Biol. Chem. **263:** 19159–19165.
16. DICKINSON, D.P. & M. THIESSE. 1995. A major human lacrimal gland mRNA encodes a new proline-rich protein family member. Invest. Ophthalmol. Vis. Sci. **36:** 2020–2031.
17. DICKINSON, D.P. & M. THIESSE. 1996. cDNA cloning of an abundant human lacrimal gland mRNA encoding a novel tear protein. Curr. Eye Res. **15:** 377–386.
18. WATSON, M.A. & T.P. FLEMING. 1996. Mammaglobin, a mammary-specific member of the uteroglobin gene family, is overexpressed in human breast cancer. Cancer Res. **56:** 860–865.
19. BECKER, R.M., C. DARROW, D.B. ZIMONJIC et al. 1998. Identification of mammaglobin B, a novel member of the uteroglobin gene family. Genomics **54:** 70–78.
20. WATSON, M.A., C. DARROW, D.B. ZIMONJIC et al. 1998. Structure and transcriptional regulation of the human mammaglobin gene, a breast cancer associated member of the uteroglobin gene family localized to chromosome 11q13. Oncogene **16:** 817–824.
21. ZHANG, Z., D.B. ZIMONJIC, N.C. POPESCU et al. 1997. Human uteroglobin gene: structure, subchromosomal localization, and polymorphism. DNA Cell Biol. **16:** 73–83.
22. LEYGUE, E., L. SNELL, H. DOTZLAW et al. 1999. Mammaglobin, a potential marker of breast cancer nodal metastasis. J. Pathol. **189:** 28–33.
23. ZACH, O., H. KASPARU, O. KRIEGER et al. 1999. Detection of circulating mammary carcinoma cells in the peripheral blood of breast cancer patients via a nested reverse transcriptase polymerase chain reaction assay for mammaglobin mRNA. J. Clin. Oncol. **17:** 2015–2019.
24. KOMIYA, A., H. SUZUKI, T. UEDA et al. 1996. Allelic losses at loci on chromosome 10 are associated with metastasis and progression of human prostate cancer. Genes Chromosomes Cancer **17:** 245–253.

25. DAHIA, P.L., D.J. MARSH, Z. ZHENG *et al.* 1997. Somatic deletions and mutations in the Cowden disease gene, PTEN, in sporadic thyroid tumors. Cancer Res. **57:** 4710–4713.
26. NELEN, M.R., G.W. PADBERG, E.A. PEETERS *et al.* 1996. Localization of the gene for Cowden disease to chromosome 10q22-23. Nat. Genet. **13:** 114–116.
27. FEILOTTER, H.E., M. NAGAI, A. BOAG *et al.* 1998. Analysis of PTEN and the 10q23 region in primary prostate carcinomas. Oncogene **16:** 1743–1748.
28. VERMA, R.S., M. MANIKAL, R.A. CONTE *et al.* 1999. Chromosomal basis of adenocarcinoma of the prostate. Cancer Invest. **17:** 441–447.
29. FOROZAN, F., R. VELDMAN, C.A. AMMERMAN *et al.* 1999. Molecular cytogenetic analysis of 11 new breast cancer cell lines. Br. J. Cancer **81:** 1328–1334.
30. ROUGEULLE, C. & M. LALANDE. 1998. Angelman syndrome: how many genes to remain silent? Neurogenetics **1:** 229–237.

Clara Cell Secretory Protein (CC16): Features as a Peripheral Lung Biomarker

F. BROECKAERT, A. CLIPPE, B. KNOOPS, C. HERMANS, AND A. BERNARD[a]

Industrial Toxicology and Occupational Medicine Unit, Faculty of Medicine, School of Public Health, Catholic University of Louvain, B-1200 Brussels, Belgium

ABSTRACT: Clara cell protein (CC16 or CC10) is a 15.8-kDa protein secreted all along the tracheobronchial tree and especially in the terminal bronchioles where Clara cells are localized. Even though the exact *in vivo* function of CC16 remains to be clarified, evidence is accumulating that CC16 plays an important protective role in the respiratory tract against oxidative stress and inflammatory response. CC16, however, presents also a major interest as a peripheral lung marker for assessing the cellular integrity or the permeability of the lung epithelium. The serum concentrations of CC16 are decreased in subjects with chronic lung damage caused by tobacco smoke and other air pollutants as a consequence of the destruction of Clara cells. By contrast, serum CC16 increases in acute or chronic lung disorders characterized by an increased airways permeability. The sensitivity of serum CC16 to an increased leakiness of the lung allows for the detection of defects of the epithelial barrier at ozone levels below current air-quality guidelines. Although the clinical significance of these early epithelial changes detected by serum CC16 remains to be determined, these results clearly show that the assay in serum of lung secretory proteins such as CC16 represents a new noninvasive approach to evaluate the integrity of the respiratory tract.

INTRODUCTION

The central activity of Clara cells, one of the most multifunctional epithelial cell type of the mammalian lung, appears to be devoted to the protection of the respiratory tract against toxic inhaled agents. Indeed, Clara cells have been shown to repair damaged epithelium, to detoxify xenobiotics, and to secrete proteins with important biological activities such as surfactant-associated proteins, leukocyte-protease inhibitors, and the 15.8-kDa Clara cell protein (CC16).[1,2] CC16 is the major protein secreted by Clara cells and one of the main secretory proteins of the lung. The protein occurs in very high concentrations in the epithelial lining fluid where it appears to play an antioxidant/inflammatory role, notably by modulating the production and/or activity of phospholipase-A2, interferon-γ, and tumor necrosis factor-α (for a review, see reference 3).

From a diagnostic point of view, CC16 is at the origin of a novel approach for assessing the integrity of the respiratory epithelium, based on the assay in serum of

---

[a]Author for correspondence: A. Bernard, Industrial Toxicology and Occupational Medicine Unit, Faculty of Medicine, School of Public Health, Catholic University of Louvain, 30.54 Clos Chapelle-aux-Champs, B-1200 Brussels, Belgium.

Bernard@toxi.ucl.ac.be

lung-specific proteins.[2–6] Most biomarkers of lung inflammation or injury described so far are constituents of the epithelial lining fluid sampled by bronchoalveolar lavage (BAL) techniques. These biomarkers consist of plasma-derived proteins or of molecules produced locally by the epithelial and inflammatory cells such as cytokines, enzymes, or growth factors. Although of great potential for evaluating the extent of inflammation or tissue damage, markers in BAL are not applicable for monitoring populations exposed to air pollutants in the environment or workplace. They also cannot be used for assessing health risks in children and patients with compromised cardiopulmonary function who are the most vulnerable to air pollution. Recent clinical and experimental studies have shown that the determination of CC16 in serum is a sensitive biomarker to detect increased leakiness of the lung epithelial barrier and/or to evaluate the integrity of Clara cells. The aim of this paper is to summarize recent advances concerning the application of CC16 as a lung peripheral marker.

FEATURES AND METABOLISM OF HUMAN CC16

The Clara cell protein family encodes a group of approximately 20 homologous proteins including the human CC16. This protein, first identified in urine of patients with renal failure[7] and purified later from lung lavage,[8] is the counterpart of rabbit uteroglobin (UG), a secretory protein of the lung (both sexes) and the uterus (during pregnancy) of lagomorphs (rabbits, hares).[9] Both proteins are homodimeric, consisting of two amino acid subunits oriented antiparallel and connected by two disulfide bonds. Although unreduced CC16 shows an apparent molecular weight of approximately 10 kDa on gel electrophoresis, its molecular mass, determined by electrospray ionization/spectrometry is exactly 15,840 Da.[10]

In humans, the organ- and tissue-specific expression of the CC16 gene has been studied by Northern blot analysis using specific probes for CC16. As shown in FIGURE 1, CC16 is mainly expressed in the lung, the trachea, and the fetal lung. Earlier studies, however, have shown that CC16 is not produced uniformly in the respiratory tract, with the protein being predominantly found in airways with an increasing density from the trachea and bronchi to the terminal bronchioles.[11,12] An extrapulmonary synthesis has also been evidenced in the prostate, the endometrium, and the kidney, but the levels of expression in these organs are on average 20 times lower than in the lung. This much higher pulmonary expression of CC16 combined with the fact that the lung offers the largest exchange surface area with blood explains why serum CC16 derives almost exclusively from the respiratory tract.[2]

This tissue distribution of CC16 is fully consistent with the pattern of CC16 concentrations in biological fluids. The highest concentrations of CC16 are found in pulmonary fluids such as the epithelial lining fluid (ELF), the bronchoalveolar lavage fluid (BALF), and the sputum. CC16 also is present in high concentrations in human amniotic fluid, where it reaches a mean value of about 1 mg/L before delivery.[13] The protein occurs also in extrapulmonary fluids such as serum and urine, but in much smaller concentrations (10–20 and 5–10 µg/L, respectively). In human BALF, the concentration of CC16 is on average between 1 and 2 mg/L, which represents 2–3% of the total soluble protein content of BALF. Since the lavage tech-

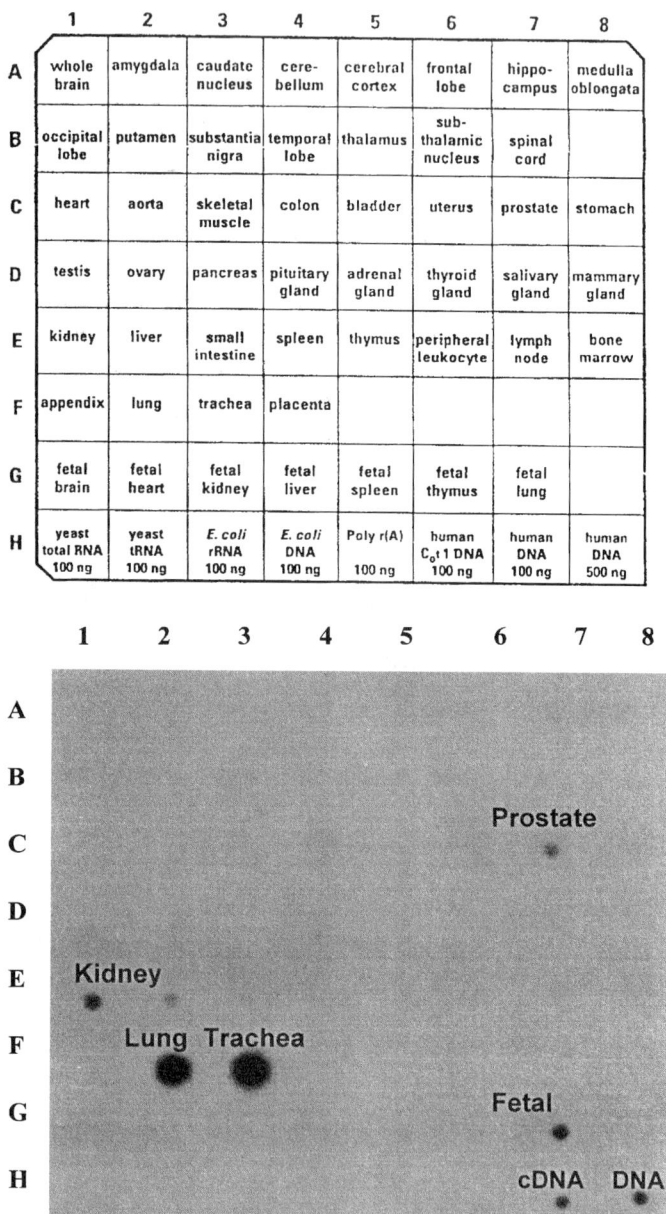

FIGURE 1. Master Blot analysis of poly(A)$^+$ RNA from approximately 50 human organs and tissues (Human RNA Master Blot™, Clontech, San Diego, CA). For hybridization, cDNA ^{32}P-radiolabeled probes specific for CC16 and β-actin (control probe) were used on the same Northern blot. From reference 2.

nique results in an approximately 100-fold dilution of the ELF, CC16 concentrations at the surface of airways can be estimated at about 100 mg/L. However, the concentrations of CC16 in BALF of healthy subjects show a great dispersion that cannot be explained solely by the variable dilution of BALF samples, suggesting the existence of interindividual variations in the pulmonary synthesis/secretion of CC16. In serum from healthy nonsmokers, CC16 concentrations are on average between 10 and 15 µg/L, that is, a value about 10,000 times lower than in ELF. This huge concentration gradient probably provides the driving force for the movement of CC16 from the lung into the blood.[2] The integrity of the bronchoalveolar/capillary membranes is an important determinant of this equilibrium as indicated by situations in which this barrier is compromised (e.g., acute lung injury, sarcoidosis). Like other low-molecular-weight proteins, CC16 is rapidly cleared from plasma by glomerular filtration with a half-life estimated at about 2 to 3 hours. As a corollary, serum CC16 rises as the glomerular filtration rate (GFR) declines.[14] The slight elevation of CC16 in serum with aging[15] is presumably due to the age-related decrease of the GFR. CC16 in serum, however, is independent of the level of lipids, body mass index, and sex and does not show nocturnal variations.[16]

Once filtered through the glomeruli, CC16 is almost completely reabsorbed by the proximal tubules where it is catabolized, explaining why urine contains normally only small amounts of CC16. In males, however, from puberty on, CC16 is secreted by the prostate directly into the urogenital tract, which increases the urinary concentrations of CC16 by a factor of 10 up to 100 depending on the mode of urine collection (only the first mL are highly contaminated due to the washout of the urogenital tract). A defect in the tubule reabsorptive capacity results in a markedly increased urinary excretion of CC16, which correlates with that of other low-molecular-weight proteins characteristic of a tubular proteinuria.[17]

In rodents, CC16 follows the same metabolic pathway as in humans, leaking from the respiratory tract into serum and increasing in urine when proximal tubular function fails.[18] While the exact mechanism by which lung CC16 reaches the bloodstream remains to be clarified, various observations support a transport by convection and/or passive diffusion. Recent pharmacokinetic studies in rats have shown that, following intratracheal administration, recombinant CC16 rapidly reaches the blood, from where it is eliminated by renal filtration with a half-life of approximately 12 hours.[19]

There is now strong evidence from *in vitro* studies and experiments on transgenic mice that CC16 has an important protective role in the respiratory tract against processes of oxidative stress, inflammation, and ultimately fibrosis. It is thus very likely that a depletion of CC16 due to Clara cell damage or intravascular leakage of the protein across the breached epithelium may play a critical role in the development of lung inflammation and injury.[3]

CC16 IN ACUTE LUNG INJURY

Because of their high biotransformation activity, Clara cells are very sensitive to lung toxicants undergoing a metabolic activation. Classic examples of compounds activated by Clara cells are the toxins, 4-ipomeanol and naphthalene, which cause

TABLE 1. Serum concentrations of Clara cell protein (CC16) in humans and rodents with short-term exposure to ozone

	Ozone level (ppm)	Duration of exposure (h)	CC16 controls (μg/L)	CC16 exposed (μg/L)	p	Ref.
Cyclists						
Male	0.076 ± 0.010	2	11.2 ± 0.8	12.3 ± 0.9	*	23
Female	idem	2	11.1 ± 0.6	11.9 ± 1.3	*	23
Mice						
$C_{57}Bl/6$	0.08 ± 0.01	4	59 ± 3	75 ± 5	*	3, 23
	idem	8	idem	88 ± 3	**	3, 23
	0.11 ± 0.02	24	62 ± 7	122 ± 5	*	3, 6
	1.82 ± 0.12	3	56 ± 4	201 ± 21	***	np
C_3H	0.11 ± 0.02	24	36 ± 5	74 ± 4	*	3, 6
	1.82 ± 0.12	3	40 ± 4	93 ± 8	*	np
Rats						
Sprague-Dawley	0.3	3	16 ± 3	19 ± 3	ns	24
	0.6	3	idem	33 ± 9	*	24
	1.0	3	idem	159 ± 22	**	24

NOTE: np, not published; statistical significance: $*p < 0.05$; $**p < 0.01$; $***p < 0.001$; all values are expressed as mean ± SE.

highly selective damage to Clara cells. Exposure of rodents to such compounds results in a rapid destruction of Clara cells, with an ensuing decrease in the abundance of CC16 mRNA and protein levels in BALF.[20,21] The acute changes induced by 4-ipomeanol are associated with a parallel elevation of CC16 in serum and of albumin in BALF due to increased permeation of these proteins across the damaged respiratory epithelium.[20] Parallel increases of CC16 in serum and of albumin in BALF have been found in rats during LPS-induced lung inflammation.[22]

The high sensitivity of serum CC16 to lung epithelium injury was demonstrated in recent studies in rodents and cyclists exposed to low concentrations of ozone (O_3) (TABLE 1). In a study on cyclists, for instance, the determination of CC16 in serum before and after a 2-h ride during episodes of photochemical smog revealed that ambient O_3 produces a dose-dependent increase of serum CC16 at exposure levels below current air-quality guidelines (with O_3 levels on average between 0.033 and 0.103 ppm[23]). This finding was reproduced in mice exposed to an O_3 level corresponding to the new national U.S. air-quality standard (0.08 ppm). These observations demonstrate that the assay of serum CC16 allows for the detection of very subtle defects in the lung epithelial barrier permeability. However, at higher O_3 levels, the increase of serum CC16 was found to correlate with the extent of lung injury as assessed by the levels of albumin, LDH, and inflammatory cells in BALF.[24] Changes in serum CC16 induced by O_3 are paralleled by an elevation of the urinary excretion of CC16 resulting from an overloading of the tubular reabsorption process.[24] Although O_3 at high exposure levels causes a significant reduction of CC16 concentrations in BALF, CC16 mRNA levels in lung tissue were not altered, suggesting that O_3 could affect the process of CC16 secretion at a posttranscriptional level. This hypothesis is supported by the study of Dodge et al.,[12] which showed that CC16 accumulates in Clara cell granules of rats exposed for 20 months to 1 ppm O_3.

In humans, a transient elevation of serum CC16 has been found in firefighters after only 20 minutes of smoke inhalation.[16] Of interest, this increase was found in the absence of any functional signs of lung impairment (spirometric tests), confirming the high sensitivity of CC16 as a marker of acute lung injury.

CC16 IN CHRONIC LUNG INJURY

The determination of serum CC16 can also be used to evaluate the chronic effects of air pollutants on the respiratory epithelium. The concentration of CC16 in serum is decreased in a dose-dependent way by tobacco smoking, with the concentration of the protein decreasing on average by 15% for each 10-pack-year.[15] This effect, which has been consistently found in several independent studies,[2,25] is a reflection of a concomitant reduction of CC16 in lung lavage due to the progressive loss of Clara cells. A significant reduction of serum CC16 has also been found in workers exposed to crystalline silica[26] or foundry dust (unpublished data). In most studies, these changes were found in asymptomatic subjects with normal or only slightly impaired lung function tests.

Changes of CC16 concentrations in BALF and/or serum have also been studied in different human lung diseases.[2] A polymorphism in the 5′-untranslated region of exon 1 of the CC16 gene has recently been associated with an increased risk of asthma in childhood, with homozygous and heterozygous subjects demonstrating a 6.9-fold and 4.2-fold increased risk of developing asthma, respectively.[27] Of interest, the mutation is associated with a significant decrease of the serum concentration of CC16.[28] This observation is in agreement with a previous work showing that the mean CC16 concentration in BALF of asthmatics was significantly decreased as compared with control healthy subjects.[29] In serum, asthmatic patients with a long duration of the disease (≥10 years) had significantly lower CC16 levels than those with a short duration of the disease (<10 years).[30] CC16 has also been found to be increased in the serum of patients with sarcoidosis, idiopathic pulmonary fibrosis, and acute respiratory distress syndrome.[2]

MECHANISMS OF CC16 CHANGES IN SERUM

Changes of CC16 concentrations in serum may result from three basic mechanisms: (i) an increased intravascular leakage of CC16 from the lung across the damaged lung epithelium barrier; (ii) a decreased production/secretion of CC16 in the respiratory tract with an ensuing decreased efflux of the protein into serum; (iii) a reduction of the clearance of CC16 from plasma associated with renal insufficiency.

The lung-blood barrier is not entirely impermeable to proteins. It behaves as a molecular sieve, allowing the bidirectional exchange of proteins between the lung and plasma. Although the exact mechanisms and pathways of this lung-blood transfer remain to be clarified, the leakage of lung proteins into the circulation or the reversed movement of plasma proteins appears to be governed by several factors, among which the most important are the epithelium permeability, the molecular fea-

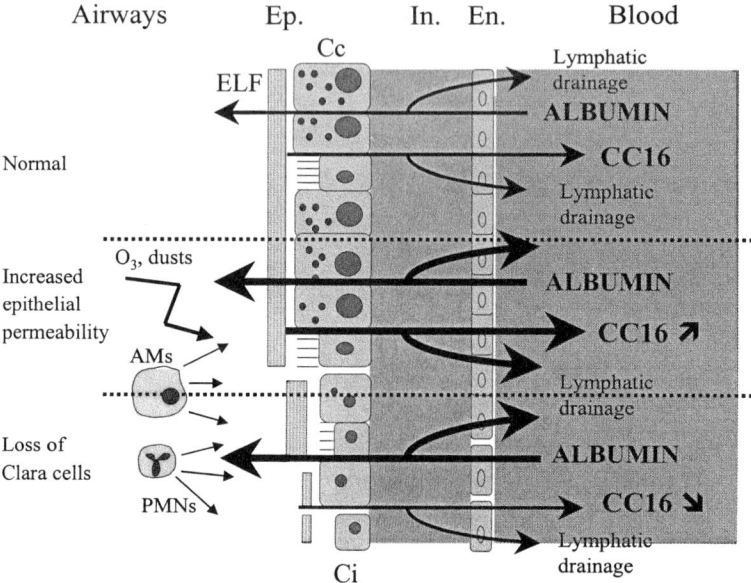

FIGURE 2. Schematic representation of the passage of albumin and CC16 across the different barriers (En., endothelium; Ep., epithelium; In., interstitium; Cc, Clara cells; Ci, ciliated cells) separating the epithelial lining fluid (ELF) from the blood at the level of a terminal bronchiole, under normal conditions, with an increased permeability of the alveolar/blood barrier or in case of Clara cell damage with a normal or slightly altered permeability of the lung epithelial barrier (as caused by tobacco smoking). The thickness of the arrows illustrates the relative fluxes of albumin or CC16 across these barriers and their clearance by lymphatic drainage. The concentrations of albumin in the ELF and blood, estimated in normal subjects, are around 3.5 g/L and 20 g/L, respectively. The corresponding values for CC16 are approximately 100 mg/L and 15 µg/L, assuming a free exchange of the protein across the endothelium. Adapted from reference 23.

tures of the proteins (size), and the transepithelial protein concentration gradient. The latter, which is probably a major driving force of the transepithelial protein transport, is under steady state conditions determined by the clearance of the protein in the compartment in which it is leaking (much faster for CC16 in serum than for albumin in BALF).

The importance of these factors is illustrated in FIGURE 2, showing schematically the passage of CC16 across the terminal airways in normal state or following acute lung epithelium damage induced by a lung irritant such as O_3. The epithelium is the main barrier hindering the bidirectional air/blood exchange of proteins, with the endothelium offering some resistance only to the passage of proteins that are the size of albumin or larger (interstitial fluid/plasma albumin ratio estimated at 0.5–0.6). Proteins that enter the lung interstitium are constantly removed by lymphatic drainage. Following acute exposure to O_3, the first barrier to be damaged is the lung epithelium and, more precisely, the tight junctions. Under these circumstances, the leakiness of CC16 across the epithelial barrier occurs before that of albumin. One explanation is the huge transepithelial concentration gradient of CC16 (around 5000) compared to that of albumin (around 2), which facilitates the movement of CC16. However, because of the very low baseline levels of CC16 in serum, the leakage is probably also more rapidly detected with CC16 than with albumin since the latter requires much larger quantities of proteins to significantly rise in BALF. These differences in gradients and baseline levels reflect the sizes of the compartments in which leaking proteins are diluted (epithelial lining fluid, 20 mL; CC16 distribution space, 40 L). Conceivably also, the small size of CC16 might make this protein more sensitive than albumin to a slight enlargement of the transepithelial protein pathways.

The second mechanism is a decreased production and secretion of CC16 in the respiratory tract that is mirrored in the serum (FIG. 2). This is the case when Clara cells are damaged and reduced in number following chronic exposure to lung toxicants such as tobacco smoke or crystalline silica. It is important to stress that a reduction of CC16 levels in the respiratory tract is paralleled by a reduction in serum only when the lung epithelial barrier is relatively well preserved. When the lung epithelium is damaged by acute toxicants such as 4-ipomeanol or by severe lung inflammation (LPS), the concentration of CC16 in serum rises despite the simultaneous decrease of the protein in BALF, indicating that in acute situations the permeability of the epithelial barrier is a more important determinant of the circulating level of CC16 than the absolute level at the surface of airways.[20]

Like all small-sized molecules circulating in plasma, CC16 is readily eliminated by glomerular filtration. As a corollary, its concentration in serum rises as the renal function fails and, like other low-molecular-weight proteins (cystatin C and β-microglobulin) in patients with renal insufficiency, the concentration of CC16 in serum is inversely correlated with the GFR. The concentration of CC16 in serum can be used as a specific marker of the lung epithelium integrity only when the renal function is normal or only moderately decreased.[2,14]

ACKNOWLEDGMENTS

This study was supported by the European Union (QKL4-1308), the program on Sustainable Development of the Belgian Federal Government (DD/MD006), and the National Fund for Scientific Research (Belgium).

REFERENCES

1. PLOPPER, C.G., D.M. HYDE & A.R. BUCKPITT. 1997. Clara cells. *In* The Lung, pp. 517–533. Lippincott-Raven. Philadelphia.
2. HERMANS, C. & A. BERNARD. 1999. Lung epithelium specific proteins: characteristics and potential application as markers—state of the art. Am. J. Respir. Crit. Care Med. **159:** 646–678.
3. BROECKAERT, F. & A. BERNARD. 2000. Clara cell secretory protein (CC16): characteristics and perspectives as lung peripheral biomarker. Clin. Exp. Allergy **30:** 469–475.
4. HERMANS, C. & A. BERNARD. 1998. Pneumoproteinaemia: a new perspective in the assessment of lung disorders. Eur. Respir. J. **11:** 801–803.
5. BERNARD, A., F.X. MARCHANDISE, S. DEPELCHIN *et al.* 1992. Clara cell protein in serum and bronchoalveolar lavage. Eur. Respir. J. **5:** 1231–1238.
6. BERNARD, A., F. BROECKAERT, C. HERMANS & B. KNOOPS. 1998. *In* Biomarkers: Medical and Workplace Applications, pp. 273–283. Joseph Henry Press. Washington, D.C.

7. JACKSON, P.J., R. TURNER, J.N. KEEN et al. 1988. Purification and partial amino acid sequence of human urine protein 1: evidence for homology with rabbit uteroglobin. J. Chromatogr. **452:** 359–367.
8. SINGH, G., J. SINGH, S.L. KATYAL et al. 1988. Identification, cellular localization, isolation, and characterization of human Clara cell–specific 10 kD protein. J. Histochem. Cytochem. **36:** 73–80.
9. NIETO, A. & M. LOMBARDERO. 1982. Uteroglobin-like antigens in species of Lagomorpha. Comp. Biochem. Physiol. **71:** 511–514.
10. BERNARD, A., X. DUMONT, H. ROELS et al. 1993. The molecular mass and concentrations of protein 1 or Clara cell protein in biological fluids: a reappraisal. Clin. Chim. Acta **223:** 189–191.
11. JENSEN, S.M., J.E. JONES, H. PASS et al. 1994. Clara cell 10 kDa protein mRNA in normal and atypical regions of human respiratory epithelium. Int. J. Cancer **58:** 629–637.
12. DODGE, D.E., R.B. RUCKER, G. SINGH & C.G. PLOPPER. 1993. Quantitative comparison of intracellular concentration and volume of Clara cell 10 kD protein in rat bronchi and bronchioles based on laser scanning confocal microscopy. J. Histochem. Cytochem. **41:** 1171–1183.
13. BERNARD, A., N. THIELEMANS, R. LAUWERYS et al. 1994. Clara cell protein in human amniotic fluid: a potential marker of fetal lung growth. Pediatr. Res. **36:** 771–775.
14. KABANDA, A., E. GOFFIN, A. BERNARD et al. 1995. Factors influencing serum levels and peritoneal clearances of low molecular weight proteins in continuous ambulatory peritoneal dialysis. Kidney Int. **48:** 1946–1952.
15. BERNARD, A., H. ROELS, J.P. BUCHET & R. LAUWERYS. 1993. Serum Clara cell protein: an indicator of bronchial cell dysfunction caused by tobacco smoking. Environ. Res. **66:** 96–104.
16. BERNARD, A., C. HERMANS & G. VAN HOUTE. 1997. Transient increase of serum Clara cell protein (CC16) after exposure to smoke. Occup. Environ. Med. **54:** 63–65.
17. BERNARD, A.M., N.O. THIELEMANS & R.R. LAUWERYS. 1994. Urinary protein 1 or Clara cell protein: a new sensitive marker of proximal tubular dysfunction. Kidney Int. Suppl. **47:** S34–S37.
18. HALATEK, T., C. HERMANS, F. BROECKAERT et al. 1998. Quantification of Clara cell protein in rat and mouse biological fluids using a sensitive immunoassay. Eur. Respir. J. **11:** 726–733.
19. PILON, A.L., J. FARROW, R. WELCH et al. 2000. Pharmacokinetics of recombinant human CC10 in Wistar rats. Poster presented at this conference.
20. HERMANS, C., B. KNOOPS, M. WIEDIG et al. 1999. Clara cell protein as a marker of Clara cell damage and bronchoalveolar blood barrier permeability. Eur. Respir. J. **13:** 1014–1021.
21. STRIPP, B.R., K. MAXSON, R. MERA & G. SINGH. 1995. Plasticity of airway cell proliferation and gene expression after acute naphthalene injury. Am. J. Physiol. **269:** L791–L799.
22. ARSALANE, K., F. BROECKAERT, B. KNOOPS et al. 2000. Clara cell specific protein (CC16) expression after acute lung inflammation induced by intratracheal lipopolysaccharide administration. Am. J. Respir. Crit. Care Med. **161:** 1624–1630.
23. BROECKAERT, F., K. ARSALANE, C. HERMANS et al. 2000. Serum Clara cell protein (CC16): a sensitive marker of epithelial damage caused by ambient ozone. Environ. Health Perspect. **108:** 533–537.
24. ARSALANE, K., F. BROECKAERT, B. KNOOPS et al. 1999. Increased serum and urinary concentrations of lung Clara cell protein in rats acutely exposed to ozone. Toxicol. Appl. Pharmacol. **159:** 169–174.
25. SHIJUBO, N., Y. ITOH, T. YAMAGUSHI et al. 1997. Serum and BAL Clara cell 10 kDa protein (CC10) levels and CC10-positive bronchiolar cells are decreased in smokers. Eur. Respir. J. **10:** 1108–1114.
26. BERNARD, A., J.M. GONZALES-LORENZO, E. SILES et al. 1994. Early decrease of serum Clara cell protein in silica-exposed workers. Eur. Respir. J. **7:** 1932–1937.
27. LAING, I.A., J. GOLDBLATT, E. EBER et al. 1998. A polymorphism of the CC16 gene is associated with an increased risk of asthma. Med. Genet. **35:** 463-467.

28. LAING, I.A., C. HERMANS, A. BERNARD et al. 2000. Association between plasma CC16 levels, the A38G polymorphism, and asthma. Am. J. Respir. Crit. Care Med. **161:** 124–127.
29. VAN VYVE, T., P. CHANEZ, A. BERNARD et al. 1995. Protein content in BAL fluid of patients with asthma and control subjects. J. Allergy Clin. Immunol. **95:** 60–68.
30. SHIJUBO, N., Y. ITOH, T. YAMAGUCHI et al. 1999. Serum levels of Clara cell 10-kDa protein are decreased in patients with asthma. Lung **177:** 45–52.

Mammaglobin, a Breast-Specific Gene, and Its Utility as a Marker for Breast Cancer

TIMOTHY P. FLEMING[a,b] AND MARK A. WATSON[c]

[b]*Department of Surgery and* [c]*Department of Pathology, Washington University School of Medicine, St. Louis, Missouri, USA*

ABSTRACT: The mammaglobin gene encodes a 10-kDa glycoprotein that is distantly related to a family of proteins that includes rat estramustine binding protein (EMBP)/prostatein and human Clara cell 10-kDa protein (CC10)/uteroglobin. Among normal adult tissues, mammaglobin mRNA expression has been detected only in the mammary gland. As an initial step to determine mammaglobin's clinical utility as a breast tumor marker, we evaluated the frequency and specificity with which mammaglobin expression could be detected in primary breast tumors, metastatic breast tumors, and breast tumor cells present in the peripheral circulation. Approximately 80% of all primary and metastatic breast tumors examined were strongly immunopositive for mammaglobin protein, and staining was independent of tumor grade. Among peripheral stem cell collections from breast cancer patients, mammaglobin mRNA could be detected in 60% of cases. Recent work has identified the secreted mammaglobin protein in the sera of some breast cancer patients using both Western blot and ELISA. This study demonstrates that the detection of mammaglobin protein and mRNA in clinical samples may be a useful marker for primary, metastatic, and occult breast cancer.

INTRODUCTION

Our laboratory is interested in identifying novel genes whose expression is altered in primary human breast tumors as compared to patient-matched normal breast tissue.[1] We believe that identification and characterization of these genes will yield clinically useful diagnostic markers for patient management and novel targets for breast cancer treatment, as well as a further understanding of breast cancer pathogenesis. Based on this premise, we have used a differential screening approach to isolate a novel human breast cancer–associated gene that we have named mammaglobin.[2]

The mammaglobin gene encodes a 10-kDa glycoprotein that is distantly related to a family of proteins that includes rat estramustine binding protein (EMBP)/prostatein[3,4] and human Clara cell 10-kDa protein (CC10)/uteroglobin.[5] Although the function of the mammaglobin protein is unknown, related family members are small, epithelial secretory proteins that can either modulate inflammatory processes[6]

---

[a]Address for correspondence: T. P. Fleming, Department of Surgery/Box 8109, Washington University School of Medicine, 660 South Euclid Avenue, St. Louis, MO 63110. Voice: 314-362-4981; fax: 314-362-3638.

fleming@vision.wustl.edu

or bind steroid ligands.[7] Structural similarities between mammaglobin and these other proteins suggest that mammaglobin may serve a related function in the human mammary gland.

Unlike these related proteins, however, mammaglobin displays two characteristics to suggest that its expression is particularly relevant to breast cancer biology. First, both Northern blot analysis and a more-sensitive reverse transcription/polymerase chain reaction (RT-PCR) analysis of several adult human tissues have demonstrated that mammaglobin expression is restricted to the mammary gland.[2] Second, mammaglobin mRNA is present at high levels in human breast tumor cell lines and primary breast tumors as compared to nonmalignant breast tissue.[2] In a preliminary survey, 5 of 10 breast carcinoma cell lines and 8 of 35 primary human breast tumors exhibited high levels of mammaglobin mRNA.

Based on its breast cancer–associated and somewhat unique breast-specific pattern of expression, we believe that mammaglobin is an excellent candidate for a novel and clinically useful breast tumor marker. The results of this study demonstrate that mammaglobin may be a useful tool for clinical breast cancer research.

MATERIALS AND METHODS

Immunohistochemical Staining for Mammaglobin

The carboxy-terminal sequence of the mammaglobin protein was synthesized as a 16-residue peptide (EVFMQLIYDSSLCDLF), conjugated to carrier, and injected into rabbits for antiserum production (Research Genetics, Huntsville, AL). Archived breast tumor specimens were chosen at random from the Vanderbilt University Department of Pathology and the Washington University Cancer Center Tumor Repository. Formalin-fixed, paraffin-embedded tissues were cut at 5 µm, mounted on charged slides, and dried.

For immunohistochemical analysis, slides were deparaffinized and rehydrated in graded solutions of ethanol and distilled water. Tissue sections were preincubated with normal goat serum (Vector Laboratories, Burlingame, CA) at a 1:100 dilution in 3% bovine serum albumin (BSA)/phosphate-buffered saline (PBS) and then with antimammaglobin rabbit antiserum at a 1:1000 dilution for 1 h at room temperature. After several rinses in PBS, sections were incubated in a solution of normal goat serum (1:1000), 3% BSA, and 6 µg/mL of biotinylated goat anti-rabbit IgG (Vector Laboratories, Burlingame, CA) in PBS for 1 h. The secondary antibody solution was rinsed four times in PBS, and tissues were then incubated with a 1:1000 dilution of streptavidin peroxidase (Boehringer Mannheim, Indianapolis, IN) also in a solution of 3% BSA/PBS. After a 30-min incubation, slides were again rinsed four times in PBS and exposed to chromagen solution containing 1 mg/mL of 3,3′-diaminobenzidine tetrahydrochloride (Dako, Carpinteria, CA) and 0.02% hydrogen peroxide for 3 min. Slides were rinsed briefly in deionized water, counterstained with Harris' hemotoxylin, and mounted under coverslips. For negative controls, tissue sections were processed identically, except a 1:500 dilution of preimmune rabbit serum was substituted for the antimammaglobin antiserum.

RNA Preparation and Northern Blot Analysis

Normalized, poly-A-enriched RNAs derived from pooled populations of normal human tissues and immobilized to a solid support (Master Blot™) were purchased from Clonetech (Palo Alto, CA). Anonymized lymph node specimens containing metastatic lesions were obtained from the Cooperative Human Tissue Network[8] and the Washington University Cancer Center Tumor Repository. Tissue specimens were pulverized and homogenized in Trizol reagent (Life Technologies, Rockville, MD) at a concentration of 100 mg of tissue per mL of reagent. RNA isolation was performed exactly as recommended in the manufacturer's protocol and the resulting RNA was resuspended at a concentration of 2 µg/µL in RNase-free water.

For Northern blot analysis, 20 µg of each RNA was subjected to formaldehyde agarose gel electrophoresis and transferred to Nytran Plus membrane (Schleicher & Schuell, Keene, NH) using 10× SSC buffer and standard methodology. To generate a mammaglobin hybridization probe, the ~450-bp mammaglobin cDNA[2] was radio-labeled with ^{32}P-α-dCTP (10 mCi/mL; >3000 Ci/mmol) and the Rediprime labeling kit (Amersham, Arlington Heights, IL) following the supplier's protocol. Forward (5'-CGCGGATCCAGGATTGTCCTGCAGAT-3') and reverse (5'-CCGGAATTC-CCATCCCTCTACCCAGA-3') keratin 19–specific oligonucleotide sequences were used in a standard PCR amplification reaction of normal human breast cDNA to generate a K19 cDNA fragment encompassing the coding region from nt. 477 to nt. 1298.[9] For Northern blot analysis of lymph node specimens, filters were first hybridized with mammaglobin cDNA probe and exposed to film. Filters were allowed to decay for several half-lives and then rehybridized with the keratin 19 cDNA probe without pretreatment of the filters. Equal counts and specific activities of mammaglobin and keratin 19 cDNA probes were used, and each hybridization was exposed to film under equivalent conditions.

Reverse Transcriptase and Polymerase Chain Reaction (RT-PCR) Assay

The mammaglobin RT-PCR assay was performed in the Washington University Cancer Center Molecular Diagnostics Core Facility. Aliquots of $\sim 1 \times 10^6$ peripheral leukocytes were obtained from leukapheresis products of breast cancer patients undergoing peripheral stem cell collection for autologous stem cell transplant. For positive controls, 10 mammaglobin-expressing MDA-MB175 human breast tumor cells[2] were mixed with 10^6 human OM431 melanoma cells to yield a $1:10^5$ breast cancer cell dilution. A pure population of OM431 cells was used as a negative control. Frozen cell aliquots were immediately lysed in 1 mL of Trizol reagent (Life Technologies, Rockville, MD), and total RNA was prepared as per the supplier's protocol. To assess the integrity of synthesized cDNA, 10% of the cDNA was subjected to a 50-µL PCR reaction containing a final concentration of 1× Taq DNA polymerase buffer, 1.5 mM $MgCl_2$, 200 µM dNTPs, 0.6 µM glyceraldehyde-3-phosphate dehydrogenase (GAPDH) forward amplification primer (5'-CCACCCATG-GCAAATTCCATGGCA-3'), 0.6 µM GAPDH reverse amplification primer (5'-TCTAGACGGCAGGTCAGGTCCACC-3'), and 2.5 units of Taq DNA polymerase (Life Technologies, Rockville, MD). Reactions were heated to 94°C for 1 min and then subjected to 30 cycles of 94°C for 30 s, 58°C for 60 s, and 72°C for 45 s. PCR products were analyzed on a 2% agarose gel and a single, uniformly intense fragment of 599 nt. indicated that each cDNA synthesis reaction had been successful. An additional 10% of the cDNA reaction was then subjected to a second 50-µL PCR reaction containing a final concentration of 1× Taq DNA polymerase buffer, 1.5 mM $MgCl_2$, 200 µM dNTPs, 0.6 µM mammaglobin forward amplification primer (5'-AGCACTGCTACGCAGGCTCT-3'), 0.6 µM mammaglobin reverse amplification primer (5'-ATAAGAAAGAGAAGGTGTGG-3'), and 2.5 units of Taq DNA polymerase (Life Technologies, Rockville, MD). Reactions were heated to 94°C for 1 min and then subjected to 45 cycles of 94°C for 30 s, 58°C for 60 s, and 72°C for 30 s. Amplification products were delivered to an alternate laboratory site, electrophoresed on a 2% agarose gel, and subjected to Southern blot analysis using 0.2 µM Nytran Plus membrane and the supplier's protocol for the neutral transfer method (Schleicher & Schuell, Keene, NH). The resulting filter was hybridized with 1×10^6 cpm/mL of mammaglobin cDNA probe and washed as described above.

RESULTS

Breast-Specific Expression

Perhaps the most provocative and clinically useful property of mammaglobin is its apparent breast-specific expression.[2] To unequivocally demonstrate this property, we reexamined mammaglobin expression in a more extensive and quantitative panel of poly-A-selected mRNAs from pooled populations of adult and fetal human tissues. FIGURE 1 demonstrates mammaglobin's breast specificity. Expression is absent in other closely related apocrine glands (e.g., salivary gland) and hormonally responsive epithelial tissues (e.g., uterus, prostate). Expression is also undetectable in peripheral leukocytes, lymph node, and bone marrow. Apart from the mammary gland, mammaglobin mRNA could not be detected in any other nonneoplastic adult or fetal tissue surveyed. This result suggests potential advantages for using mammaglobin gene expression as a very specific marker for breast cancer.

Immunohistochemical Staining in Primary Breast Tumors

High levels of mammaglobin mRNA have been previously observed in a small set of primary human breast tumors.[2] To study mammaglobin protein expression, control for cellular heterogeneity between tumor specimens, and develop a reagent that could be implemented for routine immunohistochemical analysis, we synthesized a peptide corresponding to the C-terminal mammaglobin protein sequence and generated rabbit polyclonal, antimammaglobin antibodies. This reagent was used in a larger survey of primary breast tumors of varying grades and histological types. A summary of mammaglobin protein expression in these specimens is presented in TABLE 1. Overall, 80% of ductal carcinomas examined demonstrated strong global or focal cell staining for mammaglobin protein. Interestingly, staining was equally frequent among well-differentiated (78%), moderately differentiated (67%), and poorly differentiated (63%) tumors. Strong staining was also seen in 3/3 cases of pure ductal carcinoma *in situ* (DCIS). The cellular staining pattern of mammaglobin was predominantly diffuse and cytoplasmic, although some cells also demonstrated localized staining adjacent to the nucleus. In normal breast tissue, mammaglobin staining was observed only in rare epithelial cells within small ducts and lobules. However, as has been observed with other secretory proteins, increased expression of mammaglobin coincided with features of apocrine metaplasia. In benign breast tissue with metaplastic apocrine epithelium, mammaglobin immunoreactivity was present both within the epithelium and in the apocrine cyst fluid. The specificity of these patterns of positive staining was documented by the lack of signal from iden-

TABLE 1. Summary of mammaglobin expression in primary breast cancer

	Total no.	4+	3+	2+	1+	0
Lobular	3	0	1	1	0	1
Well differentiated	14	11	0	1	0	2
Moderately differentiated	42	28	5	3	2	2
Poorly differentiated	38	24	9	2	0	3
DCIS	3	1	2	0	0	0

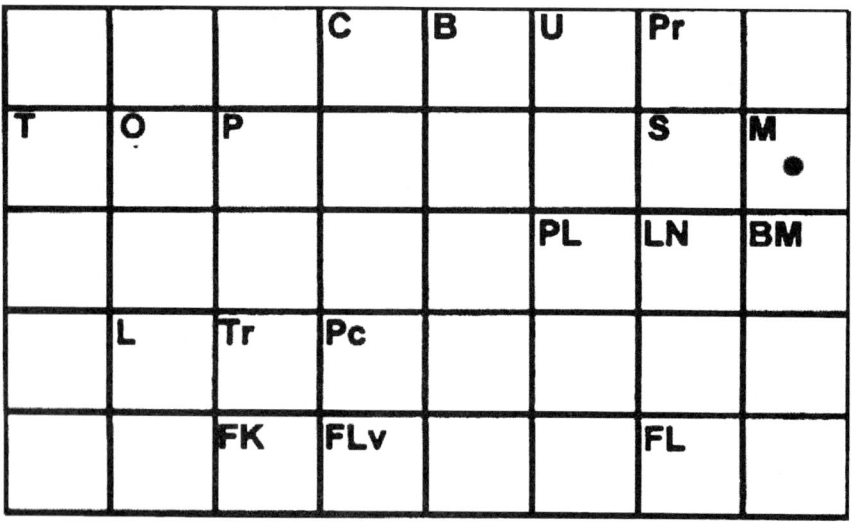

FIGURE 1. Expression of mammaglobin in human adult and fetal tissues. A commercially prepared dot-blot of human poly-A mRNAs from nonmalignant tissues was hybridized with the human mammaglobin cDNA. A portion of the hybridized blot is shown with only the most relevant mRNA samples denoted. C, colon; B, bladder; U, uterus; Pr, prostate; T, testis; O, ovary; P, pancreas; S, salivary gland; M, mammary gland; PL, peripheral leukocyte; LN, lymph node; BM, bone marrow; L, liver; Tr, trachea; Pc, placenta; FK, fetal kidney; FLv, fetal liver; FL, fetal lung. Note that several other tissue mRNAs are present on the blot (unlettered boxes) and also do not hybridize to the mammaglobin probe.

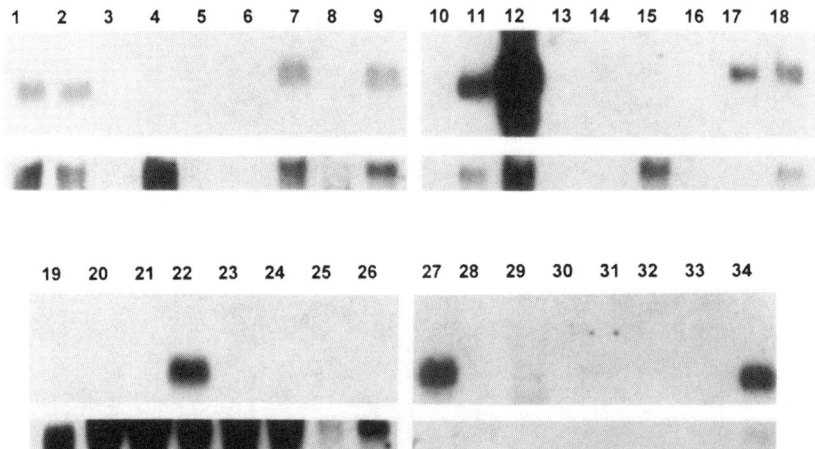

FIGURE 2. Detection of mammaglobin and keratin 19 expression in lymph nodes containing metastatic epithelial cancers. Northern blot analysis using a mammaglobin cDNA probe (top) and the identical blot reprobed with a keratin 19 cDNA probe (bottom). Lymph node specimens contain histologically documented metastases from primary breast tumors (1–20, 27), squamous cell tumors of the head and neck (21, 23, 24), endometrial cancer (22), adenocarcinoma of the colon (25), biliary carcinoma (26), and adenocarcinoma of the lung (28). Cases 29–33 are lymph node biopsies from patients without any known malignancy and case 34 is normal breast tissue.

tical specimens incubated with either preimmune rabbit serum or antimammaglobin antiserum preincubated with competing C-terminal peptide (data not shown).

Expression in Lymph Node Metastases

Mammaglobin's expression in a large percentage of primary breast tumors and its absence in lymphoid tissues suggested that it would be a sensitive and specific marker for detecting metastatic breast tumor cells within lymph nodes. To test this idea, we examined mammaglobin mRNA expression in lymph nodes containing either histologically documented breast metastases or other nonmammary metastatic tumors. Expression of mammaglobin was also compared to that of keratin 19, an epithelial cell-specific marker, to normalize mammaglobin signals to the total number of malignant epithelial cells present in each lymph node sample. FIGURE 2 shows the results of this survey. Among 21 cases of lymph nodes containing histologically documented metastatic breast cancer, mammaglobin expression could be detected in 9 (43%). However, in 6 of the 21 cases, neither mammaglobin nor keratin 19 mRNA could be detected, suggesting that the sample of lymph node may have contained too few tumor cells to be assayed by this method. When these 6 cases were excluded, 60% (9/15) of cases contained detectable mammaglobin mRNA. In at least 1 case, mammaglobin expression was detected in the absence of keratin 19 expression, suggesting that mammaglobin may be a more sensitive marker in a subset of tumors. Furthermore, keratin 19 was ubiquitously expressed in many cases of nonmammary metastases (e.g., lymph nodes with laryngeal and biliary tumor involvement), while

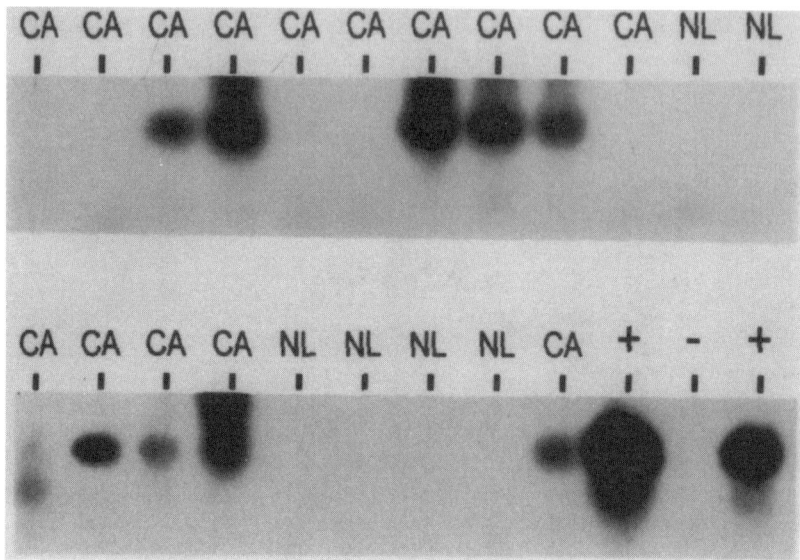

FIGURE 3. Detection of mammaglobin in circulating breast tumor cells. RNA from patients with metastatic breast cancer (CA) or normal donors (NL) was obtained from peripheral stem cell collection products and subjected to RT-PCR and Southern hybridization analysis for mammaglobin expression. Negative control RNA (−) is derived from the OM431 melanoma cell line. Positive control RNA (+) is derived from a mix of 1 in 10^5 MDA-MB-175 breast tumor cells in a background of OM431 cells.

mammaglobin expression was detected in only 1 case of a nonmammary metastatic lesion (case 22; FIG. 2). This case was an inguinal lymph node containing poorly differentiated adenocarcinoma of unknown origin. Since a previous history of endometrial cancer was documented, this lesion was presumed to be a metastatic recurrence of the disease. In all cases of lymph nodes without metastatic disease, both keratin 19 and mammaglobin expression were undetectable.

Detection of Circulating Tumor Cells

Given mammaglobin's apparent sensitivity and specificity for detecting breast tumor cells in lymph node metastases, particularly as compared to keratin 19, we investigated whether mammaglobin mRNA could be used as a marker for circulating breast tumor cells as well. Peripheral blood stem cell (PBSC) collections from 15 patients undergoing high-dose chemotherapy and autologous stem cell transplant for metastatic breast cancer were obtained and subjected to a sensitive RT-PCR assay to detect mammaglobin mRNA. The results of this assay are shown in FIGURE 3. Of 15 cases, 9 patients (60%) yielded detectable mammaglobin mRNA from their PBSCs, suggesting that the collected products contained contaminating breast tumor cells. In no cases of collections from healthy donors was mammaglobin mRNA detected. Based on the robust hybridization signal obtained from duplicated positive controls containing $1:10^5$ breast tumor cells, the limit of detection for the current assay

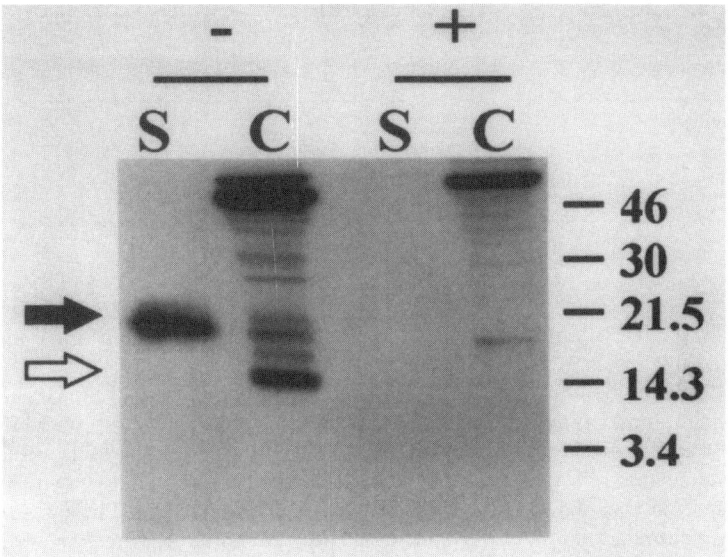

FIGURE 4. Detection of mammaglobin protein in the culture supernatant of the human breast cell line MDA-MB-415. The ~21-kDa band is detected by Western blot analysis in the culture supernatant (S) and the ~14-kDa precursor form is observed in the cell lysate (C). The right panel (+) is identical to the left panel (−), except that competing peptide was added to demonstrate specificity of the peptide antibody generated to the mammaglobin protein.

format is probably well below $1:10^6$ cells. The frequency of detection and high specificity of this assay suggest that mammaglobin may be a useful tool for studying the prognostic significance of occult circulating tumor cells in larger, prospective studies.

Detection of the Mammaglobin Protein in the Breast Cancer Cell Line MDA-MB-415

To detect physiologic mammaglobin protein, we obtained a rabbit antibody directed against the 18 amino acids comprising the carboxy-terminus of the predicted mammaglobin protein. This antibody was used to detect mammaglobin from the mammaglobin-expressing human breast tumor cell line MDA-MB-415. Culture supernatants (S) and cell lysates (C) from the MDA-MB-415 cell line were analyzed by Western blot using this rabbit peptide antibody. We were able to detect a 21-kDa secreted protein from the MDA-MB-415 cell line (FIG. 4) and the cell lysate contained immunoreactive bands of 14–21 kDa. These bands appeared to be specific to the antimammaglobin antibody as the bands in both the supernatant and the cell lysate were successfully competed with the peptide recognized by the antibody. There were bands observed above the 46-kDa molecular-weight marker, but these proteins were not blocked by the competing peptide and likely represent nonspecific proteins. Other breast tumor cell lines have been analyzed for mammaglobin protein, and the 21- and 14-kDa mammaglobin proteins, similar to those observed in the MDA-MB-415 cell line, were detected by Western blot (data not shown).

TABLE 2. Detection of serum mammaglobin in blood samples: mammaglobin-positive by ELISA

Source	No. tested	No. positive	Percentage
Normal control	25	1	4%
Primary breast cancer patients	98	32	33%
Metastatic breast cancer patients	36	16	44%

Detection of Mammaglobin in the Sera of Breast Cancer Patients

Based on our ability to detect mammaglobin in the conditioned media of mammaglobin-positive breast cancer cell lines and human breast secretions, we sought to know whether we could identify mammaglobin in sera of breast cancer patients. Using monoclonal antibodies directed against bacterially expressed mammaglobin, it was possible to demonstrate that the secreted mammaglobin protein was indeed detectable in sera of some breast cancer patients (data not shown). Of note, mammaglobin was not observed in sera from controls analyzed. These results indicate that mammaglobin may have utility as a serum marker for breast cancer.

Based on the ability to detect mammaglobin in sera of some breast cancer patients, efforts were focused to develop an ELISA assay to allow rapid screening of serum samples for the presence of mammaglobin. This screening tool would determine the feasibility of using mammaglobin as a serum marker for breast cancer detection and also would establish the sensitivity and specificity of these assays. Monoclonal and polyclonal antibodies were developed for an ELISA. We analyzed double-blinded samples of archived sera from both normal patients ($n = 25$) and patients with known breast cancer disease ($n = 135$).

The results of these experiments are summarized in TABLE 2. Only 1 normal sample yielded a positive assay (4% of the total). No data were available regarding this individual's history of breast cancer, age, or menopausal status. Moreover, 33% of the primary breast cancer sera and 44% of the metastatic breast cancer sera yielded positive assays. These results indicate that sera detection is feasible using the mammaglobin detection assay.

DISCUSSION

The mammaglobin gene encodes a novel secreted glycoprotein that is expressed in mammary epithelial cells and upregulated in a majority of breast tumors. Although the function of the mammaglobin protein and its role in the etiology of breast cancer are currently unknown, the present study demonstrates that mammaglobin expression could be a potentially useful tumor marker for the management of breast cancer patients.

Mammaglobin's tissue specificity and the frequency with which it can identify primary breast cancer suggested that mammaglobin expression would also be an effective marker for detecting metastatic and occult disease. Several "epithelial-specific" genes have been used as surrogate markers to detect occult epithelial malignancies. A nested RT-PCR assay for CEA has been used to identify metastases in histologically negative lymph nodes from breast and gastrointestinal cancer patients.[10] Many similar assays based on cytokeratin 20[11] and keratin 19[12–14] have been used to detect occult breast tumor cells in lymph nodes and the peripheral circulation. These molecular assays demonstrate enhanced sensitivity over serial sectioning and immunohistochemical approaches. However, their utility is often limited by a low level of keratin gene expression in lymphoid and other nonepithelial cells.[14,15] In our Northern blot assay, at least 1 lymph node that contained metastatic breast cancer produced no K19 hybridization signal, but was strongly positive for mammaglobin. This enhanced sensitivity may be due to the fact that the cellular expression levels of mammaglobin are higher than cytokeratins in some breast tumors. Conversely, 6 cases of K19-positive nodes were mammaglobin-negative. Since we were unable to evaluate the primary tumors in these cases, it is possible that these tumors simply did not express mammaglobin. Therefore, if only cases in which primary tumors were known to be mammaglobin-positive were considered, the sensitivity of mammaglobin for detecting metastatic lesions could be even higher than reported in this study. Not surprisingly, the K19 probe yielded strong hybridization signals from other metastatic lesions of epithelial origin, while mammaglobin expression was limited to breast cancer metastases, with the exception of 1 case of presumed endometrial cancer. Interestingly, we have recently detected mammaglobin expression in several primary endometrial cancers (unpublished data), thus confirming the idea that mammaglobin, while not expressed in normal endometrial tissue, may be expressed in a subset of endometrial as well as breast cancers.

To detect circulating breast tumor cells, the mammaglobin assay described in this report implemented a single-step RT-PCR amplification approach followed by hybridization of amplification products to a radiolabeled mammaglobin cDNA probe. This format consistently provided a sensitivity of 1 breast cancer cell in a 10^5 cell background based on the strong hybridization signal obtained from a dilution of positive control cells. At this level of sensitivity, mammaglobin expression in both lymphoid tissue and peripheral lymphocytes was undetectable. Increasing the sensitivity of our assay (e.g., by implementing "nested" PCR methodology) may reveal low levels of transcription in nonmammary tissues. However, the relatively high level of mammaglobin transcripts in breast cancer cells should preclude the need for adopting more sensitive assays where expression in lymphoid cells becomes problematic. In the current study, mammaglobin expression among circulating cells was seen in 9 of 15 cases. However, in the remaining 6 cases, we were unable to determine whether a lack of signal represented a lack of tumor cells that express mammaglobin, a true absence of tumor cells, or assay insensitivity. Limiting future studies to patients whose primary tumors definitively express mammaglobin (by either immunohistochemical or molecular assays of the primary tumor) will simplify interpretation of a negative result. The clinical significance of occult tumor cells in lymph nodes, in bone marrow, and in the circulation remains controversial. Nonetheless, mammaglobin expression may have sufficient sensitivity and specificity to be a useful tool to further explore this issue in larger clinical trials. For example, mammaglobin expression may be useful to evaluate tumor contamination of peripheral stem cell products after cytokine priming[16] or tumor cell purging protocols.[17]

The mammaglobin protein is a small, actively secreted glycoprotein. The mammaglobin protein was detected in serum of patients whose breast tumors express mammaglobin, similar to prostate-specific antigen[18] in prostate cancer patients. Serum markers currently used in clinical practice to follow breast cancer relapse, such as CA15.3, are epitopes of large glycoproteins that are passively shed from tumor cells. They often suffer from limited sensitivity and specificity.[19] Given the biochemical properties of mammaglobin, its high frequency of expression in both well-differentiated and poorly differentiated tumors, its relatively low expression in normal breast tissue, and its lack of demonstrable expression in other normal tissues, mammaglobin is an attractive candidate for a serum breast tumor marker. The assay to detect serum mammaglobin indicates that sera detection of mammaglobin may be a useful marker for the detection of primary breast cancer and tumor recurrence. It should be noted that the antibodies used in these experiments were derived from bacterially expressed mammaglobin. As mammaglobin is highly glycosylated, antibodies directed against native mammaglobin may be more sensitive in the detection of mammaglobin in sera. Other factors that may impact sera detection of mammaglobin, including pregnancy, menstrual cycle, and menopausal status, will need to be examined in detail to better define the clinical utility of circulating mammaglobin as a diagnostic marker for breast cancer.

In summary, the current study demonstrates that mammaglobin expression is a sensitive and specific marker for neoplastic breast epithelial cells and provides sufficiently promising evidence to warrant larger clinical studies using mammaglobin as a molecular marker for early detection, staging, prognosis, and/or relapse monitoring of breast cancer.

ACKNOWLEDGMENTS

This work was supported by Public Health Service Grant Nos. CA76227 (to T. P. Fleming), CA76223-01 (to M. A. Watson), and CA68485 from the National Cancer Institute, National Institutes of Health, Department of Health and Human Services. Further support was provided by the Jewish Hospital Auxiliary Fund, Barnes-Jewish Hospital, St. Louis, MO (to T. P. Fleming).

REFERENCES

1. WATSON, M.A. & T.P. FLEMING. 1994. Isolation of differentially expressed sequence tags from human breast cancer. Cancer Res. **54:** 4598–4602.
2. WATSON, M.A. & T.P. FLEMING. 1996. Mammaglobin, a mammary-specific member of the uteroglobin gene family, is overexpressed in human breast cancer. Cancer Res. **56:** 860–865.
3. FORSGREN, B., P. BJORK, K. CARLSTROM et al. 1979. Purification and distribution of a major protein in rat prostate that binds estramustine, a nitrogen mustard derivative of estradiol-17β. Proc. Natl. Acad. Sci. U.S.A. **76:** 3149–3153.
4. OSCAR, L.A., P. PETRUSZ & F.S. FRENCH. 1979. Prostatein: a major secretory protein of the rat ventral prostate. J. Biol. Chem. **254:** 6196–6202.
5. MIELE, L., E.C. MIELE, G. MANTILE et al. 1994. Uteroglobin and uteroglobin-like proteins: the uteroglobin family of proteins. J. Endocrinol. Invest. **17:** 679–692.
6. KUNDU, G.C., G. MANTILE, L. MIELE et al. 1996. Recombinant human uteroglobin suppresses cellular invasiveness via a novel class of high-affinity cell surface binding site. Proc. Natl. Acad. Sci. U.S.A. **93:** 2915–2919.

7. HEYNS, W. & D. BOSSYNS. 1983. A comparative study of estramustine and pregnenolone binding to prostatic binding protein: evidence for subunit cooperativity. J. Steroid Biochem. **19:** 1689–1694.
8. LIVOLSI, V.A., K.P. CLAUSEN, W. GRIZZLE *et al.* 1993. The Cooperative Human Tissue Network. Cancer **71:** 1391–1394.
9. STASIAK, P.C. & E.B. LANE. 1987. Sequence of cDNA coding for human keratin 19. Nucleic Acids Res. **15:** 10058.
10. MORI, M., K. MIMORI, H. INOUE *et al.* 1995. Detection of cancer micrometastases in lymph nodes by reverse transcriptase–polymerase chain reaction. Cancer Res. **55:** 3417–3420.
11. SOETH, E., I. VOGEL, C. RODER *et al.* 1997. Comparative analysis of bone marrow and venous blood isolates from gastrointestinal cancer patients for the detection of disseminated tumor cells using reverse transcription PCR. Cancer Res. **57:** 3106–3110.
12. SHINZABURO, N., A. TOMOHIKO, M. KAZUYOSHI *et al.* 1996. Histologic characteristics of breast cancers with occult lymph node metastases detected by keratin 19 mRNA reverse transcriptase–polymerase chain reaction. Cancer **78:** 1235–1240.
13. DATTA, Y.H., P.T. ADAMS, W.R. DROBYSKI *et al.* 1994. Sensitive detection of occult breast cancer by the reverse transcriptase–polymerase chain reaction. J. Clin. Oncol. **12:** 475–482.
14. SCHOENFELD, A., Y. LUQMANI, D. SMITH *et al.* 1994. Detection of breast cancer micrometastases in axillary lymph nodes by using polymerase chain reaction. Cancer Res. **54:** 2986–2990.
15. TRAWEEK, S.T., J. LIU & H. BATTIFORA. 1993. Keratin gene expression in nonepithelial tissues. Am. J. Pathol. **142:** 1111–1118.
16. PASSOS-COELHO, J.L., A.A. ROSS, D.J. KAHN *et al.* 1996. Similar breast cancer cell contamination of single-day peripheral-blood progenitor-cell collections obtained after priming with hematopoietic growth factor alone or after cyclophosphamide followed by growth factor. J. Clin. Oncol. **14:** 2569–2575.
17. PASSOS-COELHO, J., A.A. ROSS, J.M. DAVIS *et al.* 1994. Bone marrow micrometastases in chemotherapy-responsive advanced breast cancer: effect of *ex vivo* purging with 4-hydroperoxycyclophosphamide. Cancer Res. **54:** 2366–2371.
18. ARCANGELI, C.G., D.K. ORNSTEIN, D.W. KEETCH & G.L. ANDRIOLE. 1997. Prostate-specific antigen as a screening test for prostate cancer. Urol. Clin. North Am. **24:** 299–305.
19. AMERICAN SOCIETY OF CLINICAL ONCOLOGY. 1996. Clinical practice guidelines for the use of tumor markers in breast and colorectal cancer. J. Clin. Oncol. **14:** 2843–2877.

The Uteroglobin Fold

ISABELLE CALLEBAUT,[a,b] ANNE POUPON,[c] RÉNÉE BALLY,[a,d]
JEAN-PHILIPPE DEMARET,[e] DOMINIQUE HOUSSET,[f] JEAN DELETTRÉ,[a]
PAUL HOSSENLOPP,[a] AND JEAN-PAUL MORNON[a]

[a]*Systèmes Moléculaires et Biologie Structurale, Laboratoire de Minéralogie-Cristallographie Paris (LMCP), CNRS UMR 7590, Universités Paris 6 et Paris 7, Paris, France*

[c]*DIEP, CEA Saclay, 91191 Gif sur Yvette, France*

[e]*Laboratoire de Physicochimie Biomoléculaire et Cellulaire, Université Paris 6, Paris, France*

[f]*Laboratoire de Cristallographie et Cristallogénèse des Protéines, Institut de Biologie Structurale J. P. Ebel, CEA-CNRS-UJF, Grenoble, France*

> ABSTRACT: Uteroglobin (UTG) forms a fascinating homodimeric structure that binds small- to medium-sized ligands through an internal hydrophobic cavity, located at the interface between the two monomers. Previous studies have shown that UTG fold is not limited to the UTG/CC10 family, whose sequence/structure relationships are highlighted here, but can be extended to the cap domain of *Xanthobacter autotrophicus* haloalkane dehalogenase. We show here that UTG fold is adopted by several other cap domains within the α/β hydrolase family, making it a well-suited "geode" structure allowing it to sequester various hydrophobic molecules. Additionally, some data about a new crystal form of oxidized rabbit UTG are presented, completing previous structural studies, as well as results from molecular dynamics, suggesting an alternative way for the ligand to reach the internal cavity.

STRUCTURAL AND MOLECULAR DYNAMICS STUDIES OF UTEROGLOBIN

The structural study of oxidized uteroglobin (UTG)[1,2] began relatively early in the saga of Structural Biology, more than 20 years ago (FIG. 1A). Starting from several milligrams of pure protein extracted from pregnant rabbits sacrificed at day 6, several crystal forms were obtained by our group, which allowed the harvesting of data at a very high resolution (1.34 Å), taking into account the limited means at that time.[3] The C222$_1$ crystal form (PDB code 1UTG) was solved and published,[4] and further refined, using one of the first array processors.[5] The P2$_1$ crystal form was also solved and published (PDB code 2UTG).[6] Meanwhile, the well-diffracting P2$_1$2$_1$2 crystal form was investigated at 1.34-Å resolution, but remains unpublished. The reader will find hereafter a summary of this contribution (APPENDIX A) and a

---

[b]Address for correspondence: Systèmes Moléculaires et Biologie Structurale, Laboratoire de Minéralogie-Cristallographie Paris (LMCP), CNRS UMR 7590, Universités Paris 6 et Paris 7, Case 115, 4 Place Jussieu, 75252 Paris Cedex 05, France.
  Isabelle.Callebaut@lmcp.jussieu.fr
[d]Currently retired from Laboratoire de Minéralogie-Cristallographie, Paris, France.

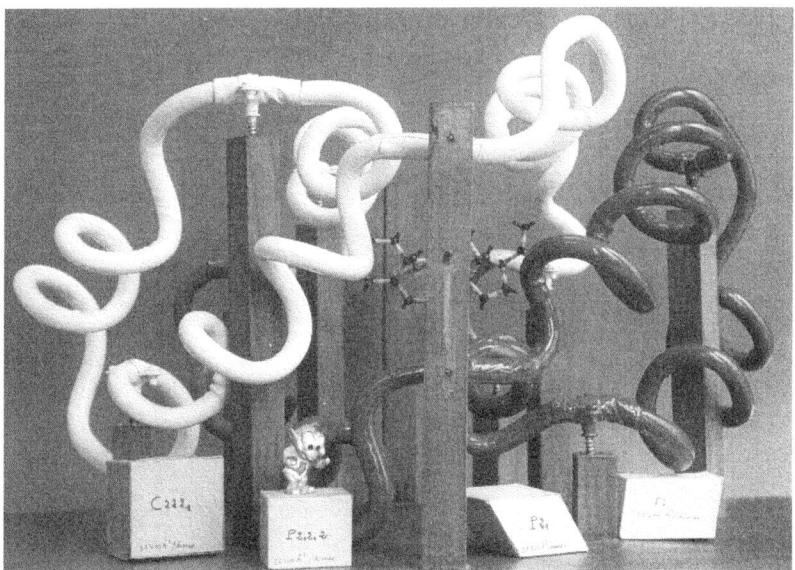

FIGURE 1A. Wire model of the rabbit uteroglobin dimer (1979; PDB 1UTG). Cells of four different crystal forms are illustrated in front of the model.

comparison at high resolution of these three structures of uteroglobin, highlighting the conservation of a very similar fold.

The uteroglobin monomer structure is composed of four α-helices, H1 to H4, that do not form a canonical four helix–bundle motif, but a "boomerang"-shaped structure in which helices H1, H3, and H4 form the interface of a dimeric structure displaying a large oblong internal cavity of 440 Å^3 (FIG. 1B). The structure can also appear as a "splinter bundle" formed by an α-hairpin (made of helices H2 and H3) wrapping onto a pair of interfacing helices H1 and H4′ (this latter helix belonging to the other subunit), whereas these interfacing helices, together with helices H1′ and H4, form a canonical "square bundle"[7] (FIG. 1B). The large cavity, at the interface of the two monomers, joined together in an antiparallel manner, is able to accommodate small- to medium-sized hydrophobic ligands, such as progesterone, but its access is sealed by two intermolecular disulfide bridges involving conserved cysteines (Cys3 to Cys69′ and Cys3′ to Cys69) (FIG. 1B). Two other smaller cavities, C2 and C3, have also been observed within each monomer of human uteroglobin, on both sides of the large cavity C1.[8] From these structural studies on rabbit uteroglobin,[4,5,9] completed by those performed on rat Clara cell 17-kDa protein,[10] it was demonstrated that Y21 and T60 are in a suitable position for forming hydrogen bonds with O-3 and O-20 of progesterone. The crucial role of these residues was experimentally demonstrated by mutational and simulation studies,[11,12] suggesting two alternative orientations of the ligand within the cavity. More recently, the structures of rat uteroglobin and human Clara cell complexed with methylsulfonyl–polychlorinated biphenyl (PCB) and phospholipid, respectively, gave further insights about ligand binding specificities.[13,14] These experiments especially stressed the crucial role of

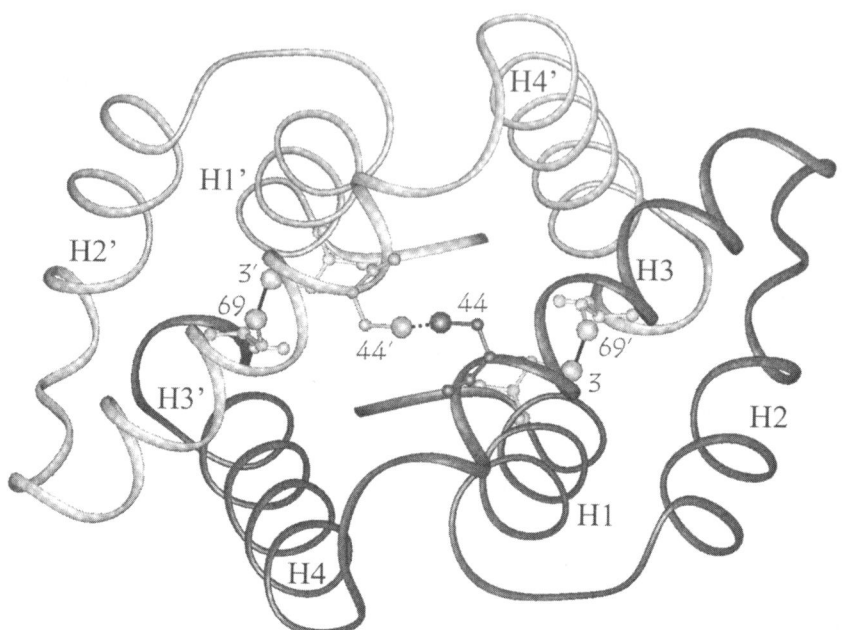

FIGURE 1B. Ribbon representation of the same dimer displayed on a silicon graphic station using InsightII (MSI Inc., San Diego) and viewed down the crystallography twofold axis, in a similar orientation as in FIGURE 1A. The two disulfide bonds (between Cys3 and Cys69' and between Cys3' and Cys69) are highlighted. Val44 and Val44' are substituted by two cysteine residues to highlight the additional disulfide bridge likely existing in many members of the uteroglobin family (see FIGS. 3 and 5C).

the hydroxyl group of Y21 (numbered Y23 in the UTG/PCB complex) in binding the sulfonyl oxygen of $(MeSO_2)_2$-PCB and the free oxygen or ester oxygen of the phospholipid phosphate group, respectively.

Further, the mechanism by which ligands gain access to the internal cavity has attracted much attention. From mutational,[15] X-ray,[4–6] and NMR[13,16,17] structural studies, it became clear that the greater mobility observed at the N- and C-termini might allow access to the cavity through a channel involving helices H1 and H4. Moreover, this mechanism was proposed to be controlled by the redox state of the disulfide bonds.[13] A molecular dynamics study performed by several of us several years ago, but still unpublished, gives some further insights into this mechanism, by proposing an unexpected alternative way for the ligand to enter the cavity between helices H3 and H3' (APPENDIX B).

UTEROGLOBIN AND THE BIRTH OF HYDROPHOBIC CLUSTER ANALYSIS

Hydrophobic cluster analysis (HCA) is a sensitive tool to analyze and compare amino acid sequences.[18–20] HCA is able to decipher structural and functional fea-

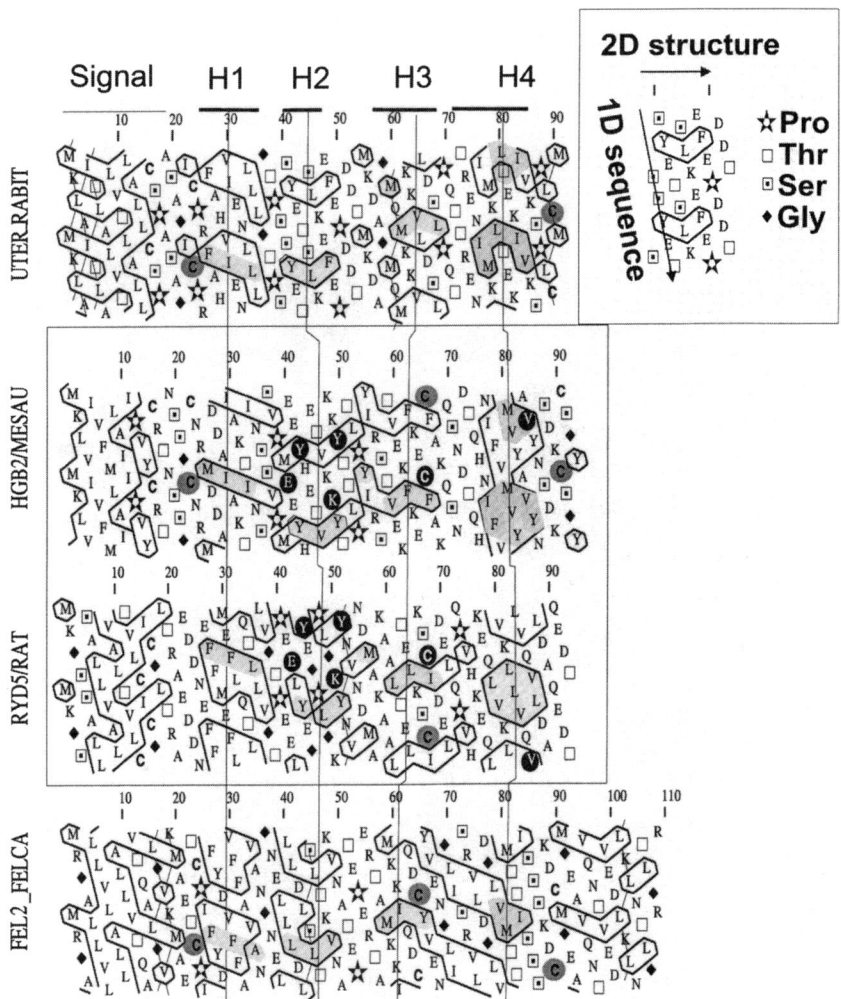

FIGURE 2. Comparison of HCA plots of four members of the uteroglobin family. Briefly, the sequences are shown on a duplicated α-helical net, in which strong hydrophobic amino acids are contoured. These do not distribute randomly, but form clusters that mainly correspond to internal faces of regular secondary structures[20] (α-helices as well as β-strands), as illustrated here by the correspondence of clusters and rabbit uteroglobin helices H1 to H4 (PDB 1UTG, UTER_RABIT). The ability of HCA to detect significant structural relationships at a low level of sequence identity is particularly well illustrated here by the comparison of *Mesocricetus auratus* heteroglobin 2 (HGB2/MESAU) and rat RYD5 (RYD5/RAT), two sequences sharing only 9.5% identity (*white letters on a black background*). A clear conservation of hydrophobic clusters (*shaded gray*) is, however, observed, as assessed by the high HCA score (70%, conserved hydrophobic amino acids/total number of hydrophobic amino acids) and its associated Z score [4.89 σ, difference (expressed in standard deviation) between the observed score and the mean score of a distribution of scores calcu-

tures of orphan genes at very low levels of sequence identity (typically in the 15% range), clearly below the so-called twilight zone where significant signals of similarity and random noise become indiscernible (FIG. 2). We recently performed many studies in various topics of cellular biology, which were based on the ability of HCA to recognize very remote relationships between protein domains (e.g., references 21–25). HCA relies on two main motors: a physical one based on the compactness of hydrophobic amino acids and a mathematical one based on topology [curvature of the sequence linear space (R^1) in the current three-dimensional space (R^3)].[20]

In the present context, it is noteworthy that it was with the uteroglobin sequence, whose HCA plot is shown on FIGURE 2, that the HCA method emerged in 1986, following a contact with Valery Lim from the Institute of Protein Science in Puschino (Moscow region). He proposed to one of us (J-P. Mornon) a rather good model of the three-dimensional structure of uteroglobin before public release of the coordinates. From an elegant tool well suited to predict small α-helical assemblies, a new facet emerged with the birth of the HCA method itself. HCA also opens new ways to delineate fundamental features of stable protein folds. An additional curvature of the sequence space in the topological R^4 space associated with, among other features, the notion of "topohydrophobic positions" (see hereafter) is now giving rise to a new technique, "hydrophobic cluster folding" (HCF), for investigating the *ab initio* prediction of protein folds using only sequence information.

Using HCA in combination with profile-based programs such as PSI-BLAST,[26] we were able to highlight some new, undescribed members belonging to the uteroglobin family, such as the chain 2 of the Fel dI major allergen (FIGS. 2 and 3). This protein, forming a heterodimeric structure with the "uteroglobin-like" Fel dI chain 1, was previously described as sharing no obvious similarity with other proteins,[27] but the conservation of all "topohydrophobic positions" (see hereafter) as well as three cysteine characteristics of the UTG family (corresponding to UTG positions 3, 44, and 69) clearly identify it as a true member of the family (FIGS. 2 and 3). Therefore, the Fel dI chain 2 probably forms a pseudohomodimeric structure with the Fel dI chain 1, exactly as lipophilins A and C.[28]

SEQUENCE/STRUCTURE RELATIONSHIPS IN THE UTG FAMILY

UTG belongs to a large family of small proteins (FIG. 3A) whose sequences are quite divergent [mean identity between sequence pairs in the multiple alignment: 25.5% (σ 14.6%) with a minimum (9.6%) observed for the HGB2/MESAU–RYD5/

---

lated by comparing HGB2 versus 1000 random permutations of RYD5 sequence]. This conservation is indicative of similar folds, and conserved hydrophobic amino acids mostly correspond to topohydrophobic positions defined on the multiple alignment of FIGURE 3. HCA scores (and associated Z scores) are 64% (4.09σ) and 69% (4.53σ) for UTER/HGB2 (15.3% identity) and UTER/RYD5 (25% identity) comparisons, respectively. The conservation of hydrophobic clusters defining the signature of the UTG family is also clearly observed on the HCA plot of *Felis silvestris catus* major allergen Fel dI chain 2 sequence (FEL2/FELCA), sharing 19.4%, 12.5%, and 12.5% identity with UTER_RABIT, HGB2/MESAU, and RYD5/RAT, respectively (HCA scores of 48%, 50%, and 64%, respectively).

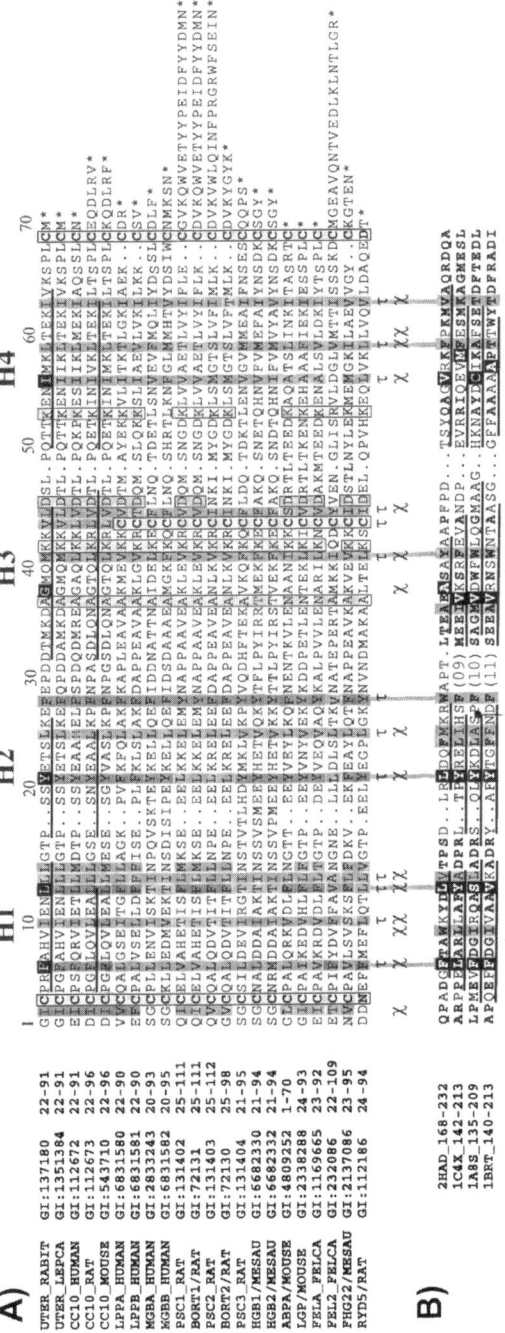

FIGURE 3. See following page for caption.

RAT comparison]. Among the 23 strong hydrophobic amino acids (V, I, L, M, F, Y, W) present in the rabbit UTG sequence (highlighted in FIG. 4), only 13 correspond to "topohydrophobic" positions (F6, I10, L13, L14, Y21, L25, F28, M41, V44, L45, I56, L59, I63; "τ" in FIG. 3). By topohydrophobic position, we mean a position always occupied in the whole family by strong hydrophobic amino acids such as V, I, L, M, F, Y, and W, while A, C, and T, which often substitute strong hydrophobic amino acids, can be considered in an extended list. In recent studies from our group, topohydrophobic positions were shown to possess remarkable properties: (1) they are deeply buried inside the domain and less dispersed between family members than other hydrophobic residues;[29,30] (2) they form a hydrophobic network that holds the core of globular domains.[31] Although most of the topohydrophobic positions in

---

FIGURE 3. (A) Multiple alignment of the sequences of representative members of uteroglobin/Clara cell 10-kDa family of proteins. Protein names and species (following the Swiss-Prot syntax, with names and species separated by "_" corresponding to true Swiss-Prot identifiers), as well as GenBank identifiers (GI), are given in front of each sequence. The position of the first and last amino acids belonging to the domain are also indicated, whereas sequence ends are indicated by *stars*. The secondary structures, as experimentally observed in rabbit uteroglobin (1UTG) and rat Clara cell 10-kDa protein (1CCD), are underlined below their sequences. Topohydrophobic positions (*see text*) and residues involved in the formation of the internal cavity are indicated by symbols τ and χ, respectively. *Names*: ABPA, androgen-binding protein subunit alpha; BORT1, prostatic steroid-binding protein C1; BORT2, prostatic steroid-binding protein C2; CC10, Clara cell 10-kDa secretory protein; FEL2, major allergen I polypeptide chain 2; FELA, major allergen I polypeptide chain 1; FHG22, FHG22 protein; HGB1, heteroglobin 1; HGB2, heteroglobin 2; LGP, lacrimal gland protein; LPPA, lipophilin A; LPPB, lipophilin B; MGBA, mammaglobin A; MGBB, mammaglobin B; PSC1, prostatic steroid-binding protein C1; PSC2, prostatic steroid-binding protein C2; PSC3, prostatic steroid-binding protein C3; RYD5: probable ligand-binding protein RYD5; UTER, uteroglobin. *Species*: FELCA, *Felis silvestris catus* (cat); HUMAN, *Homo sapiens* (human); LEPCA, *Lepus capensis* (brown hare); MESAU, *Mesocricetus auratus* (golden hamster); MOUSE, *Mus musculus* (mouse); RABIT, *Oryctolagus cuniculus* (rabbit); RAT, *Rattus norvegicus* (rat). (B) Sequence alignment of four cap domains of known 3D structures and proposed correspondence with UTG family. Sequence identities with rabbit uteroglobin are 7.6% (2HAD), 9.5% (1C4X), 7.8% (1A8S), and 6.3% (1BRT), whereas 2HAD shares 9.5%, 4.8%, and 11.1% identity with 1C4X, 1A8S, and 1BRT, respectively, and 1C4X shares 12.7% and 14.3% identity with 1A8S and 1BRT, respectively. For each helix in both alignments, one representative marker of overall 3D correspondence is shown *white on a dark background*, as observed from equivalence between topohydrophobic position networks of these two families (*see text*): H1 (1UTG F6, V9, L14/2HAD F172, W175, L179), H2 (1UTG Y21, S24, L25, F28/2HAD L187, F190, M191, W194), H3 (1UTG G38, M41, L45/2HAD A203, Y206, F210), H4 (1UTG I56, M57, L59, T60/2HAD V219, R220, F222, P223). The observed rms deviation (rmsd) between the Cα atoms of the just-described positions is minimal for the 1UTG/1UTR (1.54 Å) and 1A8S/1BRT (1.44 Å) pairs, whereas it ranges from 2.51 Å to 3.99 Å for the 13 other comparisons (involving 1UTG, 1UTR, 2HAD, 1C4X, 1Q8S, and 1BRT). Mean solvent accessibility areas for topohydrophobic (T) versus non-topohydrophobic (NT) positions are 4.5 Å^2/10.3 Å^2 for Phe (18 T and 28 NT), 3.5 Å^2/12.1 Å^2 for Leu (14 T and 60 NT), and 5.1 Å^2/14.1 Å^2 for Met (5 T and 19 NT). The mean interdistances between topohydrophobic positions is 12.2 Å (vs. 18 Å for NT) and the mean dispersion of side-chain gravity centers for topohydrophobic positions is 3.2 Å (vs. 3.9 Å for NT), illustrating the effect of global distortion of the fold. The mean distance between topohydrophobic positions and their two nearest neighbors is 5.7 Å (σ 1.0 Å), highlighting a similar contact network between UTG and UTG-like folds.

FIGURE 4. Ribbon representation of the rabbit uteroglobin dimer structure (1UTG) viewed perpendicular to the crystallographic twofold axis, in which only strong hydrophobic amino acids (V, I, L, M, F, Y, W) are shown. Three rabbit faces are hidden in this picture (for fun, only!).

UTG (11 out of 13; FIG. 3) are involved in the formation of the internal cavity (which takes the place, as it were, of the hydrophobic core of a classic "compact" structure), the number of topohydrophobic positions per secondary structure is similar to that observed for other "standard" globular structures[29,32] [4 in helix H1, 2 in helix H2, 3 in helix H3, and 3 in helix H4, with the 13th one (F28) being located in the H2–H3 loop].

The solvent accessibility values, calculated with the dimer coordinates, clearly show that the topohydrophobic positions of uteroglobin are highly buried within the dimeric structure (FIG. 5A). Among the 11 topohydrophobic positions forming the internal cavity, only 4 do not have nil accessible areas [FIG. 5A; F6 (11.5 Å^2), L13 (6.1 Å^2), M41 (14.3 Å^2), and I63 (4.9 Å^2)] and directly border the cavity (FIG. 5C). F28, corresponding to one of the two topohydrophobic positions remote from the central cavity, has a total accessibility value of 0 Å^2 in the dimer structure (FIG. 5A). This value differs from that calculated on the UTG monomer (FIG. 5A), indicating that the concerned amino acid is involved in the dimeric interface (FIG. 5C). From the alignment depicted in FIGURE 3, it is interesting to note that among topohydro-

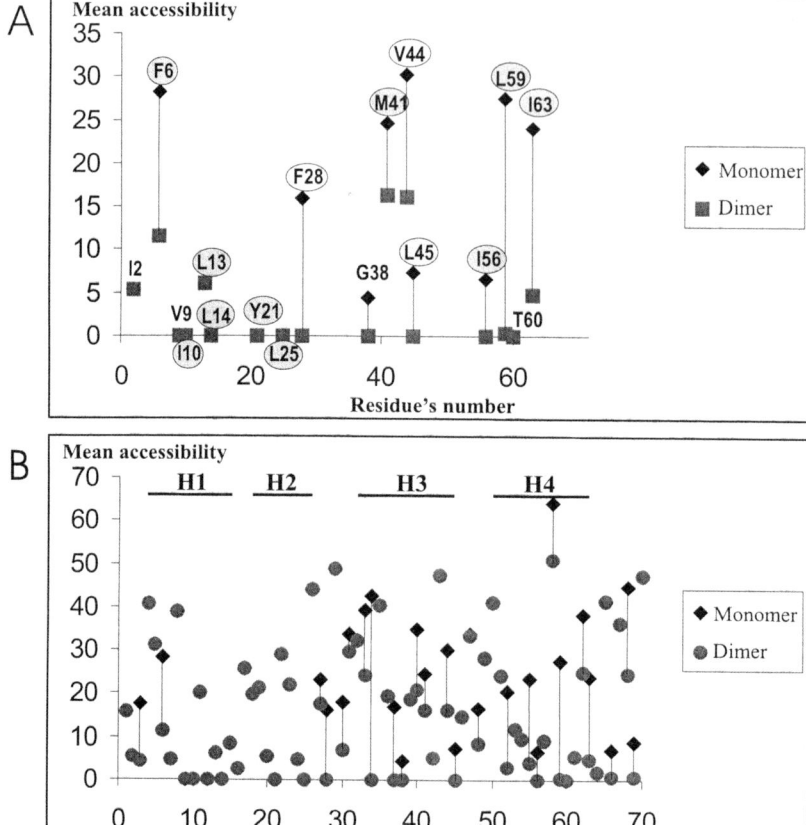

FIGURE 5A/5B. (**A**) Mean accessibility (Å^2) of topohydrophobic positions and/or residues forming the internal cavity calculated on the monomeric and dimeric structures (1UTG). Topohydrophobic positions are *encircled* and, when participating to the cavity, *shaded gray*. Residues that are not encircled are only involved in the cavity formation, but do not correspond to topohydrophobic position. (**B**) Mean accessibility (Å^2) of the 70 residues of uteroglobin, also calculated on the monomeric and dimeric structures.

phobic positions, only F28, which is involved in the dimeric interface, keeps aromatic amino acids in the whole UTG family. Conversely, F6 and Y21 are involved in ligand binding. Replacement of these positions by aliphatic residues, as observed in prostatic steroid-binding protein, should strongly influence the ligand-binding specificities of the concerned proteins. Similarly, L13, which is accessible to solvent within the hydrophobic pocket (FIGS. 5A and 5C), is often substituted by aromatic amino acids, indicating the particular importance of this position for ligand-binding specificity.

The second topohydrophobic position that does not participate in the central cavity corresponds to V44 and stays accessible in the dimeric form, suggesting that it

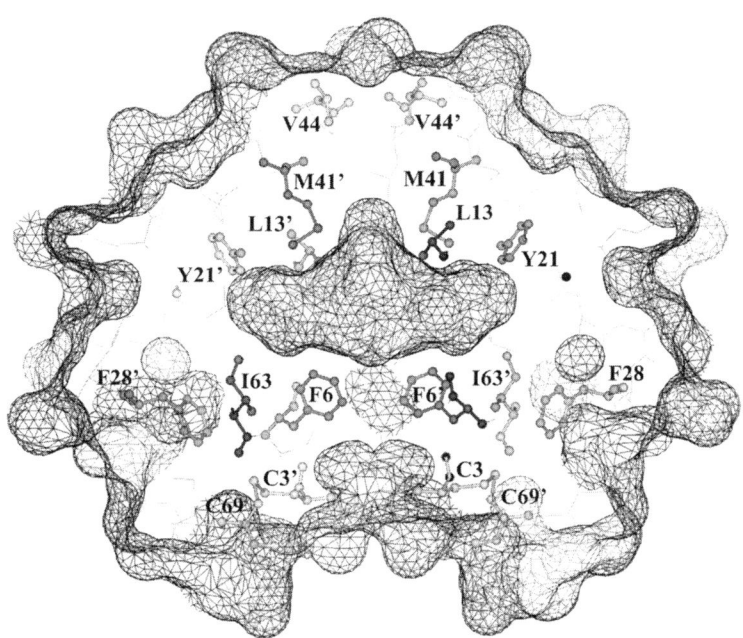

FIGURE 5C. View of the internal cavity, highlighting the importance of F6, L13, M41, and I63. Y21, which is hydrogen-bonded to the ligand, is also shown, as is F28, which is only involved in the dimeric interface without participating to the cavity. The positions of the two cysteines, and of V44 (often substituted by a cysteine), are also highlighted.

could play a functional role. This position is very often occupied by a cysteine residue that can form an intermolecular disulfide bond (linking H3 to H3′) with the same position in the other subunit (FIGS. 1B and 5C). This additional disulfide bond at the opposite sides of the two "canonical" ones should reinforce the stability of the dimeric structure. Molecular dynamic simulations also suggested that this region could play a role in ligand entry and release (see APPENDIX B). It is significant that the rat RYD5 sequence, which noticeably misses the highly conserved C3 and C69, conserves a cysteine in the V44 position.

Hydrophobic amino acids that do not correspond to topohydrophobic positions are generally characterized by much higher mean accessible areas [I2 (5.3 Å^2), L15 (8.3 Å^2), M39 (18.5 Å^2), L48 (8.4 Å^2), M57 (9.1 Å^2), L68 (24.6 Å^2), M70 (47.1 Å^2)] when only V9, M34, and V64 have values equal or almost equal to zero. The characteristics of topohydrophobic positions can thus be used as a robust signature for describing and identifying the uteroglobin fold.

In a more general way, the comparison of accessibility values of all amino acids with those calculated on the monomer structure allows us to distinguish those that are connected with the dimeric interface and/or the formation of the internal cavity, and clearly shows the involvement of helices H3 and H4 in the dimeric interface formation (FIG. 5B).

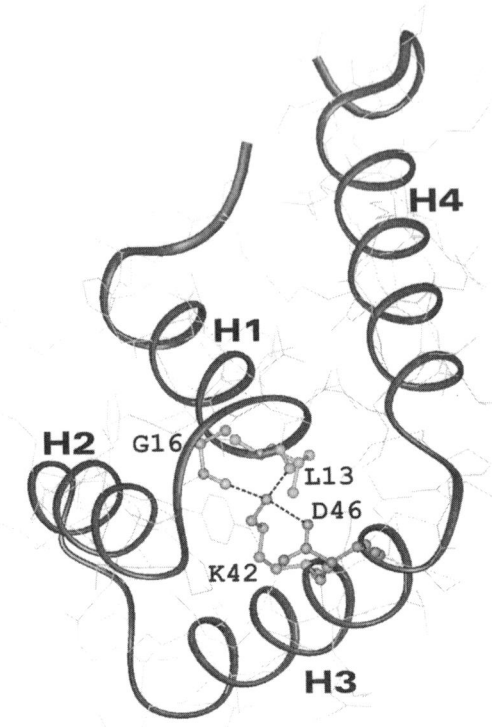

FIGURE 6. Ribbon representation of a uteroglobin monomer, where K42 is hydrogen-bonded to D46 (helix H3) and to L13 and G16 (helix H1 and loop H1-H2).

Besides topohydrophobic positions, two other amino acid positions (especially K42 and, to a lesser extent, D46) are highly conserved. The K42 Nζ atom is hydrogen-bonded with the D46 OD2 atom and also with the L13 and G16 O atoms (belonging to H1; FIG. 6), contributing to the stability of the structure. Finally, one can also note that a small amino acid (alanine) often occupies position 37 (helix H3) of the alignment, which is spatially constrained by the close proximity of helix H4′ in the uteroglobin dimer.

ENLARGEMENT OF THE STRUCTURAL FAMILY: THE UTG-LIKE FOLD

Several structural similarities of uteroglobin with other proteins have been reported previously, but only two of them are related to the entire UTG fold, harboring the typical boomerang-shaped structure of the four UTG helices. On the one hand, the uteroglobin structure has been found to be strikingly similar to four helices of the pore-forming domain of colicin A, an antibiotic protein that kills sensitive *E. coli* cells.[33] Although this similarity involves a circular permutation (the four sequential helices of uteroglobin corresponding to helices 6, 7, 3, and 4 of colicin A) and numerous deletions from colicin A, it nevertheless suggests that uteroglobin could bind to membrane like colicin A, in a monotypic fashion involving the reduction of the two disulfide bonds joining the uteroglobin monomers. It is interesting to note that, in one of the two models proposed for insertion into membranes and channel formation ("umbrella" model[34]), the UTG-like substructure corresponds to a "functional" unit, lying in the closed state flat on the membrane with the hydrophobic sides facing the lipid bilayer, whereas the central hydrophobic helical hairpin (including helices 8 and 9) is inserted perpendicular to the membrane surface. In the open state, the UTG-like structure should be directly inserted into membranes, leading to the formation of the open channel. It is noteworthy that the two aromatic amino acids of colicin A that have been shown by fluorescence spectroscopy to be exposed to the core of the lipid bilayer[34] correspond, in the UTG/colicin A structural alignment, to Y21 and T60, the two amino acids involved in uteroglobin ligand binding. The functional implications of the structural similarity with colicin A are consistent with the observed mechanism of phospholipase A2 (PLA2) inhibition by uteroglobin.[35]

Concerning phospholipase A2 inhibition, it is also noteworthy that uteroglobin shares with this enzyme a comparable surface, with surface pockets (H1H4 in uteroglobin) located similarly with respect to their central cavity.[5]

On the other hand, it has also been reported[36] that the uteroglobin structure is strikingly similar to the structure of the cap domain of *Xanthobacter autotrophicus* haloalkane dehalogenase (PDB codes: 2HAD, 2DHC[37]). Haloalkane dehalogenases catalyze the hydrolysis of a variety of haloalkanes to the corresponding alcohol, halide, and a hydrogen ion. These enzymes belong to the α/β hydrolase superfamily[38,39] (family 6 of the superfamily in the SCOP database[40]), whose common fold consists in a mixed β-sheet of eight strands connected by α-helices. In the *X. autotrophicus* haloalkane dehalogenase, an excursion from the α/β hydrolase core forms a cap domain (resembling the uteroglobin structure) lying on top of the α/β hydrolase core where the active site is located (2HAD on FIG. 7). This two-domain structure of the enzyme allows the creation of an internal hydrophobic cavity of about 40 Å^3, which contains the catalytic triad and is able to accommodate hydrophobic substrates.

Thus, although the cap domain of *X. autotrophicus* haloalkane dehalogenase does not form dimers, like uteroglobin it does form a hydrophobic cavity with an interacting partner (the large α/β hydrolase core, in a "pseudo"-heterodimeric way) that is filled by hydrophobic substrates at approximately equivalent positions (2DHC in FIG. 8).

As stated by Russel and Sternberg,[36] and as often in such situations, it is not clear whether this similarity represents a remote homology, implying a horizontal gene transfer, or a convergence to a common four-helix binding motif devoted to hydrophobic ligand binding. Nevertheless, this similarity suggests that other substructures adopting this fold would be good candidates for hydrophobic molecule binding.

Although it was previously thought that the UTG-like cap domain of *X. autotrophicus* haloalkane dehalogenase was not conserved throughout the α/β hydrolase superfamily (other enzymes are described to have different helical inserts at equivalent positions),[38,39] we carefully examined all the known structures of this super-

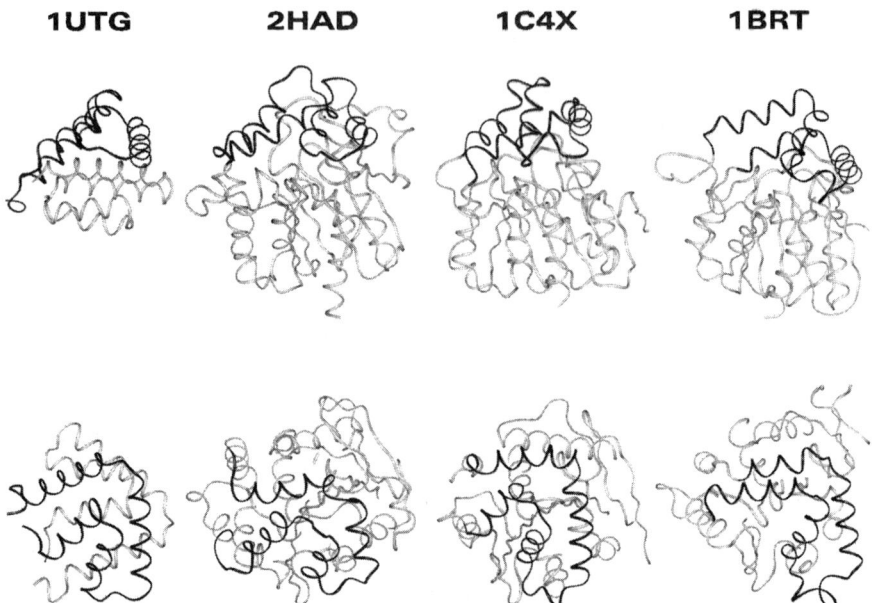

FIGURE 7. Orthogonal ribbon representations of the UTG and UTG-like structures (*dark*) in interaction with their partners [another uteroglobin subunit (1UTG) or the α/β hydrolase core (2HAD, 1C4X, 1BRT)], allowing the creation of a hydrophobic pocket for ligand binding. 2HAD (*X. autotrophicus* haloalkane dehalogenase), 1C4X (2-hydroxy-6-*oxo*-6-phenylhexa-2,4-dienoate hydrolase from *Rhodococcus sp.* strain RHA1), and 1BRT (*Streptomyces aureofaciens* bromoperoxidase A2) belong to families 6 (haloalkane dehalogenase), 8 (carbon/carbon bond hydrolase), and 10 (haloperoxidase) of the α/β hydrolase superfamily, respectively. UTG-like substructures are included between amino acids (aa) 165 and 235 (2HAD), aa 145 and 210 (1C4X), and aa 140 and 205 (1BRT), respectively.

family. Interestingly, we found that canonical UTG-like substructures are also present in at least two other families of this superfamily: namely, the carbon-carbon bond hydrolase and haloperoxidase families (families 8 and 10 in the SCOP classification, respectively; 1C4X and 1BRT on FIG. 7).

The only member of family 8 whose structure was experimentally determined is the 2-hydroxyl-6-*oxo*-6-phenylhexa-2,4-dienoic acid (HPDA) hydrolase (BPHD) from *Rhodococcus* sp. strain RHA1[41] (PDB code 1C4X), which is involved in the PCB degradation pathway. In contrast, several structures of the haloperoxidase family are known, all of which catalyze the halogenation of organic compounds in the presence of halide ions and peroxides such as H_2O_2 (PDB codes 1BRT, 1BRU, 1A7U, 1A8U, 1A8S, 1A88, 1A8Q[42,43]).

Again, the position of the ligands is very similar to the position of PCB in uteroglobin, as exemplified by the location of benzoic acid within the hydrophobic pocket of chloroperoxidase T (1A8U on FIG. 8). Although all internal pockets share the presence of aromatic amino acids involved in ligand binding, these are generally not located in similar positions in the different families.

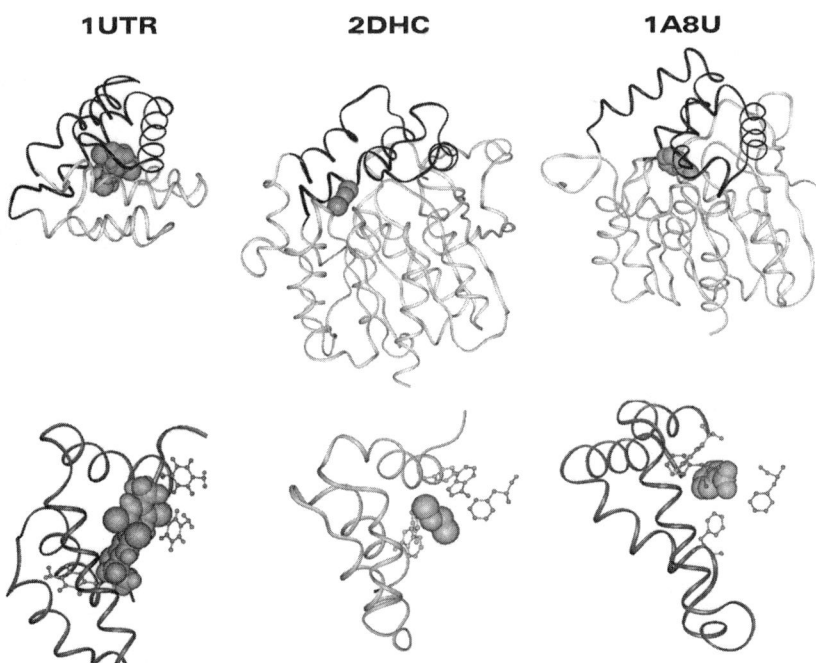

FIGURE 8. Representation of UTG and UTG-like structures complexed with ligands (CPK representations): PCB (polychlorinated biphenyl) for 1UTR (uteroglobin), DCA (1,2-dichloroethane) for 2DHC (*X. autotrophicus* haloalkane dehalogenase, family 6), and BEZ (benzoic acid) for 1A8U (*Streptomyces aureofaciens* chloroperoxidase T, family 10). **(Top)** Relative to the quaternary structure in which they are included [including another uteroglobin subunit (1UTR) or the α/β hydrolase core (2DHC, 1A8U)]. **(Bottom)** Highlighting all aromatic amino acids [belonging to the UTG or UTG-like structure (cap domains) and to the interacting partner (UTG or α/β hydrolase core)] participating in ligand binding. UTG-like substructures are included between aa 170 and 230 (2DHC) and aa 140 and 205 (1A8U).

The finding of other substructures adopting the UTG-like fold and involved in the binding of hydrophobic ligands thus strengthens the relevance of the "four-helix binding motif" designation. Moreover, it is worth noting that considerable variation certainly can occur around this four-helix binding motif, as exemplified by the structure of the cap domain of *Rhodococcus rhodochrous* haloalkane dehalogenase[44] (PDB code 1CQW, 1BN7, 1BN6). This enzyme from family 6 belongs to a different class from that of *X. autotrophicus* haloalkane dehalogenase and is especially distinguished from it by significantly different substrate specificity and halide-binding properties. Although the *Rhodococcus* and *Xanthobacter* enzymes have high structural similarity in the α/β hydrolase core, they differ considerably in the cap domain (FIG. 9). Indeed, if helices H2, H3, and H4 can be well superimposed (corresponding to helices α6, α7, and α8 in the haloalkane dehalogenase structures), helix H1 (α5 in the *Xanthobacter* enzyme) is disrupted by a turn giving two short helices (α4 and α5) almost at right angles to each other. This structural difference creates a substan-

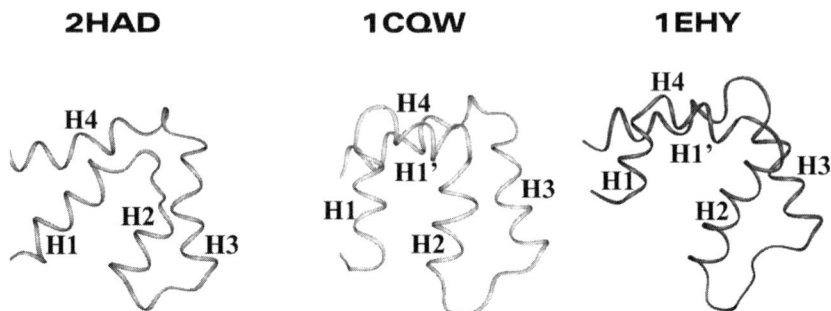

FIGURE 9. Comparison of the UTG-like substructures (cap domains) of haloalkane dehalogenases from *X. autotrophicus* and *R. rhodochrous* (PDB code 2HAD and 1CQW, respectively; family 6 of the α/β hydrolase superfamily) and of epoxide hydrolase from *Agrobacterium radiobacter* (PDB code 1EHY; family 9 of the α/β hydrolase superfamily). The first helix H1 of the UTG-like fold splits into two almost orthogonal helices H1 and H1'. UTG-like substructures are included between aa 165 and 235 (2HAD), aa 145 and 245 (1CQW), and aa 150 and 220 (1EHY).

tially larger active site cavity, which is consistent with the broader specificity of the *Rhodococcus* enzyme for primary, secondary, and cyclic haloalkanes.

This "modified" UTG-like substructure is very similar to that of the cap domain of epoxide hydrolase from *Agrobacterium radiobacter*, an enzyme that catalyzes the cofactor-independent hydrolysis of reactive and toxic epoxides and that belongs to family 9 of the α/β hydrolase superfamily[45] (PDB code 1EHY; FIG. 9). A similar organization, with helix 1 split into two almost perpendicular helices, is also observed in the recently published structure of *Aspergillus niger* epoxide hydrolase[46] (PDB code 1QO7).

The UTG fold adopted by the archetypal UTG/CC10 family and by several cap domains of the α/β hydrolase superfamily illustrates the notion that, for very distant homologous all-α folds, a unique correspondence cannot be established between sequence alignment (the "genetic memory," provided that most frequent evolutive divergence is favored instead of convergence) and 3D structural alignment (present state of the protein architecture). For such fold superfamilies, it appears that 3D distortions may be clearly more important than sequence divergence, and obvious sequence relationships can be evidenced while the compared 3D structures can be considered unrelated for a long time.[47,48] Hence, only local structural alignments, often involving supersecondary structures (e.g., two consecutive helices), may be significant in contrast to the current global ones. In this context, the definition of topohydrophic positions is tricky and can be inspired by both sequence and structural alignments, as illustrated here mainly by the third and fourth helices of the UTG fold (FIG. 3), which are consistently more concerned with sequence/structure constraints of functional evolution within the binding site. Conversely, to β-strands participating in β-sheets, the limited hydrogen bond network of helices clearly favors local winding/unwinding together with a clear propensity to establish soft distortable sets of side-chain interactions.[49] Thus, like the cradle fold encountered in small-G GTPase activating proteins (GAP),[47,48] the UTG fold may provide evidence for a better conservation of 1D sequence features rather than 3D structures.

The topohydrophobic positions of the enlarged UTG-fold superfamily nevertheless strikingly correspond to most of those defined in the first and second helices for the strict UTG family (FIG. 3), thus strengthening the pertinence of observed structural relationships and suggesting a divergent evolution from a common ancestor. As for the strict UTG family, topohydrophobic positions are much more buried within the total structure (UTG dimer or pseudoheterodimeric structure of α/β hydrolases). On average, they point in the same direction (low dispersion index—mean distance between gravity centers of their side chains in the same position within the alignment) and are close to each other (low proximity index—mean distance between gravity centers of all the amino acids), forming a hydrophobic network that defines the core or, for this particular case, the central cavity. Another common point that links UTG and UTG-like structures is the participation of helices H3 and H4 in the formation of the dimer and/or the cavity, as assessed by differences of solvent accessibility calculated for monomeric and oligomeric structures (data not shown). Finally, it is noteworthy that topohydrophobic positions that are most accessible to solvents within the internal cavity (F6 and M41; FIGS. 5A and 5C) are very often occupied by aromatic amino acids (FIGS. 3 and 10). The position of r28 is also always occupied by aromatic amino acids, highlighting its importance for the UTG-like fold.

In conclusion, the UTG-like four-helix binding motif appears much more widespread than previously thought, although it probably does not correspond to an autonomous folding unit, and it is susceptible to large structural variations. The hydrophobic pocket allowing the binding of hydrophobic ligands is indeed subordinated to its existence in a dimeric form (homodimer or homodimer-like in the case of the uteroglobin/Clara cell 10-kDa family; clearly "pseudo"-heterodimeric in the

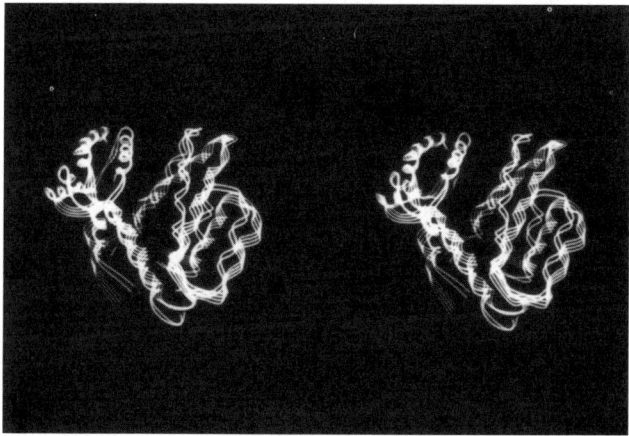

FIGURE 10. Stereo view of the superimposition of crystalline P2$_1$ uteroglobin (*two spline ribbons*) with dynamic simulations described in APPENDIX B (*four spline ribbons*), illustrating the possible opening of the uteroglobin fold in the vicinity of helices H3.

case of the "cap" domains in the α/β hydrolase superfamily). New members of this enlarged UTG-like superfamily are now being identified through sequence and structural investigations (Callebaut *et al.*, in preparation). These will certainly offer new insights for further study of the sequence/structure/function relationships associated with this particular fold and for highlighting subtle structural variations from the archetypal four-helix bundle to accommodate various hydrophobic ligands.

ACKNOWLEDGMENTS

J-P. Mornon acknowledges E. Milgrom (Kremlin-Bicêtre) for the very fruitful adventure of structural biology initiated nearly 25 years ago with rabbit uteroglobin. He also acknowledges the kind cooperation from M. Buehner (Wurzburg, Germany) and A. Nieto (Madrid, Spain), who contributed to the complete understanding of several fascinating aspects of structural biology of uteroglobin. We all thank N. Boisset for a critical reading of this manuscript. We apologize to those whose work we are unable to acknowledge due to space limitations.

REFERENCES

1. MORNON, J.-P., E. SURCOUF, R. BALLY *et al.* 1978. X-ray analysis of a progesterone-binding protein (uteroglobin): preliminary results. J. Mol. Biol. **122:** 237–239.
2. MORNON, J.-P., R. BALLY, F. FRIDLANSKY & E. MILGROM. 1979. Characterization of two new crystal forms of uteroglobin. J. Mol. Biol. **127:** 237–239.
3. BLUNDELL, T.L. & L.N. JOHNSON. 1976. Protein Crystallography. Academic Press. New York/London.
4. MORNON, J.-P., F. FRIDLANSKY, R. BALLY & E. MILGROM. 1980. X-ray crystallographic analysis of a progesterone-binding protein: the $C222_1$ crystal form of oxidized uteroglobin at 2.2 Å resolution. J. Mol. Biol. **137:** 415–429.
5. MORIZE, I. *et al.* 1987. Refinement of the $C\,222_1$ crystal form of oxidized uteroglobin at 1.34 Å resolution. J. Mol. Biol. **194:** 725–739.
6. BALLY, R. & J. DELETTRÉ. 1989. Structure and refinement of the oxidized $P2_1$ form of uteroglobin at 1.64 Å resolution. J. Mol. Biol. **206:** 153–170.
7. LIN, S.L., C.-J. TSAI & R. NUSSINOV. 1995. A study of four-helix bundles: investigating protein folding via similar architectural motifs in protein cores and in subunit interfaces. J. Mol. Biol. **248:** 151–161.
8. MUKHERJEE, A.B. *et al.* 1999. Uteroglobin: a novel cytokine? Cell. Mol. Life Sci. **55:** 771–787.
9. VANEY, M.-C. 1986. Conception d'un logiciel interactif de graphisme et de simulation moléculaires: logiciel MANOSK. Exemple d'application: modélisation de la structure tridimensionnelle de la "Prostatic Binding Protein." Ph.D. thesis, University of Paris.
10. UMLAND, T.C. *et al.* 1992. Refined structure of rat Clara cell 17-kDa protein at 3.0 Å resolution. J. Mol. Biol. **224:** 441–448.
11. PETER, W. *et al.* 1991. Identification of residues essential for progesterone binding to uteroglobin by site-directed mutagenesis. J. Steroid Biochem. Mol. Biol. **38:** 27–33.
12. DUNKEL, R., G. VRIEND, M. BEATO & G. SUSKE. 1995. Progesterone binding to uteroglobin: two alternative orientations of the ligand. Protein Eng. **8:** 71–79.
13. HÄRD, T. *et al.* 1995. Solution structure of a mammalian PCB-binding protein in complex with a PCB. Nat. Struct. Biol. **2:** 983–989.
14. UMLAND, T.C. *et al.* 1994. Structure of a human Clara cell phospholipid-binding protein-ligand complex at 1.9 Å resolution. Nat. Struct. Biol. **1:** 538–545.
15. PETER, W. *et al.* 1992. Interchain cysteine bridges control entry of progesterone to the central cavity of the uteroglobin dimer. Protein Eng. **5:** 351–359.

16. CARLOMAGNO, T. et al. 1997. Resonance assignment and secondary structure determination and stability of the recombinant human uteroglobin with heteronuclear multidimensional NMR. J. Biomol. NMR **9:** 35–46.
17. WINKELMANN, R., S. GESCHWINDNER, M. HAUN & H. RUTERJANS. 1998. Solution structure of the recombinant oxidized rabbit uteroglobin using homonuclear and heteronuclear multidimensional NMR. Eur. J. Biochem. **258:** 521–532.
18. GABORIAUD, C., V. BISSERY, T. BENCHETRIT & J.-P. MORNON. 1987. Hydrophobic cluster analysis: an efficient new way to compare and analyse amino-acid sequences. FEBS Lett. **224:** 149–155.
19. WOODCOCK, S., J.-P. MORNON & B. HENRISSAT. 1992. Detection of secondary structure elements in proteins by hydrophobic cluster analysis. Protein Eng. **5:** 629–635.
20. CALLEBAUT, I. et al. 1997. Deciphering protein sequence information through hydrophobic cluster analysis (HCA): current status and perspectives. Cell. Mol. Life Sci. **53:** 621–645.
21. HENRISSAT, B. et al. 1995. Conserved catalytic machinery and the prediction of a common fold for several families of glycosyl hydrolases. Proc. Natl. Acad. Sci. U.S.A. **92:** 7090–7094.
22. CALLEBAUT, I. & J.-P. MORNON. 1997. From BRCA1 to RAP1: a widespread BRCT module closely associated with DNA repair. FEBS Lett. **400:** 25–30.
23. CALLEBAUT, I. & J.-P. MORNON. 1997. The human EBNA-2 coactivator p100: multidomain organization and relationship to the staphylococcal nuclease fold and to the tudor protein involved in Drosophila melanogaster development. Biochem. J. **321:** 125–132.
24. CALLEBAUT, I., J.-C. COURVALIN & J.-P. MORNON. 1999. The BAH (bromo adjacent homology) domain: a link between methylation, replication, and transcriptional regulation. FEBS Lett. **446:** 189–193.
25. GIRAULT, J.-A., G. LABESSE, J.-P. MORNON & I. CALLEBAUT. 1999. The N-termini of FAK and JAKs contain divergent band 4.1 domains. Trends Biochem. Sci. **24:** 54–57.
26. ALTSCHUL, S.F. et al. 1997. Gapped BLAST and PSI-BLAST: a new generation of protein database search programs. Nucleic Acids Res. **25:** 3389–3402.
27. MORGENSTERN, J.P. et al. 1991. Amino acid sequence of Fel dI, the major allergen of the domestic cat: protein sequence analysis and cDNA cloning. Proc. Natl. Acad. Sci. U.S.A. **88:** 9690–9694.
28. LEHRER, R.I. et al. 1998. Lipophilin, a novel heterodimeric protein of human tears. FEBS Lett. **432:** 163–167.
29. POUPON, A. & J.-P. MORNON. 1998. Populations of hydrophobic amino acids within protein globular domains: identification of topohydrophobic positions. Proteins **33:** 329–342.
30. POUPON, A. & J.-P. MORNON. 1999. "Topohydrophobic positions" as key markers of globular protein folds. Theor. Chem. Acc. **101:** 2–8.
31. POUPON, A. & J.-P. MORNON. 1999. Predicting the protein folding nucleus from sequences. FEBS Lett. **452:** 283–289.
32. COLLOCH, N., A. POUPON & J.-P. MORNON. 2000. Sequence and structural features of the T-fold, an original tunneling building unit. Proteins **39:** 142–154.
33. DE LA CRUZ, X. & B. LEE. 1996. The structural homology between uteroglobin and the pore-forming domain of colicin A suggests a possible mechanism of action for uteroglobin. Protein Sci. **5:** 857–861.
34. PARKER, M.W. et al. 1992. Refined structure of the pore-forming domain of colicin A at 2.4 Å resolution. J. Mol. Biol. **224:** 639–657.
35. FACCHIANO, A., E. CORDELLA-MIELE, L. MIELE & A.B. MUKHERJEE. 1991. Inhibition of pancreatic phospholipase A2 activity by uteroglobin and antiflammin peptides: possible mechanism of action. Life Sci. **48:** 453–464.
36. RUSSEL, R.B. & M.J.E. STERNBERG. 1997. Two new examples of protein structural similarities within the structure-function twilight zone. Protein Eng. **10:** 333–338.
37. FRANKEN, S.M., H.J. ROZENBOOM, K.H. KALK & B.W. DIJKSTRA. 1991. Crystal structure of haloalkane dehalogenase: an enzyme to detoxify halogenated alkanes. EMBO J. **10:** 1297–1302.
38. OLLIS, D.L. et al. 1992. The α/β hydrolase fold. Protein Eng. **5:** 197–211.

39. HEIKINHEIMO, P., A. GOLDMAN, C. JEFFRIES & D.L. OLLIS. 1999. Of barn owls and bankers: a lush variety of α/β hydrolases. Structure **7:** R141–R146.
40. LO CONTE, L. *et al.* 2000. SCOP: a structural classification of proteins database. Nucleic Acids Res. **28:** 257–259.
41. NANDHAGOPAL, N. *et al.* 1997. Three-dimensional structure of microbial 2-hydroxyl-6-oxo-6-phenylhexa-2,4-dienoic acid (HPDA) hydrolase (BPHD enzyme) from *Rhodococcus* sp. strain RHA1, in the PCB degradation pathway. Proc. Jpn. Acad., Ser. B **73:** 154.
42. HECHT, H.J. *et al.* 1994. The metal-ion-free oxidoreductase from *Streptomyces aureofaciens* has an α/β hydrolase fold. Nat. Struct. Biol. **1:** 532–537.
43. HOFFMAN, B. *et al.* 1998. Structural investigation of cofactor-free chloroperoxidases. J. Mol. Biol. **279:** 889–900.
44. NEWMAN, J. *et al.* 1999. Haloalkane dehalogenases: structure of a *Rhodococcus* enzyme. Biochemistry **38:** 16105–16114.
45. NARDINI, M. *et al.* 1999. The X-ray structure of epoxide hydrolase from *Agrobacterium radiobacter* AD1. J. Biol. Chem. **274:** 14579–14586.
46. ZOU, J. *et al.* 2000. Structure of the *Aspergillus niger* epoxide hydrolase at 1.8 Å resolution: implications for the structure and the function of the mammalian microsomal class of epoxide hydrolases. Structure **8:** 111–122.
47. CALMELS, T.P.G. *et al.* 1998. Sequence and 3D structural relationships between mammalian Ras- and Rho-specific GTPase-activating proteins (GAPs): the cradle fold. FEBS Lett. **426:** 205–211.
48. RITTINGER, K., W.R. TAYLOR, S.J. SMERDON & S.J. GAMBLIN. 1998. Support for shared ancestry of GAPs. Nature **392:** 448–449.
49. BOWIE, J.U. 1997. Helix packing angle preferences. Nat. Struct. Biol. **4:** 915–917.
50. MUIRHEAD, H., J.M. COX, L. MAZZARELLA & M.F. PERUTZ. 1967. Structure and function of hemoglobin. 3. A three-dimensional Fourier synthesis of human deoxyhemoglobin at 5.5 angstrom resolution. J. Mol. Biol. **28:** 117–150.
51. JONES, T.A. 1985. FRODO: a graphics fitting program for macromolecules. Methods Enzymol. **115:** 157–171.
52. HENDRICKSON, W.A. & J.H. KONNERT. 1980. *In* Computing in Crystallography, pp. 13.01–13.23. Indian Institute of Science. Bangalore.
53. HENDRICKSON, W.A. & J.H. KONNERT. 1980. *In* Biomolecular Structure, Function, Conformation, and Evolution, pp. 43–47. Pergamon. Oxford.
54. IMMIRZI, A. & B. PERINI. 1977. Prediction of density in organic crystals. Acta Crystallogr. **A33:** 216–218.
55. HOUSSET, D. 1990. Dynamique moléculaire de l'utéroglobine et du complexe Utéroglobine-progestérone. Ph.D. thesis, Univ. Joseph Fourier, Grenoble.
56. BROOKS, B.R. *et al.* 1983. CHARMM: a program for macromolecular energy, minimization, and dynamics calculations. J. Comput. Chem. **4:** 187–217.
57. MOUAWAD, L. *et al.* 1990. The effects of ligands on the conformation of phosphoglycerate kinase: fluorescence anisotropy decay and theoretical interpretation. Biopolymers **30:** 1151–1160.

APPENDIX A

Structure and Refinement of $P2_12_12$ Rabbit Oxidized Uteroglobin Crystal Form at 1.34-Å Resolution (an "Archeo-crystallographic" Report)[g]

Rabbit UTG $P2_12_12$ native intensity measurements were recorded from a crystal (0.70 × 0.45 × 0.25 mm) using a Philips PW1100 single-automatic diffractometer from a ϑ resolution interval to 1.50 Å. The cell dimensions were $a = 44.52$ Å,

---

[g]This study, initiated 20 years ago by J-P. Mornon, was completed in the beginning of the 1990s with R. Bally.

$b = 37.00$ Å, $c = 32.38$ Å, $V = 53,338$ Å^3, and $B = 12.6$ Å^2. Various heavy-atom derivatives were prepared with K_2PtCl_4, $K_2PtCl_4 + K_2Pt(CN)_4$, $K_2PtCl_4 + K_2PtCl_6$ + $HAuCl_4$, $KAu(CN)_2$ + $RhCl_6$, and $PtSmCl_3$. A decay correction was applied as a function of time to 8700 reflections. Intensities were corrected from absorption, as described in reference 6. Only the first four derivatives were good and were used to solve the phases with the isomorphous replacement method running LSPHAS, a least-squares program[50] kindly given by E. Duée and G. Buisson (private communication). An electron density map calculated with the most probable native protein phases (3–10 Å, $m = 0.72$, 1200 reflections) was drawn by hand on glass sheets with a scale of 1 cm/1 Å using a local "Richard's Box"[3] and hardware "Labquick" plastic models. Only Cα atoms were picked up through a manual compass triangulation procedure. Other atoms were obtained by fitting P2$_1$-form atoms over these Cα atoms with an adaptation of the RAFMLC program (Vallino, private communication). Several density maps ($2F_O - F_C$) were calculated with FRODO software[51] on an Evans & Sutherland PS300, including only N, Cα, C, O, and Cβ atoms and the more stable hydrophobic residues. Other residues and solvent molecules were obtained by degrees. Atomic coordinates and individual isotropic parameters were refined using the PROLSQ program[52,53] (5448 reflections; 1.65- to 5.0-Å resolution).

Another colorless crystal (current UTG crystals are yellow) was used several years later, after almost completely drying, with a set of 13,780 reflections to 1.34 Å ($a = 44.58$ Å, $b = 36.87$ Å, $c = 32.31$ Å, $V = 53,104$ Å^3, $B = 14.1$ Å^2). No absorption corrections were made. The crystallographic refinements using PROLSQ software were continued with 8450 reflections, where $F_O > 2\sigma (F_O)$, and 7051 reflections, where $F_O > 3\sigma (F_O)$. The XPLOR program was also used during several cycles to regularize atomic distances.

Solvent molecules were considered as water molecules. Using increments reported by Immirzi and Perini,[54] about 297 water molecules were found per monomer (only 24% of the total volume). An additional set of 84 water molecules were progressively included, representing only 28% of the total solvent content. All the sites were refined as oxygen atoms. No occupancies were applied during refinement. The average B parameter for solvent alone is 23.4 Å^2. B values vary from 4.1 to 34.0 Å^2. Inside the cavity, there are six water positions (three by monomer) between Thr60 OG1 (A) and Thr60 OG1 (B).

Finally, the R factor was 0.21 for 7051 reflections where $F_O > 3\sigma (F_O)$ from 5-Å to 1.34-Å resolution where only 69 distances out of 1451 bonds showed greater

TABLE A1. Root-mean-square differences (Å) when superimposing two crystalline forms (fitting on main chain 3 to 69)

rms	P2$_1$2$_1$2/P2$_1$	P2$_1$2$_1$2/C222$_1$
All	1.34	1.30
Main chain	0.59	0.45
Side chain	2.04	2.06
Chain termini	2.21	1.62
Hydrophilic	2.35	2.40
Hydrophobic	1.45	1.42
Cavity (radius 11 Å)	0.60	0.46

TABLE A2. Hydrogen bonding out of helices, <3.2 Å

Arg5 NE	Glu27 OE1	2.91
His8 NE2	Glu29 O	2.60 (c)
Asn12 OD1	His8 O	2.99
ND2	Ser20 OG	2.64
Thr17 OG1	Ser20 N	2.98
Ser20 OG	Thr17 N	2.91
	Thr17 O	3.16
Tyr21 OH	Gly38 O	3.13
Glu22 OE1	Gln50 N	3.06 (d)
OE2	Gln50 N	3.10 (d)
Thr23 OG1	Ser19 O	2.95
Ser24 OG	Tyr21 O	2.89
Lys26 NZ	Ser47 O	3.09 (d)
Asp32 OD2	Asp46 OD1	2.54 (d)
Lys35 NZ	Asp46 O	2.86 (d)
Asp36 OD1	Asp46 OD1	2.75 (d)
Gln40 OE1	Thr52 OG1	2.65 (a)
Lys42 NZ	Asn12 O	3.19
	Leu13 O	2.64
	Gly16 O	2.82
	Asp46 OD2	2.89
Lys43 NZ	Leu68 O	2.98 (b)
Arg53 NH1	Leu48 O	2.98
NH2	Leu14 O	2.95
	Gln50 NE2	2.88
Asn55 OD1	Thr51 O	3.01
Thr60 OG1	Ile56 O	3.02
Lys62 NZ	Phe28 O	3.04 (a)
Ser66 OG	Leu68 N	3.05

NOTE: (a) $-x, 1-y, z$; (b) $-x, 1-y, 1-z$; (c) $0.5-x, 0.5+y, -z$; (d) $0.5-x, -0.5+y, -1-z$.

deviation than 0.04 Å from ideal values. The R factor was reduced to 0.20 if five atoms were considered to occupy two positions (Gln31 CG, Gln31 OD1, Gln31 OD2, Lys35 NZ, Ser47 OG). The Lys65 side chain, however, remained ill-defined. The average B about 548 atoms of the monomer is 11.5 Å^2 from 4.4 Å^2 to 15.7 Å^2.

When superimposing two-by-two the three crystalline refined forms of rabbit uteroglobin, they seem almost identical and their three respective internal pockets are very similar. The strongest deviations are observed for the hydrophilic residues (mean rms of 2.35 Å, while being only 1.45 Å for the hydrophobic residues). These features are illustrated in TABLE A1. TABLE A2 gives intramolecular and intermolecular hydrogen bonding out of helices. There are only two hydrogen bonds between monomers, inside the dimer (position a). The average hydrogen bonding of solvent is 2.89 Å, as calculated from 147 water bonds.

APPENDIX B

Molecular Dynamics Studies of Rabbit Uteroglobin[h]

(A) While preparing his Ph.D. thesis (1987–90), D. Housset performed an extensive study of uteroglobin through molecular dynamics simulations and normal mode analysis.[55] Several simulations were performed for the oxidized and the reduced form of uteroglobin, with and without progesterone, and with the *in silico* mutations of some key residues. The normal mode analysis of the oxidized uteroglobin was compared with molecular dynamics simulations using a quasi-harmonic approach.

(i) The molecular dynamics trajectories were computed using the program CHARMM,[56] using an explicit model of the solvent. The uteroglobin model was put in a spherical box (radius 27 Å) filled with water molecules (about 1500 water molecules for a total of 5771 atoms for the entire system). Some previous tests *in vacuo* showed the absolute necessity of explicitly taking solvent into account, despite the tremendous requirement (for the late 1980s) in CPU time and internal memory. Nevertheless, a trajectory of 80 ps for oxidized uteroglobin was computed, preceded by a thermalization stage of 30 ps in order to reach thermal equilibrium at 310 K (37°C). Uteroglobin showed a perfect stability all along the trajectory (rms atomic displacement of 0.7 Å), with a limited deviation from the crystal structure (rms difference of 0.81 Å). This stability allowed us to take this simulation as a reference to analyze the simulations [(ii) and (iii)] below. The atomic displacement derived from the simulation correlates well with the thermal motion derived from the B factor analysis of the crystal structure. During the simulation, the inner hydrophobic cavity never communicates with the outside of the protein. Moreover, the H-bond network of Lys42 showed a perfect coherence, confirming the crucial role that it probably plays for the uteroglobin fold and closely related ones.

(ii) A shorter simulation (20 ps preceded by a 30-ps thermalization stage) was calculated for reduced uteroglobin and showed few modifications outside the immediate proximity of the free cysteines. This confirms the paramount importance of van der Waals interactions for the ternary and quaternary structure stability. However, the helix H1–helix H2 loops were modified within 1-Å rms for the main chain and may be due to its inherent flexibility. At the same time, an analysis of the normal modes of vibration of the protein showed the major role of low-frequency modes in the atomic displacement (4% of the lowest frequency normal modes account for 92% of the rms atomic displacement), as in previous calculations on other well-known protein models. If the harmonic approximation used for normal mode calculation is not the most appropriate to describe protein dynamics, it is a valuable tool to investigate qualitatively some collective behavior. The comparison of normal modes with

---

[h]These studies were performed in the early 1990s by D. Housset while preparing his Ph.D. thesis.[55] He is now a specialist of protein crystallography at the Institut de Biologie Structurale J. P. Ebel (CEA-CNRS-UJF), Grenoble, France. He pursued these studies with R. Bally in partnership with J-P. Demaret, "Maître de conférences" at the Paris 6 University.

a quasi-harmonic analysis of the molecular dynamics trajectory well support this idea, especially for low-frequency modes. The lowest frequency modes show that helices behave as nearly rigid bodies, with connecting loops behaving as molecular hinges. This approach highlights some collective movements of helices that may affect the internal pocket.

(iii) A 20-ps molecular dynamics simulation of the modeled uteroglobin-progesterone complex showed a stable complex within the internal cavity where H-bonds quickly link progesterone O-3 and O-20 with the $O\eta$ of Tyr 21. In short scale times (~10 ps), however, the steroid was able to rotate along its long axis by about 90°, exemplifying the relative adaptability of this inner binding site. A gradual breaking of disulfide bridges followed by a short simulation in the reduced state (10 ps) did not suggest a way for the steroid to reach or escape the internal cavity, although the reduced uteroglobin clearly appeared to rearrange itself more freely than the oxidized form.

Studies of hypothetical complexes at the surface of the protein did not reveal any stable position. Finally, *in silico* mutations of Phe6 by Gly and Lys42 by Met showed few and rather large effects, respectively. For instance, the replacement of Lys42 by Met resulted in a movement of the helix 1–helix 2 loop up to 1.5 Å, illustrating the role of the lysine side chain for the structure stability.

(B) At the beginning of the 1990s, R. Bally and J-P. Demaret performed a new study of molecular dynamics on the asymetric uteroglobin from the $P2_1$ crystalline form ($C\alpha$ rms of 0.6 Å between the two chains, with a large difference for some side chains of up to 9 Å for K43, I2, and M70, i.e., the neighbors of the UTG-fold cysteines) to understand what eventually happens after the *in silico* reduction of disulfide bridges. CHARMM software was also used (Param19 library and Top19 topology), together with a protocol proposed by Mouawad *et al.*[57] to treat more properly exposed side chains. Thermal equilibrium to 300 K (56 ps in three steps) was followed by 100 ps of dynamics simulation at rigorously constant energy; then, the disulfide bridges were cut, this time instantaneously by suppressing covalent forces, and a further 20 ps was explored. After breaking disulfide bridges, helices H3 (H3A and H3B) moved away from each other, thereby opening the internal cavity at the opposite side of the disulfide position. The 38(A)–49(A) and 38(B)–49(B) segments of helices H3 are the more involved in this movement. This simulation performed without solvent was repeated again using a sphere of 27-Å radius containing 1500 water molecules. The same phenomenon was observed with an increase of more than 7 Å of the 41(A) $C\alpha$–45(B) $C\alpha$ distance, that is, between H3 helices, with the rest of the structure showing only more limited movements. The mean rms difference between dimers before and after simulation is near 3.0 Å. The S atoms of Cys3 and Cys69 are about 6 Å apart. This simulation suggests an opening around the twofold axis in the vicinity of the H3 helices. A disulfide bridge is likely to occur for several UTG-like dimers around the twofold axis between the 44 and 44′ positions (helix 3, FIGS. 1B and 10), supporting this hypothesis since it correlates well with the need of reducing the disulfide bridge to access the steroid binding site. This additional link may further lock the internal cavity when loaded with a hydrophobic ligand. Cysteines at opposite sides of the UTG fold (3/69 and 44) suggest that the entry or escape of ligand is not limited to one side, but perhaps may use both.

Crystal Structure Analysis of Recombinant Human Uteroglobin and Molecular Modeling of Ligand Binding

N. PATTABIRAMAN,[a,b] JOHN H. MATTHEWS,[b] KEITH B. WARD,[b,c] GIUDITTA MANTILE-SELVAGGI,[d] LUCIO MIELE,[d,e] AND ANIL B. MUKHERJEE[d]

[b]*Laboratory for the Structure of Matter, Naval Research Laboratory, Washington, District of Columbia 20375-5341, USA*

[d]*Section on Developmental Genetics, Heritable Disorders Branch, National Institute of Child Health and Human Development, National Institutes of Health, Bethesda, Maryland 20892-1830, USA*

ABSTRACT: Uteroglobin, a steroid-inducible, cytokine-like, secreted protein with immunomodulatory properties, has been reported to bind progesterone, polychlorinated biphenyls (PCB), and retinol. Structural studies may delineate whether binding of ligands is a likely physiological function of human uteroglobin (hUG). We report a refined crystal structure of uncomplexed recombinant hUG (rhUG) at 2.5-Å resolution and the results of our molecular modeling studies of ligand binding to the central hydrophobic cavity of rhUG. The crystal structure of rhUG is very similar to that of reported crystal structures of uteroglobins. Using molecular modeling techniques, the three ligands—PCB, progesterone, and retinol—were docked into the hydrophobic cavity of the dimer structure of rhUG. We undocked the progesterone ligand by pulling the ligand from the cavity into the solvent. From our modeling and undocking studies of progesterone, it is clear that these types of hydrophobic ligands could slip into the cavity between helix-3 and helix-3' of the dimer instead of between helix-1 and helix-4 of the monomer, as proposed earlier. Our results suggest that at least one of the physiological functions of UG is to bind to hydrophobic ligands, such as progesterone and retinol.

INTRODUCTION

The expression of many proteins is regulated by steroid hormones. However, the structure-function relationships of only a limited number of these proteins have been clearly elucidated. Blastokinin[1] or uteroglobin[2] (UG) is a steroid-inducible, cytokine-like protein with anti-inflammatory/immunomodulatory properties.[3,4] Although UG was first discovered in the rabbit uterus, this protein is expressed in many

---

[a]Address for correspondence: Advanced Biomedical Computing Center, NCI-Frederick/SAIC, 430 Miller Drive, Frederick, MD 21702. Voice: 301-846-5705; fax: 301-846-5762. pattabir@ncifcrf.gov

[c]Current address: Office of Naval Research, 800 North Quincy Street, Ballston Tower One, Arlington, VA 22217-5660.

[e]Current address: Department of Pathology, Cardinal Bernardin Cancer Center, Loyola University Medical Center, Maywood, IL 60153.

extrauterine tissues, including the thymus, pituitary gland, lungs, gastrointestinal tract, pancreas, mammary gland, prostate, and seminal vesicle.[5] UG is also present in the blood[6,7] and urine,[8] although it is not synthesized in the kidneys. Currently, this protein is known by several names, which are primarily derived from the organ or body fluid in which it is detectable or from the type of ligands and xenobiotics with which it interacts. Thus, it is called urine protein-1,[8] Clara cell 10-kDa protein,[9] mammaglobin,[10] polychlorinated biphenyl–binding protein,[11] progesterone-binding protein,[12] and retinol-binding protein.[13] Structurally, UGs are homodimers in which the two 70-amino-acid subunits are covalently linked in an antiparallel orientation by two interchain disulfide bonds. Each monomer consists of four α-helices and one β-turn between α-helix-2 and -3, but there is no β-sheet structure.

The UG gene consists of three exons and a 5′-flanking region containing several steroid hormone–response elements that regulate its differential expression in a tissue-specific manner. For example, while progesterone upregulates UG production in the uterus, its induction in the fallopian tubes is dependent upon estrogen. Prolactin appears to enhance the progesterone-induced UG gene expression in the uterus.[14–16] In the prostate and seminal vesicle, UG gene transcription is regulated by testosterone, while in the lungs it is stimulated by glucocorticosteroids. Recently, it has been found that treatment of mice with IFNγ stimulates UG production in the lungs. Furthermore, an IFNγ-response element has been identified in the 5′-flanking region of the mouse UG promoter.[17] In addition, disruption of the UG gene caused severe inflammatory renal glomerular disease in mice.[18] Moreover, generation and characterization of UG-antisense transgenic mice[19] revealed that UG deficiency led to an identical phenotype, as reported for UG-knockout mice.[18] Further characterization of the phenotype of these mice uncovered that both the knockout and UG-antisense transgenic mice developed IgA nephropathy, the most common primary renal glomerular disease worldwide.[20] Taken together, these results suggest a role of this protein in protecting an animal against immunological/inflammatory processes such as IgA nephropathy.

Both natural and recombinant human UG (rhUG) have an anomalous electrophoretic mobility in SDS-PAGE, under nonreducing conditions, that is similar to that of a 10-kDa protein, although its calculated molecular mass is 17.8 kDa. The availability of substantial quantities of recombinant rabbit[21] and human[22] UGs allowed comparison of their structural features by X-ray crystallography[23] and multidimensional NMR.[24] While rabbit and human UGs have only a 68% identity in amino acid sequence, their secondary, tertiary, and quaternary structures are virtually indistinguishable and are functionally identical.[22]

UG is one of the preferred substrates of transglutaminase and thrombin-activated coagulation factor XIII.[25,26] It inhibits extracellular PLA_2 (phospholipase A_2) activity[27,28] and has potent immunomodulatory,[26,29,30] anti-inflammatory,[26,31] antichemotactic,[32] and antithrombotic[33] properties. The PLA_2-inhibitory and anti-inflammatory activities appear to reside in a nonapeptide region (residues 39–47) of the α-helix-3 of UG.[28,31] The results of site-directed mutagenesis involving this region confirmed that the active site of UG utilized in PLA_2 inhibition resides, at least in part, within amino acid residues 39–47 of α-helix-3 of UG (Mantile-Selvaggi *et al.*, manuscript in preparation). Because of its secretion by uterine endometrial epithelia and its ability to bind progesterone,[12] it has been proposed that UG prevents progesterone toxicity in the developing embryo as this steroid is abundant in the uterus during implantation. Since UG is expressed at a high level in the lungs and binds xenobiotics such as polychlorinated biphenyls (PCBs),[11] it has also been suggested that UG may promote increased accumulation of PCBs in the lungs, causing serious damage to this organ. Thus, among the steroid-induced gene products, UG is one of the most exhaustively studied proteins from the standpoint of its biochemistry, molecular biology, and regulation of tissue-specific expression.[3,4] More recently, by targeted disruption of the UG gene[18] and by the generation of UG-antisense transgenic mice,[19] we uncovered a novel and critical role of this protein in preventing IgA nephropathy, the most common primary glomerular disease worldwide. The molecular mechanism by which UG, under physiological conditions, prevents this disease appears to involve its interaction with fibronectin (Fn), forming UG-Fn heteromers, which counteract Fn self-assembly and Fn-IgA complex formation that are essential for abnormal cellular deposition of Fn and IgA, respectively. Moreover, lack of UG in UG gene–disrupted mice caused higher serum PLA_2 activity, which was further enhanced when the mice were treated with bacterial lipopolysaccharide (LPS).[18] Hydrolysis of phosphatidic acid by increased PLA_2 activity in $UG^{-/-}$ mice may lead to the generation of lysophosphatidic acid, which is known to enhance the multimerization of Fn and the activation of integrins that bind Fn. These findings in $UG^{-/-}$ mice suggest that an analogous mechanism of pathogenesis may underlie human IgA nephropathy and its progression. Further, we[34] and others[35] have reported the presence of high-affinity binding sites (putative receptors) for UG on the surface of several cell types. These results raise the possibility that UG has receptor-mediated functions as well. Thus, the determination of the structural features of hUG is critical for our understanding of its interaction with other proteins such as Fn,[18] its receptor(s),[34,35] and various ligands and xenobiotics.[11–13] In a previous report,[23] we have described our initial findings on the crystallization of rhUG. Here, we present the results of further refinement of the crystal structure of rhUG at 2.5-Å resolution and that of computer modeling to delineate the interactions of UG with various ligands and xenobiotics. We also report the results of our undocking studies of progesterone using molecular dynamics simulation from the central hydrophobic cavity by pulling the ligand between the helix-3 and helix-3′ interface.

METHODS

X-ray Crystallographic Studies

Data Collection

Recombinant hUG crystals were grown from ammonium sulfate as described previously.[23] X-ray diffraction data were measured with a Siemens X-100A area-detector system and a Rigaku RU200 rotating Cu-anode X-ray source. The X-ray generator was operated at 50 kV and 100 mA, and the Cu K_α radiation was monochromatized and focused with a Siemens double-mirror focusing system.[45] For each of three crystal orientations, 440 frames of data were collected at 0.25° increments in ω. Data reduction and analysis were carried out with the XENGEN 2.1 program package.[46] The space group was C2 and the unit-cell parameters for this crystal were determined to be $a = 70.0(3)$ Å, $b = 83.5(3)$ Å, $c = 58.3(3)$ Å, and $\beta = 99.4°$. Two molecules in the asymmetric unit give a value of solvent fraction (50%) more consistent with the diffraction and mechanical strength of the crystal. From 38,506 observations, the merging of intensities for equivalent reflections and Bijvoet pairs led to 11,425 unique observations, or about 92% of the data theoretically measurable at a resolution of 2.44 Å with $R_{merge} = 0.052$.

Structure Determination

The X-PLOR program package (A. T. Brunger, *X-PLOR Manual, Version 3.0*, 1992, Yale University) was used for molecular replacement calculations, molecular dynamics calculations, and energy minimizations and structure factor least-squares refinement of the molecular model. A self-rotation search, carried out using data in the resolution range (5–15 Å), used a polyalanine model of the two-chain rabbit UG (PDB entry 2UTG[38]) for the search model. The results were essentially invariant under limited variation of the resolution range of the data. Each of the six highest peaks in the rotation function and a few of the weaker ones were further analyzed as potentially correct molecular orientations. Euler angles were tuned by PC refinement[47] to maximize the correlation between the squared structure moduli from experiment and the oriented pseudo-UG for each orientation considered. These calculations were followed with translation searches and individual rigid-body refinements.

The strongest result from cross-rotation searching scored well in subsequent rigid-body coordinate refinement for the pseudo-UG model, with convergence at $R = 0.53$ in the 3.2- to 10-Å resolution range. The noncrystallographic twofold from the self-rotation search was applied to the placed UG model to give a second orientation. The orientation of this second molecule was also confirmed in a cross-rotation search. For this second dimer alone, convergence was again at $R = 0.53$. The spacing of the two independent dimers along the polar axis was determined in a separate translation search to delineate their correct relative disposition. Refinement of the resulting two dimers (four chains) as rigid bodies gave an R factor of 0.486 in the 3.2- to 10-Å resolution range.

Refinement continued with several cycles of energy minimization in the 3.2- to 10-Å resolution range to relieve strain or bad contacts. This lowered the R factor to 0.370. Subsequent simulated annealing with an initial temperature of 3000 K, followed by conventional structure factor least-squares refinement, lowered the R factor to 0.307. Side chains and two N-terminal alanines for each chain were located from $(2|F_o| - |F_c|)$ electron density maps. Additional cycles of map-fitting and refinement gave the final model with 129 waters and $R = 0.193$ in the 2.44- to 10-Å resolution range. The overall B value was 15.8 Å. The bond lengths and bond angles showed an rms deviation from ideal values of 0.018 Å and 3.3°, respectively. All of the nonglycine residues fell inside or close to the allowed regions of the Ramachandran plot.

Modeling Studies of Ligand Binding

The initial models of the PCB-, progesterone-, and retinol-hUG complexes were built based on the solution structure of the PCB-UG complex (Protein Data Bank code is 1UTR[37,40]). The ligands were constructed using the molecular modeling tools of BioSym molecular modeling software. The coordinates of these complexes were energy-minimized using the CFF91 force field of BioSym molecular modeling software (Insight 98, a molecular modeling program of BioSym, Inc., CA). The structures were minimized to an rms value of 0.01 kcal-Å using a distance-dependent dielectric constant to mimic the solvent.

In our undocking studies, using molecular dynamics (MD) simulation techniques, the progesterone ligand docked in the cavity of the rhUG dimer was pulled slowly from the cavity by applying enough force on the center of coordinates of the ligand. The ligand was pulled along the twofold direction toward the interface between helix-3 and helix-3′ into the solvent. The center of coordinates of the ligand was first pulled by 0.1 Å and an MD simulation was carried out for 0.4 ps. The MD simulation was repeated until the force of the pull was less than 15 kcal. The preceding procedure was repeated until the ligand molecule was completely exposed to solvent. We again used the BioSym molecular modeling software (Insight 98).

RESULTS AND DISCUSSION

Structure of rhUG

The crystal form of rhUG that we analyzed is a homodimer that has a pseudodyad between the monomers. The two monomers are held together by two disulfide bonds: one between Cys3 and Cys69′ and the other between Cys3′ and Cys69. A ribbon structure drawing of the two monomers covalently linked by these disulfide

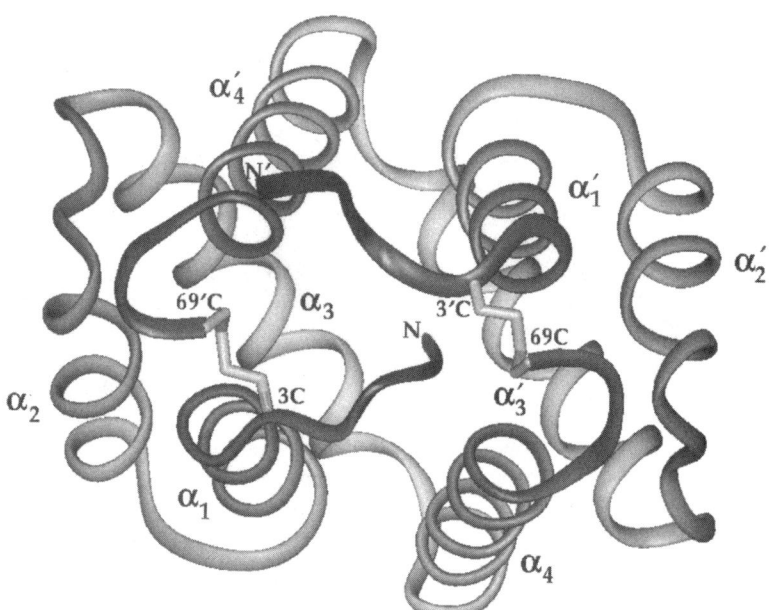

FIGURE 1. Ribbon representation of the rhUG dimer crystal structure. Two disulfide bonds (one between Cys3′ and Cys69 and the other between Cys3 and Cys69′) are shown. Alpha helices are also marked. The darker the ribbon or the bonds, the nearer they are to the reader.

bridges is shown in FIGURE 1. Each monomer is made up of four α-helices: α-1, α-2, α-3, and α-4. When these two monomers are superimposed, the rms deviation for the backbone atom coordinates from residues 2 to 70 is 0.28 Å. In this crystal form (space group C2), there are two crystallographically independent dimers in the asymmetric unit, although they are similar in structure. The rms deviation for the backbone atoms between these independent dimers is 0.4 Å. Two additional Ala residues at the N-terminus of this protein present in the rhUG do not disturb the folding of the monomers. The formation of the two disulfide bonds between the monomers and the formation of the disulfide-linked UG monomers are also not disturbed. In addition to the two interchain disulfide bridges, there are a number of hydrophobic contacts between the two monomers. Based on the occlusion surfaces[36] of one monomer residue by the other monomer, the residue pairs that have the most significant hydrophobic contacts are Leu44':Leu50, Leu44':Leu46, Val45':Leu43, Leu48':Leu46, Leu48':Gln42, Ile56':Ala39, Leu59':Met36, and Ile63':Phe8; and the corresponding pairs for the second monomer. The structure of rhUG is similar to that of the rabbit UG referred to in the Brookhaven National Laboratory Protein Data Bank[37] by the identification code 2utg.[38] The rms deviation between each of the dimers and the rabbit UG structure for the backbone atoms of residues 2 to 70 is 0.9 Å. Also, the dimer structure of rhUG is similar to that of the rat Clara cell protein structure,[39,40] with an rms deviation of 0.8 Å for the backbone atoms of residues 2 to 70. The structure of rhUG is very similar to that of a recently published crystal structure of a UG-phospholipid complex purified from human cadaver lungs.[31]

In FIGURE 2, the sequence alignment of the rhUG, rabbit UG (2utg), rat Clara cell (1ccd), and mouse Clara cell (JC2026[41]) 10-kDa protein is shown. The alignment was made using the GCG software (program manual for the Wisconsin package, Version 8, September 1994, Genetics Computer Group, Madison, WI). The secondary structures of rat Clara cell protein, rabbit UG (2utg), and rhUG were determined using the program DSSP.[42] In FIGURE 2, we have underlined the residues forming the four helices as found in the 2utg structure. The residues forming helices α-1, α-2, and α-4 are identical between the structures of rhUG and 2utg. However, α-helix-3 differs by 1 amino acid residue at the C-terminal and, as a result, the α-helix-3 of 2utg is slightly longer than that of rhUG. The residues that are in contact between helices α-1 and α-2 and between α-1 and α-4 of 2utg are marked by the letters "x" and "o", respectively, located above the sequences. Out of 7 residues (shown as "x"), 5 residues are different between rhUG and 2utg. These residues are on the surface of the protein and do not affect the interaction of helices α-1 and α-2. Between helices α-1 and α-4, the amino acid differences occur only on the surface of the protein. The dimers of both rhUG and 2utg have a hydrophobic cavity. The residues forming the internal cavity in the 2utg structure are marked by the letter "c" above the sequence in FIGURE 2. Out of 12 residues that form the internal cavity, only 3 residues (25, 41, and 60) are different between the rhUG and 2utg sequences. By superimposing the two crystal structures, we found that residue 25 is on the surface of the protein and an alteration in the side chain could occur without changing the shape of the hydrophobic cavity. However, compared to 2utg, the amino acid change in residue 41 of hUG increases the volume of its hydrophobic cavity, although the amino acid change in residue 60 of rhUG resulted in a decreased volume of the cavity.

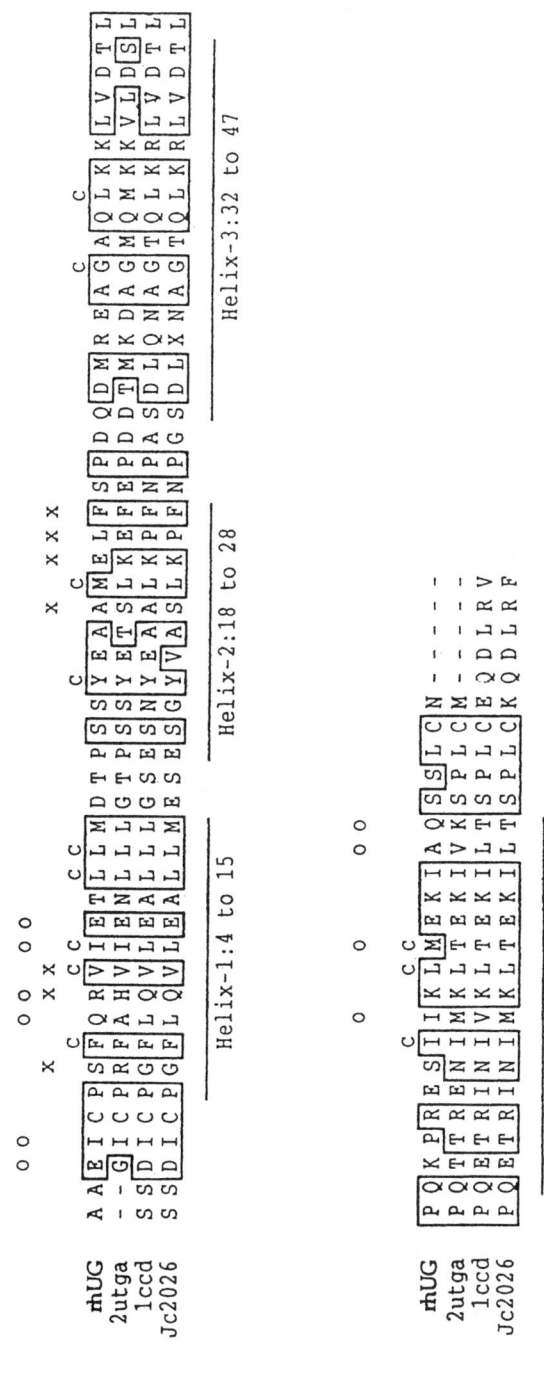

FIGURE 2. Sequence alignment of the rhUG, rabbit UG (2utg), rat Clara cell (1ccd), and mouse Clara cell (JC2026) 10-kDa protein. The sequences for rabbit UG and rat Clara cell are from the crystal structures in the Protein Data Bank. Four α-helices are marked. The residues forming the inner cavity are marked by the letter "c". The residues in contact between helix α-1 and helix α-2 are marked by the letter "x" and those between helix α-1 and helix α-4 are marked by the letter "o".

Ligand Binding

From our modeling studies, we were able to dock PCB methylsulfone, progesterone, and retinol into the cavity (as reported earlier[39,40]) of the rhUG crystal structure in order to generate energetically favorable ligand-rhUG complexes. The polar groups in PCB and progesterone form hydrogen bonds with the hydroxyl group of tyrosine residues in the dimer. In the case of retinol, only one of the hydroxyl groups of tyrosine forms a hydrogen bond with the hydroxyl group of retinol. These three molecules bind to the cavity without much change in the conformation of the amino acids that form the cavity.

It has been proposed that small hydrophobic ligands such as PCB may bind UG[11,43] and, since the UG gene is constitutively expressed in the lung, it may bring about harmful pathologic changes in this vital organ. Our molecular modeling studies revealed that the UG molecule would have to be reduced in order for such interactions to take place. Reduction of UG may induce local unfolding of the N- and C-terminal residues and may be accompanied by separation of the α-helices of UG (creating a channel) as the side chains are reoriented at the binding site. Also, during this process of channel formation, some of the side chains had to be reoriented. There are at least two problems with this proposed model. First, naturally occurring UG molecules are invariably found as homodimers (oxidized form) and reduction does not occur within the cell. In addition, even if a channel is formed, as proposed,[11,43] the ligand would have to rotate within the dimer to correctly orient itself into the binding site of the cavity. FIGURE 3 shows a ribbon representation of the model of the rhUG-progesterone complex, looking down the twofold axis from the α-3 and α-3' helices to the two disulfide bonds. Two tyrosine residues and a progesterone molecule are also shown. In this model, the two tyrosines that form a part of the hydrophobic cavity are very close to the α-3 helix. These tyrosine residues appear to be critical for the binding of small ligands by forming hydrogen bonds. Also, the orientation of the longest dimension of progesterone shown in FIGURE 3 appears to make an approximately 60° turn about the axis of the α-3 helix of UG. In addition, the possibility exists that each monomer could independently rotate around the disulfide bonds to create an opening between the two monomers near the α-3 helix region.

FIGURE 4 shows a plot of the total interaction energy of progesterone with the rhUG dimer structure versus the frame during the undocking simulation. The docked and undocked regions are marked in the figure. It is clear from FIGURE 4 that the barrier of the undocking between helix-3 and helix-3' is not huge. The ligand could find a way out of the cavity with some barrier. From FIGURE 4, the ligand goes through a number of local minima before it is undocked. Unlike other ligands binding to an open pocket in an enzyme, progesterone has to spend energy to open up the helix-3 and helix-3' interface. FIGURE 5 shows a snapshot of the undocking simulation. In this figure, only 12 frames were plotted from docking to undocking of progesterone from the rhUG dimer structure. When the ligand was undocking, initially (frames 1 to 4) the ligand was parallel to the helix-3 and helix-3' interface. Then, the ligand was rotated by 90° such that it was perpendicular to the helix-3 and helix-3' interface (frames 5 to 12) to get out of the cavity. Also, the hydroxyl of Tyr21 helps the ligand to rotate perpendicular to the interface. The docking of these ligands might be easier if the disulfide bonds are broken between the two monomers.

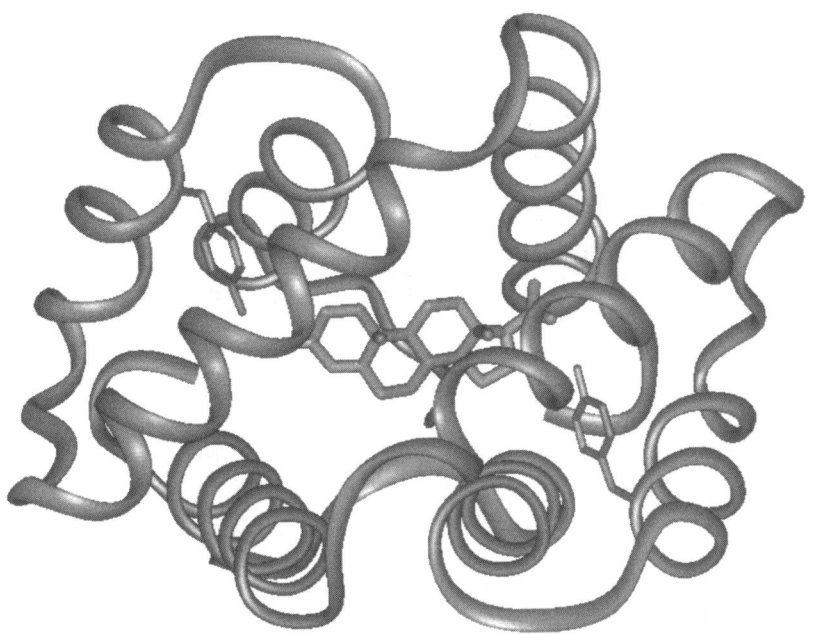

FIGURE 3. Ribbon representation of the rhUG dimer complexed with progesterone. Two tyrosines are also shown. This view is looking down the twofold axis from helix α-3 and α-3'.

In light of the previous considerations, we propose that one mechanism by which small hydrophobic ligands, such as PCB, progesterone, and retinol, can associate with UG is by interacting with the region of UG near helix-3. This may involve interactions with the hydrophobic residues of the UG dimer that form the hydrophobic cavity. This binding may open up the UG dimer near this region. Because the majority of the cavity is made up of hydrophobic residues, these ligands may further slip into the cavity in order to form favorable hydrophobic UG-ligand interactions. Finally, the reorientation of these ligands into the cavity could occur due to the presence of the two anchoring tyrosine residues. Thus, our results clearly show that these types of hydrophobic ligands could slip into the central hydrophobic cavity between helix-3 and helix-3' of the dimer instead of between helix-1 and helix-4 of the monomer, as predicted earlier.[48]

We have refined the crystal structure of rhUG at 2.5-Å resolution and have used molecular modeling to study the interactions of several hydrophobic ligands such as PCB, progesterone, and retinol. The structure of rhUG is very similar to those of rabbit UG[44] and rat Clara cell protein.[39,40] The amino acid sequence similarities between UG and the Clara cell proteins from different mammals and their virtually identical activity (i.e., PLA_2 inhibition, substrate of transglutaminase, and binding of hydrophobic ligands, including PCB) attest to the fact that these proteins belong to the same family. A similar conclusion was arrived at when recombinant expres-

122 ANNALS NEW YORK ACADEMY OF SCIENCES

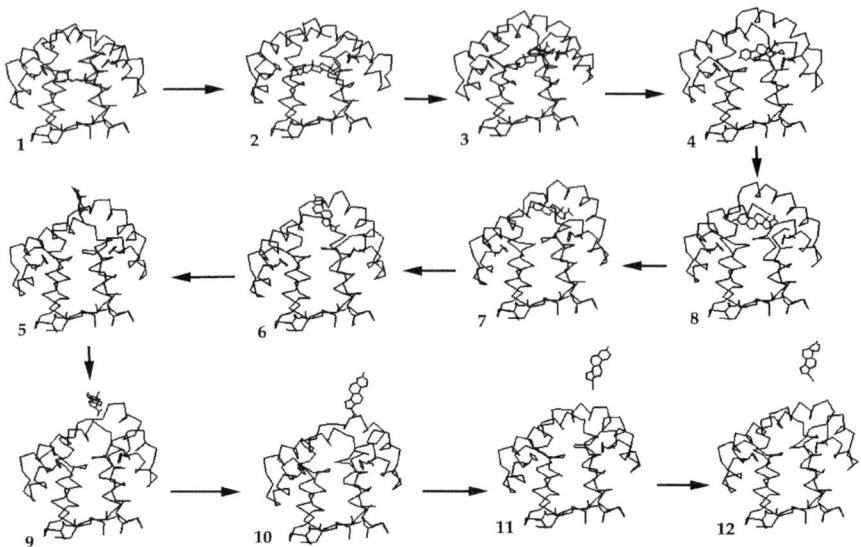

FIGURE 4. Plot of the interaction energy of progesterone with rhUG dimer during the undocking simulation.

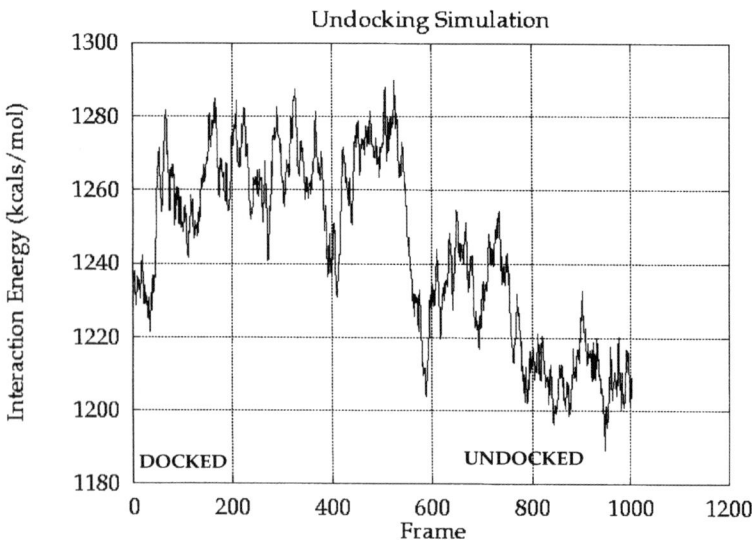

FIGURE 5. Snapshots of the undocking of progesterone from the central hydrophobic cavity of the dimer structure of rhUG.

sion and characterization of the human protein in *E. coli* was accomplished[22] and by analysis of its solution structure by heteronuclear multidimensional NMR.[24]

Despite the myriad biological properties of UG that have been described in the literature, the physiological functions of this protein have remained unknown for nearly three decades. Recently, the generation of UG-null mice by targeted disruption of the UG gene in ES cells[18] and antisense transgenic mice[19] in which UG production is inhibited by the expression of UG-antisense RNA has been performed. The results of these experiments showed that UG is essential for preventing IgA nephropathy and for maintaining normal renal function in mice. The severe renal disease developed in UG-null mice is caused by abnormal glomerular deposition of IgA, Fn, and collagen. The molecular mechanism by which UG prevents abnormal deposition of IgA in the renal glomeruli involves its high-affinity interaction with Fn, forming Fn-UG heteromers. The formation of Fn-UG heteromers prevents Fn self-assembly as well as Fn-IgA interaction required for abnormal tissue deposition of both IgA and Fn. Thus, UG appears to play a critical role in preventing abnormal deposition of multimeric Fn in the renal glomeruli and IgA. Since UG is an anti-inflammatory protein, the development of an inflammatory autoimmune kidney disease was not totally unexpected. However, the development of this disease in the kidney, an organ in which the UG gene is not expressed, was not anticipated. The presence of a UG receptor in the kidney (Kundu *et al.*, unpublished results) and the absence of UG in the plasma of our knockout mice may play a further role in the pathogenesis of this disease via an as yet unknown mechanism. Therefore, delineation of the hUG structure may facilitate our understanding of how UG interacts with various ligands and with its receptors. We recently discovered that polymorphisms (+38 A→G), $(GTTT)_m$, and $(ATTT)_n$ in the UG gene correlate with inflammatory lung disease such as asthma (Choi *et al.*, unpublished results). Since UG is constitutively expressed in the lung at a high level and it has been reported that lung inflammation in humans can cause a lower plasma level of UG, it is possible that this downregulation of UG can lead to abnormal deposition of IgA in the renal glomeruli. Thus, our present findings suggest that these polymorphisms are a predictor of either a susceptibility to IgAN or IgAN disease progression (or both). Thus, understanding the interactions of UG with its receptor may allow the development of novel therapeutic interventions in human hereditary Fn-deposit glomerulopathy. Undoubtedly, the structural data reported in the present study will be very useful in realizing this goal.

UG circulates in the blood of both rabbits[6] and humans.[7] We have demonstrated that Fn-UG heteromers could be found in the plasma of normal mice, but not in mice where the UG gene has been inactivated by targeted disruption in ES cells[18] or in transgenic mice where the production is suppressed by the expression of UG-antisense RNA.[19] The results of this study also showed that UG binds to Fn with high affinity (K_d = 13 nM) and prevents Fn-Fn interaction as the affinity of Fn for binding to itself is much lower (K_d = 176 nM).[18] Similarly, UG can also effectively compete with IgA and collagen for their binding to Fn. It has been demonstrated that Fn has binding sites not only for itself, but also for IgA and collagen. Thus, it is possible that UG (a homodimer) may simultaneously bind Fn on its own binding site as well as on that of IgA and collagen. By doing so, UG may cause conformational changes in the Fn molecule and prevent Fn self-assembly, as well as abrogate the Fn-

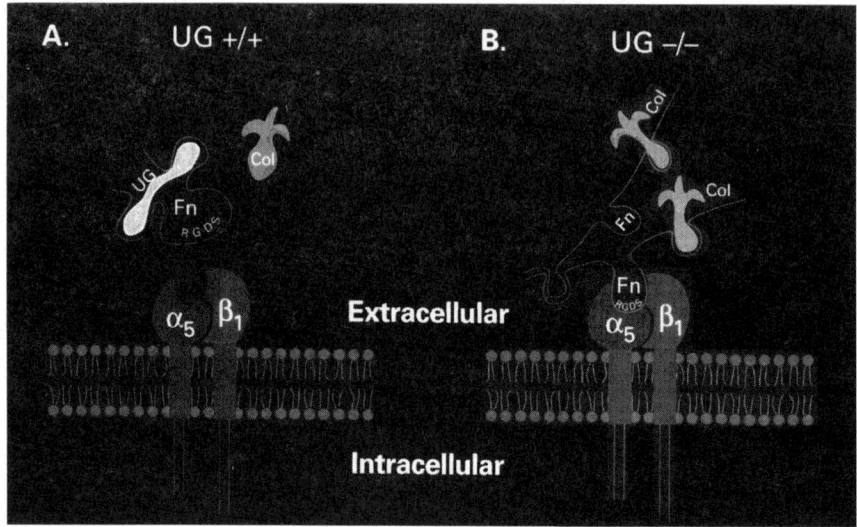

FIGURE 6. Possible mechanisms of action of UG in preventing multimerization and cellular deposition of Fn. **(A)** In $UG^{+/+}$ mice, the binding of UG to Fn may cause a conformational change in the Fn molecules. As a result of this altered conformation, Fn-Fn and Fn-collagen aggregates fail to form. Moreover, this conformational change of Fn may alter the conformation of the GRDGS cell-binding epitope of Fn, counteracting its binding with the cell-surface Fn receptor, $\alpha_5\beta_1$ integrin. **(B)** Due to a lack of UG in $UG^{-/-}$ mice, the formation of Fn-Fn and Fn-collagen complexes remains unopposed and these high-molecular-weight aggregates accumulate in the glomeruli, causing renal failure. Alternatively, it is possible that glomerular filtration fails to clear the multimeric Fn and Fn-collagen aggregates and accumulation occurs.

IgA interaction essential for abnormal cellular deposition of Fn and IgA. Since Fn also binds to cells via its receptor, $\alpha_5\beta_1$ integrin, a heterodimeric protein molecule that recognizes the GRGDS cell-binding sequence in Fn, another possibility is that this alteration of conformation, arising from UG binding to Fn, may prevent the interaction of Fn with its cell-surface receptor ($\alpha_5\beta_1$ integrin). A schematic model explaining this concept is shown in FIGURE 6.

Recently, Stripp et al.[43] described a UG-null mouse model that did not accumulate PCBs in the organs that normally express the UG gene (e.g., lungs, uterus, prostate). These investigators concluded that UG is the determinant of PCB binding and that this is one of the physiological functions of this protein. While it is possible that PCB can penetrate epithelial cells and bind reduced UG during its intracellular passage, this does not necessarily mean that binding of this ligand is a physiological function of this protein. In fact, our present crystallographic data show that binding of PCB within the hydrophobic cavity of oxidized UG dimer, while possible, is highly unlikely as this binding requires alterations in the UG molecule that are possible only if secreted UG were to be present in its reduced form. However, in nature, reduced UGs are not detectable in extracellular fluids of the animals studied so far. In fact, the recombinant rabbit and human UGs produced in *E. coli* have demonstrated that the dimerization of these proteins, with the formation of disulfide bonds, occurs within the cytoplasm of the bacteria,[21,22] supporting the notion that the extracellular existence of reduced UG under physiological conditions is highly unlikely. The structure of UGs has been essentially conserved throughout evolution, at least since the appearance of the rodents. Thus, one cannot but wonder how this family of proteins may have evolved with the specific function to bind toxic xenobiotics like PCB, substantial levels of which in the atmosphere may not have been present before the advent of the industrial revolution. Furthermore, since lungs constitutively express UG, accumulation of PCB or similar xenobiotics in this vital organ would have had a deleterious effect on the survival of air-breathing animals. This would have exerted a selective pressure to eliminate such binding properties of UG or the gene coding for this protein during evolution. Whether PCB binds UG intracellularly remains to be proven, but the present data suggest that such binding is highly unlikely to be a physiological function of this protein. Similar arguments may apply to other ligands (e.g., progesterone, retinol) that bind reduced UG. On the other hand, one cannot rule out the possibility that there exists an endogenous ligand structurally similar to these xenobiotics and that the binding of such a ligand to UG confers important physiological benefits. It is clear that UG is a highly interactive protein as it has been demonstrated that it specifically binds to Fn, a high-molecular-weight extracellular matrix protein with high affinity (K_d = 13 nM). In addition, hUG purified from human cadaver lungs has been found in association with phospholipids, and this protein binds to microsomal[35] and plasma membrane[34] proteins with high affinity and specificity. All in all, the results of our present study may allow further experiments to explain this and other interactions of UG on a rational structural basis.

ACKNOWLEDGMENTS

We thank Douglas M. Collins for valuable discussions and Connor McGrath for helping with the BCL script for undocking study using the BioSym software. We acknowledge the National Cancer Institute for allocation of computing time and staff support at the Advanced Biomedical Computing Center of NCI-Frederick.

REFERENCES

1. KRISHNAN, R.S. & J.C. DANIEL, JR. 1967. Blastokinin: inducer and regulator of blastocyst development in the rabbit uterus. Science **158:** 490–492.
2. BEIER, H.M. 1968. Uteroglobin: a hormone sensitive endometrial protein involved in blastocyst development. Biochim. Biophys. Acta **160:** 289–291.
3. MUKHERJEE, A.B. *et al.* 1998. Uteroglobin: its physiological role in normal glomerular function uncovered by UG gene knockout in mice. Am. J. Kidney Dis. **32:** 1106–1120.
4. MUKHERJEE, A.B. *et al.* 1999. Uteroglobin: a novel cytokine? Cell. Mol. Life. Sci. **55:** 771–787.
5. PERI, A. *et al.* 1993. Tissue-specific expression of the gene coding for human Clara cell 10 kDa protein, a phospholipase A_2 inhibitory protein. J. Clin. Invest. **92:** 2099–2109.
6. KIKUKAWA, T. & A.B. MUKHERJEE. 1989. Detection of a uteroglobin-like phospholipase A_2 inhibitory protein in the circulation of rabbits. Mol. Cell. Endocrinol. **62:** 177–187.

7. AOKI, A. *et al.* 1996. Isolation of human uteroglobin from blood filtrate. Mol. Hum. Reprod. **2:** 489–497.
8. JACKSON, P.J. *et al.* 1988. Purification and partial amino acid sequence of human urine protein-1: evidence for homology with rabbit uteroglobin. J. Chromatogr. **452:** 359–367.
9. SINGH, G. *et al.* 1988. Identification, cellular localization, isolation, and characterization of human Clara cell 10 kd protein. J. Histochem. Cytochem. **36:** 73–80.
10. WATSON, M.A. & T.P. FLEMING. 1996. Mammaglobin, a mammary-specific member of the uteroglobin gene family, is overexpressed in human breast cancer. Cancer Res. **56:** 860–865.
11. GILLENER, M. *et al.* 1988. The binding of methylsulfonyl polychloro-biphenyls to uteroglobin. J. Steroid Biochem. **31:** 27–33.
12. BEATO, M. & H.M. BEIER. 1975. Binding of progesterone to uteroglobin. Biochim. Biophys. Acta **392:** 346–356.
13. LOPEZ DE HARO, M.S. *et al.* 1994. Binding of retinoids by uteroglobin. FEBS Lett. **349:** 249–251.
14. CHILTON, B.S., S.K. MANI & D.W. BULLOCK. 1988. Servomechanism of prolactin and progesterone in regulating uterine gene expression. Mol. Endocrinol. **2:** 1169–1175.
15. RANDALL, G.W., J.C. DANIEL, JR. & B.S. CHILTON. 1991. Prolactin enhances uteroglobin gene expression by uteri of immature rabbits. J. Reprod. Fertil. **91:** 249–257.
16. HAYWARD-LESTER, A. *et al.* 1996. Cloning, characterization, and steroid-dependent posttranscriptional processing of RUSH-1 alpha and beta, two uteroglobin promoter-binding proteins. Mol. Endocrinol. **10:** 1335–1349.
17. MAGDALENO, S.M. *et al.* 1997. Interferon gamma regulation of Clara cell gene expression: *in vivo* and *in vitro*. Am. J. Physiol. Lung Cell. Mol. Physiol. **272:** L1142–L1151.
18. ZHANG, Z. *et al.* 1977. Severe fibronectin-deposit renal glomerular disease in mice lacking uteroglobin. Science **276:** 1408–1412.
19. ZHENG, F. *et al.* 1999. Uteroglobin is essential in preventing immunoglobin A nephropathy in mice. Nat. Med. **5:** 1018–1025.
20. D'AMICO, G. 1998. Pathogenesis of immunoglobin A nephropathy. Curr. Opin. Nephrol. Hypertens. **7:** 247–250.
21. MIELE, L., E. CORDELLA-MIELE & A.B. MUKHERJEE. 1990. High level bacterial expression of uteroglobin, a dimeric eukaryotic protein with two interchain disulfide bridges, in its natural quaternary structure. J. Biol. Chem. **265:** 6427–6435.
22. MANTILE, G. *et al.* 1993. Human Clara cell 10 kDa is the counterpart of rabbit uteroglobin. J. Biol. Chem. **268:** 20343–20351.
23. MATTHEWS, J.H. *et al.* 1994. Crystallization and characterization of the recombinant human Clara cell 10 kDa protein. Proteins Struct. Funct. Genet. **20:** 191–196.
24. CARLOMAGNO, T. *et al.* 1997. Resonance assignment and secondary structure determination and stability of the recombinant human uteroglobin with heteronuclear multi-dimensional NMR. J. Biomol. NMR **9:** 35–46.
25. MANJUNATH, R., S.I. CHUNG & A.B. MUKHERJEE. 1984. Crosslinking of uteroglobin by transglutaminase. Biochem. Biophys. Res. Commun. **12:** 400–407.
26. MUKHERJEE, A.B. *et al.* 1988. Modulation of cellular response to antigens by uteroglobin and transglutaminase. Adv. Exp. Med. Biol. **231:** 135–145.
27. LEVIN, S.W. *et al.* 1986. Uteroglobin inhibits phospholipase A2 activity. Life Sci. **38:** 1813–1819.
28. FACCHIANO, A. *et al.* 1991. Inhibition of porcine pancreatic phospholipase A2 by uteroglobin and antiflammin peptides: possible mechanism of action. Life Sci. **48:** 453–464.
29. MUKHERJEE, A.B., R.E. ULANE & A.K. AGRAWAL. 1982. Role of uteroglobin and transglutaminase in masking antigenicity of implanting rabbit embryos. Am. J. Reprod. Immunol. **2:** 135–141.
30. MUKHERJEE, A.B. *et al.* 1983. Suppression of epididymal sperm antigenicity in the rabbit by uteroglobin and transglutaminase. Science **219:** 989–991.
31. MIELE, L. *et al.* 1988. Novel anti-inflammatory peptides from the region of highest similarity between uteroglobin and lipocortin-1. Nature **335:** 726–729.
32. VASANTHAKUMAR, G. *et al.* 1988. Inhibition of phagocyte chemotaxis by uteroglobin, an inhibitor of blastocyst rejection. Biochem. Pharmacol. **37:** 389–393.

33. MANJUNATH, R. 1987. Inhibition of thrombin-induced platelet aggregation by uteroglobin. Biochem. Pharmacol. **36:** 741–745.
34. KUNDU, G.C. *et al.* 1996. Suppression of cellular chemoinvasiveness by human uteroglobin via a novel high affinity binding site. Proc. Natl. Acad. Sci. U.S.A. **93:** 2915–2919.
35. DIAZ GONZALEZ, K. & A. NIETO. 1995. Binding of uteroglobin to microsomes and plasmatic membranes. FEBS Lett. **361:** 255–258.
36. PATTABIRAMAN, N., K.B. WARD & T.J. FLEMING. 1995. Occluded molecular surface: analysis of protein packing. J. Mol. Recognit. **8:** 334–344.
37. BERNSTEIN, F.C. *et al.* 1977. The Protein Data Bank: a computer-based archival file for macromolecular structures. J. Mol. Biol. **112:** 535–542.
38. BALLY, R. & J. DELETTRE. 1989. Structure and refinement of the oxidized P21 form of uteroglobin at 1.64 Å resolution. J. Mol. Biol. **206:** 153–170.
39. UMLAND, T.C. *et al.* 1992. Refined structure of rat Clara cell 17 kDa protein at 3.0 Å resolution. J. Mol. Biol. **224:** 441-448.
40. HARD, T. *et al.* 1995. Solution structure of a mammalian PCB-binding protein in complex with a PCB. Nat. Struct. Biol. **2:** 983–989.
41. RAY, M.K. *et al.* 1993. Cloning and characterization of the mouse Clara cell specific 10 kDa protein gene: comparison of the 5′-flanking region with the human rat and rabbit gene. Biochem. Biophys. Res. Commun. **197:** 163–171.
42. KABSCH, W. & C. SANDER. 1983. Dictionary of protein secondary structure: pattern recognition of hydrogen bonded and geometrical features. Biopolymers **22:** 2577–2637.
43. STRIPP, B.R. *et al.* 1996. Clara cell secretory protein: a determinant of PCB bioaccumulation in mammals. Am. J. Physiol. **271:** L656–L664.
44. MORIZE, I. *et al.* 1987. Refinement of the C2221 crystal form of oxidized uteroglobin at 1.34 Å. J. Mol. Biol. **194:** 725–735.
45. ARNDT, U.W. & R.M. SWEET. 1977. Collimation and monochromatization. *In* The Rotation Method in Crystallography, pp. 59–63. North-Holland. Amsterdam.
46. HOWARD, A.J., C. NIELSEN & N.H. XUONG. 1985. Software for a diffractometer with multiwire area detector. Methods Enzymol. **114:** 452–472.
47. BRUNGER, A.T. 1990. Extension of molecular replacement: a new search strategy based on Patterson correlation refinement. Acta Crystallogr. **A46:** 46–57.
48. PETER, W. *et al.* 1992. Interchain cysteine bridges control entry of progesterone to the central cavity of the uteroglobin dimer. Protein Eng. **5:** 351–359.

Antiflammins

Bioactive Peptides Derived from Uteroglobin

LUCIO MIELE[a]

Cancer Immunology Program, Cardinal Bernardin Cancer Center,
Loyola University Chicago, Maywood, Illinois 60153, USA

ABSTRACT: Uteroglobin/Clara cell 10-kDa protein (UG/CC10) is a hormonally regulated small secretory protein that has a variety of *in vitro* and *in vivo* pharmacological effects. These include a potent anti-inflammatory activity and inhibitory effects on neutrophil migration, thrombin-induced platelet aggregation, *in vitro* chemoinvasion, as well as "tumor suppressor"–like effects and other properties. Several mechanisms of action have been proposed for these effects. Pharmacological properties suggest that UG itself or substances derived from it may be used as experimental drugs for several indications. The group of oligopeptides collectively known as "antiflammins" (AFs) were originally described in 1988. Their design was derived from the region of highest sequence similarity between UG and another group of proteins with anti-inflammatory properties, the lipocortins or annexins. Nanomolar concentrations of these peptides can reproduce several of the pharmacological activities of UG, including its *in vivo* anti-inflammatory effects and inhibition of platelet aggregation. The AFs have been safely and effectively used to suppress inflammation and fibrosis in several animal models. Progress in clarifying the mechanism of action of the AFs may facilitate the structure-based design of a novel class of potent anti-inflammatory, antichemotactic drugs.

BIOLOGICAL ACTIVITIES OF UTEROGLOBIN FAMILY PROTEINS

For over three decades, the proteins of the uteroglobin (UG)/Clara cell "10"-kDa (CC10) family (henceforth, UG) have presented a fascinating puzzle to the scores of investigators who have studied them. UG proteins are a highly conserved family of small-molecular-weight homodimers that are secreted by mucosal epithelia in the airways and reproductive tract and by numerous other tissues. UG expression is regulated by progestins, glucocorticoids, androgens, prolactin, and most likely other hormones and cytokines.[1,2] Despite their small size and extensive structural, biological, and genetic characterization, our understanding of the physiological role(s) of these proteins remains incomplete. This is largely due to the fact that UG proteins have a bewildering array of *in vitro* and *in vivo* biological activities, in sharp contrast with their deceptively simple structure.[1,2] These include *in vitro* inhibitory effects on platelet aggregation[3] and on neutrophil chemotaxis,[4,5] *in vivo* inhibition of carrageenan-induced acute inflammation,[6] inhibition of chemoinvasion by various

---

[a]Address for correspondence: Cancer Immunology Program, Cardinal Bernardin Cancer Center, Loyola University Chicago, 2160 South First Avenue, Maywood, IL 60153. Voice: 708-327-3362; fax: 708-327-3238.
lmiele@luc.edu

transformed cell lines,[7-9] and inhibition of anchorage-independent growth of transformed cell lines.[10,11] It is still unclear which of these multiple activities are physiologically relevant and which can be considered pharmacological effects. Targeted disruption of the mouse UG gene causes severe inflammatory nephropathy with fibronectin/IgA deposits[12-14] and increases the incidence of malignancies.[11] A different model of mouse UG gene inactivation shows increased sensitivity to oxidizing insults to the lung.[15]

To make matters even more complicated, no single mechanism has emerged so far that can explain all the biological activites of UG proteins. Their biochemical properties described so far are almost as numerous as their biological effects and include, among others, the following: (1) binding a variety of small hydrophobic molecules such as progestins, retinoids, and toxic xenobiotics;[16-21] (2) inhibition of secretory phospholipases A_2 (sPLA$_2$'s);[6,22-26] (3) binding to a saturable membrane receptor;[8,9] and (4) binding to fibronectin.[12-14] Ca^{2+}-dependent binding to negatively charged liposomes[27] and specific binding to a microsomal protein, which may or may not be related to the membrane receptor,[28] have been reported as well.

The relationship between biochemical and biological properties of UG proteins is complex and incompletely understood. The antichemoinvasion effect appears to be mediated by the UG receptor,[8,9] while fibronectin binding is likely to participate in the prevention of IgA nephropathy.[12-14] The mechanisms of the *in vivo* anti-inflammatory and tumor-suppressive activities and the possible relationship between these two properties are less clear. In principle, inhibition of sPLA$_2$'s, the UG receptor, and fibronectin binding may all participate in these effects and it is not possible to rule out additional, as yet undiscovered, mechanisms.

STRUCTURE-FUNCTION RELATIONSHIP OF UG PROTEINS: THE ANTIFLAMMIN PEPTIDES

Aside from the thorny question of physiological functions, it is apparent from this brief summary that the pharmacological activities of UG proteins make them extremely attractive candidates for therapeutic applications. Recombinant human UG is an excellent drug candidate in its own right because of its small size and exceptional stability[29] and because it can be produced in bacteria in its native, disulfide-bonded form using an expression system that we developed for that purpose[30,31] and subsequently improved for industrial use.[32] Additionally, understanding the structural basis for UG's pharmacological activities could open the way to the development of orally active, nonpeptide drugs that mimic UG activities. Based on the known information on the biological activities of UG, possible indications for UG-mimetic drugs range from inflammatory and autoimmune disorders to lung and prostate cancers.

In 1988, we[6] identified a 9-amino-acid region within α-helix-3 of rabbit UG (residues 39–47) that could reproduce the anti-inflammatory activity of the purified parent protein *in vivo*. This led to the development of a family of pharmacologically active oligopeptides collectively known as "antiflammins" (AFs) (reviewed in reference 33). These peptides were derived from the region of highest sequence similarity between UG and annexin I/lipocortin I, another glucocorticoid-induced protein orig-

TABLE 1. Sequence of AF peptides and proposed nomenclature

Name	Sequence	Parent protein	Reference
AF-1	MQMKKVLDS	Rabbit UG	6
AF-2	HDMNKVLDL	Annexin I	6
AF-2A	HDANKVLDL	Annexin I	34
AF-2N[a]	HD(nL)NKVLDL	Annexin I	34
AF-2NS[a]	HD(nL)NKVLDS	Annexin I	34
AF-3	SHLRKVFDK	Annexin V	35
AF-4	LRKVFDK	Annexin V	35
AF-5	AQLKKLVDT	Human UG	This article
Consensus[b]	XXX-FIL-FOB-FIL-LYS-LEU/VAL-VAL/LEU/PHE-ASP-FIL		

[a]nL = norleucine.
[b]XXX = indetermined; FOB = hydrophobic; FIL = hydrophilic.

inally identified because of its anti-inflammatory properties.[6] Over the past decade, the pharmacology of AFs has been explored by numerous groups and these peptides have clearly emerged as potentially useful therapeutic agents. Several AF peptide derivatives have been designed, additional pharmacological activities have been described, and mechanisms responsible for inactivation of AFs have been identified. The structures of AF peptides described to date and the putative structure of a human UG-derived antiflammin are shown in TABLE 1, together with a proposed nomenclature for this group of peptides. The consensus sequence shows a specific pattern of hydrophobic and hydrophilic residues that can be modeled as an amphipathic α-helix, with invariant Lys and Asp residues, respectively, at 4 residues and 1 residue before the C-terminus. These correspond to invariant Lys 43 and Asp 46 in the sequence of UG.

PHARMACOLOGICAL PROPERTIES OF AFs

In our original report, AFs, purified rabbit UG, and recombinant annexin/lipocortin I potently inhibited carrageenan-induced rat-paw acute inflammation.[6] Their effect was comparable to that of dexamethasone (10 µg/kg) and superior to that of indomethacin (1 mg/kg). Histologically, the most striking features were a dramatic decrease in edema and in the number of neutrophils infiltrating carrageenan-injected areas. Subsequently, Vostal *et al.* showed that AFs have a weak inhibitory effect on ADP-induced platelet aggregation *in vitro*.[33] Camussi *et al.*[34–36] demonstrated that AFs inhibit the synthesis of platelet-activating factor (PAF) induced by TNF or by phagocytosis in rat macrophages and human neutrophils or by thrombin in endothelial cells. PLA_2 activity in cell lysates and arachidonic acid (AA) release were inhibited by AFs, as was the activation (but not the catalytic activity) of acetyl-CoA-lyso-PAF acetyltransferase in intact cells. IC_{50} values for these activities were between 50 and 100 nM. Importantly, pretreatment with AFs before cell activation was more effective than treatment after application of the stimulus. These same authors also demonstrated that AFs inhibit neutrophil chemotaxis and aggregation and block intradermal Arthus reaction induced by injection of TNF or $C5_a$ in rats.[34–36] They[37] also showed that AFs inhibit TNF-induced cytoskeletal reorganization in human endothelial cells. Subsequently, Ialenti et al.[38,39] showed that AFs and annexins/lipocortins inhibit the late phase of carrageenan-induced rat-paw edema, which is mediated by eicosanoid release, but not the early phase, which is primarily mediated by vasoactive amines. Chan et al.[40] showed that AFs are as potent as dexamethasone in the treatment of experimental endotoxin-induced uveitis, a model of autoimmune anterior uveitis. Reduced PLA_2 activity in the aqueous humor was documented. Perretti et al.[41,42] described a novel AF peptide derived from a region of annexin/lipocortin V that is homologous to the "AF" region of rabbit UG. This peptide and the lipocortin I–derived AF-2 inhibited the contractions of isolated rat stomach strips induced by the application of porcine pancreatic PLA_2. These AFs also inhibited prostaglandin E_2 production induced by bradykinin in human skin fibroblasts or by opsonized zymosan in rat peritoneal macrophages. Sierra-Honigmann and Murphy[43] showed that recombinant lipocortin I and AF-2 inhibit IL-1-mediated activation of murine Th_2 cells *in vitro*. Lloret and Moreno[44] showed that AFs inhibit collagen-induced platelet activation and that they have a potent anti-inflammatory effect by local and systemic administration in carrageenan- and phorbol ester–induced inflammation, but not when inflammation was induced by treatment with AA or snake venom (*Naja naja*) $sPLA_2$. At the same time, Calderaro et al.[45] showed that AFs inhibit PGE_2 production and chloride secretion induced by Ca^{2+} ionophore A23187 in rabbit distal colonic mucosa. AA-induced PGE_2 production was inhibited by 100 nM AF-2 to the same extent as 10 µM indomethacin. Chloride secretion and cAMP accumulation induced by direct administration of PGE_2 were not inhibited, suggesting that, in this system, AFs inhibit PG production. A direct inhibitory effect on colonic mucosa cyclooxygenase with 100 nM AF-2 was observed by these authors. On the other hand, Lloret and Moreno[46] reported that AF-1 and -2 significantly inhibited carrageenan-induced inflammation, mast-cell degranulation, leukocyte infiltration, and histamine release in rat paws, but did not inhibit edema caused by snake venom (*Naja naja*) $sPLA_2$. Cabre et al.[47] showed that AFs inhibit inflammation *in vivo* caused by carrageenan, croton oil, and oxazolone, but not snake venom (*Naja naja*) $sPLA_2$. Human synovial (group II) $sPLA_2$ was not inhibited by AFs when bacterial cells were used as the substrate, but was inhibited in a mixed micellar assay. Subsequently, the same group[48] showed that topical treatment with AFs have a potent and dose-dependent inhibitory effect on edema, cell infiltration, plasma leakage, and leukotriene B_4 (LTB_4) production in phorbol ester–treated murine ears. Edema induced by topical administration of AA was not inhibited, suggesting that the AFs act in this model by inhibiting either AA release or the lipoxygenase pathway. In a different *in vitro* model of murine 3T6 fibroblasts or peritoneal macrophages, the same authors[49] did not observe effects of AF-2 on either AA release or metabolism, while a potent effect on leukocyte migration was observed. The authors suggest that this effect may be due to inhibition of leukocyte adhesion to endothelial cells and that it may participate in the *in vivo* anti-inflammatory effects of the AFs. More recently, Rodgers et al.[50] demonstrated that postoperative administration of AF-2 by mini-osmotic pumps significantly reduced peritoneal adhesions in two different animal models. Peritoneal adhesions are a consequence of inflammation induced by surgical injury and a significant source of postoperative morbidity and mortality. These data suggest that AFs are potential candidates for the prevention of this serious postoperative complication. Rayner and Van[51] reported that AF-2 strongly inhibited substance-P-induced lymphatic vessel contractions, but had no effect when a thromboxane A_2 analogue was used as the stimulus, suggesting that AF-2 acts upstream of AA metabolism in this model. Mize et al.[52] used AF-1 in a chlorpromazine-induced skin inflammation model and showed that this peptide can be delivered intradermally by iontophoresis, retaining its anti-inflammatory activity as determined by inhibition of erythema and lesion formation. This suggests that AFs may be used for local treatment in dermatological indications. PGE_2 production was not inhibited in this model, in contrast with the results of Perretti et al. in human fibroblasts and murine macrophages[41,42] and of Calderaro et al. in rabbit colonic mucosa[45] (see earlier). In an animal model of conjunctivitis, Li et al.[53] demonstrated that topical administration of AF-2 inhibits ocular inflammation caused by compound 48/80. Interestingly, the expression of $sPLA_2$'s and that of inducible NO synthase (iNOS) were inhibited by AF-2 and by dexamethasone in this model. Gao et al.[54] have recently shown that 1 nM AF-1 inhibits lymphatic vessel contraction induced by ATP. The latter is thought to be mediated by increased expression of $sPLA_2$. Very recently, the effects of AF-1 and -2 on leukocyte adhesion were studied in detail by Zouki et al.[55] These authors demonstrated that AFs and recombinant lipocortin I inhibit the changes in L-selectin and CD11/CD18 induced in human neutrophils by PAF and IL-8. The peptides and the recombinant protein had similar dose-response relationships. Moreover, AFs inhibited adhesion of human neutrophils to human coronary artery endothelial cells activated by LPS. This effect was additive with anti-E-selectin and anti-L-selectin antibodies, but not with anti-CD18 antibodies. The authors conclude that AFs modulate human neutrophil adhesion primarily by affecting CD18 expression and that these peptides are potential candidates for therapeutic indications that require modulating human neutrophil adhesion in vivo.

In summary, data from numerous groups have shown that AFs are highly potent anti-inflammatory agents that reproduce many pharmacological activities of their parent proteins, UG and annexins/lipocortins. Together, these data show that AFs:

(1) are active in vivo by parenteral (intradermal and systemic) and topical administration and can be delivered by iontophoresis and microosmotic pumps;
(2) have highly potent anti-inflammatory activity in diverse animal models in various organs and tissues, and prevent postoperative surgical adhesions;
(3) inhibit leukocyte adhesion and function, modulate the expression of adhesion molecules on neutrophils, and inhibit mast-cell degranulation, smooth muscle contraction, and lymphatic vessel contraction;
(4) inhibit the production or release of inflammatory mediators, including PAF, eicosanoids, and histamine, and possibly modulate the expression of $sPLA_2$'s and iNOS;
(5) have variable effects on AA release and the cyclooxygenase or lipooxygenase pathways, depending upon the experimental model.

These data indicate that these peptides, or structural analogues designed from them, are potentially attractive candidate drugs for a variety of anti-inflammatory uses, including dermatological and ophthalmological applications, and for the prevention of postoperative adhesions. The effect of AFs on neutrophil adhesion to human coronary endothelial cells suggests that they may be useful for such indications as prevention of reperfusion injury in myocardial ischemia and infarction. Although a formal toxicity study has not been performed, it is important to note that no apparent *in vivo* toxic effects have been reported thus far when AFs were tested at doses as high as 2 mg/kg in numerous animal models.

MECHANISM OF ACTION OF AFs

As is the case for most drugs and for UG proteins themselves, the multitude of *in vivo* activities of AFs are difficult to integrate into a straightforward model based on a simple mechanism of action. The information obtained to date is summarized below. The parent proteins, UG and annexins/lipocortins, inhibit the catalytic activity of sPLA$_2$'s *in vitro*. The mechanism(s) of this effect is still the subject of much controversy, which cannot be discussed in detail here for reasons of space (for reviews, see references 1, 26, and 33). Briefly, substrate sequestration due to Ca^{2+}-dependent binding to anionic phospholipids has been proposed as a mechanism of inhibition for annexins,[56,57] but others working with different models have questioned this conclusion.[58,59] Annexins also inhibit intracellular, cytosolic PLA$_2$ activity,[60,61] while no evidence of such intracellular activity has yet been described for UG. Similar to annexins, Ca^{2+}-dependent binding to anionic phospholipids has been described for UG.[27] However, UG also inhibits sPLA$_2$ in the absence of anionic phospholipids with phosphatidylcholine vesicles as the substrate.[22] Based upon the structural[6,62] and sequence similarity[25] between UG and sPLA$_2$'s, we originally proposed that UG proteins may interfere with the process of interfacial activation or with the interaction of sPLA$_2$ with lipid-water interfaces rather than the actual catalytic process.[1,6,24,25,63] Interestingly, UG has been suggested to have a Ca^{2+}-binding loop similar to that of sPLA$_2$'s.[64] It has been suggested that Asp 46, an invariant residue in the AF region of UG and in AF peptides, is a critical residue in this loop. These data are consistent with our hypothesis that UGs are structural analogues of sPLA$_2$'s and further identify the AF region as an area of high structural similarity between these protein families. AF peptides inhibit porcine pancreatic sPLA$_2$ with micellar substrates[6,25] and human group II sPLA$_2$ with micellar substrates,[47] as well as human neutrophil sPLA$_2$,[35,36] but like UG they have no effect when bacterial cells are used as the substrate or with snake venom sPLA$_2$'s (see earlier). Preincubation of pancreatic PLA$_2$ with AF-1 blocks its activity on mast-cell degranulation.[65] AFs interact with lipid-water interfaces[66] and inhibit the increase in fluorescence of Trp 3 in porcine pancreatic sPLA$_2$ that is normally triggered by binding to micellar substrates.[25] All in all, the most likely explanation for these observations is that, like UG, AFs also inhibit either the interfacial activation or the interaction of sPLA$_2$'s with lipid-water interfaces. PLA$_2$ inhibition by AFs had been questioned earlier.[67,68] However, in these studies, the peptides also showed no pharmacologic activity,[67] calling into question the quality of the peptide preparation used, or were used under different conditions than we used and/or with bacterial cells as the substrate.[68] Since details of peptide handling and storage were not discussed in these reports, one possible interpretation of such findings is that these authors may have been working with inactivated peptides (see below). Very recently, Chowdhury *et al.* (this volume), using site-directed mutagenesis, have definitively confirmed that the C-terminal half of α-helix-3 is the active site of UG for PLA_2 inhibition, as we originally proposed.[6]

The interpretation of *in vitro* PLA_2 inhibition studies vis-à-vis the mechanism of pharmacological activity of AFs is complicated by several factors. First, most of the studies reported so far, with the notable exception of Camussi *et al.* (see above), have not used physiologically relevant PLA_2's or substrates, relying instead on commercially available digestive or snake venom enzymes and artificial substrates. The recent isolation of several mammalian $sPLA_2$'s with structural similarities, but distinct catalytic properties,[69-71] and observations indicating that endogenous mammalian enzymes have vastly different biological activities in mammalian cells compared to snake venom enzymes[72] suggest caution in interpreting data obtained with experimental systems of questionable physiologic relevance. In general, the very nature of PLA_2 catalysis makes most assay systems highly artificial and notoriously difficult to standardize. This is especially true with nonlipid-soluble, noncompetitive inhibitors such as proteins or peptides. Thus, the role of PLA_2 inhibition as a potential *in vivo* mechanism of action of UG and AFs should be studied using mammalian nondigestive enzymes and physiologically relevant substrates. It is likely that different mammalian enzymes will show different sensitivity to inhibition, depending upon their affinities for lipid-water interfaces and mechanisms of interfacial activation, and that the type of substrate used will affect the results. The *in vivo* evidence available to date indicates that UG-deficient mice have higher plasma PLA_2 activity.[12]

A second potential source of confusion in interpreting AF studies is peptide inactivation. In our original studies,[6,25] AF peptides were stored in sealed glass vials under rigorously anhydrous conditions. Solutions were prepared with ice-cold buffers immediately before use and discarded after use. No AF solutions were stored. We[25,63] and others[34-36,73] discovered that AFs were rapidly inactivated upon storage in solution or under nonhydrous conditions. Camussi *et al.*[34-36,73] found that AFs containing Met residues were rapidly oxidized in the presence of air and that reducing agents could protect these peptides from this mechanism of inactivation. This problem was circumvented by Tetta *et al.*[73] by replacing Met with Ala or Nle residues. Additionally, these authors also discovered that AFs were completely inactivated by freezing and thawing in solution. This may be a consequence of aggregation and/or precipitation and could be partially reversed by heating solutions at 45°C. More recently, Wolfe *et al.*[74,75] demonstrated that AF-2 spontaneously hydrolyzes in aqueous solutions with pseudo-first-order kinetics. The primary hydrolysis point was at the C-terminal side of Asp 46. Under similar conditions, Met sulfoxide was the primary oxidation product.[76] In summary, there are at least three different mechanisms through which AF solutions can lose activity upon storage. One of these (freezing-induced inactivation) is not associated with chemical degradation and would not necessarily be detected by chemical methods such as mass spectrometry or HPLC. This suggests that special attention should be paid to peptide handling in AF studies and that inconsistent results are likely to ensue if this is not done.

A third possible confounding factor in interpreting pharmacological studies with AFs is that not all the *in vivo* activities of $sPLA_2$'s are mediated by their catalytic activity. High-affinity membrane receptors for $sPLA_2$'s have been isolated in recent years.[77–83] These receptors belong to the same superfamily as mannose-6-phosphate receptors (M-type receptors). M-type receptors are multifunctional molecules that bind PLA_2's via their carbohydrate recognition domains (CRDs) and have an N-terminal fibronectin type II repeat that most likely mediates cell adhesion.[82] M-type $sPLA_2$ receptors mediate a variety of cellular effects that had been previously attributed to $sPLA_2$ catalysis. These include stimulation of AA release and eicosanoid production from intact cells,[84,85] chemotaxis and chemoinvasion,[86] smooth muscle contraction,[87] and cell proliferation.[77] Through M-type receptors, group I $sPLA_2$'s increase the expression and secretion of group II $sPLA_2$.[88] Interestingly, many of these processes are inhibited by UG and/or AF peptides in various experimental models, including models in which pancreatic PLA_2 is used as the stimulus (see earlier). On the other hand, snake venom enzymes, which can readily hydrolyze mammalian cell membranes, do not appear to respond to AFs *in vivo*. These observations raise the question of whether AFs and/or UG may act *in vivo* on $sPLA_2$ M-type receptors. Theoretically, there are two different possible mechanisms for such binding: AFs and/or UG may mimic the $sPLA_2$ Ca^{2+}-binding loop,[64] which is essential for PLA_2 receptor binding,[89] and thus compete with PLA_2's for the CRD; alternatively, UG or AFs may bind the N-terminal fibronectin repeat, without competing with the PLA_2-binding site, and through this they could affect cell adhesion, trigger receptor internalization,[90,91] and thus indirectly prevent PLA_2 receptor–mediated effects and downregulate the expression of group II $sPLA_2$. UG does bind fibronectin,[12] but it is still unknown whether AFs do. This possible additional mechanism of action is at present entirely speculative. However, it is consistent with all the available evidence and could help explain several biological activities of UG and AFs, as well as the variable results on eicosanoid metabolism observed with AFs in different models. Yet another possibility is that AFs may interact with the UG receptor;[8,9] whether the latter receptor is related to M-type $sPLA_2$ receptors is still unclear.

In summary, AFs inhibit various mammalian $sPLA_2$'s *in vitro* with micellar substrates, including $sPLA_2$ activity in human neutrophil lysates, and block certain effects of group I (pancreatic) mammalian $sPLA_2$ on intact cells and organ preparations. AFs do not inhibit snake venom enzymes *in vitro* nor block their toxic effects *in vivo*. Some of the effects that are inhibited by AFs are mediated by the group I PLA_2 M-type receptor, raising the possibility that, *in vivo*, AFs and UG may also interact with these PLA_2 receptors, either at the same site bound by PLA_2's or at a different site. At present, it is impossible to rule out that other mechanisms that have been proposed, such as inhibition of PAF synthesis or of cyclooxygenase, and interaction with the UG receptor, as well as hitherto undiscovered mechanisms, may also participate in the *in vivo* pharmacological activities of the AFs.

CONCLUSIONS AND FUTURE DIRECTIONS

Results obtained by many independent groups over the past 11 years have demonstrated that AF oligopeptides derived from the C-terminal half of α-helix-3 of UG and from the homologous regions of annexins/lipocortins are extremely active antiinflammatory agents and reproduce many of the pharmacological activities of the parent proteins. AFs inhibit inflammation *in vivo* in numerous models, as well as surgical adhesion formation, smooth muscle contraction, lymphatic vessel contraction, and leukocyte adhesion and activation. Invariant residues in these peptides are Lys 5 and Asp 8 (corresponding to Lys 43 and Asp 46 in intact UG). Site-directed mutagenesis confirms that AFs correspond to the active site for this activity of UG. PLA_2 inhibition and possibly other activities such as PLA_2 receptor antagonism, cyclooxygenase inhibition, and as yet undiscovered mechanisms may contribute to the pharmacologic activities of the AFs.

A number of critical questions require further investigation: Do the AFs share the effects of UG on neoplastic cell adhesion, anchorage-independent growth, and chemoinvasion? Is PLA_2 receptor antagonism an *in vivo* mechanism of action of the AFs? Do AFs bind M-type $sPLA_2$ receptors and, if so, do they bind the CRDs or the N-terminal fibronectin repeat? Do they bind fibronectin or other fibronectin repeat–containing proteins and/or the UG receptor? Answering these questions will better delineate the spectrum of candidate therapeutic indications for AFs or their derivatives and will provide information on possible docking sites for AFs that may be used in the rational design of nonpeptide analogues.

REFERENCES

1. MIELE, L. *et al.* 1994. Uteroglobin and uteroglobin-like proteins: the uteroglobin family of proteins. J. Endocrinol. Invest. **17:** 679–692.
2. MUKHERJEE, A.B. *et al.* 1999. Uteroglobin: a novel cytokine? Cell. Mol. Life Sci. **55:** 771–787.
3. MANJUNATH, R. *et al.* 1987. Inhibition of thrombin-induced platelet aggregation by uteroglobin. Biochem. Pharmacol. **36:** 741–746.
4. SCHIFFMANN, E. *et al.* 1983. Adherence and regulation of leukotaxis. Agents Actions Suppl. **12:** 106–120.
5. VASANTHAKUMAR, G. *et al.* 1988. Inhibition of phagocyte chemotaxis by uteroglobin, an inhibitor of blastocyst rejection. Biochem. Pharmacol. **37:** 389–394.
6. MIELE, L. *et al.* 1988. Novel anti-inflammatory peptides from the region of highest similarity between uteroglobin and lipocortin I. Nature **335:** 726–730.
7. LEYTON, J. *et al.* 1994. Recombinant human uteroglobin inhibits the *in vitro* invasiveness of human metastatic prostate tumor cells and the release of arachidonic acid stimulated by fibroblast-conditioned medium. Cancer Res. **54:** 3696–3699.
8. KUNDU, G.C. *et al.* 1996. Recombinant human uteroglobin suppresses cellular invasiveness via a novel class of high-affinity cell surface binding site. Proc. Natl. Acad. Sci. U.S.A. **93:** 2915–2919.
9. KUNDU, G.C. *et al.* 1998. Uteroglobin (UG) suppresses extracellular matrix invasion by normal and cancer cells that express the high affinity UG-binding proteins. J. Biol. Chem. **273:** 22819–22824.
10. SZABO, E. *et al.* 1998. Overexpression of CC10 modifies neoplastic potential in lung cancer cells. Cell Growth Differ. **9:** 475–485.
11. ZHANG, Z. *et al.* 1999. Loss of transformed phenotype in cancer cells by overexpression of the uteroglobin gene. Proc. Natl. Acad. Sci. U.S.A. **96:** 3963–3968.
12. ZHANG, Z. *et al.* 1997. Severe fibronectin-deposit renal glomerular disease in mice lacking uteroglobin. Science **276:** 1408–1412.
13. MUKHERJEE, A.B. *et al.* 1998. Uteroglobin: physiological role in normal glomerular function uncovered by targeted disruption of the uteroglobin gene in mice. Am. J. Kidney Dis. **32:** 1106–1120.
14. ZHENG, F. *et al.* 1999. Uteroglobin is essential in preventing immunoglobulin A nephropathy in mice. Nat. Med. **5:** 1018–1025.
15. REYNOLDS, S.D. *et al.* 1999. Normal function and lack of fibronectin accumulation in kidneys of Clara cell secretory protein/uteroglobin deficient mice. Am. J. Kidney Dis. **33:** 541–551.

16. BEATO, M. & R. BAIER. 1975. Binding of progesterone to the proteins of the uterine luminal fluid: identification of uteroglobin as the binding protein. Biochim. Biophys. Acta **392**: 346–356.
17. BEATO, M. 1976. Binding of steroids to uteroglobin. J. Steroid Biochem. **7**: 327–334.
18. SAXENA, S.K. *et al.* 1983. Specific interaction of some non-steroidal compounds with the progesterone binding site of uteroglobin. J. Steroid Biochem. **18**: 303–308.
19. GILLNER, M. *et al.* 1988. The binding of methylsulfonyl-polychloro-biphenyls to uteroglobin. J. Steroid Biochem. **31**: 27–33.
20. ANDERSSON, O. *et al.* 1991. Purification and level of expression in bronchoalveolar lavage of a human polychlorinated biphenyl (PCB)-binding protein: evidence for a structural and functional kinship to the multihormonally regulated protein uteroglobin. Am. J. Respir. Cell Mol. Biol. **5**: 6–12.
21. LOPEZ DE HARO, M.S. *et al.* 1994. Binding of retinoids to uteroglobin. FEBS Lett. **349**: 249–251.
22. LEVIN, S.W. *et al.* 1986. Uteroglobin inhibits phospholipase A2 activity. Life Sci. **38**: 1813–1815.
23. MUKHERJEE, A.B. *et al.* 1988. Modulation of cellular response to antigens by uteroglobin and transglutaminase. Adv. Exp. Med. Biol. **231**: 135–152.
24. MIELE, L. *et al.* 1990. Inhibition of phospholipase A2 by uteroglobin and antiflammin peptides. Adv. Exp. Med. Biol. **279**: 137–160.
25. FACCHIANO, A. *et al.* 1991. Inhibition of pancreatic phospholipase A2 activity by uteroglobin and antiflammin peptides: possible mechanism of action. Life Sci. **48**: 453–464.
26. MUKHERJEE, A.B., E. CORDELLA-MIELE & L. MIELE. 1992. Regulation of extracellular phospholipase A2 activity: implications for inflammatory diseases. DNA Cell Biol. **11**: 233–243.
27. NORD, M., J.A. GUSTAFSSON & J. LUND. 1995. Calcium-dependent binding of uteroglobin (PCB-BP/CCSP) to negatively charged phospholipid liposomes. FEBS Lett. **374**: 403–406.
28. DIAZ, G.K. & A. NIETO. 1995. Binding of uteroglobin to microsomes and plasmatic membranes. FEBS Lett. **361**: 255–258.
29. CARLOMAGNO, T. *et al.* 1997. Resonance assignment and secondary structure determination and stability of the recombinant human uteroglobin with heteronuclear multidimensional NMR. J. Biomol. NMR **9**: 35–46.
30. MIELE, L., E. CORDELLA-MIELE & A.B. MUKHERJEE. 1990. High level bacterial expression of uteroglobin, a dimeric eukaryotic protein with two interchain disulfide bridges, in its natural quaternary structure. J. Biol. Chem. **265**: 6427–6435.
31. MANTILE, G. *et al.* 1993. Human Clara cell 10-kDa protein is the counterpart of rabbit uteroglobin. J. Biol. Chem. **268**: 20343–20351.
32. MANTILE, G. *et al.* 2000. Stable, long-term bacterial production of soluble, dimeric, disulfide-bonded protein pharmaceuticals without antibiotic selection. Biotechnol. Prog. **16**: 17–25.
33. VOSTAL, J.G. *et al.* 1989. Novel peptides derived from a region of local homology between uteroglobin and lipocortin-1 inhibit platelet aggregation and secretion. Biochem. Biophys. Res. Commun. **165**: 27–36.
34. CAMUSSI, G. *et al.* 1990. Anti-inflammatory peptides inhibit synthesis of plateletactivating factor. Prog. Clin. Biol. Res. **349**: 69–80.
35. CAMUSSI, G., C. TETTA & C. BAGLIONI. 1990. Antiflammins inhibit synthesis of platelet-activating factor and intradermal inflammatory reactions. Adv. Exp. Med. Biol. **279**: 161–172.
36. CAMUSSI, G. *et al.* 1990. Anti-inflammatory peptides (antiflammins) inhibit synthesis of platelet-activating factor, neutrophil aggregation and chemotaxis, and intradermal inflammatory reactions. J. Exp. Med. **171**: 913–927.
37. CAMUSSI, G. *et al.* 1991. Tumor necrosis factor alters cytoskeletal organization and barrier function of endothelial cells. Int. Arch. Allergy Appl. Immunol. **96**: 84–91.
38. IALENTI, A. *et al.* 1990. Anti-inflammatory effects of vasocortin and nonapeptide fragments of uteroglobin and lipocortin I (antiflammins). Agents Actions **29**: 48–49.
39. DI, R.M. & A. IALENTI. 1990. Selective inhibition of inflammatory reactions by vasocortin and antiflammin 2. Prog. Clin. Biol. Res. **349**: 81–90.

40. CHAN, C.C. *et al.* 1991. Effects of antiflammins on endotoxin-induced uveitis in rats. Arch. Ophthalmol. **109:** 278–281.
41. PERRETTI, M. *et al.* 1991. A novel anti-inflammatory peptide from human lipocortin 5. Br. J. Pharmacol. **103:** 1327–1332.
42. PERRETTI, M. 1994. Lipocortin-derived peptides. Biochem. Pharmacol. **47:** 931–938.
43. SIERRA-HONIGMANN, M.R. & P.A. MURPHY. 1992. Suppression of interleukin-1 action by phospholipase-A2 inhibitors in helper T lymphocytes. Pept. Res. **5:** 258–261.
44. LLORET, S. & J.J. MORENO. 1992. *In vitro* and *in vivo* effects of the anti-inflammatory peptides, antiflammins. Biochem. Pharmacol. **44:** 1437–1441.
45. CALDERARO, V. *et al.* 1992. Antiflammins suppress the A23187- and arachidonic acid–dependent chloride secretion in rabbit distal colonic mucosa. J. Pharmacol. Exp. Ther. **263:** 579–587.
46. LLORET, S. & J.J. MORENO. 1994. Effect of nonapeptide fragments of uteroglobin and lipocortin I on oedema and mast cell degranulation. Eur. J. Pharmacol. **264:** 379–384.
47. CABRE, F., J.J. MORENO *et al.* 1992. Antiflammins: anti-inflammatory activity and effect on human phospholipase A2. Biochem. Pharmacol. **44:** 519–525.
48. LLORET, S. & J.J. MORENO. 1995. Effects of an anti-inflammatory peptide (antiflammin 2) on cell influx, eicosanoid biosynthesis, and edema formation by arachidonic acid and tetradecanoyl phorbol dermal application. Biochem. Pharmacol. **50:** 347–353.
49. MORENO, J.J. 1996. Antiflammin-2, a nonapeptide of lipocortin-1, inhibits leukocyte chemotaxis, but not arachidonic acid mobilization. Eur. J. Pharmacol. **314:** 129–135.
50. RODGERS, K.E. *et al.* 1997. Reduction of adhesion formation by intraperitoneal administration of anti-inflammatory peptide 2. J. Invest. Surg. **10:** 31–36.
51. RAYNER, S.E. & H.D. VAN. 1997. Evidence that the substance P–induced enhancement of pacemaking in lymphatics of the guinea-pig mesentery occurs through endothelial release of thromboxane A2. Br. J. Pharmacol. **121:** 1589–1596.
52. MIZE, N.K. *et al.* 1997. Antiflammin 1 peptide delivered non-invasively by iontophoresis reduces irritant-induced inflammation *in vivo*. Exp. Dermatol. **6:** 181–185.
53. LI, Q. *et al.* 1998. Suppressive effect of antiflammin-2 on compound 48/80–induced conjunctivitis: role of phospholipase A2s and inducible nitric oxide synthase. Ocul. Immunol. Inflamm. **6:** 65–73.
54. GAO, J. *et al.* 1999. Evidence that the ATP-induced increase in vasomotion of guinea-pig mesenteric lymphatics involves an endothelium-dependent release of thromboxane A2. Br. J. Pharmacol. **127:** 1597–1602.
55. ZOUKI, C., S. OUELLET & J.G. FILEP. 2000. The anti-inflammatory peptides, antiflammins, regulate the expression of adhesion molecules on human leukocytes and prevent neutrophil adhesion to endothelial cells. FASEB J. **14:** 572–580.
56. BUHL, W.J. 1992. Annexins and phospholipase A2 inhibition. Eicosanoids 5(suppl.): S26–S28.
57. BUCKLAND, A.G. & D.C. WILTON. 1998. Inhibition of secreted phospholipases A2 by annexin V: competition for anionic phospholipid interfaces allows an assessment of the relative interfacial affinities of secreted phospholipases A2. Biochim. Biophys. Acta **1391:** 367–376.
58. KIM, K.M. *et al.* 1994. Annexin-I inhibits phospholipase A2 by specific interaction, not by substrate depletion. FEBS Lett. **343:** 251–255.
59. SPEIJER, H. *et al.* 1997. Partial coverage of phospholipid model membranes with annexin V may completely inhibit their degradation by phospholipase A2. FEBS Lett. **402:** 193–197.
60. BUCKLAND, A.G. & D.C. WILTON. 1998. Inhibition of human cytosolic phospholipase A2 by human annexinV. Biochem. J. **329:** 369–372.
61. MIRA, J.P. *et al.* 1997. Inhibition of cytosolic phospholipase A2 by annexin V in differentiated permeabilized HL-60 cells: evidence of crucial importance of domain I type II Ca^{2+}-binding site in the mechanism of inhibition. J. Biol. Chem. **272:** 10474–10482.
62. MORIZE, I. *et al.* 1987. Refinement of the C222(1) crystal form of oxidized uteroglobin at 1.34 Å resolution. J. Mol. Biol. **194:** 725–739.
63. MUKHERJEE, A.B. & L. MIELE. 1994. Design of immunomodulatory peptides based on active site structures. *In* Chemical and Structural Approach to Rational Drug Design, pp. 237–261. CRC Press. Boca Raton, Florida.

64. BARNES, H.J. et al. 1996. Structural basis for calcium binding by uteroglobins. J. Mol. Biol. **256:** 392–404.
65. NAGAI, H. et al. 1991. Extracellular phospholipase A2 and histamine release from rat peritoneal mast cells. Int. Arch. Allergy Appl. Immunol. **96:** 311–316.
66. NEWMAN, R.H. et al. 1990. Interaction of synthetic peptides from annexin I and uteroglobin with lipid monolayers and their effect on phospholipase A2 activity. Biochem. Soc. Trans. **18:** 1233–1234.
67. MARKI, F. et al. 1990. "Antiflammins": two nonapeptide fragments of uteroglobin and lipocortin I have no phospholipase A2–inhibitory and anti-inflammatory activity. FEBS Lett. **264:** 171–175.
68. HOPE, W.C., B.J. PATEL & D.R. BOLIN. 1991. Antiflammin-2 (HDMNKVLDL) does not inhibit phospholipase A2 activities. Agents Actions **34:** 77–80.
69. CHEN, J. et al. 1994. Cloning and recombinant expression of a novel human low molecular weight Ca(2+)-dependent phospholipase A2. J. Biol. Chem. **269:** 2365–2368.
70. HAN, S.K., B.I. LEE & W. CHO. 1997. Bacterial expression and characterization of human pancreatic phospholipase A2. Biochim. Biophys. Acta **1346:** 185–192.
71. HAN, S.K., E.T. YOON & W. CHO. 1998. Bacterial expression and characterization of human secretory class V phospholipase A2. Biochem. J. **331:** 353–357.
72. MOUNIER, C. et al. 1994. Platelet secretory phospholipase A2 fails to induce rabbit platelet activation and to release arachidonic acid in contrast with venom phospholipases A2. Biochim. Biophys. Acta **1214:** 88–96.
73. TETTA, C., G. CAMUSSI et al. 1991. Inhibition of the synthesis of platelet-activating factor by anti-inflammatory peptides (antiflammins) without methionine. J. Pharmacol. Exp. Ther. **257:** 616–620.
74. WOLFE, J.L. et al. 1994. Degradation of antiflammin 2 in aqueous solution [letter]. J. Pharm. Sci. **83:** 1762–1764.
75. YE, J.M. et al. 1996. Degradation of antiflammin 2 under acidic conditions. J. Pharm. Sci. **85:** 695–699.
76. YE, J.M. & J.L. WOLFE. 1996. Oxidative degradation of antiflammin 2. Pharm. Res. **13:** 250–255.
77. ARITA, H. et al. 1991. Novel proliferative effect of phospholipase A2 in Swiss 3T3 cells via specific binding site. J. Biol. Chem. **266:** 19139–19141.
78. HANASAKI, K. & H. ARITA. 1992. Characterization of a high affinity binding site for pancreatic-type phospholipase A2 in the rat: its cellular and tissue distribution. J. Biol. Chem. **267:** 6414–6420.
79. HANASAKI, K. & H. ARITA. 1992. Purification and characterization of a high-affinity binding protein for pancreatic-type phospholipase A2. Biochim. Biophys. Acta **1127:** 233–241.
80. ISHIZAKI, J. et al. 1993. Receptor-binding capability of pancreatic phospholipase A2 is separable from its enzymatic activity. FEBS Lett. **324:** 349–352.
81. ISHIZAKI, J. et al. 1994. Molecular cloning of pancreatic group I phospholipase A2 receptor. J. Biol. Chem. **269:** 5897–5904.
82. ANCIAN, P., G. LAMBEAU & M. LAZDUNSKI. 1995. Multifunctional activity of the extracellular domain of the M-type (180 kDa) membrane receptor for secretory phospholipases A2. Biochemistry **34:** 13146–13151.
83. ANCIAN, P. et al. 1995. The human 180-kDa receptor for secretory phospholipases A2: molecular cloning, identification of a secreted soluble form, expression, and chromosomal localization. J. Biol. Chem. **270:** 8963–8970.
84. XING, M., L. MIELE & A.B. MUKHERJEE. 1995. Arachidonic acid release from NIH 3T3 cells by group-I phospholipase A2: involvement of a receptor-mediated mechanism. J. Cell. Physiol. **165:** 566–575.
85. KISHINO, J. et al. 1995. Pancreatic-type phospholipase A2 activates prostaglandin E2 production in rat mesangial cells by receptor binding reaction. J. Biochem. (Tokyo) **117:** 420–424.
86. KUNDU, G.C. & A.B. MUKHERJEE. 1997. Evidence that porcine pancreatic phospholipase A2 via its high affinity receptor stimulates extracellular matrix invasion by normal and cancer cells. J. Biol. Chem. **272:** 2346–2353.

87. KANEMASA, T. *et al.* 1992. Contraction of guinea pig lung parenchyma by pancreatic type phospholipase A2 via its specific binding site. FEBS Lett. **303:** 217–220.
88. KISHINO, J. *et al.* 1994. Pancreatic-type phospholipase A2 induces group II phospholipase A2 expression and prostaglandin biosynthesis in rat mesangial cells. J. Biol. Chem. **269:** 5092–5098.
89. LAMBEAU, G. *et al.* 1995. Structural elements of secretory phospholipases A2 involved in the binding to M-type receptors. J. Biol. Chem. **270:** 5534–5540.
90. ROSSINI, G.P. *et al.* 1996. Binding and internalization of extracellular type-I phospholipase A2 in uterine stromal cells. Biochem. J. **315:** 1007–1014.
91. ZVARITCH, E., G. LAMBEAU & M. LAZDUNSKI. 1996. Endocytic properties of the M-type 180-kDa receptor for secretory phospholipases A2. J. Biol. Chem. **271:** 250–257.

Therapeutic Applications of Antiflammin Peptides in Experimental Ocular Inflammation

CHI-CHAO CHAN,[a,b] NADINE TUAILLON,[a] QIAN LI,[b,c] AND DE FEN SHEN[b]

[b]*Section of Immunopathology, Laboratory of Immunology, National Eye Institute, National Institutes of Health, Bethesda, Maryland, USA*

[c]*Wilmer Ophthalmic Institute, The Johns Hopkins University, Baltimore, Maryland, USA*

ABSTRACT: Antiflammins are synthetic peptides derived from the region of highest local similarity between uteroglobulin and lipocortin. These peptides have shown anti-inflammatory activity on carrageenan-induced rat footpad edema. They are potent inhibitors for phospholipase A_2 activation both *in vitro* and *in vivo*. Previously, we have demonstrated the effectiveness of topical antiflammins in suppressing acute ocular inflammation and allergic response in rodent endotoxin-induced uveitis and murine allergic conjunctivitis. The mechanisms by which antiflammins protect against inflammation and allergy in these ocular models may involve inhibition of phospholipase A_2 and inducible nitric oxide synthase (iNOS) as well as the production of proinflammatory cytokine, interleukin-6.

INTRODUCTION

Glucocorticoids are potent anti-inflammatory agents with selective inhibitory action on phospholipase A_2 (PLA_2).[1] Two proteins induced by glucocorticoids, lipocortin I and uteroglobin, have regions of high similarity.[2] Antiflammins are synthetic peptides derived from the region of highest local similarity between human lipocortin I and rabbit uteroglobulin.[3] These peptides also inhibit PLA_2 *in vitro* and have potent anti-inflammatory activity *in vivo* on carrageenan-induced rat footpad edema.[3,4] It has been suggested that the anti-inflammatory effect may be on PLA_2 activation rather than on the enzyme or enzyme-substrate interaction.[5]

The P2 or antiflammin-2 peptide (HDMNKVLDL; corresponding to lipocortin I residues 247–255) inhibits the synthesis of platelet-activating factor (PAF) induced by TNF or phagocytosis in rat macrophages and human neutrophils, and by thrombin in vascular endothelial cells.[6] The P1 or antiflammin-1 peptide (MQMKKVLDS; corresponding to the 9 C-terminal amino acid residues 39–47 of α-helix 3 in uteroglobin) is less inhibitory than antiflammin-2 for macrophages and not inhibitory for neutrophils after a 5-min preincubation. This finding suggests that antiflammin-1 is inactivated by neutrophil secretory products, possibly oxidizing agents. Synthesis of PAF is inhibited by antiflammin-2 without an appreciable lag, but this inhibition is reversed when neutrophils or macrophages are washed and incubated in fresh medium. Therefore, antiflammins must be continuously present to inhibit PAF syn-

---

[a]Address for correspondence: Chi-Chao Chan, NIH/NEI, 10 Center Drive, Building 10, Room 10N103, Bethesda, MD 20892-1857. Voice: 301-496-0417; fax: 301-402-8664.
ccc@helix.nih.gov

thesis. Antiflammins block activation of the acetyltransferase required for PAF synthesis, suggesting that this enzyme is another target for the inhibitory activity of antiflammins.

Antiflammins inhibit neutrophil aggregation and chemotaxis.[6,7] Antiflammin-2 suppresses the increase in vascular permeability and the leukocyte infiltration induced in rats by an Arthus reaction or by intradermal injection of TNF and complement component C5a.[4] Both synthesized oligopeptides are water-soluble and without the known side effects of corticosteriods.

We have examined the effects of antiflammins in two acute ocular inflammatory models of endotoxin-induced uveitis in rats and allergic conjunctivitis in mice.[8–10] In both models, antiflammins protect against inflammatory responses. The mechanism involves inhibition of not only PLA_2, but also other inflammatory mediators and cytokines.

ENDOTOXIN-INDUCED UVEITIS

Endotoxin, when injected at sites far from the eye, elicits an acute ocular inflammation. Ayo first demonstrated that a single intravenous injection of endotoxin induced ocular inflammation.[11] Endotoxin-induced uveitis (EIU) is an experimental model for anterior uveitis of the eye by footpad injection of sublethal dose of endotoxin, the lipopolysaccharide (LPS) component of gram-negative bacterial cell wall.[12] EIU in the rat usually peaks at 16–20 hours and subsides at 48 hours after endotoxin injection. Histopathology of EIU is characterized by acute inflammatory cells, including neutrophils and macrophages in the anterior chamber, iris, and ciliary body, as well as protein transudate in the anterior chamber. Various inflammatory cytokines are also released in the eyes during EIU.[13] It has been suggested that activation of PLA_2, rather than a direct action of endotoxin into the eye, is one of the initial events during EIU.[14] Inhibitors of PLA_2 such as disodium D,L-α-tocopheryl L-ascorbate 2-O-phosphate diester (EPC) and dexamethasone have been reported to be most effective in treating EIU.

We previously reported the efficacy of topical antiflammins in the treatment of EIU in rats.[8,9] Antiflammin-1 and -2 were dissolved in Dacriose sterile ophthalmic irrigating solution (Cooper Vision Pharmaceuticals Inc., Puerto Rico). They were administered topically or in combination with intramuscular injection. EIU rats receiving either topical or systemic antiflammins showed reduced inflammatory cell counts in the anterior chamber, less ocular inflammation, and lower PLA_2 levels in the aqueous humor (FIGS. 1A and 1B). The effects were compatible with corticosteroid treatment. Therefore, antiflammins could have applications in the treatment of acute anterior uveitis.

ALLERGIC CONJUNCTIVITIS

Seasonal allergic conjunctivitis, the most common allergic ocular disease, is an IgE-mediated hypersensitivity reaction in the eye induced by airborne allergens.[15] The typical symptoms, including itching, eyelid swelling, conjunctival hyperemia, chemosis (conjunctival edema), and mucous discharge, are mostly caused by degran-

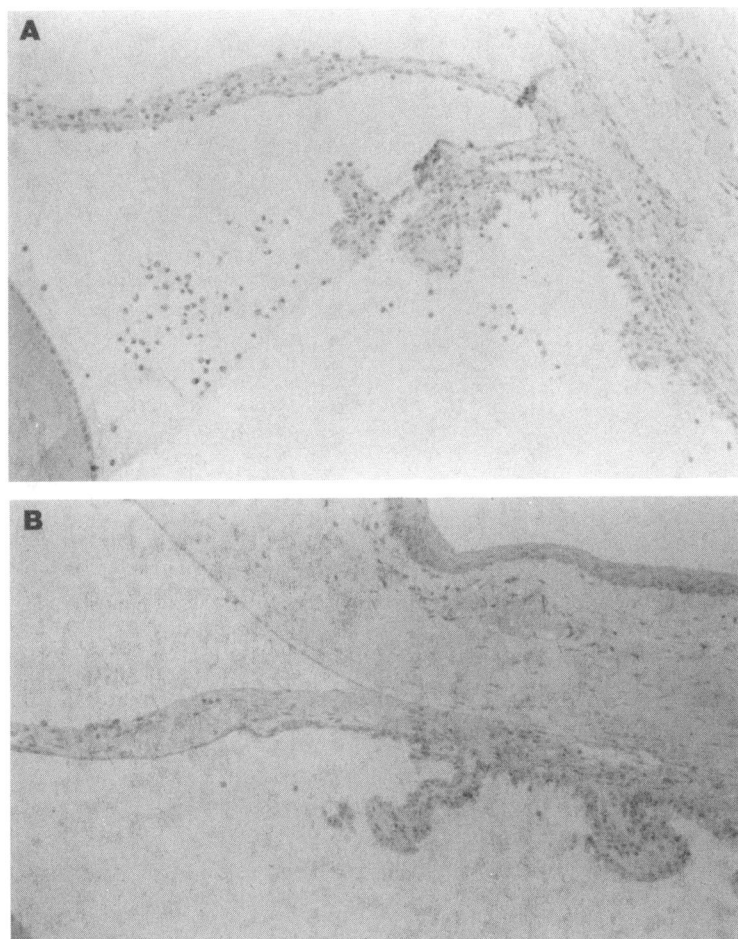

FIGURE 1. Photomicrographs showing inhibition of ocular inflammation in rat EIU eyes treated with antiflammin. **(A)** Many inflammatory cells are present in the anterior and posterior chambers of a control eye with EIU. **(B)** Only a few inflammatory cells are present in the anterior and posterior chambers of an eye that received topical antiflammin. (Hematoxylin & eosin, ×100.)

ulating mast cells and by the mediators they release in the conjunctiva.[16] Topical instillation of a mast cell degranulating agent, compound 48/80 (C48/80), onto the conjunctiva induces allergic conjunctivitis in the mouse.[17] Fifteen minutes after C48/80 exposure, the mice develop conjunctival chemosis. Six to 24 hours later, neutrophils, macrophages, T lymphocytes, and few eosinophils infiltrate the edematous and congested conjunctiva. Inducible nitric oxide synthase (iNOS) is produced in the allergic conjunctival tissue.

Low levels of PLA_2 and their messages are detected in normal conjunctival epithelium.[10] In C48/80-induced allergic conjunctivitis, PLA_2 protein and mRNA are

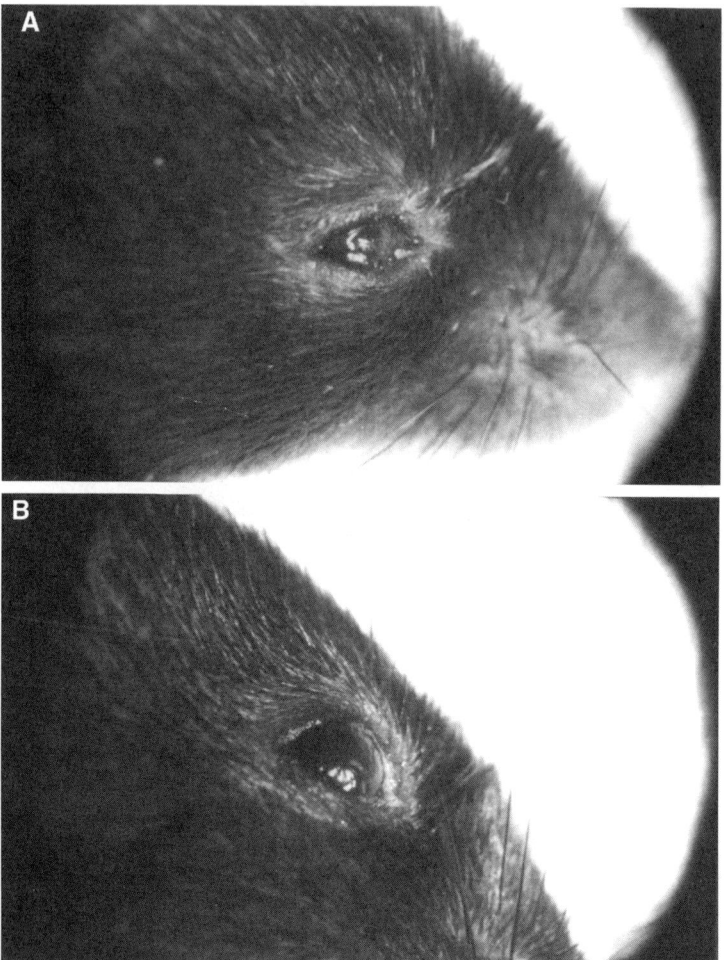

FIGURE 2. Photographs showing lower inflammatory reaction in mouse C48/80-induced allergic conjunctivitis treated with antiflammin. (**A**) Discharge and tearing in an eye with allergic conjunctivitis. (**B**) Minimal discharge in an allergic eye treated with antiflammin.

also located in conjunctival substantia propria. Application of antiflammins onto the eyes abets allergic response characterized by diminished eyelid discharge and swelling (FIGS. 2A and 2B), decreased numbers of neutrophils and macrophages, and suppressed iNOS as well as PLA_2 in the conjunctiva. The mechanisms by which antiflammins protect against allergy may involve the inhibition of PLA_2 and iNOS.

A linkage between PLA_2 and a proinflammatory cytokine, interleukin-1 (IL-1), has been reported.[18] We studied the effect of antiflammins on another proinflammatory cytokine, IL-6, in murine C48/80-induced allergic conjunctivitis. Induction of IL-6 production is an early event in acute inflammatory response, leading to vascular

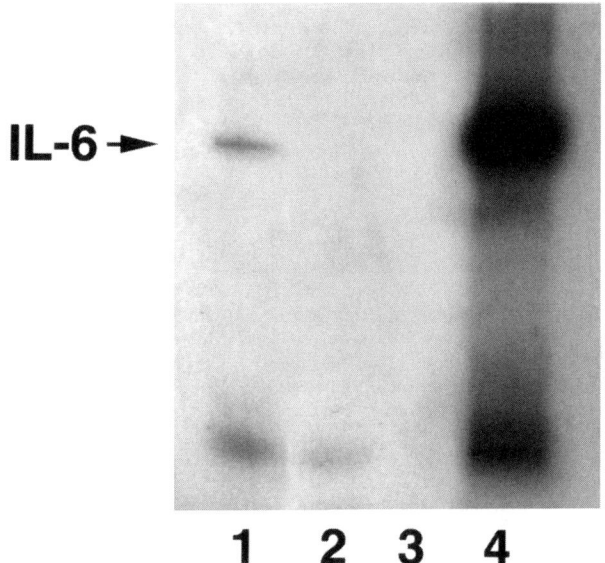

FIGURE 3. RT-PCR detection of murine conjunctival IL-6 mRNA. Lane 1, C48/80 and control ophthalmic solution; lane 2, C48/80 and antiflammin 2; lane 3, negative control; lane 4, positive control.

leakage and acute inflammatory cellular infiltration.[19–21] IL-6 is the first cytokine message to be detected in the murine eyes with endotoxin-induced uveitis.[22]

Allergic conjunctiva cells were selectively captured by microdissection. IL-6 mRNA in these cells was analyzed using RT-PCR technique. IL-6 mRNA was found in conjunctival epithelial cells and fibroblasts treated with C48/80 and control ophthalmic solution, but not in cells treated with C48/80 and antiflammins (FIG. 3). This result suggests that antiflammins are capable of downregulating the production of IL-6.

SUMMARY

Antiflammins are potent inhibitors for PLA_2 activation both *in vitro* and *in vivo*. The effectiveness of topical antiflammins in suppressing acute ocular inflammation and allergic response in rodent endotoxin-induced uveitis and murine allergic conjunctivitis makes them alternative therapeutic medications for ocular inflammatory diseases. The mechanisms by which antiflammins protect against inflammation and allergy may involve inhibition of PLA_2 and inducible nitric oxide synthase (iNOS) as well as the production of proinflammatory cytokines, IL-6 and IL-1.

REFERENCES

1. DI ROSA, M., R.J. FLOWER, F. HIRATA *et al.* 1984. Anti-phospholipase proteins. Prostaglandins **28**(4): 441–442.

2. HUANG, K.S., B.P. WALLNER, R.J. MATTALIANO et al. 1986. Two human 35 kd inhibitors of phospholipase A2 are related to substrates of pp60v-src and of the epidermal growth factor receptor/kinase. Cell **46**(2): 191–199.
3. MIELE, L., E. CORDELLA-MIELE, A. FACCHIANO & A.B. MUKHERJEE. 1988. Novel antiinflammatory peptides from the region of highest similarity between uteroglobin and lipocortin I. Nature **335**(6192): 726–730.
4. CAMUSSI, G., C. TETTA & C. BAGLIONI. 1990. Antiflammins inhibit synthesis of platelet-activating factor and intradermal inflammatory reactions. In Biochemistry, Molecular Biology, and Physiology of Phosphalipase A_2 and Its Regulatory Factors, pp. 161–172. Plenum. New York.
5. LLORET, S. & J.J. MORENO. 1992. In vitro and in vivo effects of the anti-inflammatory peptides, antiflammins. Biochem. Pharmacol. **44**(7): 1437–1441.
6. CAMUSSI, G., C. TETTA, F. BUSSOLINO & C. BAGLIONI. 1990. Anti-inflammatory peptides (antiflammins) inhibit synthesis of platelet-activating factor, neutrophil aggregation and chemotaxis, and intradermal inflammatory reactions. J. Exp. Med. **171**(3): 913–927.
7. MORENO, J.J. 1996. Antiflammin-2, a nonapeptide of lipocortin-1, inhibits leukocyte chemotaxis, but not arachidonic acid mobilization. Eur. J. Pharmacol. **314**(1–2): 129–135.
8. CHAN, C.C., M. NI, L. MIELE et al. 1991. Effects of antiflammins on endotoxin-induced uveitis in rats. Arch. Ophthalmol. **109**(2): 278–281.
9. CHAN, C.C., M. NI, L. MIELE et al. Antiflammins: inhibition of endotoxin-induced uveitis in Lewis rats. In Ocular Immunology Today, pp. 467–470. Elsevier. Amsterdam/New York.
10. LI, Q., D. LUYO, D.M. MATTESON & C.C. CHAN. 1998. Suppressive effect of antiflammin-2 on compound 48/80–induced conjunctivitis: role of phospholipase A2s and inducible nitric oxide synthase. Ocul. Immunol. Inflamm. **6**(2): 65–73.
11. AYO, C. 1943. A toxic ocular reaction. I. New property of Schwartzman toxins. J. Immunol. **46**: 113–132.
12. ROSENBAUM, J.T., H.O. MCDEVITT, R.B. GUSS & P.R. EGBERT. 1980. Endotoxin-induced uveitis in rats as a model for human disease. Nature **286**(5773): 611–613.
13. DE VOS, A.F., V.N. KLAREN & A. KIJLSTRA. 1994. Expression of multiple cytokines and IL-1RA in the uvea and retina during endotoxin-induced uveitis in the rat. Invest. Ophthalmol. Visual Sci. **35**(11): 3873–3883.
14. HERBORT, C.P., A. OKUMURA & M. MOCHIZUKI. 1988. Endotoxin-induced uveitis in the rat: a study of the role of inflammation mediators. Graefe's Arch. Clin. Exp. Ophthalmol. **226**(6): 553–558.
15. FOSTER, C.S. 1995. The pathophysiology of ocular allergy: current thinking. Allergy **50**(21): 6–9; 34–38 (discussion).
16. ANDERSON, D.F., J.D. MACLEOD, S.M. BADDELEY et al. 1997. Seasonal allergic conjunctivitis is accompanied by increased mast cell numbers in the absence of leukocyte infiltration. Clin. Exp. Allergy **27**(9): 1060–1066.
17. LI, Q., D. LUYO, N. HIKITA et al. 1996. Compound 48/80–induced conjunctivitis in the mouse: kinetics, susceptibility, and mechanism. Int. Arch. Allergy Immunol. **109**(3): 277–285.
18. SIERRA-HONIGMANN, M.R. & P.A. MURPHY. 1992. Suppression of interleukin-1 action by phospholipase-A2 inhibitors in helper T lymphocytes. Pept. Res. **5**(5): 258–261.
19. MAY, L.T., U. SANTHANAM, S.B. TATTER et al. 1988. Phosphorylation of secreted forms of human beta 2–interferon/hepatocyte stimulating factor/interleukin-6. Biochem. Biophys. Res. Commun. **152**(3): 1144–1150.
20. HIERHOLZER, C., J.C. KALFF, L. OMERT et al. 1998. Interleukin-6 production in hemorrhagic shock is accompanied by neutrophil recruitment and lung injury. Am. J. Physiol. **275**: L611–L621.
21. SAWA, Y., H. ICHIKAWA, K. KAGISAKI et al. 1998. Interleukin-6 derived from hypoxic myocytes promotes neutrophil-mediated reperfusion injury in myocardium. J. Thorac. Cardiovasc. Surg. **116**(3): 511–517.
22. SHEN, D.F., M.A. CHANG, D.M. MATTESON et al. 2000. Biphasic ocular inflammatory response to endotoxin-induced uveitis (EIU) in the mouse. Arch. Ophthalmol. **118**(4): 521–527.

Antiflammin Peptides in the Regulation of Inflammatory Response

JUAN J. MORENO[a]

Department of Physiology, Faculty of Pharmacy, Barcelona University, E-08028 Barcelona, Spain

> ABSTRACT: This review focuses on the role of antiflammins in the regulation of the inflammatory response, in particular acute inflammation. The results show that antiflammins were effective on several classical models of inflammation. Preliminary data suggest that antiflammin action may be due to their ability to suppress leukocyte trafficking to the lesion.

INTRODUCTION

Glucocorticosteroids (GS) are a classic treatment for inflammatory disorders. However, they induce numerous adverse side effects resulting from inadequate selectivity of action. These hormones/drugs are known to produce several biological effects by modifying gene expression. As much as 1% of the tools genome may be altered by GS in their target cells, resulting in changes in the expression of large numbers of enzymes and other proteins. Several studies indicate that their anti-inflammatory action is mediated, at least in part, by the induction of regulatory proteins such as lipocortins and uteroglobins. Thus, in the late 1970s, an important step in understanding the molecular mechanism of action of GS was the identification of a protein involved in the glucocorticoid-induced inhibition of the release of eicosanoids.[1] Lipocortin-1 (LC-1) was another protein induced by GS.[2] It was recognized that the gene for LC-1 contains a number of control factors, including at least one GS response element.[3]

Several biological roles have been attributed to lipocortins. However, most members of the lipocortin family are known to possess antiphospholipase A_2 (PLA_2) activity. This action had been described on the 14-kDa secretory PLA_2 ($sPLA_2$) and the proposed mechanism accepted was substrate depletion.[4] The discovery of cytosolic PLA_2 ($cPLA_2$) and our understanding of its sequence and regulation have permitted a new approach to the lipocortin effect. Using pure $cPLA_2$, Kim *et al.*[5] showed that this enzyme was inhibited through a specific interaction by LC-1. Moreover, recombinant human LC-1 mimics a variety of anti-inflammatory properties of GS both *in vivo* and *in vitro*.[6] Some biological effects of LC-1 related with its anti-inflammatory action are summarized in TABLE 1.

---

[a]Address for correspondence: Department of Physiology, Faculty of Pharmacy, Barcelona University, Avda. Joan XXIII s/n, E-08028 Barcelona, Spain. Voice: 3493 4024505; fax: 3493 4021896.

moreno@farmacia.far.ub.es

TABLE 1. Some anti-inflammatory effects of lipocortin-1

Parameter studied	Experimental model	Reference
Edema	Carrageenan paw edema	7
Neutrophil influx	Zymosan pleurisy	8
	Air-pouch	9
	Gel granuloma	10
Monocyte influx	Zymosan pleurisy	8
TNF/PGE$_2$ release	LPS/IL-1/mononuclear cells	11
NOS levels/activity	Lung exposure to LPS	12
NOS activity	Synovial macrophages	13
Adhesion neutrophils	Intravital microscopic studies	14

NOTE: TNF = tumor necrosis factor; PGE$_2$ = prostaglandin E$_2$; LPS = lipopolysaccharides; IL-1 = interleukin-1; NOS = nitric oxide synthase.

Rabbit blastokinin[15] or uteroglobin (UG)[16] are other GS-dependent, multifunctional, cytokine-like proteins with potent anti-inflammatory activity.[17] UG was first discovered in the rabbit uterus during early pregnancy, but was subsequently found in many other organs, such as thymus, pituitary gland, respiratory and gastrointestinal tracts, pancreas, mammary gland, prostate, and seminal vesicle.[18]

UG inhibits low-molecular-weight group I and II sPLA$_2$, enzymes that could play critical roles in the production of proinflammatory lipid mediators.[19,20] This effect was attributed to the fact that UG sequesters calcium ions, an essential cofactor for sPLA$_2$ activity.[21] However, Mukherjee et al.[22] observed that mutations in a critical region of the UG molecule can abrogate its PLA$_2$ inhibitory activity without affecting sequestration of calcium. Furthermore, UG is a potent inhibitor of neutrophil and monocyte chemotaxis.[23] Its biological activities directly related with its anti-inflammatory action are summarized in TABLE 2.

ANTIFLAMMINS

Lipocortins constitute a family of at least 13 structurally related cytoplasmic proteins widely distributed in mammals. All members have two different regions: the

TABLE 2. Some anti-inflammatory effects of uteroglobin

Parameter studied	Experimental model	Reference
Platelet aggregation	Thrombin-induced aggregation	24
PLA$_2$ activity	Pancreatic PLA$_2$/RAW 264.7	19
Phagocyte chemotaxis	Chemotaxis induced by formyl peptides	23
Endometrial PGE$_2$ levels	Rabbits treated with progesterone	25

NOTE: PLA$_2$ = phospholipase A$_2$; PGE$_2$ = prostaglandin E$_2$.

N-terminal domain, constituted by 32 amino acids of LC-1, and the C-terminal domain, named the core, which is composed of four repeats of a highly conserved 70- to 80-amino-acid sequence.[26,27]

UG is a homodimeric protein in which the 70 amino acid subunits, in an antiparallel orientation, are connected by two disulfide bonds.[28,29] The dimer structure has three cavities: cavity C1 could accommodate small molecules such as progesterone or retinol; cavities C2 and C3 are located within each monomer and are formed by the α-helices 1, 2, and 3.[29]

The identification of small molecules that may mimic the effect of these proteins is an attractive and practical approach to develop a molecule with the biological effects of LC-1/UG.

On the basis of computer analysis, Miele et al.[30] designed several synthetic peptides corresponding to the region of highest similarity between UG and LC-1: the nonapeptides corresponding to UG residues 39–47 (MQMKKVLDS) and LC-1 residues 246–254 (HDMNKVLDL). Both peptides inhibited pancreatic PLA_2 in vitro and were effective on carrageenan-induced rat footpad edema. From these results, these peptides were named antiflammins (AFs): AF-1 for UG-derived peptide and AF-2 for LC-1-derived peptide.

ANTIFLAMMINS IN THE REGULATION OF INFLAMMATORY RESPONSE

Vostal et al.[31] observed that AFs inhibited platelet aggregation induced by thrombin and ADP, and they also inhibited PLA_2 from human polymorphonuclear leukocytes and the synthesis of platelet-activated factor (PAF).[32] However, the ability of AFs to inhibit PLA_2 and their anti-inflammatory activity were questioned. Thus, several authors observed that AFs do not inhibit porcine pancreatic PLA_2 activity in vitro or carrageenan-induced rat paw edema.[33–35] In an attempt to clarify these preliminary results and the later controversy, we studied the effects of AFs on the inflammatory process. It is important to consider that AFs were stored below 0°C in anhydrous conditions and dissolved in ice-cold buffer immediately before use. Solutions were never stored and unused portions were discarded. AFs are unstable and are also degradable in acidic conditions.[36,37] Furthermore, Camussi et al.[32] observed that frozen AFs are readily inactivated by the oxidation of methionine residues.

Our first experiments were performed with phospholipid-deoxycholate mixed micelles and E. coli biomembranes as substrate, and porcine pancreatic PLA_2, N. naja naja PLA_2, and human synovial fluid PLA_2. Our experimental conditions avoided the aggregation of AFs and their oxidation. In these conditions, no significant inhibitory effect on these PLA_2's was observed.[38] In contrast, with these results in vitro, subplantar administration of AFs inhibited carrageenan-induced rat paw edema during the early and late phases of the process.[38,39] These results are in agreement with Miele et al.,[30] but only partially in accordance with Ialenti et al.,[40] who found that AF-2 was ineffective in the early phase (0–1 h), which is developed mainly by histamine and serotonin release. However, when experimental inflammation was induced directly by PLA_2 (human synovial fluid PLA_2 or N. naja naja PLA_2) injection, AFs did not display any significant anti-inflammatory effect, indicating that the anti-inflammatory effects of AFs do not involve a direct interaction between

AFs and these PLA_2's. Furthermore, AFs were effective in other models of acute inflammation induced by different agents. Thus, we reported that AFs were able to inhibit ear edema induced by croton oil or oxazolone in presensitized mice.[38]

Subsequent studies were performed to clarify the mechanism of the antiinflammatory activity of AFs. AFs inhibited platelet aggregation and thromboxane B_2 (TxB_2) synthesis induced by collagen, but did not affect the aggregation and thromboxane production stimulated by arachidonic acid (AA).[41] Similar results were obtained when AFs were used to treat ear edema induced by AA or croton oil. Nonapeptides significantly reduced inflammation induced by croton oil.[41] These results suggested that AFs have no significant effect on the cyclooxygenase pathway when AA is added exogenously, whereas they could inhibit PLA_2 activation and the subsequent AA release, an essential step in collagen-induced platelet aggregation or croton oil ear edema.

Acute inflammatory reactions are characterized by changes in vascular permeability and vasodilatation, resulting in edema. We examined whether AFs inhibit the increase in vascular permeability induced by histamine, bradykinin, PAF, and C5a *in vivo*. AFs had no significant effect on the action of these autacoids. Thus, a direct antihistaminic effect on the early phase of carrageenan-induced paw edema seems unlikely, although an indirect action of AFs through inhibition of mast cell degranulation should be considered.[42]

The migration and accumulation of neutrophils and mononuclear cells is another characteristic feature of inflammation. AA application on skin produced a short-lived edema response with rapid onset associated with marked increases in prostaglandin E_2 (PGE_2) synthesis and minimal cellular influx, whereas 12-O-tetradecanoylphorbol 13-acetate (TPA) produced a longer-lasting edema associated with marked influx of neutrophils and mononuclear cells, as well as predominant formation of leukotriene B_4 (LTB_4). Thus, AA-induced ear edema appeared to be dependent on prostaglandins, whereas TPA-induced edema could be dependent on the cell influx and the subsequent LTB_4 release by these cells. AF-2 dose-dependently reduced plasma leakage, cell influx, edema, and LTB_4 levels in response to TPA, but had no effect when inflammation was induced by AA.[43] These results also suggest that AFs do not modify the cyclooxygenase pathway, which is in contradiction with Calderano *et al.*,[44] who observed that AF-2 inhibits cyclooxygenase activity in rabbit distal colonic mucosa. On the other hand, these findings suggest that the antiedematous effect and the inhibitory effect on cell influx and eicosanoid production of AF-2 could be related to an inhibitory action of nonapeptides on AA mobilization and/or AA metabolism by lipoxygenases in the TPA model. However, we must consider an alternative mechanism to explain the anti-inflammatory effect of AF-2, which may involve an antichemotactic effect.

Probably, the best way of assessing whether a substance is a PLA_2 inhibitor *in vivo* is to determine the effect on AA release in whole cells. Recently, we demonstrated that AF-2 does not significantly reduce AA release stimulated by TPA or calcium ionophore A23187 in murine 3T6 fibroblasts or murine resident peritoneal macrophages, whereas dexamethasone was effective.[42] However, AFs inhibited AA release and prostaglandin production induced by thrombin in endothelial cells (personal observation). Recent advances in the understanding of PLA_2 have revealed that, in general, several PLA_2's are involved in cellular regulation and lipid messen-

TABLE 3. Some anti-inflammatory effects of antiflammins

Parameter studied	Experimental model	Reference
Edema	Carrageenan rat paw edema	30, 41
Edema	TPA ear edema	43
Edema	Contact hypersensitivity	38
Cell influx	TPA ear edema	43
PGE_2/LTB_4 levels	TPA ear edema	43
Platelet aggregation	Collagen-induced platelet aggregation	41
TxB_2 release	Collagen-induced platelet aggregation	41
Histamine release	Mast cell degranulation	39
Erythema	Skin irritation	46

NOTE: TPA = 12-O-tetra-decanoylphorbol 13-acetate; PGE_2 = prostaglandin E_2; LTB_4 = leukotriene B_4; TxB_2 = thromboxane B_2.

ger formation. The importance of each PLA_2 enzyme depends on the cell type or stimuli used. This could explain these apparent contradictory results.

The accumulation of neutrophils and monocyte/macrophages is a characteristic feature of the inflammatory response, as mentioned above. This leukocyte extravasation into inflamed areas involves a complex interaction of leukocytes with the endothelium through regulated expression of surface adhesion molecules. Recently, we reported that sialidase treatment, which affects the structure of selectins and inhibits leukocyte influx, significantly reduced eicosanoid and edema in TPA-induced ear edema.[45] Our results showed that AF-2 significantly reduced leukocyte adhesion to the endothelium stimulated by N-formyl-Met-Leu-Phe.[42] These data were correlated with the inhibitory effect of AFs on cell influx and the development of the inflammatory process. Some effects of AFs related with their anti-inflammatory action are shown in TABLE 3.

Considering these findings, we proposed that the effects of AFs on the inflammatory process may be due to their ability to suppress leukocyte trafficking to the lesion. However, additional experiments should be performed to clarify whether antiflammins interfere with the expression or activation/affinity of the adhesion molecules.

REFERENCES

1. BLACKWELL, G.J., R. CARNUCCIO, M. DI ROSA *et al.* 1980. Macrocortin: a polypeptide causing the anti-phospholipase effect of glucocorticoids. Nature **287:** 147–149.
2. FLOWER, R.J. 1988. Lipocortin and the mechanism of action of glucocorticoids. Br. J. Pharmacol. **94:** 987–1015.
3. BROWNING, J.L., M.P. WARD, B.P. WALLNER & R.B. PEPINSKY. 1990. Studies on the structural properties of lipocortin-1 and the regulation of its synthesis by steroids. *In* Cytokines and Lipocortins in Inflammation and Differentiation, pp. 27–45. Wiley-Liss. New York.
4. DAVIDSON, F.F., E.A. DENNIS, M. POWELL & J.R. GLENNEY. 1987. Inhibition of phospholipase A_2 by lipocortins and calpectins: an effect of binding to substrate phospholipids. J. Biol. Chem. **262:** 1698–1705.

5. KIM, K.M., D.K. KIM, Y.M. PARK et al. 1994. Annexin-I inhibits phospholipase A_2 by specific interactions, not by substrate depletion. FEBS Lett. **343:** 251–255.
6. FLOWER, R.J. & N.J. ROTHWELL. 1994. Lipocortin-1: cellular mechanisms and clinical relevance. Trends Pharmacol. Sci. **15:** 71–76.
7. CIRINO, G., S.H. PEERS, R.J. FLOWER et al. 1989. Human recombinant lipocortin 1 has acute local anti-inflammatory properties in the rat paw edema test. Proc. Natl. Acad. Sci. U.S.A. **86:** 3428–3432.
8. BECHERUCCI, C., M. PERRETTI, E. SOLITO et al. 1993. Conceivable difference in the antiinflammatory mechanisms of lipocortin 1 and 5. Mediators Inflammation **2:** 109–113.
9. PERRETTI, M. & R.J. FLOWER. 1993. Modulation of interleukin-1 induced neutrophil migration by dexamethasone and lipocortin-1. J. Immunol. **150:** 992–999.
10. ERRASFA, M. & F. RUSSO-MARIE. 1989. A purified lipocortin shares the anti-inflammatory effect of glucocorticoids *in vivo* in mice. Br. J. Pharmacol. **97:** 1051–1080.
11. SUDLOW, A.W., F. CAREY, R. FORDER & N.J. ROTHWELL. 1996. The role of lipocortin-1 in dexamethasone-induced suppression of PGE_2 and TNF alpha release from human peripheral blood mononuclear cells. Br. J. Pharmacol. **117:** 1449–1455.
12. BRYANT, C.E., M. PERRETTI & R.J. FLOWER. 1998. Suppression by dexamethasone of inducible nitric oxide synthase protein expression *in vivo*: a possible role for lipocortin-1. Biochem. Pharmacol. **55:** 279–285.
13. YANG, Y.H., P. HUTCHINSON, L.L. SANTOS & E.F. MORAND. 1998. Glucocorticoid inhibition of adjuvant arthritis synovial macrophage nitric oxide production: role of lipocortin-1. Clin. Exp. Immunol. **111:** 117–122.
14. LIM, L.H.K., E. SOLITO, F. RUSSO-MARIE et al. 1998. Promoting detachment of neutrophils adherent to murine postcapillary venules to control inflammation: effect of lipocortin-1. Proc. Natl. Acad. Sci. U.S.A. **95:** 14535–14539.
15. KRISHNAN, R.S. & J.C. DANIEL. 1967. Blastokinin inducer and regulator of blastocyst development in the rabbit uterus. Science **158:** 490–492.
16. BEIER, H.M. 1968. Uteroglobin: a hormone-sensitive endometrial protein involved in blastocyst development. Biochim. Biophys. Acta **160:** 289–291.
17. MIELE, L., E. CORDELLA-MIELE, G. MANTILE et al. 1994. Uteroglobin and uteroglobin-like proteins: the uteroglobin family of proteins. J. Endocrinol. Invest. **8:** 679–692.
18. PERI, A., E. CORDELLA-MIELE, L. MIELE & A.B. MUKHERJEE. 1993. Tissue-specific expression of the gene coding for human Clara cell 10 Kd protein, a phospholipase A_2–inhibitory protein. J. Clin. Invest. **92:** 2099–2109.
19. LEVIN, S.W., J.D. BUTLER, V.K. SCHUMACHER et al. 1986. Uteroglobin inhibits phospholipase A_2 activity. Life Sci. **38:** 1813–1819.
20. MUKHERJEE, A.B., L. MIELE & N. PATTABIRAMAN. 1994. Phospholipase A_2 enzymes: regulation and physiological role. Biochem. Pharmacol. **48:** 1–10.
21. ANDERSON, O., L. NORDLUND-MULLER, H.J. BARNES & J. LUND. 1994. Heterologous expression of human uteroglobin polychlorinated biphenyl–binding protein: determination of ligand binding parameters and mechanisms of phospholipase A_2 inhibition *in vitro*. J. Biol. Chem. **267:** 19081–19087.
22. MUKHERJEE, A.B., G.L. KUNDU, G. MANTILE-SELVAGGI et al. 1999. Uteroglobin: a novel cytokine? CMLS Cell. Mol. Life Sci. **55:** 771–787.
23. VASANTHAKUMAR, G., R. MANJUNATH, A.B. MUKHERJEE et al. 1988. Inhibition of phagocyte chemotaxis by uteroglobin, an inhibitor of blastocyst rejection. Biochem. Pharmacol. **37:** 389–394.
24. MANJUNATH, R., S.W. LEVIN, K.K. KUMAROO et al. 1987. Inhibition of thrombin-induced platelet aggregation by uteroglobin. Biochem. Pharmacol. **36:** 741–746.
25. KIKUKAWA, T., B.D. COWAN, R.I. TEJADA & A.B. MUKHERJEE. 1988. Partial characterization of a uteroglobin-like protein on the human uterus and its temporal relationship to prostaglandin levels in this organ. J. Clin. Endocrinol. Metab. **67:** 315–321.
26. PEPINSKY, R.B., R. TIZARD, R.J. MATTALIANO et al. 1988. Five distinct calcium and phospholipid binding proteins share homology with lipocortin 1. J. Biol. Chem. **263:** 10799–10811.
27. GEISOW, M.J., V. FRITSCHE, J.M. HEXHAM et al. 1986. A consensus amino-acid sequence repeat in Torpedo and mammalian Ca^{2+} dependent membrane-binding proteins. Nature **320:** 636–638.

28. PONSTINGL, H., A. NIETO & M. BEATO. 1978. Amino acid sequence of progesterone-induced rabbit uteroglobin. Biochemistry **17:** 3908–3912.
29. NIETO, A., H. PONSTINGL & M. BEATO. 1997. Purification and quaternary structure of the hormonally induced protein uteroglobin. Arch. Biochem. Biophys. **180:** 82–92.
30. MIELE, L., E. CORDELLA-MIELE, A. FACCHIANO & A.B. MUKHERJEE. 1988. Novel antiinflammatory peptides from the region of highest similarity between uteroglobin and lipocortin-1. Nature **335:** 726–730.
31. VOSTAL, J.G., A.B. MUKHERJEE, L. MIELE & N.R. SHULMAN. 1989. Novel peptides derived from a region of local homology between uteroglobin and lipocortin-1 inhibit platelet aggregation and secretion. Biochem. Biophys. Res. Commun. **165:** 27–36.
32. CAMUSSI, G., C. TETTA, F. BUSSOLINO & C. BAGLIONI. 1990. Anti-inflammatory peptides (antiflammins) inhibit synthesis of platelet-activating factor, neutrophil aggregation and chemotaxis, and intradermal inflammatory reactions. J. Exp. Med. **171:** 913–927.
33. MARKI, F., J. PFEILSCHIFTER, H. RINK & I. WIESENBERG. 1990. Antiflammins: two nonapeptide fragments of uteroglobin and lipocortin-1 have no phospholipase A_2-inhibitory and anti-inflammatory activity. FEBS Lett. **264:** 171–175.
34. HOPE, W.C., B.J. PATEL & D.R. BOLIN. 1991. Antiflammin 2 (HDMNKVLDL) does not inhibit phospholipase A_2 activities. Agents Actions **34:** 77–80.
35. MARASTONI, M., S. SALVADONI, G. BALDONI et al. 1991. Studies on the anti-phospholipase A_2 and anti-inflammatory activities of a synthesis nonapeptide from uteroglobin. Arzneim. Forsch. **41:** 240–243.
36. YE, J.M., G.E. LEE, G.K. POTTI et al. 1996. Degradation of antiflammin-2 under acidic conditions. J. Pharm. Sci. **85:** 695–699.
37. YE, J.M. & J.L. WOLFE. 1996. Oxidative degradation of antiflammin-2. Pharm. Res. **13:** 250–255.
38. CABRE, F., J.J. MORENO, A. CARABAZA et al. 1992. Antiflammins: anti-inflammatory activity and effect on human phospholipase A_2. Biochem. Pharmacol. **44:** 519–525.
39. LLORET, S. & J.J. MORENO. 1994. Effect of nonapeptide fragments of uteroglobin and lipocortin-1 on oedema and mast cell degranulation. Eur. J. Pharmacol. **264:** 379–384.
40. IALENTI, A., P.M. DOYLE, G.N. HARDY et al. 1990. Anti-inflammatory effects of vasocortin and nonapeptide fragments of uteroglobin and lipocortin 1 (antiflammins). Agents Actions **29:** 48–49.
41. LLORET, S. & J.J. MORENO. 1992. In vitro and in vivo effects of the anti-inflammatory peptides, antiflammins. Biochem. Pharmacol. **44:** 1437–1441.
42. MORENO, J.J. 1996. Antiflammin-2, a nonapeptide of lipocortin-1, inhibits leukocyte chemotaxis, but not arachidonic acid mobilization. Eur. J. Pharmacol. **314:** 129–135.
43. LLORET, S. & J.J. MORENO. 1995. Effect of an anti-inflammatory peptide (antiflammin-2) on cell influx, eicosanoid biosynthesis, and edema formation by arachidonic acid and tetradecanoyl phorbol dermal application. Biochem. Pharmacol. **50:** 347–353.
44. CALDERANO, V., C. PARRILLO, A. GIOVANE et al. 1992. Antiflammins suppress the A23187- and arachidonic acid–dependent chloride secretion in rabbit distal colonic mucosa. J. Pharmacol. Exp. Ther. **263:** 579–587.
45. SANCHEZ, T. & J.J. MORENO. 1999. Role of leukocyte influx in tissue prostaglandin H synthase-2 overexpression induced by phorbol ester and arachidonic acid in skin. Biochem. Pharmacol. **58:** 877–879.
46. MIZE, N.K., M. BUTTERY, N. RUIS et al. 1997. Antiflammin 1 peptide delivered noninvasively by iontophoresis reduces irritant-induced inflammation in vivo. Exp. Dermatol. **6:** 181–185.

Regulation of the Clara Cell Secretory Protein/Uteroglobin Promoter in Lung

MAGNUS NORD,[a,b] TOBIAS N. CASSEL,[b] HARALD BRAUN,[c] AND GUNTRAM SUSKE[c]

[b]*Department of Medical Nutrition, NOVUM, Karolinska Institute, Huddinge, Sweden*

[c]*Institut für Molekularbiologie und Tumorforschung, Philipps-Universität Marburg, Marburg, Germany*

ABSTRACT: Clara cell secretory protein/uteroglobin (CCSP/UG) is specifically expressed in the conducting airway epithelium of the lung in a differentiation-dependent manner. The proximal promoter region of the rodent CCSP/UG gene directs Clara cell specificity. Previously, it was shown that the forkhead transcription factors HNF-3α and β and the homeodomain factor TTF-1 are important transcription factors acting through this region, suggesting that they contribute to cell specificity of the CCSP/UG gene. Members of the C/EBP family of transcription factors can also interact with elements of the proximal rat and mouse CCSP/UG promoters. The onset of C/EBPα expression in Clara cells correlates with the strong increase of CCSP/UG expression. Thus, C/EBPα may play a crucial role for differentiation-dependent CCSP/UG expression. Transfection studies demonstrate that C/EBPα and TTF-1 can synergistically activate the murine CCSP/UG promoter. Altogether, these results suggest that C/EBPα, TTF-1, and HNF-3 determine the Clara cell–specific, differentiation-dependent expression of the CCSP/UG gene in murine lung. The relative importance of these three transcription factors, however, differs in rabbits and humans.

INTRODUCTION

Regional differentiation of the lung epithelium results in the functionally distinct conducting and respiratory portions of the lung, with specialized cell types serving the different subfunctions of the respiratory system.[1] The bronchioles constitute the most distal part of the conducting airways. Here, the predominant epithelial cells are ciliated cells and nonciliated Clara cells. Clara cells represent a well-differentiated cell type with a high secretory activity. Numerous proteins are secreted from these cells, with the major secretory product being the Clara cell secretory protein or uteroglobin (CCSP/UG).[2] In normal lung, the CCSP/UG gene is specifically expressed at high levels in the bronchiolar Clara cells, as well as in similar cells in bronchi and trachea.[3,4] CCSP/UG has been estimated to account for 40% of proteins secreted from rabbit Clara cells[5] and constitutes 2–12% of human bronchoalveolar lavage fluid proteins.[6,7]

[a]Address for correspondence: Magnus Nord, Department of Medical Nutrition, Karolinska Institute, NOVUM, Huddinge University Hospital, SE-141 86 Huddinge, Sweden. Voice: +46 8 5858 37 25; fax: +46 8 711 66 59.

magnus.nord@mednut.ki.se

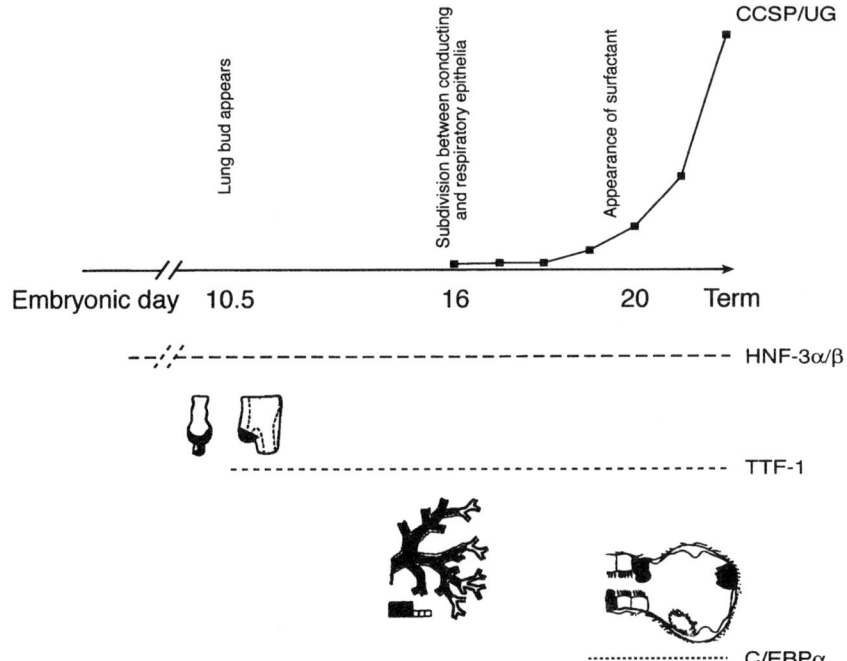

FIGURE 1. Expression of CCSP/UG mRNA in the developing rat lung. CCSP/UG expression is shown in relation to major events in lung development and to the expression pattern of transcription factors known to activate the CCSP/UG promoter (HNF-3α/β, TTF-1, and C/EBPα). The drawing is based on references 9–11, 35, 50, 55, and 72.

Several lines of evidence suggest that the expression of CCSP/UG is tightly coupled to a high degree of differentiation of the bronchiolar Clara cells. During development, low-level CCSP/UG gene expression is first detected at the time of the first distinction between the conducting and respiratory portions of the lung (FIG. 1). Expression is confined to the prospective conducting airway epithelium and thus correlates both spatially and temporally with the subdivision between conducting and respiratory epithelia.[8–11] Expression levels do not increase until later during embryonic development, in rat lung at day 19, corresponding to the initiation of Clara cell differentiation (see FIG. 1). CCSP/UG levels continue to increase as the Clara cells differentiate postnatally.[9,12–15] Studies on prematurely born human infants show a marked increase in CCSP/UG levels in the last weeks of gestation,[16] and CCSP/UG in amniotic fluid has been shown to reflect fetal lung growth.[17] Studies in mice have shown that, after injury of Clara cells by the selective cytotoxicant naphthalene, expression of CCSP/UG rapidly disappears. It does not reappear until the repopulating cells assume a mature differentiated phenotype.[18,19] Examination of lung tissues from patients with non–small cell lung cancer has demonstrated a loss of expression of CCSP/UG in airway regions undergoing changes of cell type.[20] Finally, in primary cultures of isolated rat Clara cells, CCSP/UG expression is rapidly lost as the cells dedifferentiate *in vitro*.[21]

In this review, we will focus on transcription factors that determine the Clara cell–specific, differentiation-dependent expression of CCSP/UG in lung. We will not discuss the influence of hormones and other signaling factors as this already has been addressed in several previous reviews[22,23] (also see articles in this volume).

REGULATION OF LUNG AND CLARA CELL–SPECIFIC EXPRESSION OF CCSP/UG

Studies in transgenic mice have demonstrated that *cis*-acting elements directing lung tissue and Clara cell–specific expression of CCSP/UG reside within 2.3 kb of the 5'-flanking region of the rat gene.[24,25] Experiments utilizing the Clara cell–like human lung adenocarcinoma cell line NCI-H441 as a model have demonstrated that *cis*-acting elements within the first 175 bp of the rat CCSP/UG promoter can confer cell specificity.[24] An *in vivo* 5'-deletion analysis of the murine CCSP/UG gene in transgenic mice revealed that 166 bp of the mouse promoter is sufficient for directing Clara cell–specific expression.[26] Within this region of the rodent promoters, cell and lung specific interactions of nuclear proteins occur.[24,27–29] Two binding sites for HNF-3/forkhead, or winged helix, transcription factors contribute to the activity of this promoter region in lung epithelial cells[30,31] (see FIG. 2). In addition, binding sites for the lung-enriched homeodomain transcription factor TTF-1 are present within this region[32–34] (FIG. 2). These studies suggest that factors from the HNF-3/forkhead transcription factor family, along with TTF-1, are crucial for the cell-specific expression of CCSP/UG in lung.

The HNF-3/forkhead transcription factor, hepatocyte nuclear factor-3α (HNF-3α), is first expressed in the murine developing endoderm from the mid- to late-primitive streak stage of gastrulation (around 8 days postcoitum). Subsequently, it is found in all endoderm-derived structures, including lung[35,36] (FIG. 1). In the adult lung, HNF-3α expression is confined to the conducting airway epithelium.[28,37] This suggests that HNF-3α could be a determinant of bronchiolar cell-specific expression of CCSP/UG. In accordance, HNF-3α activates the CCSP/UG promoter through two HNF-3/forkhead sites in the proximal promoter[27,28,30,31] (FIG. 2). Two additional HNF-3 factors exist, HNF-3β and HNF-3γ, of which HNF-3β is also expressed in lung. HNF-3β expression is turned on already at the onset of gastrulation (FIG. 1). Subsequently, it is also expressed in all endoderm-derived structures.[35,36] This has lead to the proposition that HNF-3α and β are involved in the specification of the endoderm. This hypothesis is supported by the results of HNF-3β gene targeting in mice. In HNF-3β(–/–) embryos, gut morphogenesis is severely affected, and orga-

FIGURE 2. Transcription factors acting through the proximal murine CCSP/UG promoter region. Positions of transcription factor binding sites in the murine CCSP/UG promoter (numbering is based on reference 29).

nized node and notochord formation is absent.[38,39] In adult lung, immunohistochemical studies have revealed expression of HNF-3β in bronchiolar cells, including Clara cells, as well as in alveolar type II cells.[40] *In situ* hybridization studies for HNF-3β mRNA, however, have produced conflicting results regarding the expression in alveolar and bronchiolar cells.[28,37] Similar to HNF-3α, HNF-3β regulates CCSP/UG expression through two HNF-3/forkhead sites in the proximal CCSP/UG promoter.[27,28,30,41]

Several other lung-enriched HNF-3/forkhead transcription factors exist, most notably the HNF-3/forkhead homologues (HFH) 4 and 8. HFH-4 is expressed in the bronchiolar epithelium. Initial studies in liver cell lines showed that HFH-4 could activate the CCSP/UG promoter.[42] However, more recent studies in other cell models have failed to demonstrate activation by HFH-4 (reference 41 and unpublished observations). During development, the onset of HFH-4 expression spatially and temporally correlates with the distinction between the prospective conducting and respiratory epithelia[43] (FIG. 1). However, HFH-4 is expressed in ciliated cells only, and knockout and transgenic experiments have uncovered a key role for this factor in ciliogenesis.[44–47] The human homologue of HFH-8, FREAC-1, activates the CCSP/UG promoter in lung epithelial cell lines.[48] However, in adult mice, HFH-8 expression is confined to the endothelium and connective fibroblasts of the alveolar region.[49] Thus, it seems unlikely that HFH-4 and -8 are involved in conferring Clara cell specificity of the CCSP/UG gene. However, the possibility that another yet uncharacterized lung or Clara cell–enriched HNF-3/forkhead transcription factor is of importance for the cell-specific expression of CCSP/UG still has to be considered.

The homeodomain transcription factor TTF-1 (thyroid transcription factor 1, also known as NKx2.1 or T/EBP) is expressed in the prospective lung epithelium from the onset of lung development, that is, the appearance of the lung bud on the ventral foregut (at day 10.5 of rat development; see FIG. 1).[50] As development proceeds, expression of TTF-1 continues in the pulmonary epithelium. In the adult murine lung, both conducting and respiratory epithelial cells express TTF-1, excluding alveolar type I cells.[40] The occurrence of TTF-1, however, has been described in type I cell–like cell lines.[51] That TTF-1 plays a crucial role in lung development was demonstrated by gene targeting experiments. In TTF-1-deficient mice, the respiratory system is severely malformed, with virtually absent lungs.[52,53] Together, this suggests a key role for TTF-1 in specification of the lung epithelial cell lineage. The murine CCSP/UG promoter contains two TTF-1 binding sites within the 166-bp proximal promoter region that confers Clara cell–specific expression (see FIG. 2). In addition, one major and two minor TTF-1 regulatory sites are present in upstream promoter regions. Transgenic studies suggest that the upstream sites are important for high-level CCSP/UG expression *in vivo*.[26,32] In the rat CCSP/UG promoter, TTF-1 regulatory sites have been identified at positions corresponding to the two proximal sites and the major distal TTF-1 site.[33,34]

In summary, the data outlined above suggest that the transcription factors HNF-3 and TTF-1 are crucial for Clara cell–specific expression of the rodent CCSP/UG genes. Both transcription factors are also important for the expression of other lung-specific genes, such as the surfactant apoproteins (recently reviewed in, e.g., reference 54). Taken together with the results of gene targeting, this makes evident that HNF-3 and TTF-1 play a key role for the specification of the pulmonary epithelium.

REGULATION OF DIFFERENTIATION-DEPENDENT EXPRESSION OF CCSP/UG IN LUNG

It seems plausible that HNF-3 and TTF-1 regulate Clara cell–specific expression of CCSP/UG. However, TTF-1 expression in the developing lung is first detectable at the time that the lung bud forms, and HNF-3α and β are expressed from earlier developmental stages.[35,36,50] This is in contrast to the onset of CCSP/UG expression, which occurs later during development with levels increasing as Clara cells start to differentiate (FIG. 1). Certainly, other factors need to be involved in the regulation of CCSP/UG expression to account for its developmental timing.

The basic region–leucine zipper transcription factor C/EBPα is expressed in the developing rat lung. Expression is first detected between day 18 and 20,[55] in close correlation to the increase in CCSP/UG expression[9,12] (see FIG. 1). In the adult rat and mouse lung, C/EBPα is expressed at high levels in alveolar type II cells and at lower levels in the bronchiolar Clara cells.[21,55,56] As primary cultures of isolated rat Clara cells are grown *in vitro*, CCSP/UG expression is rapidly lost as the cells dedifferentiate. In contrast, HNF-3α and TTF-1 expression continues. C/EBPα levels, however, parallel the disappearance of CCSP/UG.[21] Taken together, this suggests that C/EBPα could be involved in controlling the differentiation-dependent expression of CCSP/UG. In this context, it should be noted that C/EBP factors, and especially C/EBPα, play an important role in controlling differentiation and differentiation-dependent processes in other tissues. These tissues include liver (which, similar to lung, is an endodermal organ of foregut origin), adipose tissue, and white blood cells. In liver and fat, C/EBPα is an important regulator of proliferation, cell cycle arrest, and gene expression (recently reviewed in references 57–61).

In accordance with a role for C/EBPα in the differentiation-dependent expression of CCSP/UG, C/EBPα activates the rat CCSP/UG gene in transient transfections of lung cell lines. Activation occurs through a C/EBP response element in the proximal promoter.[21] In mice, expression of C/EBPδ, another C/EBP family member, is highest in lung.[62] Here, C/EBPδ is expressed at high levels in bronchiolar epithelial cells, including Clara cells, and at lower levels in alveolar epithelial cells.[56] Studies in the developing rabbit lung have revealed that, like C/EBPα, C/EBPδ also exhibits a differentiation-dependent expression pattern.[63] In transfections, the murine CCSP/UG promoter is activated by both C/EBPα and δ. Activation is mediated via a response element residing in the proximal CCSP/UG promoter. The C/EBP response element is located 60 to 100 bp upstream of the transcriptional start site and consists of two C/EBP-binding sites (FIG. 2). The integrity of both sites is necessary for full function of the element, indicating that the two sites form a compound response element.[56] C/EBPδ is a more potent activator of the CCSP/UG promoter than C/EBPα. In accordance with this, C/EBPδ strongly interacts with both binding sites in the compound response element, whereas C/EBPα interaction with the proximal site is weaker. On both C/EBP-binding sites, heterodimers between C/EBPα and δ are preferentially formed. Cotransfection of C/EBPα and C/EBPδ together resulted in a more than additive induction as compared to the factors alone. This indicates a regulatory role for the C/EBPα–C/EBPδ heterodimers.[56]

The finding that C/EBPα and δ can activate the CCSP/UG promoter as well as the correlation between CCSP/UG expression and C/EBP factors during differentiation suggest that these C/EBP factors play an important role in differentiation-dependent processes in the bronchiolar epithelium of the lung. Such a hypothesis is further supported by findings showing that another Clara cell differentiation marker, the P450-enzyme CYP2B1, is also regulated by C/EBPα and δ in lung epithelial cells.[64] A more general role for C/EBPα in regulating pulmonary epithelial differentiation is suggested by histological examinations of lungs from C/EBPα(–/–) knockout mice. C/EBPα-deficient lungs demonstrate alveolar abnormalities with hyperproliferation of epithelial cells.[65] No abnormalities in the bronchiolar epithelium were reported in these mice. A compensation for C/EBPα deficiency by C/EBPδ may occur in bronchioles of C/EBPα knockout mice since expression of C/EBPδ is high in this part of the lung. In the alveolar epithelium, where C/EBPδ expression is lower, such compensation may not be possible.[56] That such compensations can occur has been demonstrated in the liver of C/EBPα(–/–) mice, where C/EBPβ can substitute for the loss of C/EBPα.[66]

DIFFERENCES IN PULMONARY REGULATION OF THE RODENT, RABBIT, AND HUMAN CCSP/UG GENES

It is clear that TTF-1, HNF-3, and C/EBP are important transcription factors that determine expression of the CCSP/UG gene in rodent lung. However, whether these transcription factors are also major players in other mammals seems questionable. The role of TTF-1 for expression of the CCSP/UG gene in rabbit and human lung is especially unclear. Studies of the rabbit CCSP/UG gene in transgenic mice show that 600 bp of the 5′-flanking sequence is sufficient to confer lung expression. Similar to what has been described for the rodent genes, additional sequences between 0.6 and 2.3 kb are necessary for full high-level pulmonary expression.[67] Transfection studies using the Clara cell–like NCI-H441 cell line as a model have demonstrated six regions containing *cis*-acting elements of importance for expression within the first 250 bp of the rabbit CCSP/UG promoter.[68] Two of these elements are regulated by the ubiquitous transcription factors, Sp1 and Sp3,[69] and two are HNF-3/forkhead regulatory sites.[41] In contrast to these HNF-3/forkhead-binding sites and the distal C/EBP-binding site, the TTF-1 sites in the rat and mouse CCSP/UG promoters are not conserved in the corresponding rabbit and human promoter regions. The rabbit CCSP/UG promoter contains recognition sites for TTF-1 at different positions that bind TTF-1 *in vitro* (unpublished data). In transfection experiments, however, neither the rabbit nor the human CCSP/UG promoters are activated by TTF-1.[41] Moreover, in immunohistochemical studies of human term lung, TTF-1 was detected primarily in alveolar type II cells and was only rarely seen in distal conducting airway cells.[70]

Both HNF-3α and HNF-3β can activate not only the rodent, but also the rabbit and human CCSP/UG promoters. However, binding of HNF-3α and β on their own is not sufficient for activation of these promoters. The action of both transcription factors is strongly dependent on the presence of the ubiquitous transcription factors Sp1 or Sp3.[41] Whether a similar combinatorial action of Sp1/Sp3 transcription factors and HNF-3α and HNF-3β occurs in rat and mouse is unclear. The rat gene contains Sp1/Sp3-binding sites in the proximal promoter region. However, in transfection experiments, Sp1/Sp3 factors do not seem to activate these promoters (reference 33 and unpublished observations). In summary, it appears that the mechanisms that determine Clara cell–specific expression of the CCSP/UG gene in lung are similar, but not identical in rodents, rabbits, and humans.

COOPERATIVE INTERACTIONS IN THE REGULATION OF THE CCSP/UG PROMOTER

As discussed above, HNF-3, TTF-1, and C/EBP control the Clara cell–specific, differentiation-dependent expression of the rodent CCSP/UG genes through interactions with promoter proximal elements (FIG. 2). It seemed as an attractive hypothesis that these factors synergistically act together to produce the high-level CCSP/UG expression characteristic for the differentiated Clara cell. To investigate this possibility, we performed cotransfection studies in *Drosophila* Schneider SL-2 cells. SL-2 cells lack many mammalian transcription factors and are thus well suited for cooperativity studies.[71] Transient transfection of a 172-bp murine CCSP/UG promoter fragment along with expression plasmids for HNF-3α, HNF-3β, TTF-1, and C/EBPα activated the CCSP/UG promoter by 1.5-, 1.9-, 3.7-, and 1.6-fold, respectively (see FIG. 3). When TTF-1 and C/EBPα expression plasmids were transfected together, a strong synergy resulting in a 40-fold induction of the reporter gene was seen (FIG. 3). No synergism was observed between C/EBPα and HNF-3 nor between HNF-3 and TTF-1. Synergistic activation by C/EBPα and TTF-1 was further increased by HNF-3α or β (FIG. 3). No synergism was observed between TTF-1 and C/EBPδ (unpublished results).

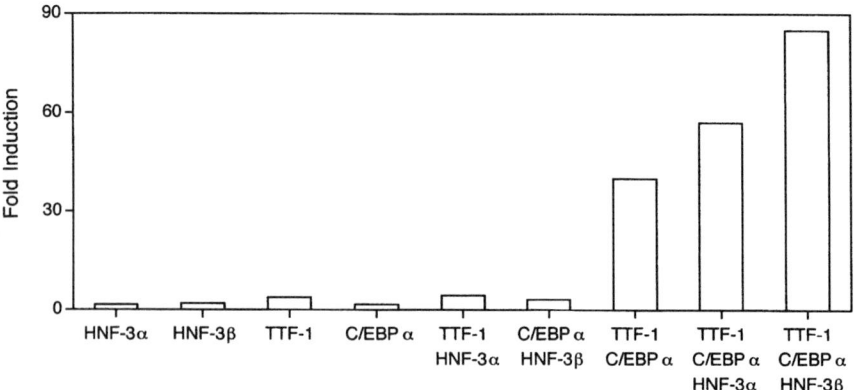

FIGURE 3. Activation of the CCSP/UG promoter by various transcription factors. *Drosophila* SL-2 cells were transfected with a 172-bp murine CCSP/UG-promoter-reporter gene construct (4 µg) along with expression plasmids for TTF-1 (2 µg), C/EBPα (2 µg), HNF-3α (20 ng), and HNF-3β (20 ng). The amount of DNA was kept constant by inclusion of empty pPac expression plasmid. Cells also received a β-galactosidase expression plasmid (2 µg) to allow normalization for variations in transfection efficiency. Assays were carried out in duplicates. Cell culture methods, transfections, and reporter gene assays have been described previously.[71]

Additional TTF-1 binding sites exist upstream of the 172-bp CCSP/UG promoter fragment.[32] When a 2.1-kb fragment containing these upstream TTF-1 sites was used instead, stronger activation by TTF-1 was observed. Synergistic activation by TTF-1 and C/EBPα, however, did not increase. This result indicates that the cooperative action of TTF-1 and C/EBPα is mediated via TTF-1-binding sites within the 172 most-proximal base pairs of the promoter (unpublished results). At this stage, the molecular mechanisms behind the synergistic activity of C/EBPα and TTF-1 and the additional effects of HNF-3 are unclear. Several possibilities exist, including cooperative binding of C/EBPα and TTF-1 mediated by protein-protein interactions and/or changes of the local DNA conformation. Another possibility would be cooperative recruitment of coactivators. In summary, these results show that C/EBPα acts in concert with TTF-1 and HNF-3 to regulate CCSP/UG gene expression. Our data suggest that C/EBPα expression is essential for high-level expression of CCSP/UG in differentiated Clara cells and that the onset of C/EBPα expression in the developing lung is crucial for the increase in CCSP/UG expression (FIG. 1). These findings also support a role for C/EBP factors in the control of lung epithelial differentiation.

CONCLUSIONS

In lung, the Clara cell secretory protein/uteroglobin is specifically expressed in Clara cells of the conducting airway epithelium in a differentiation-dependent manner. Studies during the last decade have advanced our understanding of the regulation of this gene and also started to uncover the molecular determinants of cell specification and differentiation in the developing lung. Collectively, these studies support a hypothesis for the regulation of the rodent CCSP/UG genes in which TTF-1 and HNF-3 act in concert with C/EBPα to give cell-specific and differentiation-dependent expression, respectively.

ACKNOWLEDGMENTS

This work was supported by the Swedish Medical Research Council (Grant No. 13115); the Swedish Heart-Lung Foundation (Grant No. 199941330); Swedish Match (Grant No. 199914); the Swedish Medical Society; the Research Foundations "Tore Nilssons stiftelse för medicinsk forskning", "Stiftelsen Lars Hiertas minne", "Sigurd och Elsa Goljes minne", "Magnus Bergvalls stiftelse", and "Robert Lundbergs stiftelse"; the Research Foundations of the Karolinska Institute (M. Nord and T. N. Cassel); and the BMBF and Deutsche Forschungsgemeinschaft (G. Suske).

REFERENCES

1. GAIL, D.B. & C.J.M. LENFANT. 1983. Cells of the lung: biology and clinical applications. Am. Rev. Respir. Dis. **127:** 366–387.
2. MASSARO, G.D., G. SINGH, R. MASON et al. 1994. Biology of the Clara cell: conference report. Am. J. Physiol. **266:** L101–L106.
3. BEDETTI, C.D., J. SINGH, G. SINGH et al. 1987. Ultrastructural localization of rat Clara cell 10 kD secretory protein by the immunogold technique using polyclonal and monoclonal antibodies. J. Histochem. Cytochem. **35:** 789–794.

4. LUND, J., T. DEVEREUX, H. GLAUMANN & J-Å. GUSTAFSSON. 1988. Cellular and subcellular localization of a binding protein for polychlorinated biphenyls in rat lung. Drug Metab. Dispos. **16:** 590–599.
5. PATTON, S.E., L.B. GILMORE, A.M. JETTEN *et al.* 1986. Biosynthesis and release of proteins by isolated pulmonary Clara cells. Exp. Lung Res. **11:** 277–294.
6. BERNARD, A., H. ROELS, R. LAUWERYS *et al.* 1992. Human urinary protein 1: evidence for identity with the Clara cell protein and occurrence in respiratory tract and urogenital secretions. Clin. Chim. Acta **207:** 239–249.
7. ANDERSSON, O., T.N. CASSEL, R. GRONNEBERG *et al.* 1999. *In vivo* modulation of glucocorticoid receptor mRNA by inhaled fluticasone propionate in bronchial mucosa and blood lymphocytes in subjects with mild asthma. J. Allergy Clin. Immunol. **103:** 595–600.
8. STRUM, J.M., R.S. COMPTON, S.L. KATYAL & G. SINGH. 1992. The regulated expression of mRNA for Clara cell protein in the developing airways of the rat, as revealed by tissue *in situ* hybridization. Tissue Cell **24:** 461–471.
9. NORD, M., O. ANDERSSON, M. BRÖNNEGÅRD & J. LUND. 1992. Rat lung polychlorinated biphenyl–binding protein: effect of glucocorticoids on the expression of the Clara cell–specific protein during fetal development. Arch. Biochem. Biophys. **296:** 302–307.
10. TEN HAVE–OPBROEK, A.A. 1981. The development of the lung in mammals: an analysis of concepts and findings. Am. J. Anat. **162:** 201–219.
11. OTTO-VERBERNE, C.J. & A.A. TEN HAVE–OPBROEK. 1987. Development of the pulmonary acinus in fetal rat lung: a study based on an antiserum recognizing surfactant-associated proteins. Anat. Embryol. **175:** 365–373.
12. SINGH, G., S.L. KATYAL & C.M. WONG. 1986. A quantitative assay for a Clara cell–specific protein and its application in the study of development of pulmonary airways in the rat. Pediatr. Res. **20:** 802–805.
13. CARDOSO, W.V., L.G. STEWART, K.E. PINKERTON *et al.* 1993. Secretory product expression during Clara cell differentiation in the rabbit and rat. Am. J. Physiol. **264:** L543–L552.
14. MASSARO, G.D., L. DAVIS & D. MASSARO. 1984. Postnatal development of the bronchiolar Clara cell in rats. Am. J. Physiol. **247:** C197–C203.
15. PLOPPER, C.G., D.M. HYDE & A.R. BUCKPITT. 1997. *In* The Lung: Scientific Foundations, pp. 517–533. Lippincott-Raven. Philadelphia.
16. ANDERSSON, O., G. NOACK, B. ROBERTSSON *et al.* 1994. Ontogeny of a human polychlorinated biphenyl binding protein: level of expression in tracheal aspirates in broncopulmonary dysplasia. Chest **105:** 17–22.
17. BERNARD, A., N. THIELEMANS, R. LAUWERYS *et al.* 1994. Clara cell protein in human amniotic fluid—a potential marker of fetal lung growth. Pediatr. Res. **36:** 771–775.
18. STRIPP, B.R., K. MAXSON, R. MERA & G. SINGH. 1995. Plasticity of airway cell proliferation and gene expression after acute naphthalene injury. Am. J. Physiol. **269:** L791–L799.
19. VAN WINKLE, L.S., A.R. BUCKPITT, S.J. NISHIO *et al.* 1995. Cellular response in naphthalene-induced Clara cell injury and bronchiolar epithelial repair in mice. Am. J. Physiol. **269:** L800–L818.
20. JENSEN, S.M., J.E. JONES, H. PASS *et al.* 1994. Clara cell 10 kDa protein mRNA in normal and atypical regions of human respiratory epithelium. Int. J. Cancer **58:** 629–637.
21. NORD, M., M. LÅG, T.N. CASSEL *et al.* 1998. Regulation of CCSP (PCB-BP/uteroglobin) expression in primary cultures of lung cells—involvement of C/EBP. DNA Cell Biol. **17:** 481–492.
22. MIELE, L., M.E. CORDELLA & A.B. MUKHERJEE. 1987. Uteroglobin: structure, molecular biology, and new perspectives on its function as a phospholipase A2 inhibitor. Endocr. Rev. **8:** 474–490.
23. MUKHERJEE, A.B., G.C. KUNDU, G. MANTILE-SELVAGGI *et al.* 1999. Uteroglobin: a novel cytokine? Cell. Mol. Life Sci. **55:** 771–787.
24. STRIPP, B.R., P.L. SAWAYA, D.S. LUSE *et al.* 1992. *Cis*-acting elements that confer lung epithelial cell expression of the CC10 gene. J. Biol. Chem. **267:** 14703–14712.
25. HACKETT, B.P. & J.D. GITLIN. 1992. Cell-specific expression of a Clara cell secretory protein–human growth hormone gene in the bronchiolar epithelium of transgenic mice. Proc. Natl. Acad. Sci. U.S.A. **89:** 9079–9083.

26. RAY, M.K., S.W. MAGDALENO, M.J. FINEGOLD & F.J. DE MAYO. 1995. Cis-acting elements involved in the regulation of mouse Clara cell–specific 10-kDa protein gene—in vitro and in vivo analysis. J. Biol. Chem. **270:** 2689–2694.
27. BINGLE, C.D. & J.D. GITLIN. 1993. Identification of hepatocyte nuclear factor-3 binding sites in the Clara cell secretory protein gene. Biochem. J. **295:** 227–232.
28. BINGLE, C.D., B.P. HACKETT, M. MOXLEY et al. 1995. Role of hepatocyte nuclear factor-3α and hepatocyte nuclear factor-3β in Clara cell secretory protein gene expression in the bronchiolar epithelium. Biochem. J. **308:** 197–202.
29. STRIPP, B.R., J.A. HUFFMAN & R.J. BOHINSKI. 1994. Structure and regulation of the murine Clara cell secretory protein gene. Genomics **20:** 27–35.
30. SAWAYA, P.L., B.R. STRIPP, J.A. WHITSETT & D.S. LUSE. 1993. The lung-specific CC10 gene is regulated by transcription factors from the AP-1, octamer, and hepatocyte nuclear factor-3 families. Mol. Cell. Biol. **13:** 3860–3870.
31. SAWAYA, P.L. & D.S. LUSE. 1994. Two members of the HNF-3 family have opposite effects on a lung transcriptional element; HNF-3α stimulates and HNF-3β inhibits activity of region I from the Clara cell secretory protein (CCSP) promoter. J. Biol. Chem. **269:** 22211–22216.
32. RAY, M.K., C.Y. CHEN, R.J. SCHWARTZ & F.J. DE MAYO. 1996. Transcriptional regulation of a mouse Clara cell–specific protein (mCC10) gene by the NKx transcription factor family members thyroid transcription factor 1 and cardiac muscle–specific homeobox protein (CSX). Mol. Cell. Biol. **16:** 2056–2064.
33. TOONEN, R.F.G., S. GOWAN & C.D. BINGLE. 1996. The lung enriched transcription factor TTF-1 and the ubiquitously expressed proteins Sp1 and Sp3 interact with elements located in the minimal promoter of the rat Clara cell secretory protein gene. Biochem. J. **316:** 467–473.
34. ZHANG, L.Q., J.A. WHITSETT & B.R. STRIPP. 1997. Regulation of Clara cell secretory protein gene transcription by thyroid transcription factor-1. Biochim. Biophys. Acta Gene Struct. Expr. **1350:** 359–367.
35. MONAGHAN, A.P., K.H. KAESTNER, E. GRAU & G. SCHUTZ. 1993. Postimplantation expression patterns indicate a role for the mouse forkhead/HNF-3 alpha, beta, and gamma genes in determination of the definitive endoderm, chordamesoderm, and neuroectoderm. Development **119:** 567–578.
36. ANG, S.L., A. WIERDA, D. WONG et al. 1993. The formation and maintenance of the definitive endoderm lineage in the mouse: involvement of HNF3/forkhead proteins. Development **119:** 1301–1315.
37. CLEVIDENCE, D.E., D.G. OVERDIER, R.S. PETERSON et al. 1994. Members of the HNF-3/forkhead family of transcription factors exhibit distinct cellular expression patterns in lung and regulate the surfactant protein B promoter. Dev. Biol. **166:** 195–209.
38. ANG, S.L. & J. ROSSANT. 1994. HNF-3 beta is essential for node and notochord formation in mouse development. Cell **78:** 561–574.
39. WEINSTEIN, D.C., A. RUIZ I ALTABA, W.S. CHEN et al. 1994. The winged-helix transcription factor HNF-3 beta is required for notochord development in the mouse embryo. Cell **78:** 575–588.
40. ZHOU, L., L. LIM, R.H. COSTA & J.A. WHITSETT. 1996. Thyroid transcription factor-1, hepatocyte nuclear factor-3 beta, surfactant protein B, C, and Clara cell secretory protein in developing mouse lung. J. Histochem. Cytochem. **44:** 1183–1193.
41. BRAUN, H. & G. SUSKE. 1998. Combinatorial action of HNF3 and Sp family transcription factors in the activation of the rabbit uteroglobin/CC10 promoter. J. Biol. Chem. **273:** 9821–9828.
42. LIM, L., H. ZHOU & R.H. COSTA. 1997. The winged helix transcription factor HFH-4 is expressed during choroid plexus epithelial development in the mouse embryo. Proc. Natl. Acad. Sci. U.S.A. **94:** 3094–3099.
43. HACKETT, B.P., S.L. BRODY, M. LIANG et al. 1995. Primary structure of hepatocyte nuclear factor/forkhead homologue 4 and characterization of gene expression in the developing respiratory and reproductive epithelium. Proc. Natl. Acad. Sci. U.S.A. **92:** 4249–4253.
44. CHEN, J., H.J. KNOWLES, J.L. HEBERT & B.P. HACKETT. 1998. Mutation of the mouse hepatocyte nuclear factor/forkhead homologue 4 gene results in an absence of cilia and random left-right asymmetry. J. Clin. Invest. **102:** 1077–1082.

45. TICHELAAR, J.W., L. LIM, R.H. COSTA & J.A. WHITSETT. 1999. HNF-3/forkhead homologue-4 influences lung morphogenesis and respiratory epithelial cell differentiation *in vivo*. Dev. Biol. **213:** 405–417.
46. TICHELAAR, J.W., S.E. WERT, R.H. COSTA *et al.* 1999. HNF-3/forkhead homologue-4 (HFH-4) is expressed in ciliated epithelial cells in the developing mouse lung. J. Histochem. Cytochem. **47:** 823–832.
47. BLATT, E.N., X.H. YAN, M.K. WUERFFEL *et al.* 1999. Forkhead transcription factor HFH-4 expression is temporally related to ciliogenesis. Am. J. Respir. Cell Mol. Biol. **21:** 168–176.
48. HELLQVIST, M., M. MAHLAPUU, L. SAMUELSSON *et al.* 1996. Differential activation of lung-specific genes by two forkhead proteins, FREAC-1 and FREAC-2. J. Biol. Chem. **271:** 4482–4490.
49. PETERSON, R.S., L. LIM, H. YE *et al.* 1997. The winged helix transcriptional activator HFH-8 is expressed in the mesoderm of the primitive streak stage of mouse embryos and its cellular derivatives. Mech. Dev. **69:** 53–69.
50. LAZZARO, D., M. PRICE, M. DE FELICE & R. DI LAURO. 1991. The transcription factor TTF-1 is expressed at the onset of thyroid and lung morphogenesis and in restricted regions of the fetal brain. Development **113:** 1093–1104.
51. RAMIREZ, M.I., A.K. RISHI, Y.X. CAO & M.C. WILLIAMS. 1997. TGT3, thyroid transcription factor I, and Sp1 elements regulate transcriptional activity of the 1.3-kilobase pair promoter of T1 alpha, a lung alveolar type I cell gene. J. Biol. Chem. **272:** 26285–26294.
52. KIMURA, S., Y. HARA, T. PINEAU *et al.* 1996. The t/ebp null mouse: thyroid-specific enhancer-binding protein is essential for the organogenesis of the thyroid, lung, ventral forebrain, and pituitary. Genes Dev. **10:** 60–69.
53. MINOO, P., G. SU, H. DRUM *et al.* 1999. Defects in tracheoesophageal and lung morphogenesis in Nkx2.1(–/–) mouse embryos. Dev. Biol. **209:** 60–71.
54. WHITSETT, J.A. & S.W. GLASSER. 1998. Regulation of surfactant protein gene transcription. Biochim. Biophys. Acta **1408:** 303–311.
55. LI, F., E. ROSENBERG, C.I. SMITH *et al.* 1995. Correlation of expression of transcription factor C/EBP alpha and surfactant protein genes in lung cells. Am. J. Physiol. **269:** L241–L247.
56. CASSEL, T.N., L. NORDLUND-MÖLLER, O. ANDERSSON *et al.* 2000. C/EBP alpha and C/EBP delta activate the Clara cell secretory protein gene through activation with two adjacent C/EBP-binding sites. Am. J. Respir. Cell Mol. Biol. **22:** 469–480.
57. LEKSTROM-HIMES, J. & K.G. XANTHOPOULOS. 1998. Biological role of the CCAAT/enhancer-binding protein family of transcription factors. J. Biol. Chem. **273:** 28545–28548.
58. DIEHL, A.M. 1998. Roles of CCAAT/enhancer-binding proteins in regulation of liver regenerative growth. J. Biol. Chem. **273:** 30843–30846.
59. CRONIGER, C., P. LEAHY, L. RESHEF & R.W. HANSON. 1998. C/EBP and the control of phosphoenolpyruvate carboxykinase gene transcription in the liver. J. Biol. Chem. **273:** 31629–31632.
60. DARLINGTON, G.J., S.E. ROSS & O.A. MACDOUGALD. 1998. The role of C/EBP genes in adipocyte differentiation. J. Biol. Chem. **273:** 30057–30060.
61. POLI, V. 1998. The role of C/EBP isoforms in the control of inflammatory and native immunity functions. J. Biol. Chem. **273:** 29279–29282.
62. CAO, Z., R.M. UMEK & S.L. MCKNIGHT. 1991. Regulated expression of three C/EBP isoforms during adipose conversion of 3T3-L1 cells. Genes Dev. **5:** 1538–1552.
63. BREED, D.R., L.R. MARGRAF, J.L. ALCORN & C.R. MENDELSON. 1997. Transcription factor C/EBP delta in fetal lung: developmental regulation and effects of cyclic adenosine 3′,5′-monophosphate and glucocorticoids. Endocrinology **138:** 5527–5534.
64. CASSEL, T.N., J-Å. GUSTAFSSON & M. NORD. 2000. *CYP2B1* is regulated by C/EBP alpha and C/EBP delta in lung epithelial cells. Mol. Cell. Biol. Res. Commun. **3:** 42–47.
65. FLODBY, P., C. BARLOW, H. KYLEFJORD *et al.* 1996. Increased hepatic cell proliferation and lung abnormalities in mice deficient in CCAAT/enhancer binding protein α. J. Biol. Chem. **271:** 24753–24760.

66. CRONIGER, C., M. TRUS, K. LYSEK-STUPP et al. 1997. Role of the isoforms of CCAAT/enhancer-binding protein in the initiation of phosphoenolpyruvate carboxykinase (GTP) gene transcription at birth. J. Biol. Chem. **272:** 26306–26312.
67. DE MAYO, F.J., S. DAMAK, T.N. HANSEN & D.W. BULLOCK. 1991. Expression and regulation of the rabbit uteroglobin gene in transgenic mice. Mol. Endocrinol. **5:** 311–318.
68. SUSKE, G., W. LORENZ, J. KLUG et al. 1992. Elements of the rabbit uteroglobin promoter mediating its transcription in epithelial cells from the endometrium and lung. Gene Expr. **2:** 339–352.
69. DENNIG, J., G. HAGEN, M. BEATO & G. SUSKE. 1995. Members of the Sp transcription factor family control transcription from the uteroglobin promoter. J. Biol. Chem. **270:** 12737–12744.
70. STAHLMAN, M.T., M.E. GRAY & J.A. WHITSETT. 1996. Expression of thyroid transcription factor-1 (TTF-1) in fetal and neonatal human lung. J. Histochem. Cytochem. **44:** 673–678.
71. SUSKE, G. 2000. Transient transfection of Schneider cells in the study of transcription factors. Methods Mol. Biol. **130:** 175–187.
72. ADAMSON, I.Y.R. 1997. *In* The Lung: Scientific Foundations, pp. 993–1001. Lippincott-Raven. Philadelphia.

Uteroglobin Gene Transcription: What's the RUSH?

BEVERLY S. CHILTON,[a] AVELINE HEWETSON, JERRY DEVINE, ERICKA HENDRIX, AND MALINI MANSHARAMANI

Department of Cell Biology and Biochemistry, Texas Tech University Health Sciences Center, Lubbock, Texas 79430, USA

ABSTRACT: Prolactin enhances progesterone-dependent transcription of the rabbit uteroglobin gene. RUSH transcription factors are implicated in the signal transduction pathway. The RUSH acronym identifies key features of these nuclear phosphoproteins, that is, RING-finger motif, binds the uteroglobin promoter, structurally related to the SWI/SNF family of transcription factors, and helicase-like. Cloned by recognition site screening, RUSH proteins bind to an 85-bp region (−170/−85) of the uteroglobin promoter that was subsequently identified as a novel prolactin-responsive region by promoter deletion analysis. Gel shift and linker-scanning assays further reduced the RUSH target site to −160/−110. A hexameric core of MCWTDK was identified as the RUSH-specific DNA-binding site (−126/−121) by CASting. This site overlaps authentic HNF3β and OCT-1 binding sites. A unique Type IV P-type ATPase that is embedded in the inner nuclear membrane binds the RING domain of RUSH. The conformationally flexible loop portion of this RING-finger binding protein (RFBP) extends into the nucleoplasm to contact euchromatin. The physical association of RFBP with transcriptionally active chromatin supports the speculation that RFBP targets RUSH transcription factors to the active uteroglobin promoter.

INTRODUCTION

Prolactin has been implicated in a vast number of physiological events; however, the actual number of prolactin target genes is very small. Although the mammary gland has been the focus of most studies, it is clear that prolactin is endometriotrophic. Studies with knockout mice have shown that prolactin null females are completely infertile,[1] and homozygous prolactin receptor null females have complete failure of implantation.[2] This work supports the need to define the role of prolactin in the regulation of gene expression in the endometrium.

The steroid-inducible blastokinin[3] or uteroglobin[4] gene has provided a very compelling window for viewing the combinations of regulatory sequences and transcription factors that mediate its expression. Recent studies suggest that the uteroglobin promoter is controlled by the interactions of hepatic nuclear factor 3 (HNF3) and the ubiquitous factor Sp1[5–7] along with C/EBP factors.[8] RUSH proteins are the first

---

[a]Address for correspondence: Beverly S. Chilton, Department of Cell Biology and Biochemistry, Texas Tech University Health Sciences Center, 3601 4th Street, Lubbock, TX 79430. Voice: 806-743-2709; fax: 806-743-2990.
beverly.chilton@ttmc.ttuhsc.edu

members of the SWI/SNF family of factors known to bind the uteroglobin promoter directly.[9] Prolactin augments progesterone-dependent transcriptional activation of the uteroglobin gene in the rabbit endometrium. As potential chromatin remodeling factors, RUSH proteins have been implicated in the cross talk between prolactin and progesterone in transcriptional activation of the uteroglobin gene.[10]

UTEROGLOBIN PROMOTER

In the rabbit endometrium, uteroglobin is a progesterone-dependent preimplantation uterine protein with peak availability on day 5 of pregnancy. Progesterone directs transcription of the uteroglobin gene via two strong and two weak progesterone receptor binding sites located between positions −2.7 kb and −2.3 kb (FIG. 1). DNase I protection assays, promoter deletion, and linker-scanning analysis have been used to identify seven regulatory regions[5] in the proximal promoter of the uteroglobin

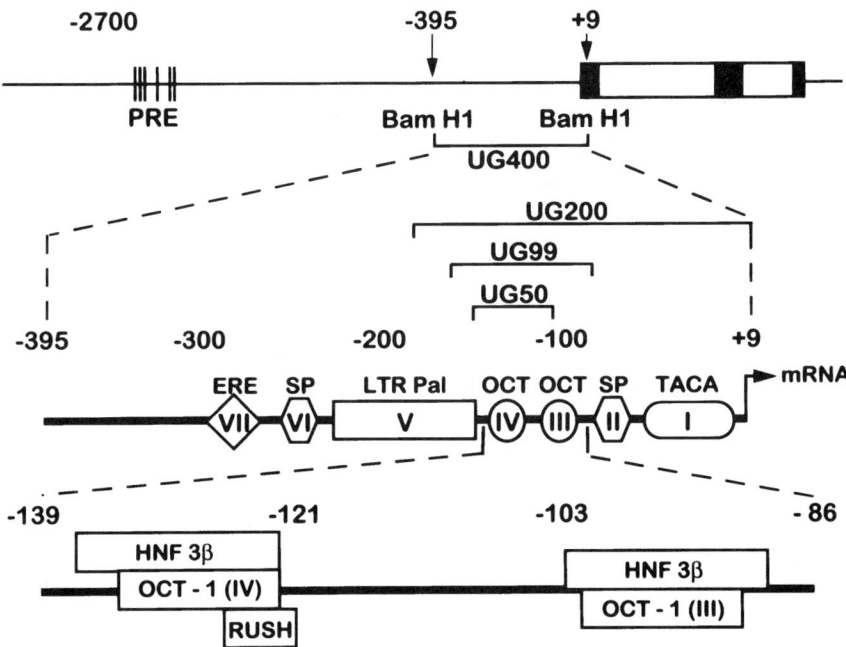

FIGURE 1. Organization of the uteroglobin (UG) gene. Exons are shown by black boxes and introns by open boxes. PRE refers to the progesterone receptor binding sites located between −2700 and −2300. The 400-bp BamH1 promoter region (−395/+9) is enlarged to show seven distinct regions characterized by an estrogen-response element (ERE), binding sites for the Sp family of transcription factors (SP), the palindromic sequence found in the long terminal repeat of murine leukemia viruses (LTR Pal), octamer factor binding sites (OCT), and a TACA element (noncanonical TATA box). The positions of UG200 (−194/+9), UG99 (−170/−85), and UG50 (−160/−110) are shown. The position of the RUSH binding domain relative to HNF3β and OCT-1 binding sites is also depicted.

gene (−395/+9). As shown in FIGURE 1, region VII is defined by the presence of an estrogen-response element (−265/−252). The ubiquitous octamer-binding protein OCT-1 binds to regions III and IV, whereas Sp1 and Sp3 bind to regions II and VI. Other regions of the promoter include a motif that is 92% identical to the GT-I motif in the simian virus 40 (SV40) enhancer (−258/−220) and a 7-bp identical inverted repeat (CAGTTTC) that is found in the long terminal repeat (−171/−148) of all murine leukemia viruses and proviruses.[11] Two HNF3β-response elements centered at −130 and −95 completely overlap the OCT-1 binding sites in regions III and IV (FIG. 1).

PROLACTIN AUGMENTS PROGESTERONE-DEPENDENT TRANSCRIPTION

In the endometrium, progesterone directs transcription of the uteroglobin gene in estrous rabbits and in short-term (3 days to 4 weeks) ovariectomized rabbits. However, with increased time postovariectomy, the uterus becomes increasingly refractory to progesterone challenge[12] such that, after 12 weeks, uteroglobin synthesis is negligible. The treatment of long-term ovariectomized (>12 weeks) rabbits with either prolactin or progesterone resulted in a dramatic increase in the receptor for the other hormone.[13–16] Sequential treatment with prolactin plus progesterone increased the endometrial uteroglobin mRNA content and stimulated uteroglobin production to a concentration equal to that found on day 5 of pregnancy.[15] One possible explanation for this observation was the existence of a direct link between prolactin and progesterone signaling that enhanced the rate of uteroglobin gene transcription.

Rider and Bullock[17] used gel shift assays to identify progesterone-dependent nuclear proteins that bound to the uteroglobin promoter (−194/+9). Then, Rider and Peterson[18] provided indirect evidence that transcriptional activation of the uteroglobin gene by progesterone requires the binding of two proteins to the uteroglobin promoter. We[19] subsequently used gel shift assays, Southwestern blots, and UV cross-linking experiments to show that prolactin augments the binding of four

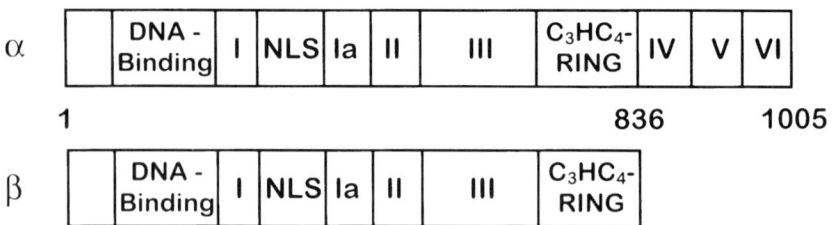

FIGURE 2. Structure of the RUSH-1α and RUSH-1β proteins. **(Top)** RUSH-1α contains the seven consecutive domains (I through VI) that are typical of ATPases and DNA helicases. The DNA-binding domain is in the N-terminus of the protein (amino acids 40–231) and the RING-motif is in the C-terminus (amino acids 757–800). The putative nuclear localization signal (NLS; amino acids 383–386) is positioned between domains I and Ia. **(Bottom)** Because RUSH-1β terminates just after the RING motif, it does not contain domains IV, V, and VI.

progesterone-dependent proteins to an 85-bp 5'-flanking region (−170/−85) of the uteroglobin gene. The cDNAs for two of these uteroglobin promoter–binding proteins, RUSH-1α (113 kDa) and β (95 kDa), were cloned by recognition site screening.[9] The RUSH acronym identifies the key characteristics of these nuclear phosphoproteins, that is, RING-finger motif, binds the uteroglobin promoter, structurally related to the SWI/SNF family of transcription factors, and helicase-like. RUSH-1α and β proteins (FIG. 2) are products of alternatively spliced messages. Competitive reverse transcriptase–polymerase chain reaction (RT-PCR) and an ion-pair reversed phase HPLC product detection system showed that RUSH-1α mRNA is the progesterone-dependent splice variant. The choice of acronym was thus influenced by the significance of alternative splicing as an adjunct to regulation of the uteroglobin promoter. When this posttranscriptional level of regulation is invoked, the response of the target cell to a physiological challenge is accelerated or RUSHed.

HUMAN HOMOLOGUES

Three human homologues of RUSH-1α were isolated by recognition site screening of λgt-11 cDNA expression libraries derived from HeLa cells. HIP116,[20] human helicase-like transcription factor (hHLTF),[21] and Zbu1[22] encode identical 1009-amino-acid proteins (116 kDa). The human RUSH gene was mapped to a single locus on human chromosome 3q25.1-q26.1.[23] Identified as sequence-specific DNA-binding proteins, activation of the PAI-1 gene by hHLTF[21] and P113, the mouse homologue of RUSH-1α,[24] provides *in vivo* evidence that RUSH proteins are transcription factors. RUSH proteins also provide a new perspective on the transcriptional activation of the uteroglobin gene.

NOVEL ELEMENTS IN THE UTEROGLOBIN PROMOTER MEDIATE THE ACTION OF PROLACTIN

To directly test the effects of prolactin ± progesterone on uteroglobin gene expression,[10] the full-length uteroglobin gene, pUG3.1, and a deletion mutant designated pUG3.1ΔRUSH were directionally subcloned into the promoterless, luciferase reporter plasmid, pGL3-Basic. As shown in FIGURE 3, the pUG3.1-luciferase construct contains all of the regions that are important to uteroglobin expression, especially the progesterone receptor binding sites and the proximal promoter region known to bind RUSH. The mutant is missing the 85-bp region (−170/−85) of the proximal uteroglobin promoter that contains the palindromic sequence found in the long terminal repeat of murine leukemia viruses, two OCT-1 binding motifs, the HNF3β binding sites, and RUSH binding sites as determined by footprinting. HRE-H9 cells were cotransfected with one of the uteroglobin-luciferase constructs and a transfection efficiency control vector that contains the herpes simplex virus thymidine kinase promoter upstream of *Renilla* luciferase (pRL-TK). The pUG-luciferase activities were normalized to pRL-TK-luciferase activities and expressed as ratios. HRE-H9 cells are an SV40-transformed cell line derived from the uterine epithelium of hCG-treated pseudopregnant rabbits.[25] These epithelial cells express receptors for

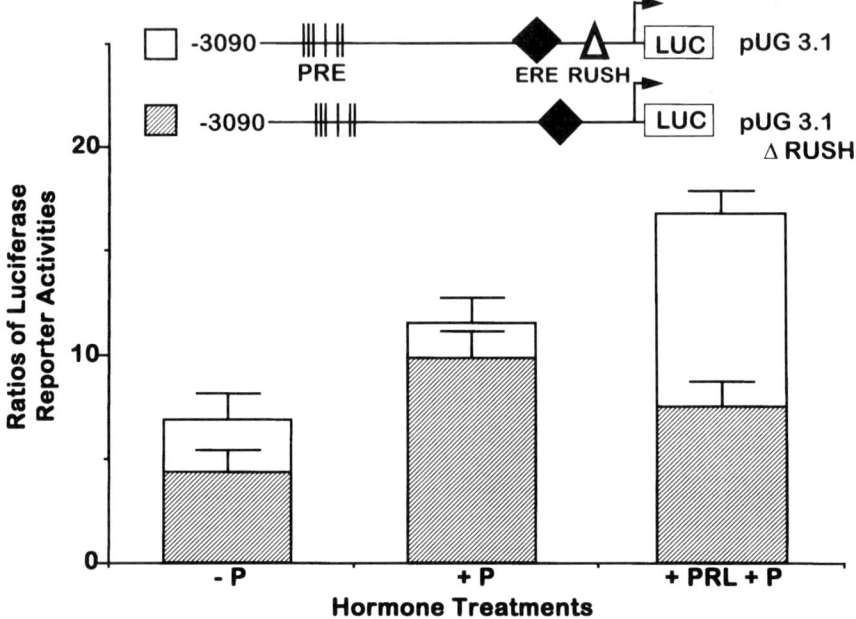

FIGURE 3. Deletion of the RUSH binding site mediates the prolactin effect. The full-length construct, pUG3.1-luciferase, and the deletion mutant, pUG3.1ΔRUSH-luciferase, were tested in transfection assays with HRE-H9 cells. Deletion of the proximal promoter regions −170/−85 had no effect on the progesterone (P)–dependent increase in uteroglobin gene transcription; however, its removal eliminated the prolactin effect. PRE, site of progesterone receptor binding; ERE, estrogen-response element. Promoter activity was normalized to the internal control (pRL-TK luciferase activity). Data are expressed as the mean ± SEM of two to five independent experiments performed in triplicate. Data were analyzed by ANOVA followed by Student-Newman-Keul's multiple range test ($P < 0.05$ significance level).

progesterone and prolactin;[10,26] however, endogenous uteroglobin mRNA is undetectable by RT-PCR.

HRE-H9 cells were pretreated with estrogen (10^{-7} M) and prolactin (10^{-9} M or 10^{-8} M) for 48 h prior to transfection. After transfection, the cells were either used as untreated controls or treated with prolactin ± progesterone (10^{-7} M) for 48 h. Prolactin levels were tested at 10^{-9} M, the value for normal estrous[27] and pregnant animals,[28–30] and at 10^{-8} M, the maximum value recorded in response to pregnancy,[29] parturition,[29] and stress.[31] As shown in FIGURE 3, progesterone alone increased ($P < 0.05$) the transcriptional activity of pUG3.1-luciferase. Transcription was further increased ($P < 0.05$) when cells were treated with prolactin + progesterone. These data are consistent with our previous *in vivo* observations[15] showing that prolactin plus progesterone increased ($P < 0.05$) the uterine content of uteroglobin mRNA over the value for progesterone alone. Prolactin alone had no effect. Progesterone alone also increased ($P < 0.05$) the transcriptional activity of pUG3.1ΔRUSH-luciferase. However, deletion of the proximal promoter region −170/−85 was highly deleterious to the prolactin effect (FIG. 3). Using this mutational approach, it was possible to retain the progesterone-dependent increase ($P < 0.05$) in transcription, but eliminate ($P > 0.05$) the prolactin-mediated increase in uteroglobin gene transcription.

Prolactin receptor signal transduction generally involves activation of the Janus family of tyrosine kinases and phosphorylation of specific members of the STAT (signal transducers and activators of transcription) family of transcription factors that homodimerize, translocate to the nucleus, and bind to the palindromic gamma-interferon-activated (GAS) sequence $TTC(N)_mGAA$, where $m = 3$ (or sometimes 4 for STAT6). Although there are subtle differences in the binding site specificities for STAT proteins,[32] no binding element for STAT5a, the family member responsible for prolactin signal transduction, is present in the −170/−85 region of the uteroglobin promoter or for that matter in the entire uteroglobin promoter (−394/+9). Therefore, the promoter deletion analysis has delineated a novel prolactin-responsive region between −170 and −85 in the uteroglobin promoter.[10,33] These data also support the speculation that RUSH proteins play an important regulatory role in prolactin signal transduction.

CASTING TO IDENTIFY THE RUSH BINDING SITE

RUSH proteins were initially identified[9] based on their ability to bind to UG200 (−194/+9) and UG99 (−170/−85). The relationship of these regions to the general organization of the uteroglobin (UG) gene is shown in FIGURE 1. The first step in mapping sequences important for RUSH binding was to further reduce the target search. Gel shift, supershift, and immunodepletion assays were used to show that RUSH proteins bind to UG50 (−160/−110), and UG50 competes for RUSH-specific binding. Protein sources for these experiments included nuclear extracts from endometrium and truncated RUSH translated with the TNT-coupled reticulocyte lysate system. The second step was to perform competition gel shift assays with overlapping 20-bp oligonucleotides that span the UG50 sequence (FIG. 4A). However, results from these experiments were ambiguous. To conclusively identify a subregion responsible for RUSH binding, native RUSH was presented with a choice of mutated templates. First, UG50 was mutagenized such that changes were introduced at 10-bp intervals (FIG. 4B). Each interval was changed into the same sequence to minimize variability, to maximize sequence disruption (i.e., a minimum of 8 out of 10 bases were mutated), to provide a consistent degree of mutation at each site, and to maintain spatial relationships. Moreover, the mutation sequence (CCTCCGGACT) does not resemble any known functional element.

The second step was to generate a set of 10 double-stranded, 147-bp probes, one for each linker-scanning (LS) mutant and a wild-type control, by PCR. Third, mutant-specific binding was evaluated in competition assays with UG50. Although some competition was noted for LS2 through LS6, binding was nearly negligible in assays with LS2 and LS5. To determine whether there is a functional link between binding site perturbation and gene transcription, transient transfection assays with mutant-luciferase constructs were performed with rabbit HRE-H9 and human Ishikawa cells as described.[10,33] Cells were cotransfected with mixtures of plasmid

```
A. Alignment of oligonucleotides with UG50 (-160/-110) and each other.
-175 AGTTCAGTTTCAATAGGGATGGAAACTGGATTGAGAAAAGGGAATATTTACTTATCCCACCAAGTC -110
     TCAGTTTCAATAGGGATGGA Oligo-1 (-172/-163)
              TAGGGATGGAAACTGGATTG Oligo-2 (-162/-143)
                       AACTGGATTGAGAAAAGGGA Oligo-3 (-153/-133)
                                AGAAAAGGGAATATTTACTT Oligo-4 (-143/-123)
                                         ATATTTACTTATCCCACCAA Oligo-5 (-133/-113)

B. Alignment of linker-scanning mutants with UG50 (-160/-110) and each other.
              (LS2)              (LS4)              (LS6)              (LS8)
            CCTCCGGACT         CCTCCGGACT         CCTCCGGACT         CCTCCGGACT
            **********         ******|***         ***||*****         **********
-175 AGTTCAGTTTCAATAGGGATGGAAACTGGATTGAGAAAAGGGAATATTTACTTATCCCACCAAGTCAATGCCCAAGTAAATAATGCAGTCAAGTAA -80
     *|*********         *****|****         **********|         ******|*|*         *******|**
     CCTCCGGACT         CCTCCGGACT         CCTCCGGACT         CCTCCGGACT         CCTCCGGACT
       (LS1)              (LS3)              (LS5)              (LS7)              (LS9)
```

FIGURE 4. Schematic for the preparation of **(A)** oligonucleotide competitors and **(B)** linker-scanning (LS) mutants. The 85-bp sequence identified as UG99 (−170/−85) in FIGURE 1 is shown here. UG50 (−160/−110), a subregion of UG99, is underlined. Top-strand sequences of double-stranded oligonucleotides used in the gel mobility shift assays are shown in panel A. A 20-bp sliding window with 10-bp overlaps was used in the synthesis of five single-stranded oligonucleotides (20 bp each) that span region −172/−113. Note that each 20-bp oligonucleotide overlaps two LS mutants shown in panel B, that is, oligo-1 overlaps LS1 and LS2, oligo-2 overlaps LS2 and LS3, etc. For the preparation of LS mutants, substitutions were introduced into pUG3.1, a 3.1-kb SacI/HindIII fragment (−3090/+37) of UG 5′-flanking DNA that was directionally subcloned into the luciferase reporter plasmid, pGL3-Basic. Mutagenesis primers were 32 bases in length and each encoded a 10-base substitution that was flanked by unaltered bases. When comparing the LS mutants with authentic promoter sequence, mutated bases are indicated with an asterisk (*) and unaltered nucleotides are indicated with a vertical line (|). The identity of each mutant was confirmed by sequence analysis. Luciferase constructs were used in transient transfection assays and as templates in the synthesis of probes for gel shift assays. A set of double-stranded, 147-bp probes, one for each LS mutant and a wild-type control, was generated by PCR. Each LS mutant template was paired with forward (−206/−174) and reverse (−91/−60) primers designed to uteroglobin flanking sequence. Thus, the wild-type control (−206/−60) contained 147 bp of unaltered uteroglobin promoter sequence that included UG99 (−170/−85).

DNA containing each pUG-luciferase mutant and the internal control plasmid pRL-TK. Luciferase assays were performed at 18–24 h posttransfection. As shown in FIGURE 5, internal deletions along the length of UG50 dramatically altered the transcriptional activity of the full-length pUG3.1 construct. Destroying the A/T-rich LS5 (−132/−123) region that was previously identified as a binding domain for members of the octamer and HNF3 families[7,34] reduced transcriptional activity to negligible levels. Fine mapping of the −160/−110 region of the uteroglobin promoter by site-directed mutagenesis revealed sequences whose integrity is essential for promoter activity. Also, these studies support the idea that the combinatorial processes of protein-protein and protein-DNA interactions play a central role in the regulation of uteroglobin gene expression.

Thus, what is the role of RUSH proteins? Are RUSH proteins relatively promiscuous in their DNA affiliation despite the fact that they were cloned by recognition site screening, that is, based on sequence-specific DNA binding? All RUSH family members are SWI/SNF-related proteins characterized by domains that are typical of ATPases,[35] and strong DNA-dependent ATPase activity has been demonstrated for

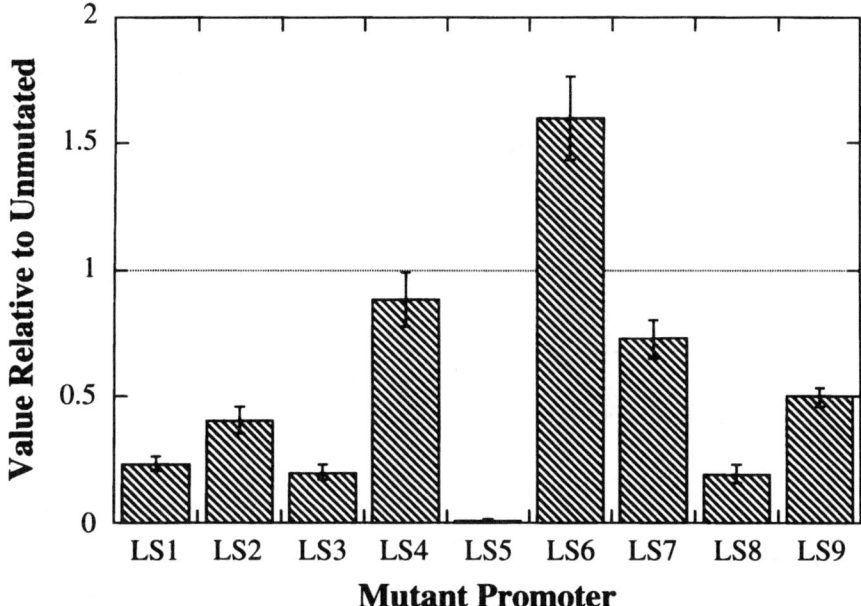

FIGURE 5. Results of the transfection experiments in HRE-H9 cells. Promoter activity was normalized to the internal control (pRL-TK luciferase activity) and these values were normalized to activity for the intact pUG3.1 control (value = 1 on this graph). Data are expressed as the mean ± SEM of three independent experiments performed in triplicate. Similar results were obtained with Ishikawa cells.

HIP116, Zbu1, and P113. Therefore, RUSH proteins may use the energy of ATP hydrolysis to disrupt chromatin structure. Evidence that SWI/SNF is an architectural protein comes from *in vitro* studies in which SWI/SNF preferentially binds synthetic four-way (Holliday) junction DNAs that may resemble the entry/exit point of the DNA in the nucleosome.[36] Thus, SWI/SNF binding might alter the helical twist of the DNA and destabilize the histone-DNA affiliation, resulting in either the loss of the H2A-H2B dimer or eviction of the entire octamer. Whatever the proposed architectural role *in vivo*, it must also account for incidences of sequence-specific binding reported for SWI/SNF.

Cyclic Amplification and Selection of Targets (CASTing)[37] was used to identify the putative RUSH target in the uteroglobin promoter. Nuclear extract from the endometrium of a progesterone-treated rabbit was mixed with a double-stranded oligonucleotide that contained 26 randomized bases centered between PCR priming sequences and affinity-purified anti-RUSH-$2_{(370-387)}$ antibodies. Immunoprecipitated protein-DNA complexes were amplified by PCR. A hexameric core sequence of A/CCA/TTNG/T (MCWTNK) that is at times palindromic (ACATGT) was identified after five cycles of selection. The same consensus sequence was identified by CASTing with an antibody (RUSH-1α) to an epitope unique to RUSH-1α protein and by CASTing with fusion protein rather than with nuclear extract. Preliminary analysis of DNA-binding-site selectivity with gel shift assays indicates the optimal consensus binding sequence is A/CCA/TTA/T/GG/T (MCWTDK) and that the affinity of protein-DNA binding varies over one order of magnitude. Although not previously identified as the binding site for any of the human (HIP116, hHLTF, Zbu1) or mouse (P113) RUSH homologues, this consensus sequence is present in each of the probes used for library screens and/or gel mobility shift assays.[35] As shown in FIGURE 1, the putative RUSH binding site (−126/−121) is a subelement of the HNF3β and OCT-1 binding sites.

A RING-FINGER BINDING PROTEIN

A presumptive nuclear localization signal and a C_3HC_4 RING-finger are characteristic of all of the RUSH family members,[35] including the RUSH-1α isologue recently cloned in *Arabidopsis thaliana*. The RING motif is the common feature of a superfamily of nearly 200 otherwise nonhomologous proteins,[38,39] many of which are transcription factors and some of which have oncogenic potential. When the RING-finger was first identified, some thought it might play a role in DNA binding because zinc fingers are known as DNA-binding motifs.[38,40] However, in the absence of any data to support this idea, many now believe that the RING-finger plays a role in unique protein-protein interactions.[39,41–43]

To test this hypothesis, we used the RUSH cDNA that encodes this motif to isolate a cDNA for a protein that complexes with it. The amino acid sequence derived from the composite cDNA sequence (4286-bp) identifies this protein as a new member of an ancient and previously unrecognized family (Type IV) of P-type ATPase enzymes. This RING-Finger Binding Protein (RFBP) has consensus sequences for the conformationally flexible loop, seven of eight core segments, the ATP-binding domain, the highly conserved phosphorylation site [DKTGT(L/I)T], and nine transmembrane domains potentially involved in substrate binding and translocation.

P-type ATPases comprise an evolutionarily diverse superfamily of ATP-driven ion pumps.[44] Phylogenetic analysis reveals five main groups of P-type ATPases, designated Type I (heavy metal pumps), Type II (Ca^{2+}, Na^{2+}/K^{2+}, and H^+/K^+ pumps), Type III (H^+ and Mg^{2+} pumps), Type IV (phospholipid pumps), and Type V (no assigned substrate specificity). Type IV is the most recently discovered family, found only in eukaryotic cells, and it appears to be very distinct from the metal ion transporters that dominate this class of enzymes. Because P-type ATPases show very little similarity to each other (<15%), any P-type ATPase is more homologous (80–90%) to others of its class in distantly related organisms than to other ATPases in the same organism. Recent interpretations of cDNA sequences and hydropathy analyses have led to the conclusion that P-type ATPases can be divided into two groups, one with eight transmembrane domains and the other with ten transmembrane domains.

The working model (FIG. 6) of membrane topology for both groups locates the NH_2- and COOH-terminal ends of the proteins on the same side of the plasma membrane and therefore in the same subcellular compartment.[45] As a result, the small, strongly hydrophilic loop (regions A–C) between transmembrane domains 2–3 and the large hydrophilic head or conformationally flexible loop (regions D–H) between transmembrane domains 4–5 are preferentially exposed to the cytoplasm (FIG. 6). Regions A–C have been assigned a role in energy transduction. Although this role

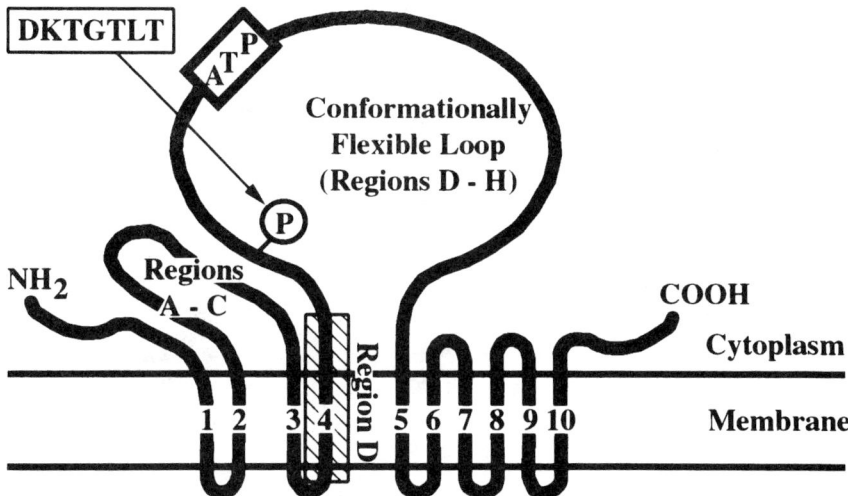

FIGURE 6. Topology of a Type IV P-type ATPase. Consensus sequences for the phosphorylation (P) site [DKTGT(L/I)T] and the ATP-binding site (ATP) are located in the conformationally flexible loop. Transmembrane domains are numbered 1–10, and region D is highlighted.

remains undefined, it may be as simple as maintaining the stability of the ATPase structure. Transmembrane domain 4 (TM4) may be directly involved in energy transduction. The sequences that link TM4 to the phosphorylation site, that is, regions D and E, are highly conserved regions in ATPases. The conformationally flexible loop contains the highly conserved phosphorylation and ATP-binding sites. The function of the phosphorylation site in all P-type ATPases is the same, that is, the transfer of γ-phosphate from ATP to the aspartate residue in the phosphorylation site.

The newly identified RFBP differs from all previously described Type IV P-type ATPases in two important ways. The most striking feature of RFBP is an odd number of transmembrane domains resulting from the absence of region D and TM4. All working models of membrane topology for P-type ATPases (FIG. 6) contain an even number of transmembrane passes with four highly conserved passes at the NH$_2$-terminus.[44,46] Collectively, these passes anchor the protein in the lipid bilayer, orient the hydrophilic domains of the protein at the membrane surface, and assemble into channel-like structures to regulate transport. Thus, the absence of the fourth membrane-embedded domain from this region of the protein alters both the orientation of the adjacent hydrophilic domains and the determinants of transport specificity. Our first thought was that the endometrial RFBP might be a splice variant. Region D and its included TM4 are encoded by 72 nucleotides that could easily participate in the simplest of alternative splicing events, that is, splice/don't splice. However, genomic cloning confirmed that region D and TM4 are absent from the RFBP gene. These results support a working model for membrane topology (FIG. 7) that orients the NH$_2$- and COOH-terminal ends of RFBP on opposite sides of the membrane. The second, equally striking feature of RFBP is its presence in the nuclear membrane.

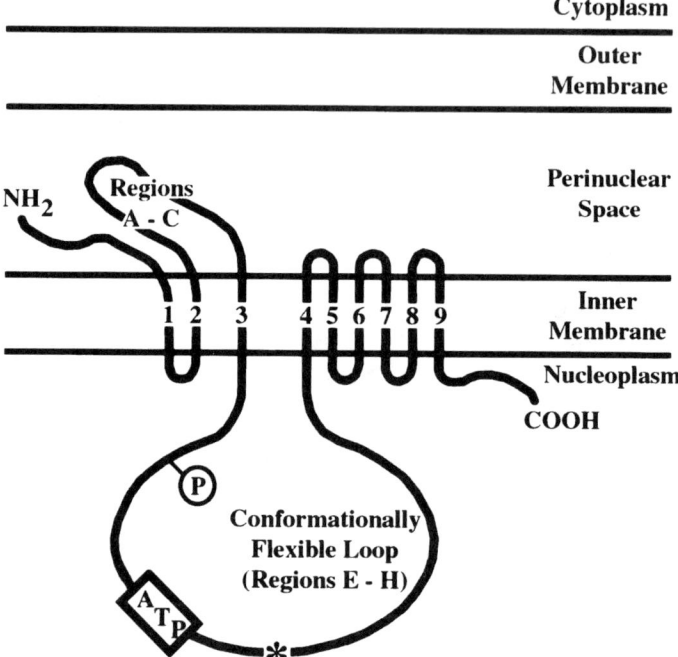

FIGURE 7. Topology of the RFBP. Consensus sequences for the phosphorylation (P) site and the ATP-binding site (ATP) are located in the conformationally flexible loop. Based on the cloning strategy, the conformationally flexible loop also contains the RING-finger binding domain. The asterisk (*) designates the antigenic region that was identified by the Peptide Structure Program (Genetics Computer Group Software) and used for antipeptide antibody production. Although region D and the fourth NH_2-terminal transmembrane domain are missing, the transmembrane domains are numbered consecutively (1–9) to avoid confusion.

Nicotera et al.[47] and Lanini et al.[48] have identified a nuclear Type II P-type Ca^{2+}-ATPase whose structure is that of a transporting pump. However, no Type IV P-type ATPases have been localized to the nuclear membrane. The use of immunoelectron microscopy to show the RFBP is located in the inner nuclear membrane provided additional insights to its atypical structure. As shown in FIGURE 7, the small loop (regions A–C) extends into the perinuclear space that is contiguous with the lumen of the endoplasmic reticulum, and the large conformationally flexible loop (regions E–H) extends into the nucleoplasm. Physical contact between the large loop and the euchromatin was also confirmed with immunoelectron microscopy.

Very little is known about the potential for hormones to regulate the expression of ATPases in general and P-type ATPases specifically. The few studies that exist consider the effects of hormones on ATPase activity. An example is the demonstration by Zylinska et al.[49] that physiologically relevant concentrations of estradiol caused a dose-dependent increase in the hydrolytic activity of rat cortical Ca^{2+}-ATPase. Two human P-type ATPases (Menkes- and Wilson-ATPases) that contain metal-binding motifs have been implicated in genetic disorders of copper homeostasis. In the case of the Menkes P-type ATPase, treatment of the human breast carcinoma cells, PMC42, with lactational hormones increased perinuclear (Golgi) and punctate (endosome) protein as measured by indirect immunofluorescence.[50] However, Northern analysis failed to show an effect of hormones (steroids or prolactin) on mRNA expression. By comparison, we quantified competitive RT-PCR reactions by ion-pair reversed phase HPLC to show that message expression for the RFBP in rabbit endometrium is increased ($P < 0.05$) in response to progesterone treatment. This is the first demonstration of an effect of hormones on the expression of any P-type ATPase. Moreover, prolactin plus progesterone further increased ($P < 0.05$) the amount of message over the value for progesterone alone. We have previously shown that prolactin enhances progesterone-dependent expression of uteroglobin[15] and Muc-1[51] in the endometrium.

A WORKING MODEL

Although we have some insight into the structure and hormonal regulation of this novel P-type ATPase, intriguing questions regarding its function and specificity remain unanswered. As a Type IV P-type ATPase, RFBP is a putative phospholipid pump. The presence of phospholipids in nuclei is well documented. Gurr et al.[52] reported that most of the nuclear phospholipids were confined to the nuclear envelope. However, Rose and Frenster[53] found that levels of intranuclear phospholipids were

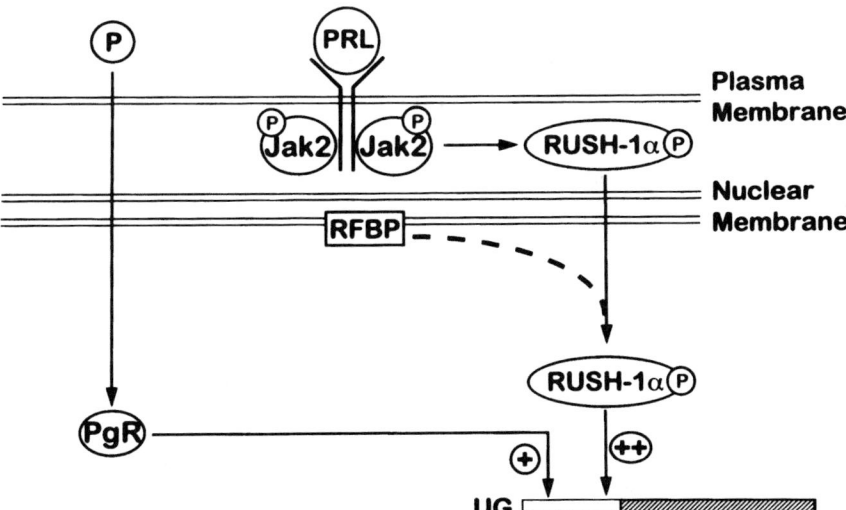

FIGURE 8. The interaction between progesterone (P) and prolactin (PRL) and their receptors in the regulation of gene expression. Prolactin enhances (++) progesterone-dependent transcriptional activation (+) of the uteroglobin gene. The role of RUSH proteins as an alternative to the Jak/Stat pathway is highlighted. Jak2 and RUSH-1α are labeled with "P" to indicate they are phosphoproteins. The location of the RING-finger binding protein (RFBP) in the inner nuclear membrane is indicated.

higher for active chromatin than for repressed chromatin. Thus, the putative phospholipid transporter[54] might be an integral membrane protein that regulates nuclear phospholipid composition.[55] Alternatively, as shown in FIGURE 8, the unique structure of the RFBP coupled to its ability to bind the RING domain and contact euchromatin supports the idea that it may have a very different function. Because gene expression requires the colocalization of transcription factors, coactivators, and RNA polymerase II at nuclear sites of RNA transcription, this novel P-type ATPase might be part of a molecular mechanism that targets regulatory proteins, in this case, RUSH transcription factors to the uteroglobin promoter.

ACKNOWLEDGMENTS

This work was supported by NIH Grant No. HD29457 to B. S. Chilton.

REFERENCES

1. HORSEMAN, N.D., W. ZHAO, E. MONTECINO-RODRIGUEZ et al. 1997. Defective mammopoiesis, but normal hematopoiesis, in mice with a targeted disruption of the prolactin gene. EMBO J. **16:** 6926–6935.
2. ORMANDY, C.J., A. CAMUS, J. BARRA et al. 1997. Null mutation of the prolactin receptor gene produces multiple reproductive defects in the mouse. Genes Dev. **11:** 167–178.
3. KRISHNAN, R.S. & J.C. DANIEL, JR. 1967. "Blastokinin": inducer and regulator of blastocyst development in the rabbit uterus. Science **158:** 490–492.
4. BEIER, H.M. 1968. Uteroglobin: a hormone-sensitive endometrial protein involved in blastocyst development. Biochim. Biophys. Acta **160:** 289–291.
5. DENNIG, J., G. HAGEN, M. BEATO & G. SUSKE. 1995. Members of the Sp transcription factor family control transcription from the uteroglobin promoter. J. Biol. Chem. **270:** 12737–12744.
6. SCHOLZ, A., T. MATHIAS & M. BEATO. 1998. Hormone-induced recruitment of Sp1 mediates estrogen activation of the rabbit uteroglobin gene in endometrial epithelium. J. Biol. Chem. **273:** 4360–4366.
7. BRAUN, H. & G. SUSKE. 1998. Combinatorial action of HNF3 and Sp family transcription factors in the activation of the rabbit uteroglobin/CC10 promoter. J. Biol. Chem. **273:** 9821–9828.
8. NORD, M., M. LAG, T.N. CASSEL et al. 1998. Regulation of CCSP (PCB-BP/uteroglobin) expression in primary cultures of lung cells: involvement of C/EBP. DNA Cell Biol. **17:** 481–492.
9. HAYWARD-LESTER, A., A. HEWETSON, E.G. BEALE et al. 1996. Cloning, characterization, and steroid-dependent posttranscriptional processing of RUSH-1α and β, two uteroglobin promoter binding proteins. Mol. Endocrinol. **10:** 1335–1349.
10. HEWETSON, A. & B.S. CHILTON. 1997. Novel elements in the uteroglobin promoter are a functional target for prolactin signaling. Mol. Cell. Endocrinol. **136:** 1–6.
11. WOLF, M., J. KLUG, R. HACKENBERG et al. 1992. Human CC10, the homologue of rabbit uteroglobin: genomic cloning, chromosomal localization, and expression in endometrial cell lines. Hum. Mol. Genet. **1:** 371–378.
12. DANIEL, J.C., JR. 1980. Factors influencing uteroglobin synthesis. *In* Steroid Induced Uterine Proteins, pp. 87–103. Elsevier/North-Holland. Amsterdam/New York.
13. CHILTON, B.S. & J.C. DANIEL, JR. 1987. Differences in the rabbit uterine response to progesterone as influenced by growth hormone or prolactin. J. Reprod. Fertil. **79:** 581–587.
14. DANIEL, J.C., JR., S.C. JUNEJA, S.P. TAYLOR et al. 1988. Variability in the response of the rabbit uterus to progesterone as influenced by prolactin. J. Reprod. Fertil. **84:** 13–21.

15. CHILTON, B.S., S.K. MANI & D.W. BULLOCK. 1988. Servomechanism of prolactin and progesterone in regulating uterine gene expression. Mol. Endocrinol. **2:** 1169–1175.
16. RANDALL, G.W., J.C. DANIEL, JR. & B.S. CHILTON. 1991. Prolactin enhances uteroglobin gene expression by uteri of immature rabbits. J. Reprod. Fertil. **91:** 249–257.
17. RIDER, V. & D.W. BULLOCK. 1988. Progesterone-dependent binding of a *trans*-acting factor to the uteroglobin promoter. Biochem. Biophys. Res. Commun. **156:** 1368–1375.
18. RIDER, V. & C.J. PETERSON. 1991. Activation of uteroglobin gene expression by progesterone is modulated by uterine-specific promoter-binding proteins. Mol. Endocrinol. **5:** 911–920.
19. KLEIS-SANFRANCISCO, S., A. HEWETSON & B.S. CHILTON. 1993. Prolactin augments progesterone-dependent uteroglobin gene expression by modulating promoter-binding proteins. Mol. Endocrinol. **7:** 214–223.
20. SHERIDAN, P.L., M. SCHORPP, M.L. VOZ & K.A. JONES. 1995. Cloning of an SNF2/SWI2-related protein that binds specifically to the SPH motifs of the SV40 enhancer and to the HIV-1 promoter. J. Biol. Chem. **270:** 4574–4587.
21. DING, H., K. DESCHEEMAEKER, P. MARYNEN *et al.* 1996. Characterization of a helicase-like transcription factor involved in the expression of the human plasminogen activator inhibitor-1 gene. DNA Cell Biol. **15:** 429–442.
22. GONG, X., S. KAUSHAL, E. CECCARELLI *et al.* 1997. Developmental regulation of Zbu1, a DNA-binding member of the SWI2/SNF2 family. Dev. Biol. **183:** 166–182.
23. LIN, Y., P.L. SHERIDAN, K.A. JONES & G.A. EVANS. 1995. The HIP116 SNF2/SWI2-related transcription factor gene (SNF2L3) is located on human chromosome 3q25.1-q26.1. Genomics **27:** 381–382.
24. ZHANG, Q., D. EKHTERAE & K.H. KIM. 1997. Molecular cloning and characterization of P113, a mouse SNF2/SWI2-related transcription factor. Gene **202:** 31–37.
25. LI, W.I., C.L. CHEN & J.Y. CHOU. 1989. Characterization of a temperature-sensitive β-endorphin-secreting transformed endometrial cell line. Endocrinology **125:** 2862–2867.
26. FLISS, A.E., F.J. MICHEL, C-L. CHEN *et al.* 1991. Regulation of the uteroferrin gene promoter in endometrial cells: interactions among estrogen, progesterone, and prolactin. Endocrinology **129:** 697–704.
27. MUCCIOLI, G., D. LANDO, G. BELLUSSI & R. DICARLO. 1983. Physiological and pharmacological variations in rabbit prolactin plasma levels. Life Sci. **32:** 703–710.
28. DURAND, P. & J. DJIANE. 1977. Lactogenic activity in the serum of rabbits during pregnancy and early lactation. J. Endocrinol. **75:** 33–42.
29. MCNEILLY, A.S. & H.G. FRIESEN. 1978. Prolactin during pregnancy and lactation in the rabbit. Endocrinology **102:** 1548–1554.
30. FUCHS, A-R., L. CUBILE & M.Y. DAWOOD. 1981. Effects of mating on levels of oxytocin and prolactin in the plasma of male and female rabbits. J. Endocrinol. **90:** 245–253.
31. MCNEILLY, A.S. & H.G. FRIESEN. 1978. Heterologous radioimmunoassay for rabbit prolactin. Endocrinology **102:** 1539–1547.
32. SCHINDLER, U., P. WU, M. ROTHE *et al.* 1995. Components of the Stat recognition code: evidence for two layers of molecular selectivity. Immunity **2:** 689–697.
33. CHILTON, B.S. & A. HEWETSON. 1998. Zinc finger proteins RUSH in where others fear to tread. Biol. Reprod. **58:** 285–294.
34. SAWAYA, P.L., B.R. STRIPP, J.A. WHITSETT & D.S. LUSE. 1993. The lung-specific CC10 gene is regulated by transcription factors from the AP-1, octamer, and hepatocyte nuclear factor 3 families. Mol. Cell. Biol. **13:** 3860–3871.
35. DEVINE, J.H., A. HEWETSON, V.H. LEE & B.S. CHILTON. 1999. After chromatin is SWItched-on can it be RUSHed? Mol. Cell. Endocrinol. **151:** 49–56.
36. ZLATANOVA, J. & K. VAN HOLDE. 1998. Binding to four-way junction DNA: a common property of architectural proteins? FASEB J. **12:** 421–431.
37. POLLOCK, R. & R. TREISMAN. 1990. A sensitive method for the determination of protein-DNA binding specificities. Nucleic Acids Res. **18:** 6197–6204.
38. FREEMONT, P.S. 1993. The RING finger: a novel protein sequence motif related to the zinc finger. Ann. N.Y. Acad. Sci. **684:** 174–192.
39. SAURIN, A.J., K.L.B. BORDEN, M.N. BODDY & P.S. FREEMONT. 1996. Does this have a familiar RING? Trends Biochem. Sci. **21:** 208–214.

40. BARLOW, P.N., B. LUISI, A. MILNER et al. 1994. Structure of the C_3HC_4 domain by H-nuclear magnetic resonance spectroscopy: a new structural class of zinc-finger. J. Mol. Biol. **237:** 201–211.
41. BORDEN, K.L.B. & P.S. FREEMONT. 1996. The RING finger domain: a recent example of a sequence-structure family. Curr. Opin. Struct. Biol. **6:** 395–401.
42. MACKAY, J.P. & M. CROSSLEY. 1998. Zinc fingers are sticking together. Trends Biochem. Sci. **23:** 1–4.
43. BORDEN, K.L.B. 2000. RING domains: master builders of molecular scaffolds. J. Mol. Biol. **295:** 1103–1112.
44. AXELSEN, K.B. & M.G. PALMGREN. 1998. Evolution of substrate specificities in the P-type ATPase superfamily. J. Mol. Evol. **46:** 84–101.
45. MOLLER, J.V., B. JULL & M. LE MAIRE. 1996. Structural organization, ion transport, and energy transduction of P-type ATPases. Biochim. Biophys. Acta **1286:** 1–51.
46. LINGREL, J.B. & T. KUNTZWEILER. 1994. Na^+, K^+-ATPase. J. Biol. Chem. **269:** 19659–19662.
47. NICOTERA, P., D.J. MCCONKEY, D.P. JONES & S. ORRENIUS. 1989. ATP stimulates Ca^{2+} uptake and increases the free Ca^{2+} concentration in isolated rat liver nuclei. Proc. Natl. Acad. Sci. U.S.A. **86:** 453–457.
48. LANINI, L., O. BACHS & E. CARAFOLI. 1992. The calcium pump of the liver nuclear membrane is identical to that of endoplasmic reticulum. J. Biol. Chem. **267:** 11548–11552.
49. ZYLINSKA, L., E. GROMADZINSKA & L. LACHOWICZ. 1999. Short-time effects of neuroactive steroids on rat cortical Ca^{2+}-ATPase activity. Biochim. Biophys. Acta **1437:** 257–264.
50. ACKLAND, M.L., E.J. CORNISH, J.A. PAYNTER et al. 1997. Expression of Menkes disease gene in mammary carcinoma cells. Biochem. J. **328**(part 1): 237–243.
51. CHILTON, B.S. 2001. Prolactin action in the uterine endometrium. *In* The Endometrium. Harwood Academic Pub. Reading, U.K. In press.
52. GURR, M.I., J.B. FINEAN & J.N. HAWTHORNE. 1963. The phospholipids of liver-cell fractions. I. The phospholipid composition of the liver-cell nucleus. Biochim. Biophys. Acta **70:** 406–416.
53. ROSE, H.G. & J.H. FRENSTER. 1965. Composition and metabolism of lipids within repressed and active chromatin of interphase lymphocytes. Biochim. Biophys. Acta **106:** 577–591.
54. TANG, X., M.S. HALLECK, R.A. SCHLEGEL & P. WILLIAMSON. 1996. A subfamily of P-type ATPases with aminophospholipid transporting activity. Science **272:** 1495–1497.
55. SIEGMUND, A., A. GRANT, C. ANGELETTI et al. 1998. Loss of Drs2p does not abolish transfer of fluorescence-labeled phospholipids across the plasma membrane of *Saccharomyces cerevisiae*. J. Biol. Chem. **273:** 34399–34405.

Physiological Regulation of Uteroglobin/CCSP Expression

ALBERT CHANG,[a] PATRICIA RAMSAY,[b,c] BIHONG ZHAO,[c] MOON PARK,[b] SUSAN MAGDALENO,[d] MICHAEL J. REARDON,[a] STEPHEN WELTY,[b] AND FRANCESCO J. DEMAYO[b,c,e]

[a]*Department of Surgery,* [b]*Department of Pediatrics,* [c]*Department of Molecular and Cellular Biology, Baylor College of Medicine, Houston, Texas 77030, USA*

[d]*Department of Developmental Neurobiology, St. Jude Children's Research Hospital, Memphis, TN 38105, USA*

ABSTRACT: Uteroglobin/CCSP is expressed specifically in the Clara cells. This allows the gene to be used as a marker to identify the elements regulating the physiologic and cell-specific expression of this gene. The regulation of UG/CCSP by IFN-γ was shown to be at the level of the proximal promoter by the upregulation of HNF3β. This has allowed the determination of the factors responsible for the expression of UG/CCSP.

INTRODUCTION

Uteroglobin was first discovered in the late 1960s as a secretory product of the rabbit reproductive tract and lung (for review, see reference 1). This protein is also known as blastokinin,[2] uteroglobin,[3] urinary protein 1,[4] polychlorobiphenyl binding protein,[5] Clara cell 10-kDa protein (CC10),[6] Clara cell 16-kDa protein (CC16),[7] and Clara cell secretory protein (CCSP).[8] For sake of simplicity, this protein will be referred to as uteroglobin/Clara cell secretory protein (UG/CCSP). Although expressed in a variety of tissues, the major site of expression of this protein in all species investigated, with the exception of lagomorphs, is the Clara cells of the lungs.[9] The mature form of this protein consists of two identical polypeptides held together in antiparallel orientation by disulfide bridges.[10,11] The antiparallel orientation of these peptides forms a hydrophobic pocket that allows this protein to bind compounds, such as polychlorinated biphenyl compounds,[12] progesterone,[1] and retinol.[13] Among the many biological properties identified with UG/CCSP, several anti-inflammatory properties have been attributed to this protein. This protein has been shown to inhibit phospholipase A_2,[14] the production and function of interferon gamma (IFN-γ),[15] and the neutrophil inflammation and chemotaxic response.[5] Additional evidence that supports this protein playing an important anti-inflammatory role lies in its location within the genome. The human UG/CCSP gene is located on chromosome 11q12.3-q13.1.[16] In this region, other genes are involved in the regula-

[e]Corresponding author: Francesco J. DeMayo, Department of Molecular and Cellular Biology, Baylor College of Medicine, 1 Baylor Plaza, M725, Houston, TX 77030. Voice: 713-798-6241; fax: 713-790-1275.

fdemayo@bcm.tmc.edu

tion of inflammation. This implies that UG/CCSP plays some role in modulating the inflammatory process.

Although the function of UG/CCSP has not been fully defined, ablation of UG/CCSP by homologous recombination in embryonic stem cells has confirmed the protective properties of UG/CCSP. Two groups, each using different targeting strategies to disrupt the expression of this gene, have ablated the mouse UG/CCSP gene (mUG/CCSP). Stripp and coworkers inserted the selection marker, PGKneo, into the first exon of this gene to disrupt expression.[17] The phenotype observed in the unchallenged mouse was a failure of the lungs to accumulate biphenyl compounds and a disruption of Clara cell morphology at the ultrastructural level. There was no disruption in pulmonary function or animal viability observed by these investigators. However, when the UG/CCSP knockout mice were challenged with a hyperoxic insult, these mice showed significantly reduced survival time when compared to wild-type mice. Surveying the lungs of the UG/CCSP knockout mice after hyperoxic challenge demonstrated elevated levels of the proinflammatory cytokines and the interleukins, IL-6, IL-1b, and IL-3, as compared to wild-type mice simultaneously exposed to hyperoxia. This analysis demonstrated a protective role for mUG/CCSP in response to hyperoxic insult.[18] The protective role may be due to maintenance of Clara cell integrity provided by UG/CCSP and/or its anti-inflammatory activity. Zhang and coworkers ablated mUG/CCSP expression by deleting part of the second exon and intron.[19] Again, these investigators observed no obvious disruption in pulmonary function in the unchallenged mouse. The UG/CCSP knockout mice showed a significant increase in serum phospholipase A activity, supporting the observation that UG/CCSP is a potent inhibitor of phospholipase A_2. However, the major discrepancy between the model generated by Zhang and coworkers and that by Stripp and coworkers is that the UG/CCSP –/– mice generated by Zhang and coworkers died from renal failure secondary to fibronectin accumulation in the glomeruli. The latter phenotype was confirmed by a transgenic antisense approach.[20] Despite the difference in phenotype between the two knockout mouse models, there is no doubt that UG/CCSP provides an important protective role in the lungs. The protective function of UG/CCSP as well as the cellular specificity in pulmonary epithelium makes UG/CCSP an ideal model to investigate the physiological regulation of CCSP in the Clara cells of the lungs. The understanding of the molecular regulation of UG/CCSP will allow for not only an understanding of UG/CCSP function, but also an understanding of the biology and differentiation of the pulmonary Clara cell.

CLARA CELLS

First identified as a distinct cell type by Kolliker in 1881, credit was given to Max Clara, who described the same cell type in 1937.[21] Clara concluded that the cells were exocrine secretory cells releasing a substance that was later identified by Kuhn as protein and not mucous in nature.[22] The Clara cells are nonciliated, secretory cells lining the bronchioles of the lung.[21] The distribution and abundance of Clara cells throughout the respiratory tree vary in a species-specific manner.[9,23] Clara cells can be identified in the airways by their distinct dome-shaped morphology and abundance of secretory granules.[23] Clara cells contain abundant levels of cytochrome P450 mixed-function oxidases, and they secrete surfactant proteins A, B, and D, a leukocyte protease inhibitor, a trypsin-like protease, and a 16-kDa protein, Clara cell secretory protein, UG/CCSP.[24] Clara cells have an important protective role in airway physiology. The cytochrome P450 mixed-function oxidases serve to metabolize xenobiotics in the lungs and UG/CCSP serves to protect the lungs against hyperoxic damage[18] and inflammation.[19] Due to the cellular preference and high levels of expression of this protein, UG/CCSP has been the subject of numerous investigations to define its regulation and functional significance. Molecular analysis of UG/CCSP gene expression will define how interactions between the transcription factors regulating UG/CCSP gene expression function to coordinate UG/CCSP gene expression in a cell-specific manner, as well as in response to external stimuli. Investigation of the murine gene for UG/CCSP is especially important because the mouse can be genetically manipulated to allow for the investigation of factors that regulate UG/CCSP gene expression.

REGULATION OF UG/CCSP GENE EXPRESSION

Initial investigations of the rabbit UG/CCSP gene showed that steroid hormones regulated this gene. This steroid hormone regulation of the UG/CCSP gene depends upon the tissue in which this gene is expressed. In the female reproductive tract, this gene is regulated by the ovarian steroids, estrogen and progesterone.[25–27] In the male reproductive tract, this gene is regulated by androgens.[28] In the lungs, UG/CCSP gene expression is influenced by glucocorticoids,[25] IFN-γ,[29] interleukin-4 (IL-4),[30] and hyperoxia.[31] Since most species studied express UG/CCSP predominantly in the lung, the investigation of UG/CCSP in the lung is of particular physiological significance. Glucocorticoids have been shown to increase UG/CCSP expression *in vivo* by administration of dexamethasone to adrenalectomized mice[32,33] and *in vitro* by administering dexamethasone to transformed mouse lung epithelial cells.[34] IFN-γ has been shown to increase UG/CCSP levels both *in vivo* by intratracheal administration of IFN-γ to mice and *in vitro* by the culture of mouse transformed Clara cells in the presence of IFN-γ.[29] Transgenic mice overexpressing IL-4 in the airways have decreased expression of UG/CCSP[30] and mice exposed to chronic hyperoxia have been shown to display a significant decrease in UG/CCSP mRNA.[31] Interestingly, hyperoxia, which increases the expression of other pulmonary epithelial secretory proteins (i.e., surfactant proteins A, B, and C), represses UG/CCSP.[35] Since UG/CCSP protects the lungs from hyperoxic damage, the repression of UG/CCSP levels by hyperoxia may worsen the severity of oxygen damage to the lungs. Identification of the mechanism by which O_2 represses UG/CCSP expression may lead to therapeutic approaches to reduce the severity of hyperoxic damage to the lungs. Molecular analysis of the regulation of the expression of the mouse UG/CCSP gene will allow for an elucidation of these mechanisms.

CHARACTERIZATION OF THE mUG/CCSP GENE

The cloning of the mouse cDNA for the UG/CCSP gene revealed that 288 bp of sequence coded for the mUG/CCSP protein. Sequence analysis revealed that the mUG/CCSP amino acid sequence was 90%, 52%, and 51% homologous to the rat, human, and rabbit UG/CCSP proteins, respectively.[36] The mUG/CCSP cDNA was expressed in bacteria and the recombinant protein was used to generate antibodies to the mUG/CCSP protein. *In situ* hybridization and immunohistochemical analyses were used to determine the developmental expression of mUG/CCSP. *In situ* hybridization analysis detected the mRNA for mUG/CCSP no earlier than on fetal day 17 and, specifically, in the Clara cells. However, immunohistochemical analysis using antibodies either to the rabbit UG/CCSP or to the mUG/CCSP protein could not detect any expression until day 18.[37]

The mUG/CCSP cDNA was also used to clone the genomic sequences of the mUG/CCSP gene. Three genomic clones ranging in size from 12 to 15 kb were isolated from a 129/SV mouse genomic library. The clone with the longest 5′-flanking sequences, that is, 5.1 kb, was characterized and 7.7 kb of this clone was sequenced.[38] The region sequenced included 3.3 kb of 5′-flanking DNA, the entire coding region, and 0.2 kb of 3′-flanking DNA. The mUG/CCSP gene consists of three exons of 114, 187, and 125 bp separated by two introns of 2434 and 1373 bp. This gene structure is conserved between all species where the UG/CCSP gene has been cloned, with differences in the sizes of the intervening sequences.[39] The cloning of the gene for UG/CCSP has been important and has allowed for an extensive analysis of the molecular regulation of transcription of this gene.

REGULATION OF UG/CCSP GENE TRANSCRIPTION

The elements that regulate UG/CCSP gene expression in the lung have been extensively studied *in vitro* by transient transfection analysis in a Clara cell–like adenocarcinoma cell line, the H441 cells, and *in vivo* by transgenic mouse analysis.[40–45] Transgenic mouse analysis demonstrated that 800 base pairs (bp) of the mUG/CCSP promoter were sufficient for directing expression of a human growth hormone (hGH) reporter gene to the Clara cells of the lungs at a level comparable to that of the endogenous mUG/CCSP mRNA.[43] Furthermore, a transgene containing a deletion of the mUG/CCSP promoter that contained only 166 bp of 5′-flanking DNA was able to direct expression of the reporter gene to the lungs of transgenic mice. However, the frequency and level of expression of the hGH reporter transgene were significantly reduced in this transgenic model. The level of expression of hGH under the control of the proximal promoter region (166 bp) was, at maximum, 10% of that of the endogenous mUG/CCSP gene. Therefore, it was demonstrated that there are two promoter regions in the mUG/CCSP 5′-flanking region: a proximal promoter region located between −166 and the start of transcription and a distal promoter region located between −800 and −166. The distal promoter region is responsible for 90% of mUG/CCSP gene expression. A diagram of the promoter elements for the mouse UG/CCSP gene is shown in FIGURE 1. Transient transfection, cotransfection, electrophoretic mobility shift (EMS), and DNase 1 footprint analysis have identified the *cis*-acting elements that regulate transcription of the rat and mouse UG/CCSP gene. A summary of the elements postulated to regulate the UG/CCSP gene is illustrated in FIGURE 1. It has been demonstrated by mutation analysis and transfection analysis that hepatocyte nuclear factors (HNF), HNF3α and HNF3β,[40,44,45] AP-1,[44] and NKx 2.1[42,46] are important for regulating basal transcription of UG/CCSP. Other

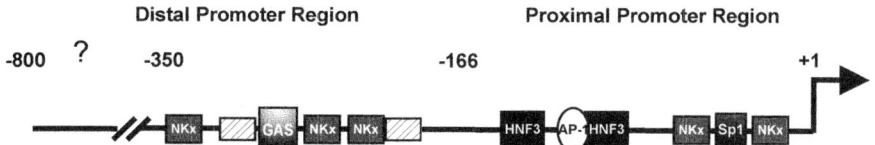

FIGURE 1. A schematic representation of the proximal and distal promoter regions of UG/CCSP. Hatched bars indicate a region where DNA/protein interaction has been observed, but no transcription factor has been identified.

transcription factors indicating that there is indirect evidence for regulating UG/CCSP include SP1, SP3,[47,48] and STAT1.[29] However, no functional analyses of the role of these latter transcription factors have been demonstrated. Also, DNase 1 footprint analysis has identified DNA-protein interactions at −236 to −213[45] and at −314 to −295.[29] These elements in the distal promoter region of the UG/CCSP gene have not been thoroughly investigated due to the limitation of available reagents to analyze the UG/CCSP promoter.

MOUSE TRANSFORMED CLARA CELLS

The majority of the analysis of the UG/CCSP promoter has been conducted using the H441 cell line. The use of the H441 cell line may not be entirely appropriate because this cell line does not express the endogenous human UG/CCSP gene. Transient transfection analysis has demonstrated that H441 cells do not recognize the contribution of the distal promoter region of UG/CCSP toward UG/CCSP gene expression.[43] Identification of the elements in the distal promoter important for UG/CCSP expression has been accomplished by searches for sequence homology, cotransfection experiments in heterologous cell types, and transgenic analysis.[42,45] Although this last approach has been used to identify NKx 2.1 as a potent distal promoter regulator of UG/CCSP, this approach is time-consuming, expensive, and not feasible to investigate the interactions of elements in the distal promoter that are responsible for the regulation of 90% of the expression of UG/CCSP. Thus, an attempt was made to develop an alternative cell line that continued to express UG/CCSP and could serve as a reagent for functional binding studies involving the UG/CCSP gene. This was accomplished by targeting the SV40 large T antigen (TAg) to the Clara cells of mice and isolating transformed Clara cells that retained the expression of the endogenous mUG/CCSP gene.

Transgenic mouse lines were generated with the SV40 TAg under the control of either the rabbit or the mouse UG/CCSP promoter.[47,48] Tumors were formed in mice expressing each transgene and 10 clonal cell lines were derived from these tumors. The cell lines were named mTCC for mouse transformed Clara cells.[29] The mTCC lines exhibited an epithelial phenotype in culture. Immunohistochemical, Northern blot, and Western blot analyses were used to characterize UG/CCSP expression in these cells. All clones isolated expressed the UG/CCSP gene. However, the level of expression of UG/CCSP in these cell lines was reduced to 0.7% of that found in total lung mRNA. The reduction in UG/CCSP expression was limited not only in the mTCC cell lines, but also in the tumors *in situ*. The cell lines were analyzed by Northern blot analysis for their ability to express other lung-specific differentiation markers. These cells expressed SP-B, NKx 2.1, HNF3α, and HNF3β, but did not express SP-A, SP-C, or SP-D. Interestingly, unlike H441 cells, these cells expressed UG/CCSP and the transcription factor HNF3. The expression of the HNF3β protein is important if these cells are to be used in defining the regulation of NKx 2.1 gene expression.

UTILITY OF mTCC LINES

The usefulness of the mouse transformed Clara cells to investigate the molecular regulation of UG/CCSP was made apparent when an investigation was initiated into the potential regulation of UG/CCSP by cytokines. UG/CCSP has been recently shown to inhibit both the action and expression of IFN-γ in the lungs.[15] If the physiological role of UG/CCSP is to regulate IFN-γ, then a negative feedback regulatory loop may exist between UG/CCSP and IFN-γ. In this loop, IFN-γ would stimulate UG/CCSP expression and, in turn, UG/CCSP would inhibit IFN-γ. Using the mTCC lines, it was shown that IFN-γ stimulated the expression of UG/CCSP mRNA in these cells. This regulation of UG/CCSP by IFN-γ was verified *in vivo* when a similar induction of CC10 mRNA was observed after intratracheal administration of IFN-γ to mouse lungs.[29]

POTENTIAL MECHANISMS OF CYTOKINE REGULATION OF UG/CCSP

Examination of the promoter elements identified to date shows several points where cytokines can regulate the expression of UG/CCSP. IFN-γ has been shown to regulate UG/CCSP expression in both the proximal and distal promoter regions.[29] IFN-γ activates gene expression by binding to its receptor located in the cell plasma membrane. Ligand binding of IFN-γ to its receptor triggers the activation of the Janus kinase (JAK)–signal transducer and activator of transcription (STAT) pathway. Activation of the JAK-STAT pathway can activate gene expression by two routes. First, IFN-γ activation of the STAT protein will cause STAT1 to bind to a gamma activation sequence (GAS) in the promoter region of the gene and will activate transcription.[49-53] Second, activation of this JAK-STAT pathway can induce the synthesis of another transcription factor, the interferon response factor, IRF-1, which can activate gene transcription by binding its own response element in the promoter region of a gene.[54] Both pathways are possible for the regulation of mUG/CCSP gene expression. A GAS site has been located by footprint analysis in the distal promoter region of the UG/CCSP promoter.[29] Also, it has been demonstrated that HNF3β is regulated by IFN-γ through activation of the IRF pathway.[55] Therefore, cytokines have the potential to activate UG/CCSP gene expression by signaling through the GAS site in the distal promoter of UG/CCSP and/or by stimulating through HNF3β activity on the proximal promoter. A summary of the potential regulatory pathways for UG/CCSP by cytokines is shown in FIGURE 2.

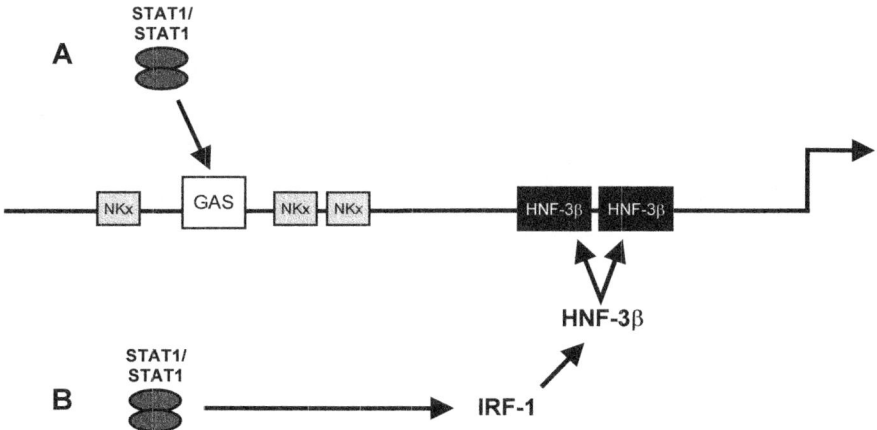

FIGURE 2. The potential mechanisms of IFN-γ stimulation of UC/CCSP expression. (**A**) One potential mechanism is IFN-γ activation of STAT1, which binds directly to the GAS site in the distal promoter of the UG/CCSP gene. (**B**) A second possible mechanism for IFN-γ stimulation is the indirect increase in HNF3β levels after STAT activation through increased transcription of HNF3β by IRF-1.

POTENTIAL MECHANISMS OF O_2 REPRESSION OF UG/CCSP

Examination of the promoter elements of mUG/CCSP also shows two potential mechanisms for O_2 repression of the mUG/CCSP promoter based on how hyperoxia regulates the activity of known transcription factors. Oxygen has been shown to regulate gene expression of antioxidant response enzymes by activating a signaling cascade that results in the activation of transcription factors, such as nuclear factor kappa B (NFκB) and AP-1. These factors bind their response element and activate gene transcription.[56–59] It has been shown that AP-1 binds the UG/CCSP promoter, making it a prime candidate for mediating O_2 regulation of UG/CCSP.[44] AP-1 is a complex of two families of transcription factors: the cJun and cFos families. The cJun family consists of vJun, cJun, JunB, and JunD. The cFos family consists of vFos, cFos, FosB, Fra-1, and Fra-2.[60] Examination of the composition of AP-1 in H441 cells has demonstrated that the AP-1 complex that binds to the UG/CCSP promoter consists of a heterodimer of the transcription factors JunB and Fra-2.[44] While Jun-Fos heterodimers activate gene transcription, Fra-JunB heterodimers do not.[60] Also, the AP-1 site in the UG/CCSP promoter overlaps with one of the HNF3β binding sites. It has been shown that AP-1 and HNF3β cannot simultaneously occupy the same site in the UG/CCSP promoter.[44] Therefore, one possible mechanism for O_2 repression of UG/CCSP gene expression is that O_2 regulation of UG/CCSP gene expression induces AP-1 to the site and subsequent competition of AP-1 with HNF3β on the UG/CCSP promoter.

The second possible mechanism of O_2 repression of UG/CCSP may involve O_2 impairing the activity of NKx 2.1 to activate UG/CCSP gene transcription. It has been shown that oxidation of NKx 2.1 causes the formation of concatamers of this

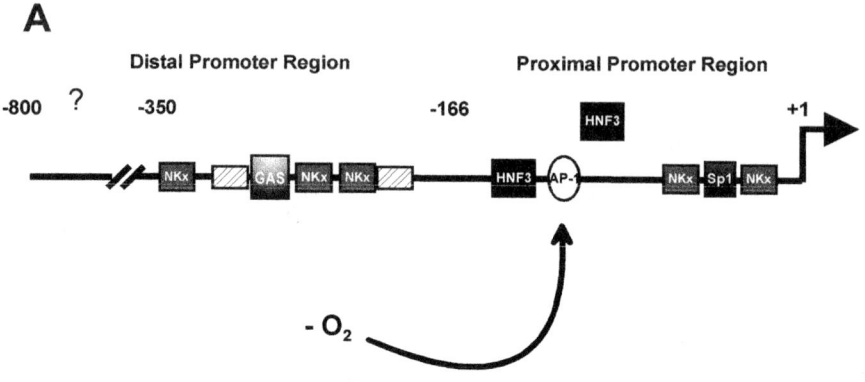

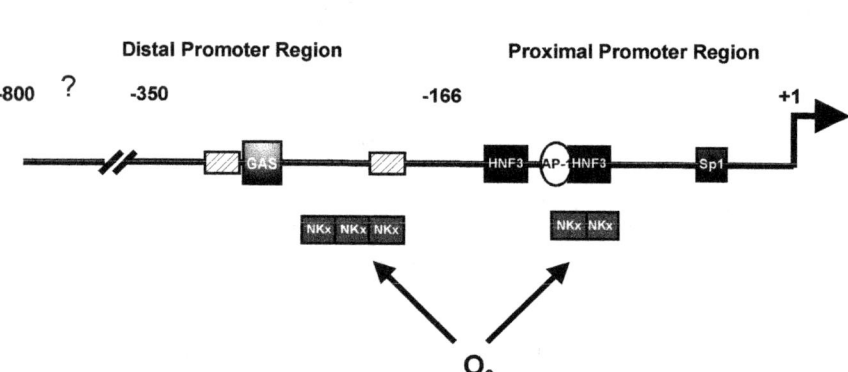

FIGURE 3. The potential mechanisms of hyperoxic repression of UG/CCSP gene expression. **(A)** Oxygen stimulates AP-1 binding to the UG/CCSP proximal promoter region, which displaces the HNF3β binding. **(B)** Oxygen causes NKx 2.1 to no longer bind the UG/CCSP promoter.

transcription factor.[61] Multimers of NKx 2.1 do not bind DNA as well as monomers and therefore are not as transcriptionally active. If O_2 causes oxidation of NKx 2.1, then this may be a possible mechanism of O_2 repression of UG/CCSP. Although O_2 inactivation of NKx 2.1 is a possible mechanism, it is not likely. While hyperoxia represses UG/CCSP gene expression, transcription of surfactant proteins A, B, and C is elevated.[31,35] Since the transcription of the surfactant protein genes is also dependent on NKx 2.1,[62] O_2 repression of NKx 2.1 activity would result in the repression of these genes and not activation. A summary of these potential mechanisms is shown in FIGURE 3.

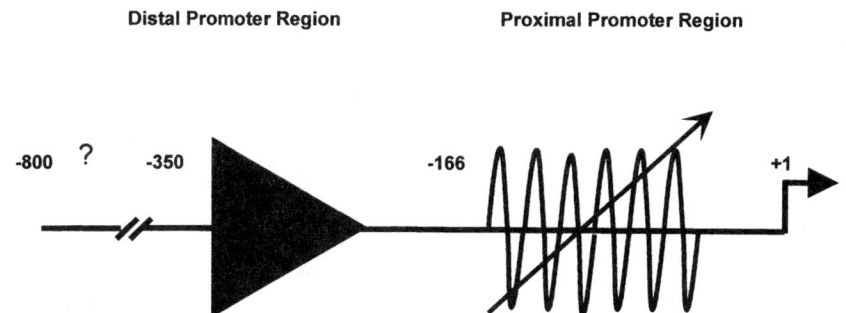

FIGURE 4. A schematic representation of the functional regions of the UG/CCSP promoter. The distal promoter regulates 90% of the transcriptional activity and is portrayed as an amplifier (triangle). The proximal promoter is important for cell-specific expression of UG/CCSP and, possibly, a key regulatory point is represented as a tuner.

DISCUSSION

Although the precise physiological significance of UG/CCSP expression and secretion has not been totally defined, this molecule has served as a marker for the investigation of the cell-specific and physiological regulation of pulmonary gene expression. Identification of the UG/CCSP promoter has identified two promoter regions regulating gene expression. Current analysis of these promoter regions has demonstrated that the distal promoter region is responsible for regulating 90% of this gene activity *in vivo*. However, the proximal promoter regions are important for governing the cell-specific regulation of this gene. These promoter elements are analogous to the circuitry of a radio, with the proximal promoter being the tuner and the distal promoter region being the amplifier (FIG. 4). Current investigations have identified the transcription factors found to be important in regulating UG/CCSP gene expression. These investigations have demonstrated that lung-specific expression of UG/CCSP is not due to the presence of lung-specific transcription factor(s), but due to the appropriate combination of non-tissue-specific transcription factors. However, these studies have not identified elements responsible for the preference of the UG/CCSP gene to be expressed in the Clara cells. Many, if not all, of the elements identified to be important for UG/CCSP gene expression have been demonstrated to be expressed in the alveolar type II cell. Future studies must be conducted to increase our understanding of how these elements coordinate the cell-specific preference of the UG/CCSP promoter.

ACKNOWLEDGMENTS

This manuscript was produced with the technical help of John Ellsworth and the editorial assistance of Janet DeMayo. This work was supported by NIH Grant No. HL61406.

REFERENCES

1. MIELE, L., E. CORDELLA-MIELE & A. MUKHERJEE. 1987. Uteroglobin: structure, molecular biology, and new perspectives on its function as a phospholipase A2 inhibitor. Endocr. Rev. **8:** 474–490.
2. KRISHNAN, R.S. & J.C. DANIEL, JR. 1967. Blastokinin: inducer and regulator of blastocyst development in rabbit uterus. Science **158:** 490–494.
3. BEIER, H.M. 1968. Uteroglobin: a hormone sensitive endometrial protein involved in blastocyst development. Biochim. Biophys. Acta **160:** 289–291.
4. BERNARD, A., R. LAUWERYS, A. NOEL et al. 1989. Urine protein 1: a sex-dependent marker of tubular or glomerular dysfunction. Clin. Chem. **35:** 2141–2142.
5. ANDERSSON, O., L. NORDLUND-MOLLER, M. BRONNEGARD et al. 1991. Purification and level of expression in bronchoalveolar lavage of a human polychlorinated biphenyl (PCB)–binding protein: evidence for a structural and functional kinship to the multihormonally regulated protein uteroglobin. Am. J. Respir. Cell Mol. Biol. **5**(1): 6–12.
6. KATYAL, S.L., G. SINGH, W.E. BROWN et al. 1990. Clara cell secretory (10 kDaltons) protein: amino acid and cDNA nucleotide sequence and developmental expression. Prog. Respir. Res. **25:** 29–35.
7. KABANDA, A., M. JADOUL, J.M. POCHET et al. 1994. Determinants of the serum concentrations of low molecular weight proteins in patients on maintenance hemodialysis. Kidney Int. **45**(6): 1689–1696.
8. STRIPP, B.R., J.A. HUFFMAN & R.J. BOHINSKI. 1994. Structure and regulation of the murine Clara cell secretory protein gene. Genomics **20:** 27–35.
9. PLOPPER, C.G., A.T. MARIASSY & L.H. HILL. 1980. Ultrastructure of the nonciliated bronchiolar epithelial (Clara) cell of mammalian lung. I. A comparison of rabbit, guinea pig, rat, hamster, and mouse. Lung Res. **1:** 139–154.
10. MORIZE, I., E. SURCOUF, M.C. VANEY et al. 1987. Refinement of the C222(1) crystal form of oxidized uteroglobin at 1.34 Å resolution. J. Mol. Biol. **194**(4): 725–739.
11. MORNON, J.P., F. FRIDLANSKY, R. BALLY & E. MILGROM. 1980. X-ray crystallographic analysis of a progesterone-binding protein: the C2221 crystal form of oxidized uteroglobin at 2.2 Å resolution. J. Mol. Biol. **127**(2): 237–239.
12. LUND, J., L. NORDLUND & J-A. GUSTAFSSON. 1988. Partial purification of a binding protein for polychlorinated biphenyls from rat lung cytosol: physiochemical and immunochemical characterization. Biochemistry **27:** 7895–7901.
13. LOPEZ DE HARO, M.S., M. PEREZ MARTINEZ, C. GARCIA & A. NIETO. 1994. Binding of retinoids to uteroglobin. FEBS Lett. **349**(2): 249–251.
14. MIELE, L., E. CORDELLA-MIELE, A. FACCHIANO & A.B. MUKHERJEE. 1988. Novel antiinflammatory peptides from the region of highest similarity between uteroglobin and lipocortin I. Nature **335:** 730–736.
15. DIERYNCK, I., A. BERNARD, H. ROELS & M. DE LEY. 1995. Potent inhibition of both human interferon-γ production and biologic activity by the Clara cell protein CC16. Am. J. Respir. Cell Mol. Biol. **12:** 205–210.
16. ZHANG, Z., D.B. ZIMONJIC, N.C. POPESCU et al. 1997. Human uteroglobin gene: structure, subchromosomal localization, and polymorphism. DNA Cell Biol. **16**(1): 73–83.
17. STRIPP, B.R., J. LUND, G.W. MANGO et al. 1996. Clara cell secretory protein: a determinant of PCB bioaccumulation in mammals. Am. J. Physiol. (Lung Cell. Mol. Physiol.) **271**(15): L656–L664.
18. JOHNSTON, C.J., G.W. MANGO, J.N. FINKELSTEIN & B.R. STRIPP. 1997. Altered pulmonary response to hyperoxia in Clara cell secretory protein deficient mice. Am. J. Respir. Cell Mol. Biol. **17:** 147–155.
19. ZHANG, Z., G.C. KUNDU, C-J. YUAN et al. 1997. Severe fibronectin-deposit renal glomerular disease in mice lacking uteroglobin. Science **276:** 1408–1412.
20. ZHENG, F., G.C. KUNDU, Z. ZHANG et al. 1999. Uteroglobin is essential in preventing immunoglobulin A nephropathy in mice. Nat. Med. **5**(9): 1018–1025.
21. CLARA, M. 1937. Zur histobiologie des bronchalepithels. Z. Mikrosk. Anat. Forsch. **41:** 321–347.
22. KUHN, C., III, L.A. CALLAWAY & F.B. ASKIN. 1974. The formation of granules in the bronchiolar Clara cells of the rat. 1. Electron microscopy. J. Ultrastruct. Res. **49**(3): 387–400.

23. PLOPPER, C., A. MARIASSY, D. WILSON et al. 1983. Comparison of nonciliated tracheal epithelial cells in six mammalian species: ultrastructure and population densities. Exp. Lung Res. **5**(4): 281–294.
24. MASSARO, G.D., G. SINGH, R. MASON et al. 1994. Biology of the Clara cell. Am. J. Physiol. **266**: L101–L106.
25. MIELE, L., E. CORDELLA-MIELE & A.B. MUKHERJEE. 1987. Uteroglobin: structure, molecular biology, and new perspectives in its function on a phospholipase A2 inhibitor. Endocr. Rev. **8**: 474–490.
26. HEINS, B. & M. BEATO. 1981. Hormonal control of uteroglobin secretion and preuteroglobin mRNA content in rabbit endometrium. Mol. Cell. Endocrinol. **21**: 139.
27. SHEN, X.Z., M-J. TSAI & D.W. BULLOCK. 1983. Hormonal regulation of rabbit uteroglobin gene transcription. Endocrinology **112**: 871–876.
28. LOPEZ DE HARO, M.S., C. GARCIA & A. NIETO. 1990. Localization of an estrogen receptor binding site near the promoter of the uteroglobin gene. FEBS Lett. **256**: 20–22.
29. MAGDALENO, S., G. WANG, K. JACKSON et al. 1997. Interferon-gamma regulation of Clara cell gene expression: in vivo and in vitro. Am. J. Physiol. (Lung Cell. Mol. Physiol.) **272**(16): L1142–L1151.
30. JAIN-VORA, S., S.E. WERT, U-A. TEMANN et al. 1997. Interleukin-4 alters epithelial cell differentiation and surfactant homeostasis in the postnatal mouse lung. Am. J. Respir. Cell Mol. Biol. **17**: 541–551.
31. WIKENHEISER, K.A., S.E. WERT, J.R. WISPE et al. 1992. Distinct effect of oxygen on surfactant protein B expression in bronchiolar and alveolar epithelium. Am. J. Physiol. **262**: L32–L39.
32. DEMAYO, F.J., S. DAMAK, T.N. HANSEN & D.W. BULLOCK. 1991. Expression and regulation of the rabbit uteroglobin gene in transgenic mice. Mol. Endocrinol. **5**: 311–318.
33. SANDMOLLER, A., A.K. VOSS, J. HAHN et al. 1991. Cell-specific, developmentally and hormonally regulated expression of the rabbit uteroglobin transgene and the endogenous mouse uteroglobin gene in transgenic mice. Mech. Dev. **34**: 57–68.
34. WIKENHEISER, K.A., D.K. VORBROKER, W.R. RICE et al. 1993. Production of immortalized distal respiratory epithelial cell lines from surfactant protein C/simian virus 40 large tumor antigen transgene mice. Proc. Natl. Acad. Sci. U.S.A. **90**: 11029–11033.
35. MINOO, P., R.J. KING & J.J. COALSON. 1992. Surfactant proteins and lipids are regulated independently during hyperoxia. Am. J. Physiol. **263**(2, pt. 1): L291–L298.
36. MARGRAF, L.R., M.J. FINEGOLD, L.A. STANLEY et al. 1993. Cloning and tissue-specific expression of the cDNA for mouse Clara cell 10 kDa protein: comparison of endogenous expression of rabbit uteroglobin promoter-driven transgene expression. Am. J. Respir. Cell Mol. Biol. **9**: 231–238.
37. RAY, M.K., G.Y. WANG, J. BARRISH et al. 1996. Immunohistochemical localization of mouse Clara cell 10 kDa protein using antibodies against the recombinant protein. J. Histochem. Cytochem. **44**(8): 919–927.
38. RAY, M.K., S. MAGDALENO, B.W. O'MALLEY & F.J. DEMAYO. 1993. Cloning and characterization of mouse Clara cell specific 10 kDa protein gene: comparison of 5′-flanking region with the human, rat, and rabbit gene. Biochem. Biophys. Res. Commun. **197**: 163–171.
39. SNEAD, R., L. DAY, M. MACE et al. 1981. Mosaic structure and mRNA precursor of uteroglobin, a hormone regulated mammalian gene. J. Biol. Chem. **98**: 503–517.
40. BINGLE, C.D., B.P. HACKETT, M. MOXLEY et al. 1995. Role of hepatocyte nuclear factor-3 alpha and hepatocyte nuclear factor-3 beta in Clara cell secretory protein gene expression in the bronchiolar epithelium. Biochem. J. **308**(pt. 1): 197–202.
41. HACKETT, B.P. & J.D. GITLIN. 1992. Cell-specific expression of a Clara cell secretory protein–human growth hormone gene in the bronchiolar epithelium of transgenic mice. Proc. Natl. Acad. Sci. U.S.A. **89**: 9079–9083.
42. RAY, M., C. CHEN, R. SCHWARTZ & F. DEMAYO. 1996. Transcription regulation of a mouse Clara cell-specific protein (mCC10) gene by the NKx transcription factor family members thyroid transcription factor 1 and cardiac muscle-specific homeobox protein (CSX). Mol. Cell. Biol. **16**(5): 2056–2064.

43. Ray, M., S. Magdaleno, M. Finegold & F.J. DeMayo. 1995. Cis-acting elements involved in the regulation of mouse Clara cell-specific 10 kDa protein gene. J. Biol. Chem. **270**(6): 2689–2694.
44. Sawaya, P.L., B.R. Stripp, J.A. Whitsett & D.S. Luse. 1993. The lung-specific CC10 gene is regulated by transcription factors from the AP-1, octamer, and hepatocyte nuclear factor-3 families. Mol. Cell. Biol. **13**(7): 3860–3871.
45. Stripp, B.R., P.L. Sawaya, D.S. Luse et al. 1992. Cis-acting elements that confer lung epithelial cell expression of the CC10 gene. J. Biol. Chem. **261**: 14703–14712.
46. Toonen, R., S. Gowan & C. Bingle. 1996. The lung enriched transcription factor TTF-1 and the ubiquitously expressed protein Sp1 and Sp3 interact with elements located in the minimal promoter of the rat Clara cell secretory protein gene. Biochem. J. **316**: 467–473.
47. DeMayo, F.J., M.J. Finegold, T.N. Hansen et al. 1991. Expression of SV40 T antigen under control of rabbit uteroglobin promoter in transgenic mice. Am. J. Physiol. **261**(2, pt. 1): L70–L76.
48. Magdaleno, S., G. Wang, V. Mireles et al. 1997. Cyclin dependent kinase inhibitor (CDKI) expression in pulmonary Clara cells transformed with SV40 large T antigen in transgenic mice. Cell Growth Differ. **8**: 145–155.
49. Ihle, J. 1996. STATs: signal transducers and activators of transcription. Cell **84**: 331–334.
50. David, M. 1995. Transcription factors in interferon signaling. Pharmacol. Ther. **65**: 149–161.
51. Tanaka, M. & T. Taniguchi. 1992. Cytokine gene regulation: regulatory cis-elements and DNA binding factors involved in the interferon system. Adv. Immunol. **52**: 263–281.
52. Ihle, J.N. 1994. The Janus kinase family and signaling through members of the cytokine receptor superfamily (43757). PSEBM **206**: 268–272.
53. Darnell, J., Jr., I. Kerr & G. Stark. 1994. Jak-STAT pathways and transcriptional activation in response to IFNs and other extracellular signalling protein. Science **264**: 1415–1421.
54. Miyamoto, M., T. Fujita, Y. Kimura et al. 1988. Regulated expression of a gene encoding a nuclear factor, IRF-1, that specifically binds to IFN-beta gene regulatory elements. Cell **54**(6): 903–913.
55. Samadani, U., A. Porcella, L. Pani et al. 1995. Cytokine regulation of the liver transcription factor hepatocyte nuclear factor-3b is mediated by the C/EBP family and interferon regulatory factor 1. Cell Growth Differ. **6**: 879–890.
56. Camhi, S.L., P. Lee & A.M.K. Choi. 1995. The oxidative stress response. New Horizons **3**(2): 170–182.
57. Jaiswal, A.K. 1994. Antioxidant response element. Biochem. Pharmacol. **48**: 439–444.
58. Sen, C.K. & L. Packer. 1996. Antioxidant and redox regulation of gene transcription. FASEB J. **720**(10): 709–720.
59. Sun, Y. & L.W. Oberley. 1996. Redox regulation of transcriptional activators. Free Radical Biol. Med. **21**(3): 335–348.
60. Suzuki, T., H. Okuno, T. Yoshida et al. 1991. Difference in transcriptional regulatory function between c-Fos and Fra-2. Nucleic Acids Res. **19**: 5537–5542.
61. Arnone, M.I., M. Zannini & R. Di Lauro. 1995. The DNA binding activity and the dimerization ability of the thyroid transcription factor 1 are redox regulated. J. Biol. Chem. **270**(20): 12048–12055.
62. Bohinski, R.J., R. Di Lauro & J.A. Whitsett. 1994. The lung-specific surfactant protein B gene promoter is a target for thyroid transcription factor 1 and hepatocyte nuclear factor 3, indicating common factors for organ-specific gene expression along the foregut axis. Mol. Cell. Biol. **14**(9): 5671–5681.

Tumor Necrosis Factor Alpha Stimulation of Human Clara Cell Secretory Protein Production by Human Airway Epithelial Cells

M. J. COWAN, X. HUANG, X. L. YAO, AND J. H. SHELHAMER[a]

Critical Care Medicine Department, Clinical Center, National Institutes of Health, Bethesda, Maryland, USA

ABSTRACT: Clara cell secretory protein (CCSP) or uteroglobin/CC10 is a product of epithelial cells in a variety of organs including the lung. CCSP has antiinflammatory properties and may act as an inhibitor of secretory phospholipase A_2's. Tumor necrosis factor alpha (TNF-α) is capable of inducing the expression of gene products including a variety of cytokines and chemokines in the airway epithelium that may upregulate the airway inflammatory response. Therefore, it was of interest to determine whether this proinflammatory cytokine might also induce the production of a counterregulatory protein such as CCSP, which might modulate the inflammatory response in the airway. Normal human tracheobronchial epithelial cells in primary culture and a human bronchial epithelial cell line (BEAS-2B) were studied. CCSP mRNA levels in BEAS-2B cells were detected by ribonuclease protection assay. CCSP mRNA levels increased in response to TNF-α (20 ng/mL) stimulation after 8–36 h, with the peak increase at 18 h. Immunoblotting of CCSP released from BEAS-2B cells into the culture media demonstrated that TNF-α induced the synthesis and secretion of CCSP over 8 to 18 h. Similarly, TNF stimulated the release of CCSP from human tracheobronchial epithelial cells in primary culture at 8 and 18 h. The CCSP reporter gene including 801 bases 5' of the transcription start site did not increase transcriptional activity in response to TNF-α stimulation. A CCSP mRNA half-life assay indicated that TNF-α induced increases in CCSP mRNA at least in part at a posttranscriptional level. Therefore, TNF-α induces airway epithelial cell expression of human CCSP and may modulate airway inflammatory responses in this manner.

INTRODUCTION

Clara cell secretory protein (CCSP) or CC 10-kDa protein (CC10) was first named from the apparent molecular mass of this protein in nonreducing SDS-PAGE.[1,2] It is also called PCB (polychlorinated biphenyl) binding protein[3,4] and is identical to the urinary protein 1 (P1).[5] CCSP may have immunosuppressive, antiinflammatory, antiproteinase, antiphospholipase A_2, and progesterone binding activities.[6–8]

The airway epithelium may play an active role in initiating, amplifying, and modulating airway inflammation.[9,10] Epithelial cells of airways may release cytokines, chemokines, and growth factors that may initiate or amplify inflammatory pro-

---

[a]Author for correspondence: J. H. Shelhamer, Critical Care Medicine Department, Clinical Center, NIH, Building 10, Room 7-D-43, Bethesda, MD 20892.

cesses. These cells may also produce proteins such as solubilized TNF receptor or IL-1 receptor antagonist that may have immunomodulatory/anti-inflammatory properties.[11,12]

The tumor necrosis factor (TNF) "family" includes two structurally and functionally related proteins, TNF-α and TNF-β. TNF-α is a multifunctional cytokine produced by monocytes, macrophages, lymphocytes, and other cells[13] that has a wide range of activities in various cell types.[14] TNF may be a stimulant of the local and systemic inflammatory processes.[15,16] TNF induces the production of prostaglandins, leukotrienes, and platelet-activating factor, which serve as potent lipid inflammatory mediators in many types of cells.[17] TNF-α also stimulates PLA_2 enzyme activation and gene expression[17–23] in the epithelium. Because TNF-α can induce gene expression in airway epithelial cells and may modulate the inflammatory response in the airway, it was of interest to study the effect of this cytokine on epithelial cell CCSP mRNA expression and CCSP synthesis and secretion.[24]

TNF SIGNALING

Two different TNF-α receptor families have been described. The TNFR-I family, including TNFR-I and Fas, contain a "death domain" and interact with intracellular molecules such as TRADD, MADD, and FADD, which also contain death domains. The TNFR-II group of receptors include TNFR-II and CD40. These interact via a less well characterized motif with the TNF receptor–associated factor (TRAF) family of proteins. TNF-α is active as a homotrimer in two different forms: membrane-bound and extracellular. Signal transduction results from ligand-stimulated receptor clustering into dimers or trimers. There is little preference for binding of free TNF-α to TNFR-I or TNFR-II; however, membrane-bound TNF-α preferentially activates TNFR-II. TNF-α receptor binding initiates two broad intracellular responses: activation of gene expression and apoptosis (FIG. 1).[28] This discussion will focus on events other than the induction of apoptosis.

TNFR-I activation results in accessory protein binding to the intracellular domain of the receptor. Through TRADD-TRAF2 interaction, JHK and p38 result in activation of the AP-1 pathway. Similarly, NFkB may be activated via TRADD, TRAF2, and TANK. In addition, association of RIP with the TNFR-I trimer activates sphingomyelinase, resulting in the activation of MAP kinase pathways. TNFR-II activation with TRAF2, TRAF1, and cIAP binding results in activation of the AP-1 and NFkB pathways.[25]

As previously described, the best-characterized transcription factors activated by TNF-α are NFkB and AP-1. NFkB is a member of the Rel family of activators and normally resides in the cytosol bound to members of the IkB family. Within minutes of TNF stimulation of the cell, IkB becomes phosphorylated by the IkB kinase complex and becomes the target for ubiquitin degradation. Free NFkB translocates to the nucleus as a dimer where it can activate promoters containing the kB site.[28]

AP-1 is formed from homo- or heterodimers of the Jun, Fos, and ATF families of proteins. Different cell types may preferentially form different AP-1 dimers. The activity of these dimers is controlled both at the level of transcription and by phosphorylation.[27]

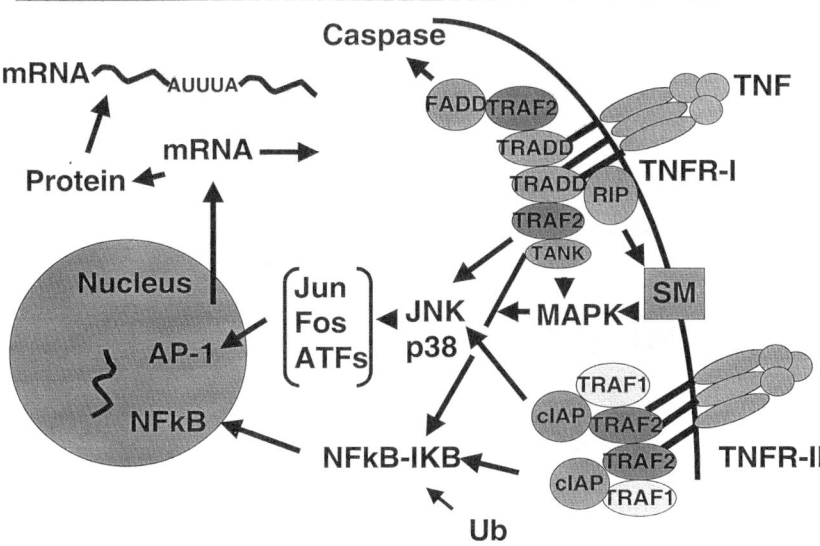

FIGURE 1. TNF signaling pathways. TNFR-I, TNFR-II: TNF receptor I, II; SM: sphingomyelinase; RIP: receptor interacting protein; TRADD: TNF receptor activated death domain; TRAF1, TRAF2: TNF receptor associated factor 1, 2; FADD: Fas associated death domain; TANK: TRAF-binding protein; cIAP: cellular inhibitor of apoptosis; MAPK: MAP kinase; JNK: Jun kinase; ATFs: activating transcription factors; NFkB: nuclear factor kappa B; IkB: inhibitors of NFkB; Ub: ubiquitin. Compiled from references 25–29.

EFFECT OF TNF ON CCSP EXPRESSION IN HUMAN LUNG EPITHELIAL CELLS

A comparison of the proximal portions of the 5′ promoter sequences for the murine and human CCSP genes is presented in TABLE 1. Motifs for sequences activated by proinflammatory mediators such as interferon-γ and TNF are shown. Neither gene has a clear NFkB sequence. Two incomplete AP-1 sites are present in the human promoter at −536 and −396. Therefore, it is possible that AP-1 activation could regulate CCSP expression in human cells. Two lines of evidence suggest that human airway epithelial cells in culture produce CCSP. First, the human airway epithelial cell line, BEAS-2B, contains detectable CCSP mRNA as assayed by ribonuclease protection assay using a 367-base riboprobe (FIG. 2). Second, a Western blot of concentrated supernatant media from BEAS-2B cells demonstrated production and release of CCSP into the supernatant media. Yao *et al.* demonstrated that TNF-α treatment of BEAS-2B cells increased steady state levels of CCSP mRNA over 8 to 36 h (FIG. 2).[24] Similarly, TNF exerts this effect on steady state mRNA levels in a dose-dependent manner (10–40 ng/mL at 24 h) (FIG. 2).[24] TNF-α also increases the production and secretion of CCSP into culture media by BEAS-2B cells in culture

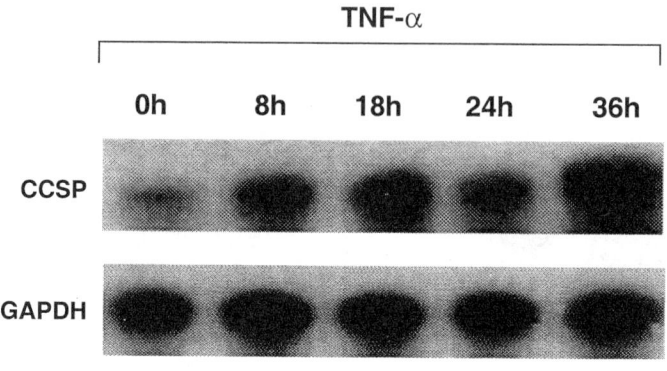

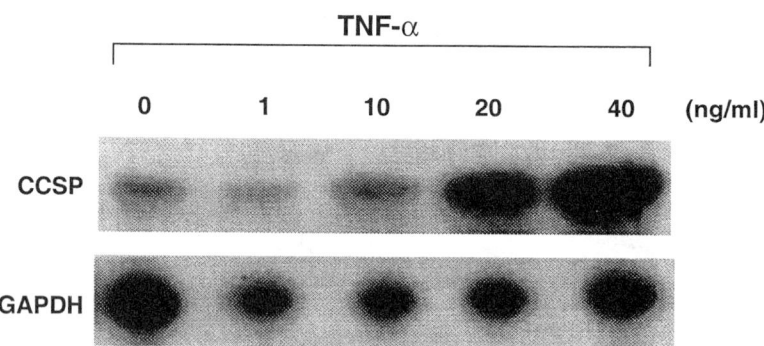

FIGURE 2. (Top) The effect of TNF-α on steady state CCSP mRNA levels. BEAS-2B cells, a human airway epithelial cell line, were treated with TNF-α (20 ng/mL) for 8, 18, 24, and 36 h before total RNA was extracted. Ten or 50 µg of the total RNA was hybridized to GAPDH and CCSP-specific radiolabeled RNA probes and assayed by RPA. The protected fragments of CCSP (367 bp) were visualized by autoradiography. **(Bottom)** The effect of TNF-α on steady state CCSP mRNA levels. BEAS-2B cells were treated with TNF-α (1, 10, 20, and 40 ng/mL) for 24 h before total RNA was extracted. Ten or 50 µg of the total RNA was hybridized to GAPDH and CCSP-specific radiolabeled RNA probes and assayed by RPA. The protected fragments of CCSP (367 bp) were visualized by autoradiography. Reproduced from reference 24 with permission.

TABLE 1. CCSP promoter subsequences

Subsequence	Murine	Human
ISRE (RGAAANNGAAACT)	−160 (12/14)	−872 (13/14)
γ-IRE (CWKKANNY)	5 sites	2 sites
GAS (TTMCNNNAA)	−301	—
AP-1 (TKAGTCA)	—	−536 (6/7) −396 (6/7)
NFkB (GGGRNYYCC)	—	—

NOTE: Bases are numbered from the presumed transcription start site at 30 bases 3′ of the TATA box.

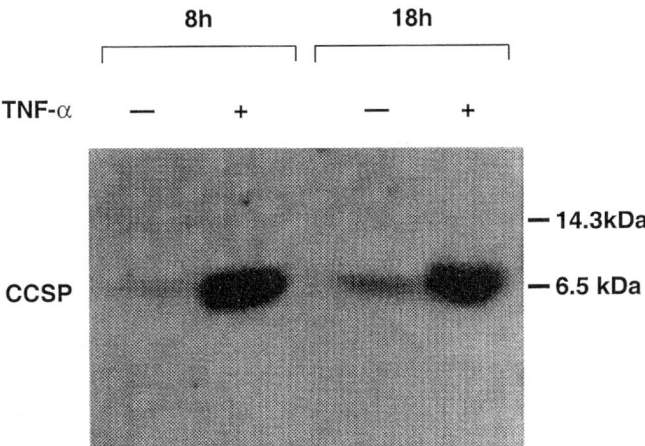

FIGURE 3. The effect of TNF-α on the secretion of CCSP in BEAS-2B cells. BEAS-2B cells were grown to near confluence and incubated with TNF-α (20 ng/mL) for 8 and 18 h. Culture media from treated and untreated cells were concentrated and subjected to gel electrophoresis and immunoblotting. CCSP expression was detected by using 1:500 dilution rabbit–anti-human CC10 polyclonal antibody (a gift from G. Singh, VA Medical Center, Pittsburgh). Reproduced from reference 24 with permission.

at 8 and 18 h of exposure (FIG. 3).[24] Similar results have been found from human tracheobronchial epithelial cells in primary culture.

As noted in FIGURE 1, TNF activation of TNF receptors may result in alteration of transcriptional activity of a variety of genes by signaling through activation of NFkB and Jun/Fos among others. A survey of the murine and human CCSP 5′ promoter regions suggests a variety of sites at which transcriptional activity of the gene might be altered by inflammatory mediators such as TNF-α, interferon-α/β, and interferon-γ (TABLE 1). There is variation between the murine and human promoters in the presence and location of these sites. A portion of the 5′ promoter region of the

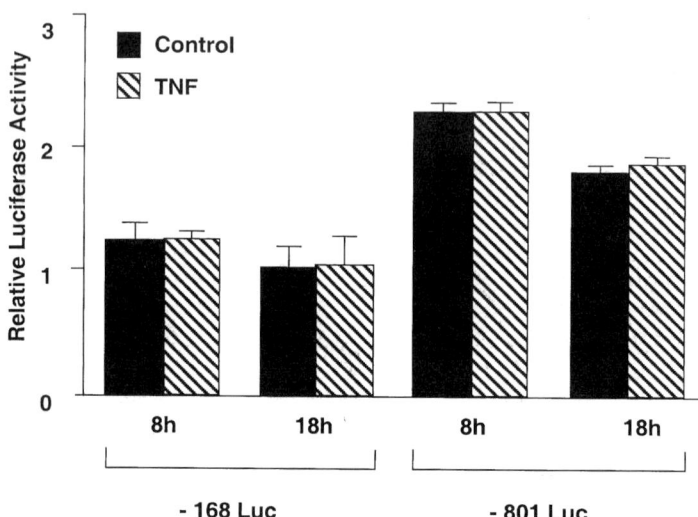

FIGURE 4. The effect of TNF-α treatment on luciferase activity from two reporter gene constructs containing regions of 5' promoter sequences from the CCSP promoter. BEAS-2B cells were transfected with luciferase reporter constructs containing CCSP promoter sequences from +31 to −168 or to −801. Cells were transfected with a plasmid expressing secretory alkaline phosphatase as a control for transfection efficiency. After transfection for 2 h, medium was changed and cells were incubated for 16 h. The transfected cells were then treated for 8 or 18 h with TNF-α (20 ng/mL). The culture medium was assayed for secreted alkaline phosphatase activity. The cells were lysed and the cell lysate was assayed for luciferase activity. The relative luciferase activity is the ratio of the luciferase activity from the cell lysate and the secreted alkaline phosphatase activity from the medium. Solid bars indicate control activity. Hatched bars indicate activity from cells treated with TNF-α. Each point represents a mean of three different experimental determinations, each done in duplicate. Reproduced from reference 24 with permission.

human CCSP gene (−801 to +31 or −168 to +31) has been cloned into a luciferase reporter plasmid. Reporter genes were transfected into BEAS-2B cells, along with a secreted alkaline phosphatase gene as a control for transfection efficiency. After transfection, some cells were treated with TNF-α for 8 or 18 h. TNF-α treatment had no clear effect on the transcriptional activity of these reporter genes (FIG. 4). Changes in steady state mRNA levels in response to TNF stimulation may also be due to posttranscriptional regulation. Toward this end, an mRNA half-life assay has been used to study posttranscriptional regulation. BEAS-2B cells were treated with TNF for 18 h prior to the addition of actinomycin D. Steady state levels of CCSP mRNA and of GAPDH mRNA were studied over the next 24 h. The results are reproduced in FIGURE 5 and demonstrate a more prolonged mRNA half-life in the TNF-treated cells. Together, these results suggest that TNF increases CCSP mRNA and protein production in this human airway epithelial cell line. This increase in steady state mRNA level appears to be due at least in part to posttranscriptional regulation.

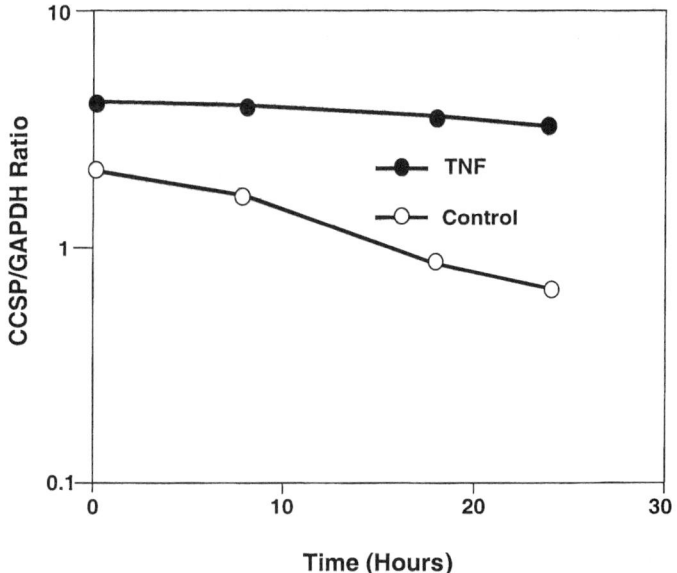

FIGURE 5. Determination of the stability of CCSP mRNA from TNF-α treated cells. BEAS-2B cells were grown to near confluence and treated with TNF-α (20 ng/mL) for 18 h prior to the addition of actinomycin D. Total cellular RNA was isolated after 0, 8, 18, or 24 h. RPA was performed using CCSP and GAPDH cRNA probes. After autoradiography, ratios of the densities of CCSP and GAPDH images were calculated. Reproduced from reference 24 with permission.

The effect of TNF on cellular function is exerted to a significant degree at the transcriptional level.[13,14,22,30] Although the induction of many TNF-responsive genes is mediated at the transcriptional level, the expression of some genes is regulated by TNF at least in part at a posttranscriptional level.[31,32] TNF induction of GLUT1 expression in adipocytes is mediated perhaps by the induction of mRNA binding proteins that protect the AUUUA destabilization motif in the 3′ untranslated region of the mRNA.[33] CCSP mRNA also contains an AUUUA motif in the 3′ untranslated region.

It is interesting to speculate that proinflammatory cytokine mediators such as TNF may also induce the expression of molecules that may modulate over time the inflammatory response as well. This mechanism could allow for the local control of the inflammatory response and protection of the airway from excessive or chronic inflammation.

REFERENCES

1. HACKETT, B.P., N. SCHIMIZU & J.D. GITLIN. 1992. Clara cell secretory protein gene expression in bronchiolar epithelium. Am. J. Physiol. **262:** 2399–2404.
2. SINGH, G., J. SINGH, S.L. KATYAL *et al.* 1988. Identification, cellular localization, isolation, and characterization of human Clara cell–specific 10 kD protein. J. Histochem. Cytochem. **36:** 73–78.

3. NORDLUND-MOLLER, L., O. ANDERSON, R. AHLGREN *et al.* 1990. Cloning, structure, and expression of a rat binding protein for polychlorinated biphenyls. J. Biol. Chem. **265:** 12690–12693.
4. LUND, J., L. NORDLUND & J. GUSTAFSSON. 1988. Partial purification of a binding protein for polychlorinated biphenyls from rat lung cytosol: physiochemical and immunochemical characterization. Biochemistry **27:** 7895–7901.
5. BERNARD, A., H. ROELS, R. LAUWERYS *et al.* 1992. Human urinary protein 1: evidence for identity with the Clara cell protein and occurrence in respiratory tract and urogenital secretions. Clin. Chim. Acta **207:** 239–249.
6. KATYAL, S.L., G. SINGH, W.E. BROWN *et al.* 1990. Clara cell secretory (10 kDa) protein: amino acid and cDNA nucleotide sequences and developmental expression. Prog. Respir. Res. **25:** 29–35.
7. DAVIDSON, F.F. & E.A. DENNIS. 1989. Biological relevance of lipocortins and related proteins as inhibitors of phospholipase A2. Biochem. Pharmacol. **38:** 3645–3651.
8. MANTILE, G., L. MIELE, E. CORDELLA-MIELE *et al.* 1993. Human Clara cell 10-kDa protein is the counterpart of rabbit uteroglobin. J. Biol. Chem. **268:** 20343–20351.
9. THOMPSON, A.B., R.A. ROBBINS & D.J. ROMBERGER. 1995. Immunological functions of the pulmonary epithelium. Eur. Respir. J. **8:** 127–149.
10. RENNARD, S.I., J.D. BECKMANN & R.A. ROBBINS. 1991. Biology of airway epithelial cells. *In* The Lung: Scientific Foundations, pp. 157–167. Raven Press. New York.
11. LEVINE, S., C. LOGUN, D. CHOPRA *et al.* 1996. Protein kinase C, interleukin-1β, and corticosteroids regulate type 1, 55 kDa soluble TNF receptor shedding from human airway epithelial cells. Am. J. Respir. Cell Mol. Biol. **14:** 254–261.
12. LEVINE, S., T. WU & J. SHELHAMER. 1997. Extracellular release of the type I intracellular interleukin-1 receptor antagonist from human airway epithelial cells: differential regulation by IL-4, IL-13, interferon-γ, and corticosteroids. J. Immunol. **158:** 5949–5957.
13. PFIZENMAIER, K., A. HIMMLER, S. SCHUTZE *et al.* 1992. TNF receptors and TNF signal transduction. *In* Tumor Necrosis Factor: The Molecules and Their Emerging Role in Medicine, pp. 439–472. Raven Press. New York.
14. FIERS, W. 1991. Tumor necrosis factor: characterization at the molecular, cellular, and *in vivo* level. FEBS Lett. **285:** 199–221.
15. CAMMUSSI, G., E. ALBANO, C. TETTA & F. BUSSOLIO. 1991. The molecular action of tumor necrosis factor-α. Eur. J. Biochem. **202:** 3–14.
16. PALOMBELLA, V.J. & J. VILCEK. 1989. Mitogenic and cytotoxic actions of tumor necrosis factor in Balb/c 3T3 cells. J. Biol. Chem. **264:** 18128–18136.
17. KNAUER, M.F., K.J. LONGMUIR, R.S. YAMAMOTO *et al.* 1990. Mechanism of human lymphotoxin and tumor necrosis factor induced destruction of cells *in vitro*: phospholipase activation and deacylation of specific membrane phospholipids. J. Cell. Physiol. **142:** 469–479.
18. CLARK, M.A., M.J. CHEN, S.T. CROOKE & J.S. BOMALAASK. 1988. Tumor necrosis factor (cachectin) induces phospholipase A_2 activity and synthesis of a phospholipase A_2–activating protein in endothelial cells. Biochem. J. **250:** 125–132.
19. HORI, T., S. KASHIYAMA, M. HAYAKAWA *et al.* 1989. Possible role of prostaglandins as negative regulators in growth stimulation by tumor necrosis factor and epidermal growth factor in human fibroblasts. J. Cell. Physiol. **141:** 275–280.
20. HOECK, W.G., C.S. RAMESHA, D.J. CHANG *et al.* 1993. Cytoplasmic phospholipase A_2 activity and gene expression are stimulated by tumor necrosis factor: dexamethasone blocks the induced synthesis. Biochemistry **90:** 4475–4479.
21. WU, T., C.W. ANGUS & J.H. SHELHAMER. 1996. TNF-alpha induces cytosolic phospholipase A2 gene expression in human bronchial epithelial cells. Biochim. Biophys. Acta **1310:** 175–184.
22. OKA, S. & H. ARITA. 1991. Inflammatory factors stimulate expression of group II phospholipase A_2 in rat cultured astrocytes: two distinct pathways of the gene expression. J. Biol. Chem. **266:** 9956–9960.
23. CROWL, R.M., T.J. STOLLER, R.R. CONROY & C.R. STONER. 1991. Induction of phospholipase A_2 gene expression in human hepatoma cells by mediators of the acute phase response. J. Biol. Chem. **266:** 2647–2651.

24. YAO, X.L., S.J. LEVINE, M.J. COWAN et al. 1998. Tumor necrosis factor-α stimulates human Clara cell secretory protein production by human airway epithelial cells. Am. J. Respir. Cell Mol. Biol. **19:** 629–635.
25. GOEDDEL, D.V. 1999. Signal transduction by tumor necrosis factor. Chest **116:** 69S–73S.
26. DARNAY, B.G. & B.B. AGGARWAL. 1999. Signal transduction by tumour necrosis factor and tumour necrosis factor related ligands and their receptors. Ann. Rheum. Dis. **58**(suppl. 1): I2–I13.
27. SZATMÁRY, Z. 1999. Tumor necrosis factor-α: molecular-biological aspects. Neoplasma **46:** 257–266.
28. LEDGERWOOD, E.C., J.S. POBER, J.R. BRADLEY et al. 1999. Recent advances in the molecular basis of TNF signal transduction. Lab. Invest. **79:** 1041–1050.
29. POMERANZ, J.L. & D. BALTIMORE. 1999. NF-kB activation by a signaling complex containing TRAF2, TANK, and TBK1, a novel IKK-related kinase. EMBO J. **18:** 6694–6704.
30. KRONKE, M., S. SCHUTZE, P. SCHEURISCH & K. PFIZENMAIER. 1992. *In* Tumor Necrosis Factors: Structure, Function, and Mechanism of Action, pp. 189–216. Dekker. New York.
31. KOEFFLER, H.P., J. GASSON & A. TOBLER. 1988. Transcriptional and post-transcriptional modulation of myeloid colony-stimulating factor expression by tumor necrosis factor and other agents. Mol. Cell. Biol. **8:** 3432–3438.
32. SHERMAN, M.L., B.L. WEBER, R. DATTA & D.W. KUFE. 1990. Transcriptional and post-transcriptional regulation of macrophage-specific colony stimulating factor gene expression by tumor necrosis factor: involvement of arachidonic acid metabolites. J. Clin. Invest. **85:** 442–447.
33. MCGOWAN, K., S. POLICE, J. WINSLOW & P. PEKALA. 1997. Tumor necrosis factor-α regulation of glucose transporter (GLUT1) mRNA turnover. J. Biol. Chem. **272:** 1331–1337.

Pulmonary Phenotype of CCSP/UG Deficient Mice: A Consequence of CCSP Deficiency or Altered Clara Cell Function?

BARRY R. STRIPP,[a,b] SUSAN D. REYNOLDS,[b] CHARLES G. PLOPPER,[c] INGER-MARGRETHE BØE,[d] AND JOHAN LUND[d]

[b]*Department of Environmental Medicine, University of Rochester, Rochester, New York 14642, USA*

[c]*Department of Anatomy and Cell Biology, University of California at Davis, Davis, California, USA*

[d]*Department of Anatomy and Cell Biology, University of Bergen, Bergen, Norway*

ABSTRACT: Clara cell secretory protein (CCSP) is the most abundant secreted protein within airways of the lung. Moreover, CCSP levels are modulated in human lung disease, supporting a potentially important role for CCSP and/or Clara cells in lung homeostasis. However, *in vivo* roles for CCSP remain elusive. A popular hypothesis is that CCSP is a regulator of the inflammatory response. The purpose of this review is to provide an overview of the phenotype of CCSP null mice and relate this phenotype to proposed functions for the protein. Phenotypic analysis of mice homozygous for the CCSP-1 null allele of the CCSP gene (CCSP–/–1) revealed susceptibility to inhaled oxidant gases. Sensitivity of CCSP–/–1 mice to inhaled ozone is unrelated to alterations in antioxidant defenses, but is associated with increased cellular injury. Additional studies investigating inflammatory control in CCSP deficient mice found no differences between wild-type and CCSP–/–1 mice in their inflammatory response to low-dose inhaled endotoxin exposure, arguing against a role for CCSP in regulation of pulmonary inflammation. The findings among CCSP–/–1 mice of ultrastructural alterations to Clara cell secretory apparatus, with associated changes in airway lining fluid protein composition, demonstrate that the CCSP–/–1 genotype results in more complex changes to airways than CCSP deficiency per se. It can be concluded that CCSP does not regulate endotoxin-induced pulmonary inflammation. Moreover, CCSP–/–1 mice represent a valuable tool for probing functional roles for Clara cells in regulation of airway lining fluid composition and lung pollutant susceptibility.

INTRODUCTION

Clara cell secretory protein (CCSP) is a 16-kDa homodimeric protein that represents the most abundant secreted protein within airways of many mammalian species.[1] In addition to its abundant expression within the lung, CCSP is also expressed

---

[a]Address for correspondence: Barry R. Stripp, Department of Environmental Medicine, University of Rochester, Box EHSC, 575 Elmwood Avenue, Rochester, NY 14642. Voice: 716-275-3685; fax: 716-256-2631.

barry_stripp@urmc.rochester.edu

within the genitourinary tract at levels that are highly species-dependent. Many activities have been ascribed to this protein based primarily upon its *in vitro* properties and have led to widespread speculation that CCSP may serve to regulate local inflammatory responses.[2–5] Other functions that have been proposed for CCSP have been largely based upon the structural conservation of a hydrophobic pocket within the homodimeric protein that serves as the binding site for lipophilic ligands such as progestins and pollutants including methyl-sulfonyl derivatives of polychlorinated biphenyls.[6,7] The finding that CCSP levels are modulated in a variety of human lung diseases such as asthma and chronic obstructive pulmonary disease (COPD) suggests a potentially important role for CCSP and/or Clara cell functions in lung pathobiology.[8–10]

To determine the validity of these proposed functions, mice have been generated that are homozygous for null alleles of the CCSP gene.[11,12] Interestingly, the phenotypes of two independently generated lines of CCSP–/– mice differ quite remarkably. The purpose of this review is to detail the phenotype of the first line of CCSP null mice to be established, termed CCSP–/–1, and to discuss the significance of this phenotype in relation to postulated *in vivo* functions for CCSP.

GENERATION AND PHENOTYPIC CHARACTERIZATION OF CCSP–/–1 MICE

Establishment of CCSP–/–1 Mice

The CCSP-1 allele was generated through insertion of DNA sequences coding for neomycin phosphotransferase under the regulatory control of the phosphoglycerate kinase promoter into position +97 of the first exon, in a reverse orientation relative to the CCSP gene. This was accomplished using genomic sequences cloned from strain 129J mice.[13] Establishment of mice homozygous for the CCSP-1 allele (CCSP–/–1) is associated with the complete lack of detectable CCSP expression as assessed using a sensitive ELISA (FIG. 1). Deficiency of CCSP in CCSP–/–1 mice was not associated with either pulmonary or systemic pathology and did not result in altered growth or breeding performance among either the inbred 129 or congenic C57Bl/6 strains.[11,14]

Altered Susceptibility to Environmental Agents

The principal site of CCSP expression in the mouse is the nonciliated (Clara) cell of the conducting airway epithelium.[15,16] Whereas the genitourinary tract represents a secondary site for expression of CCSP in many mammalian species, little CCSP is expressed within the genitourinary tract of either male or female mice.[16] CCSP is also found at low levels within serum, a consequence of leakage from the pulmonary compartment, and localized to the proximal convoluted tubule of the kidney through the active reuptake of CCSP from the glomerular filtrate.[17,18] Based upon this expression pattern, CCSP–/–1 mice have been challenged by agents capable of causing lung injury and compromised function in an attempt to reveal phenotypic abnormalities associated with CCSP deficiency. Altered susceptibility to both inhaled oxidants and microorganisms has been described.[19–21] This review will focus on the oxidant sensitive phenotype and associated changes in pulmonary inflammation.

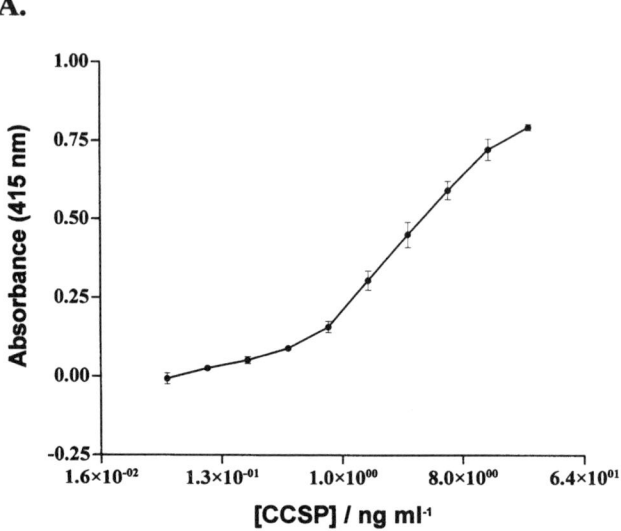
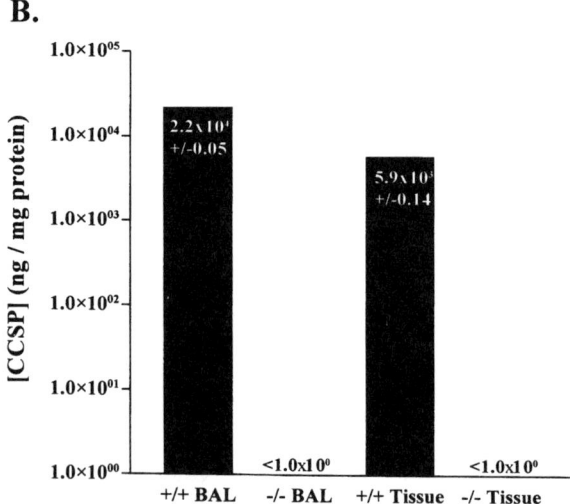

FIGURE 1. CCSP−/−1 mice express undetectable quantities of CCSP using a highly sensitive enzyme-linked immunosorbent assay. A sandwich ELISA was established using goat anti-rat CCSP affinity-purified against immobilized 4x-His-rat CCSP as capture antibody, rabbit anti-rat CCSP as secondary antibody, and anti-rabbit horseradish peroxidase (HRP) conjugate as detection antibody. Immobilized HRP was detected using ABTS substrate. A standard curve generated using recombinant rat CCSP standard (**A**) was used for the estimation of CCSP abundance within bronchoalveolar lavage (BAL) or homogenized lung tissue (Tissue) from either wild-type (+/+) or CCSP knockout (−/−) mice (**B**). CCSP levels were below the limit of detection for BAL and tissue extracts from CCSP−/−1 mice.

Oxidants

Inhalation of oxidants, in the form of either therapeutic gases such as supplemental oxygen or pollutants such as ozone, results in direct lung injury through the production of reactive oxidant species.[22–24] The direct effects of oxidant gases are further compounded by activation of infiltrating inflammatory cells and production of secondary reactive oxygen species. Exposure of CCSP–/–1 mice to either hyperoxia (>95% O_2) or 1.0 ppm ozone was associated with more severe lung injury than that observed in coexposed wild-type counterparts.[19,20] Within the airways of ozone-exposed CCSP–/–1 mice, necrotic cells were detected prior to inflammatory cell influx. This dramatic increase in the magnitude and kinetics of epithelial cell death was accompanied by a corresponding decrease in the abundance of normal ciliated cells and, surprisingly, Clara cells within airways[20] (Plopper *et al.*, submitted). Increased ozone susceptibility was further characterized by an earlier and more pronounced induction of inflammatory and stress response gene expression.[20] However, increased susceptibility of CCSP–/–1 mice to oxidant-induced lung injury was not related to alterations in the abundance of either enzymatic or nonenzymatic antioxidants or to alterations in the regional deposition of inhaled ozone (Mango *et al.*, submitted). Based upon these studies, it is clear that CCSP deficiency is associated with functional alterations to airways that contribute to oxidant susceptibility. However, the nature of these changes and whether this is either a direct or an indirect consequence of CCSP deficiency remain to be determined.

Inflammatory Stimuli

The response of CCSP–/–1 mice to inflammatory stimuli has been investigated using exposure to bacterial endotoxin to elicit a potent inflammatory response without direct cellular injury[25] (Mango *et al.*, submitted). Parameters investigated included analysis of cytokine and chemokine gene expression, and recruitment of polymorphonuclear leukocytes and total inflammatory cells to the lung. When a dose-response study was performed using nebulized endotoxin delivered via inhalation, none of the inflammatory parameters showed significant differences between endotoxin-exposed CCSP–/–1 or wild-type mice. These findings argue against a role for CCSP in regulation of pulmonary inflammatory responses and support the hypothesis that differences between wild-type and CCSP–/–1 mice in hyperoxia-induced inflammatory responses are secondary to differences in susceptibility to oxidant-induced injury.

Structural and Functional Changes to Clara Cells

Even though adult CCSP–/–1 mice do not exhibit gross phenotypic anomalies associated with CCSP deficiency in the steady state, examination of cellular ultrastructure did reveal alterations to the airway epithelium.[11] Most notable within the epithelium was a dramatic reorganization of Clara cell secretory apparatus, including a >95% reduction in abundance of rough endoplasmic reticulum and absence of secretory granules. Unique to the Clara cells of CCSP–/–1 mice were large intracytoplasmic membranous whorls that frequently contained mitochondria within their center (FIG. 2). Ultrastructural changes to Clara cells are associated with changes in the protein constituents of airway lining fluid (Stripp *et al.*, submitted). These

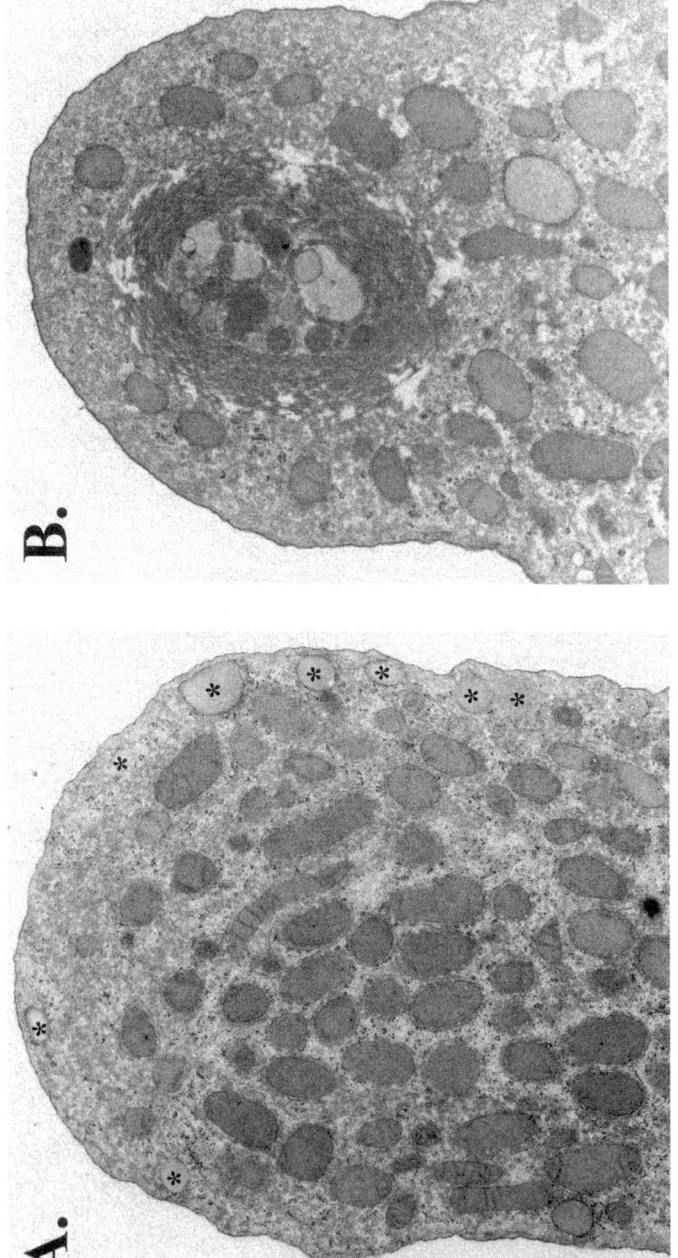

FIGURE 2. Ultrastructural features of Clara cells from wild-type and CCSP−/−1 mice. Shown are the apical portions of Clara cells from (**A**) wild-type or (**B**) CCSP−/−1 mice. Secretory granules decorating the peripheral cytoplasm of Clara cells from wild-type mice (asterisks) are absent among Clara cells from CCSP−/−1 mice. Moreover, Clara cells from CCSP−/−1 mice frequently contain membranous inclusions within their apical cytoplasm.

findings suggest biochemical changes to airways of CCSP deficient mice that may be related to altered processing and/or secretion of Clara cell–derived secretory proteins or to changes in the secretion or stability of non-Clara cell–derived proteins normally present within the airway lining fluid. It has not been determined whether these alterations are directly related to a lack of CCSP expression or are a result of altered Clara cell function.

DISCUSSION

Relationships between the Phenotype of CCSP–/–1 Mice and In Vivo Roles for CCSP

An important question that stems from the results summarized above is whether the phenotype of CCSP–/–1 mice is consistent with proposed physiological functions for CCSP that have been based largely upon the results of *in vitro* studies. CCSP is most frequently cited as a regulator of the inflammatory response.[2–5] Susceptibility of CCSP–/–1 mice to inhaled oxidants cannot be used as a model to either support or refute this hypothesis due to the combination of altered epithelial, inflammatory, and stress-response parameters. A more convincing test of this hypothesis can be made through stimulation of the inflammatory response using agents such as endotoxin at doses that do not cause direct epithelial cell injury. Failure to observe differences in inflammatory response between CCSP–/–1 and wild-type mice across a range of endotoxin doses argues against a role for either CCSP or altered Clara cell function observed in CCSP–/–1 mice in regulation of inflammatory responses.

The phenotype of CCSP–/–1 mice indicates that CCSP is required for normal Clara cell secretory function and maintenance of normal airway lining fluid composition. Furthermore, the sensitivity of Clara cells within the airways of these mice to inhaled oxidant pollutants reveals a critical requirement for normal Clara cell functions in defense of these cells from environmental insult. Although the pathways important to this process have yet to be elucidated, it is unlikely that CCSP-dependent processes modulate the inflammatory response or cell nonautonomous antioxidant defense. The alternatives to these hypotheses that may account for the oxidant susceptibility of CCSP–/–1 mice are that the phenotype may be either (1) the direct result of CCSP deficiency or (2) an indirect effect of alterations in the function of either Clara cells or airway lining fluid. The pleiotropic effects of CCSP deficiency prohibit establishment of a direct link between CCSP itself and the phenotype of CCSP–/–1 mice. As such, oxidant sensitivity of CCSP–/–1 mice is not necessarily a direct reflection of functional roles for CCSP.

Other functions for CCSP have been proposed based upon the phenotype of an independently generated line of CCSP–/– mice (termed CCSP–/–2 in this review). CCSP–/–2 mice exhibit a lethal phenotype with variable penetrance, wasting, and multiple organ damage that is associated with fibronectin accumulation within glomeruli of the kidney.[12] Moreover, CCSP–/–2 mice that survive into adulthood show an increased incidence of neoplastic disease. CCSP–/–1 mice exhibit none of these systemic anomalies.[11,14] The basis for these differences in phenotype has not been determined. However, in light of the moderate pulmonary phenotype of CCSP–/–1 mice, the severe phenotype of CCSP–/–2 mice most likely represents a more complex scenario than deficiency of CCSP per se. Studies investigating the basis for differences in the phenotype of CCSP–/–1 and CCSP–/–2 mice may yield new insights into systemic roles for CCSP.

The phenotype of CCSP–/–1 mice is particularly relevant when considering cellular and molecular mechanisms contributing to the pathogenesis of lung diseases in which CCSP levels are altered. Reduced CCSP levels seen in chronic lung diseases such as asthma and COPD[8–10] are likely to be associated with changes in Clara cell abundance and function, either of which, by analogy with phenotypic consequences of CCSP deficiency among CCSP–/–1 mice, may contribute to establishment and/or exacerbation of the disease. In contrast, a positive correlation between recovery from ARDS and elevations in serum and airway CCSP levels highlights a possible role for Clara cells and Clara cell function in the resolution of disease.[10] Importantly, as is the case for CCSP deficient mice, it is not possible to distinguish between potential roles played by CCSP versus associated changes in Clara cell function in the pathogenesis of these diseases.

REFERENCES

1. HERMANS, C. & A. BERNARD. 1999. Lung epithelium-specific protein characteristics and potential applications as markers. Am. J. Respir. Crit. Care Med. **159:** 646–678.
2. GUY, J., R. DHANIREDDY & A.B. MUKHERJEE. 1992. Surfactant-producing rabbit pulmonary alveolar type II cells synthesize uteroglobin. Biochem. Biophys. Res. Commun. **189:** 662–669.
3. PERI, A., N.H. DUBIN, R. DHANIREDDY & A.B. MUKHERJEE. 1995. Uteroglobin gene expression in the rabbit uterus throughout gestation and in the fetal lung: relationship between uteroglobin and eicosanoid levels in the developing fetal lung. J. Clin. Invest. **96:** 343–353.
4. DIERYNCK, I., A. BERNARD, H. ROELS & M. DE LEY. 1996. The human Clara cell protein: biochemical and biological characterization of a natural immunosuppressor. Multiple Sclerosis **1:** 385–387.
5. MUKHERJEE, A.B., G.C. KUNDU, G. MANTILE-SELVAGGI et al. 1999. Uteroglobin: a novel cytokine? Cell. Mol. Life Sci. **55:** 771–787.
6. LUND, J., I. BRANDT, L. POELLINGER et al. 1985. Target cells for polychlorinated biphenyl metabolite 4,4′-bis(methylsulfonyl)-2,2′,5,5′-tetrachlorobiphenyl: characterization of high affinity binding in rat and mouse lung cytosol. Mol. Pharmacol. **27:** 314–323.
7. GILLNER, M., J. LUND, C. CAMBILAU et al. 1988. The binding of methylsulfonyl-polychloro-biphenyls to uteroglobin. J. Steroid Biochem. **31:** 27–33.
8. BERNARD, A., F.X. MARCHANDISE, S. DEPELCHIN et al. 1992. Clara cell protein in serum and bronchoalveolar lavage. Eur. Respir. J. **5:** 1231–1238.
9. VAN VYVE, T., P. CHANEZ, A. BERNARD et al. 1995. Protein content in bronchoalveolar lavage fluid of patients with asthma and control subjects. J. Allergy Clin. Immunol. **95:** 60–68.
10. JORENS, P.G., Y. SIBILLE, N.J. GOULDING et al. 1995. Potential role of Clara cell protein, an endogenous phospholipase A2 inhibitor, in acute lung injury. Eur. Respir. J. **8:** 1647–1653.
11. STRIPP, B.R., J. LUND, G.W. MANGO et al. 1996. Clara cell secretory protein: a determinant of PCB bioaccumulation in mammals. Am. J. Physiol. **271:** L656–L664.
12. ZHANG, Z., G.C. KUNDU, C.J. YUAN et al. 1997. Severe fibronectin-deposit renal glomerular disease in mice lacking uteroglobin. Science **276:** 1408–1412.
13. STRIPP, B.R., R.J. BOHINSKI & J.A. HUFFMAN. 1994. Structure and regulation of the murine Clara cell secretory protein gene. Genomics **20:** 27–35.
14. REYNOLDS, S.D., G.W. MANGO, R. GELEIN et al. 1999. Normal function and lack of fibronectin accumulation in kidneys of Clara cell secretory protein/uteroglobin deficient mice. Am. J. Kidney Dis. **33:** 598–600.

15. SANDMOLLER, A., A.K. VOSS & J. HAHN. 1991. Cell-specific, developmentally and hormonally regulated expression of the rabbit uteroglobin transgene and the endogenous mouse uteroglobin gene in transgenic mice. Mech. Dev. **34:** 57–67.
16. MARGRAF, L.R., M.J. FINEGOLD, L.A. STANLEY *et al.* 1993. Cloning and tissue-specific expression of the cDNA for the mouse Clara cell 10 kD protein: comparison of endogenous expression to rabbit uteroglobin promoter-driven transgene expression. Am. J. Respir. Cell Mol. Biol. **9:** 231–238.
17. BERNARD, A., C. HERMANS & G. VAN HOUTE. 1997. Transient increase of serum Clara cell protein (CC16) after exposure to smoke. Occup. Environ. Med. **54:** 63–65.
18. BERNARD, A. & R. LAUWERYS. 1995. Low-molecular-weight proteins as markers of organ toxicity with special reference to Clara cell protein. Toxicol. Lett. **77:** 145–151.
19. JOHNSTON, C.J., G.W. MANGO, J.N. FINKELSTEIN & B.R. STRIPP. 1997. Altered pulmonary response to hyperoxia in Clara cell secretory protein deficient mice. Am. J. Respir. Cell Mol. Biol. **17:** 147–155.
20. MANGO, G.W., C.J. JOHNSTON, S.D. REYNOLDS *et al.* 1998. Clara cell secretory protein deficiency increases oxidant stress response in conducting airways. Am. J. Physiol. **275:** L348–L356.
21. HARROD, K.S., A.D. MOUNDAY, B.R. STRIPP & J.A. WHITSETT. 1998. Clara cell secretory protein decreases lung inflammation after acute virus infection. Am. J. Physiol. **275:** L924–L930.
22. PRYOR, W.A. & C.F. CHURCH. 1991. Aldehydes, hydrogen peroxide, and organic radicals as mediators of ozone toxicity. Free Radical Biol. Med. **11:** 41–46.
23. PRYOR, W.A., B. DAS & C.F. CHURCH. 1991. The ozonation of unsaturated fatty acids: aldehydes and hydrogen peroxide as products and possible mediators of ozone toxicity. Chem. Res. Toxicol. **4:** 341–348.
24. WILBORN, A.M., L.B. EVERS & A.T. CANADA. 1996. Oxygen toxicity to the developing lung of the mouse: role of reactive oxygen species. Pediatr. Res. **40:** 225–232.
25. JOHNSTON, C.J., J.N. FINKELSTEIN, G. OBERDORSTER *et al.* 1999. Clara cell secretory protein–deficient mice differ from wild-type mice in inflammatory chemokine expression to oxygen and ozone, but not to endotoxin. Exp. Lung Res. **25:** 7–21.

Insight into the Physiological Function(s) of Uteroglobin by Gene-Knockout and Antisense-Transgenic Approaches

ZHONGJIAN ZHANG,[a] GOPAL C. KUNDU,[a,b] FENG ZHENG,[a,c] C-J. YUAN,[a,d] ERIC LEE,[e] HEINER WESTPHAL,[e] JERROLD WARD,[f] FRANCESCO DEMAYO,[g] AND ANIL B. MUKHERJEE[a,h]

[a]*Section on Developmental Genetics, Heritable Disorders Branch, National Institute of Child Health and Human Development, National Institutes of Health, Bethesda, Maryland 20892-1830, USA*

[e]*Laboratory of Mammalian Genetics and Development, National Institute of Child Health and Human Development, National Institutes of Health, Bethesda, Maryland 20892-1830, USA*

[f]*Veterinary and Tumor Pathology Section, Office of Laboratory Animal Sciences, National Cancer Institute, National Institutes of Health, Frederick, Maryland 21702-1201, USA*

[g]*Department of Cell Biology, Baylor College of Medicine, Houston, Texas 77030, USA*

ABSTRACT: To determine the physiological function(s) of uteroglobin (UG), a steroid-inducible, homodimeric, secreted protein, we have generated transgenic mice that either are completely UG-deficient due to UG gene-knockout (UG-KO) or are partially UG-deficient due to the expression of UG antisense RNA (UG-AS). Both the UG-KO and UG-AS mice develop immunoglobulin A (IgA) nephropathy (IgAN), characterized by microhematuria, albuminuria, and renal glomerular deposition of IgA, fibronectin (Fn), collagen, and C3 complement. This phenotype of both UG-KO and UG-AS mice is virtually identical to that of human IgAN, the most common primary glomerulopathy worldwide. The molecular mechanism by which UG prevents this disease in mice appears to center around UG's interaction with Fn. Since Fn, IgA, and UG are present in circulation and high plasma levels of IgA-Fn complex have been reported in human IgAN, we sought to determine whether UG interacts with Fn and prevents Fn-Fn and/or IgA-Fn interactions, essential for abnormal tissue deposition of Fn and IgA. Our coimmunoprecipitation studies uncovered the formation of Fn-UG heteromers *in vitro* and these heteromers are detectable in the plasma of normal mice, but not UG-KO mice. Further, high plasma levels of IgA-Fn complex, a characteristic of human IgAN patients, were also found in UG-KO mice. Finally, coadministration of UG + Fn or UG + IgA to

[b]Present address: National Center for Cell Science, NCCS Complex, Pune 411007, India.

[c]Present address: Laboratory of Renal Cell Biology, Department of Medicine, University of Miami School of Medicine, P.O. Box 016910 (R126), Miami, FL 33101.

[d]Present address: Department of Biological Sciences and Technology, National Chiao Tung University, 75 Po-Ai Street, 300 Hsinchu, Taiwan.

[h]Author for correspondence: A. B. Mukherjee, National Institutes of Health, Building 10, Room 9S241, Bethesda, MD 20892-1830. Voice: 301-496-7213; fax: 301-402-6632.
mukherja@exchange.nih.gov

UG-KO mice prevented glomerular deposition of Fn and IgA, respectively. Our results define a possible molecular mechanism of IgAN and provide insight into at least one important physiological function of UG in maintaining normal renal function in mice.

It has been more than three decades since the laboratories of Joseph Daniel[1] in the United States and Henning Beier[2] in Germany independently discovered a low-molecular-weight (14.8 kDa), steroid-inducible, secreted protein in the rabbit uterus during early pregnancy. Daniel named it blastokinin[1] and Beier coined the term uteroglobin (UG).[2] It is now clear that this protein is secreted by virtually all mucosal epithelia of all mammals studied so far (for review, see references 3 and 4). In fact, an antigen, cross-reactive to rabbit UG, is detectable in representative species from phylum Amphibia to Mammalia.[5] Besides ovalbumin, UG is probably the most thoroughly studied protein from the standpoint of its structure, molecular biology, and biochemistry, although the physiological function(s) of this protein remained unknown until recently, when UG-deficient mouse models were generated by targeted gene disruption[6] and UG antisense RNA expression[7] in transgenic mice and their phenotypes were characterized. Compelling evidence suggests that UG is a multifunctional protein with protective effects especially on the kidneys[6,7] and lungs.[8]

The physiological importance of this protein is attested not only by the phenotypes that developed in UG-deficient mice, but also by the fact that the UG gene is conserved from rodents to humans during evolution. In fact, there is evidence to suggest that all vertebrates may express a UG-like protein.[3–5] As stated above, UG is a multifunctional secretory protein that has putative receptors via which it appears to regulate vital cellular functions such as motility, invasiveness, and differentiation.[9–11] Taken together, these properties of UG suggest that this protein may be a novel cytokine.[4]

TARGETED DISRUPTION OF THE UG GENE

The isolation and characterization of the UG gene from 129/SVJ mice have been previously reported.[12] This gene was used to generate the gene-targeting construct (FIG. 1A) and was introduced into ES R1 cells[13] by electroporation. The UG gene disruption was achieved by homologous recombination and confirmed in three ES R1 clones by Southern blot analyses (FIG. 1B). These clones were injected into the blastocysts of the C57BL/6J mouse strain, yielding several chimeric founders. Two separate lines of mice, each descended from a chimeric founder, were generated from an independent ES clone carrying UG gene disruption. Mice heterozygous for UG gene disruption (UG$^{+/-}$) were mated and the genotypes of the progeny were analyzed by PCR (FIG. 1C) and Southern blot analyses (FIG. 1D), respectively.

The gene-targeted mice were tested for UG mRNA and UG protein expression in several organs including the lungs, in which UG is constitutively expressed at a high level. Using RT-PCR, UG mRNA expression in the lungs of UG$^{+/+}$ and UG$^{+/-}$ littermates was readily detectable, whereas the lungs of UG$^{-/-}$ mice had no UG mRNA

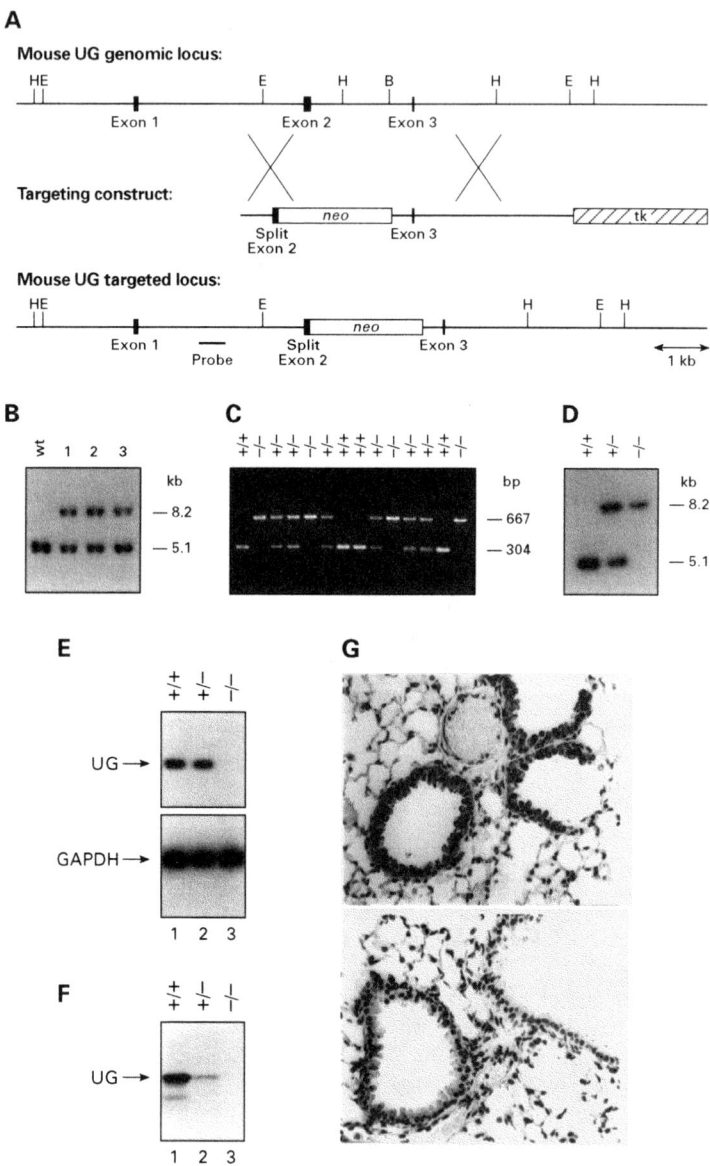

FIGURE 1. Targeting of the UG gene. (**A**) Diagrammatic representation of the UG gene structure, targeting construct, and resulting UG-targeted locus; B, Bam H1; E, Eco R1; H, Hind III. (**B**) Southern blot analysis of the targeted ES-R1 cell clones; wt, wild type. (**C**) Representative PCR analyses of genomic DNA from tail biopsies of offspring. The genotypes and their corresponding PCR products are as follows: $UG^{+/+}$, base pairs (bp); $UG^{+/-}$, 304 and 667 bp; $UG^{-/-}$, 667 bp. (**D**) Southern blot of mouse tail genomic DNA. (**E**) RT-PCR analyses of total RNA from lung tissues of littermates with $UG^{+/+}$, $UG^{+/-}$, and $UG^{-/-}$ geno-

signal (FIG. 1E). Immunoprecipitation and Western blot analyses of UG protein in the lungs showed that $UG^{-/-}$ mice had no detectable UG, while the lungs of $UG^{+/-}$ mice had a drastically reduced level of UG protein compared with $UG^{+/+}$ littermates (FIG. 1F). The results of histopathological analyses of the lungs of $UG^{-/-}$ mice also confirmed that they lacked UG-specific immunostaining in bronchiolar epithelial cells (FIG. 1G). The prostate and uteri of $UG^{-/-}$ mice, but not of $UG^{+/+}$ and $UG^{+/-}$ mice, lacked UG mRNA and protein, respectively.

The Phenotype of UG-Knockout Mice

Mice homozygous for UG gene disruption ($UG^{-/-}$) developed a progressive illness manifesting albuminuria, microhematuria, and profound weight loss. Histopathological examination of affected $UG^{-/-}$ animals revealed a renal disease characterized by massive eosinophilic deposits in the glomeruli (FIG. 2). $UG^{+/-}$ mice manifested a milder form of the renal disease compared with that observed in $UG^{-/-}$ mice. Those $UG^{-/-}$ mice that initially appeared to be healthy also had focal glomerular eosinophilic deposits at 2 months of age (late-onset disease). However, at around 10 months of age, many of these apparently healthy mice had extreme cachexia similar to that of the mice dying at 4 to 5 weeks of age (early-onset disease). Histopathological studies not only revealed that mice with the late-onset disease had severe glomerulopathy like the early-onset disease, but their kidneys also had marked fibrosis of the renal parenchyma with tubular hyperplasia (FIG. 2). Although the predominant pathology in $UG^{-/-}$ mice was found in the kidneys, histopathological studies also uncovered focal necrosis in the pancreas.

Since UG is an anti-inflammatory/immunomodulatory protein, its deficiency may cause reactive amyloidosis. Therefore, we stained kidney sections from $UG^{+/+}$ and $UG^{-/-}$ mice with Congo red, a specific test for the detection of amyloid deposits. It should be noted that, in amyloidosis, the tissue sections stained with Congo red yielded a positive birefringence when examined under polarized microscopy. Our results showed that the glomeruli of UG-null mice were negative for birefringence, indicating that the glomerular deposits are not amyloid. Immunofluorescence studies for the presence of immunocomplexes in the glomeruli of $UG^{-/-}$ mice and immunohistochemical analyses for the presence of major immunocomplexes were also negative. It should be noted, however, that immunofluorescence of immunocomplexes is more readily detectable when fresh frozen tissues are used versus formaldehyde-fixed specimens. This may explain the negative results obtained in our initial characterization of UG-knockout mice using formaldehyde-fixed kidneys.[5] These are discussed in further detail under UG-AS mice.

---

types. A 273-bp RT-PCR product was detected in the lungs of $UG^{+/+}$ and $UG^{+/-}$ mice, but was completely lacking in $UG^{-/-}$ mice. **(F)** Western blot analyses. Proteins (30 µg each) from lung lysates were resolved by electrophoresis on 4–20% gradient SDS-polyacrylamide gels under nonreducing conditions and were immunoblotted with UG antibody. **(G)** Immunohistochemical localization of UG in bronchiolar epithelial cells. The dark staining over the bronchiolar epithelial cells of the $UG^{+/-}$ mouse (*upper panel*) indicates UG immunoreactivity. Note the absence of immunoreactivity in $UG^{-/-}$ mouse lungs (*lower panel*). Magnification: ×100. Reprinted with permission from reference 6. [Figure reduced to 65%.]

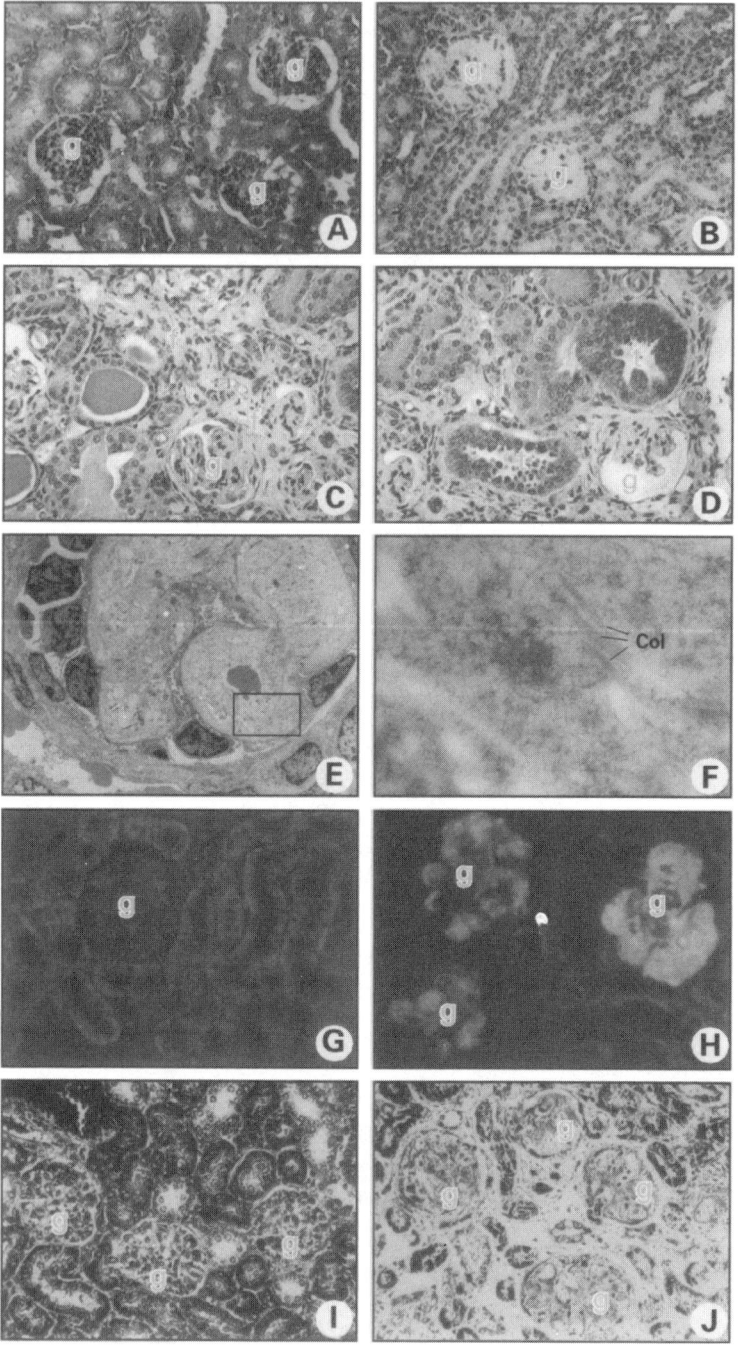

FIGURE 2. *See following page for caption.*

In order to characterize the glomerular deposits more thoroughly, we examined the kidneys of $UG^{-/-}$ mice by transmission electron microscopy. The results showed that the glomerular deposits contained primarily two types of fibrillar structures: (i) long and striated fibrils, which are relatively infrequent, and (ii) fibers that are short and diffuse, which are more abundant (FIGS. 2E and 2F). Since extracellular matrix (ECM) proteins, such as collagen and fibronectin (Fn), produce fibrillar structures, a possibility is raised that these glomerular deposits contain ECM proteins. Analysis of the glomerular deposits by immunofluorescence, using murine Fn antibody, revealed that, compared with the glomeruli of $UG^{+/+}$ mice (FIG. 2G), those of the $UG^{-/-}$ littermates had intense Fn immunoreactivity (FIG. 2H). Masson's trichrome staining, which detects collagen, uncovered the following: while the glomeruli of $UG^{+/+}$ mice were negative (FIG. 2I), those of $UG^{-/-}$ (FIG. 2J) mice were highly positive for collagen. It is well known that Fn interacts with other ECM proteins. Accordingly, we tested for the presence of laminin, vitronectin, and osteopontin in the glomeruli of $UG^{+/+}$ and $UG^{-/-}$ mice by immunohistochemistry. The results failed to show a significant glomerular accumulation of these proteins.

One possible mechanism of ECM protein deposition in the glomeruli may be overproduction of these proteins (e.g., Fn) in the kidney. In consideration of this possibility, we determined the relative amounts of Fn mRNA expression using total RNA from the kidneys, lungs, and liver of $UG^{-/-}$ and $UG^{+/+}$ mice by semiquantitative RT-PCR. The results indicated that relative amounts of Fn mRNA in $UG^{+/+}$ and $UG^{-/-}$ animals are virtually identical. Thus, we concluded that overproduction of Fn mRNA in the kidneys was not a likely cause of Fn deposition in the glomeruli of $UG^{-/-}$ mice, although we did not rule out the possibility that overexpression may have occurred locally in the glomeruli. Comparison of the Fn protein levels in the plasma, kidneys, and liver of $UG^{-/-}$ and $UG^{+/+}$ mice by SDS-PAGE under reducing conditions and by Western blotting showed that plasma as well as the kidneys and liver of wild-type mice contained only a 220-kDa Fn species, whereas in those of the $UG^{-/-}$ mice, in addition to the 220-kDa Fn band, the kidney lysates contained a high-molecular-weight, apparently multimeric Fn band (FIG. 3A). It should be noted here that Fn self-aggregation is one of the critical initial steps in Fn matrix assembly and

---

FIGURE 2. Histopathological and immunohistochemical analyses of the kidneys of UG-deficient mice. H & E staining of the kidney sections from (**A**) $UG^{+/+}$ and (**B**) $UG^{-/-}$ littermates. Note the heavy eosinophilic material in the glomeruli of the $UG^{-/-}$ mouse with the severe renal disease. (**C**) A kidney section from a "late-onset", 10-month-old mouse that has severe parenchymal fibrosis. (**D**) A region of the same mouse kidney section shown in panel C showing renal tubular hyperplasia. Magnification: ×40; g, glomerulus; f, fibrosis; t, tubule. (**E**) Transmission electron microscopy of the glomerular deposit of a $UG^{-/-}$ mouse with severe renal disease. Magnification: ×6000. (**F**) The inset in panel E is magnified ×60,000 to show the presence of long striated fibrillar structures consistent with the presence of collagen (col) and short diffuse structures consistent with Fn fibrils. (**G**) Fn immunofluorescence of a kidney section from a $UG^{+/+}$ mouse using murine Fn antibody. Note the absence of Fn immunofluorescence in glomeruli "g" of a $UG^{+/+}$ mouse and (**H**) Fn immunofluorescence in a kidney section from a $UG^{-/-}$ mouse with severe renal disease. Note intense glomerular Fn immunofluorescence. (**I**) Masson's trichrome staining of a kidney section from a $UG^{+/+}$ mouse and (**J**) that of a $UG^{-/-}$ mouse. Note the presence of staining (indicative of collagen) over the glomeruli (g) of $UG^{-/-}$, but not $UG^{+/+}$ mouse kidney sections. Reprinted with permission from reference 6. [Figure reduced to 80%.]

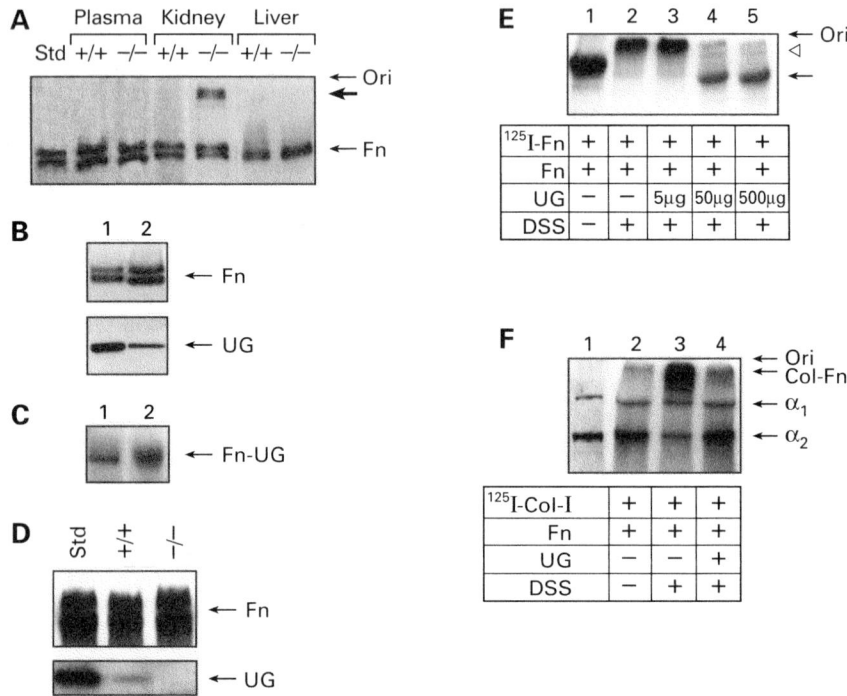

FIGURE 3. Detection of multimeric Fn in UG$^{-/-}$ mice and the effect of UG on Fn-Fn and Fn-collagen interactions. **(A)** Immunoprecipitation and immunoblotting of Fn from plasma, kidney, and liver of UG$^{+/+}$ and UG$^{-/-}$ mice. Immunoprecipitates were resolved by electrophoresis on 4–20% gradient (liver) and 6% (kidney and plasma) SDS-polyacrylamide gels, respectively, under reducing conditions. The bold arrow indicates the multimeric Fn band present only in the kidney lysate of UG$^{-/-}$ mice. **(B)** Binding of Fn with UG followed by coimmunoprecipitation and detection by immunoblotting. The immunoprecipitated proteins were resolved by electrophoresis on either 6% or 4–20% gradient SDS-polyacrylamide gels under reducing and denaturing conditions for Fn and UG, respectively. Note that the immunoprecipitates contain both Fn (lane 2, upper panel) and UG (lane 2, lower panel). Lane 1 of both panels represents corresponding standards stained with Coomassie blue. **(C)** The Fn-^{125}I-UG complex was immunoprecipitated with Fn antibody, and the immunoprecipitates were resolved by electrophoresis on 6% polyacrylamide gels under nonreducing and nondenaturing conditions. Lane 1, Coomassie blue–stained Fn-UG heteromer; lane 2, its autoradiogram. Note that there is no appreciable difference in the electrophoretic mobilities of the heteromer compared with Fn alone as the slight increase in molecular mass of the Fn-UG heteromer cannot be appreciated from that of **(D)** Fn (upper panel) or UG (lower panel). Std, standards for UG and Fn. **(E)** Affinity cross-linking of ^{125}I-Fn with unlabeled Fn in the absence of UG and DSS. Open arrowhead, multimeric Fn; lower thin arrow, 220-kDa Fn. **(F)** Affinity cross-linking of ^{125}I-collagen I with unlabeled Fn in the absence (lane 3) and presence (lane 4) of UG. Lane 1, Coomassie blue–stained collagen I; α_1, α_1-chain of collagen I; α_2, α_2-chain of collagen I. Lane 2, ^{125}I-collagen I and unlabeled Fn in the absence of UG and DSS. Reprinted with permission from reference 6.

fibrillogenesis, and these processes are currently thought to involve integrin activation by lysophosphatidic acid (LPA),[14] a by-product of phospholipase A_2 (PLA_2) catalysis. We have previously reported for the first time[15] that UG is a potent inhibitor of soluble PLA_2 ($sPLA_2$) (for review, see references 3 and 4), a key enzyme that catalyzes the hydrolysis of glycerophosphocholine, yielding arachidonic acid that is further metabolized to generate proinflammatory lipid mediators. We also reported that the α-helix 3 of the UG molecule harbors amino acid residues that are critical for PLA_2 inhibition.[16,17] In fact, the synthetic peptides from this region are potent PLA_2 inhibitors and are anti-inflammatory.[16] Thus, the level of LPA could possibly increase as a result of a lack of UG in $UG^{-/-}$ mice and may contribute to the development of glomerulonephritis, an inflammatory renal disease.[18] Accordingly, we measured the specific activities (μmol/min/mg protein) of serum $sPLA_2$. Compared with the specific activities of serum $sPLA_2$ in $UG^{+/+}$ mice [18 ± 2.8 (SEM)], those in $UG^{-/-}$ mice [36 ± 3.3 (SEM)] were significantly higher ($p < 0.05$). These results raised the possibility that higher PLA_2 activity may have caused increased LPA production and, as a result, promoted integrin activation[14] in $UG^{-/-}$ mice, causing Fn fibrillogenesis to occur.

To further understand how UG may prevent Fn self-aggregation, we determined whether this protein (UG) is capable of disrupting the Fn-Fn interaction *in vitro*. Accordingly, equimolar concentrations of UG and Fn were incubated and immunoprecipitated with Fn antibody. The immunoprecipitates were resolved by SDS-PAGE under reducing conditions and by Western blotting using either Fn or UG antibody. The results showed that Fn coimmunoprecipitated with UG (FIG. 3B). To further confirm these results, we also incubated ^{125}I-UG with Fn and resolved the complexes by electrophoresis, using a 6% polyacrylamide gel under nondenaturing and nonreducing conditions (FIG. 3C). We detected a radioactive Fn-UG heteromer (lane 2) that clearly indicated that Fn interacts with UG. We rationalized that the Fn-UG heteromers would be detectable in plasma of normal animals if our *in vitro* observations are accurate. Thus, we immunoprecipitated the plasma of $UG^{+/+}$ and $UG^{-/-}$ mice with a murine Fn antibody that has no cross-reactivity with UG and found that the Fn antibody coprecipitated both Fn and UG from plasma of $UG^{+/+}$ mice, but not $UG^{-/-}$ mice (FIG. 3D). These results suggest that Fn-UG heteromers are present in the circulation of $UG^{+/+}$ mice, but not $UG^{-/-}$ mice.

To determine the specificity and affinity of UG binding to Fn, we incubated ^{125}I-Fn with unlabeled Fn in the presence and absence of UG. Affinity cross-linking was performed using disuccinimidyl suberate (DSS), a chemical cross-linking agent. In the absence of UG, ^{125}I-Fn formed a radioactive, high-molecular-weight complex with unlabeled Fn; however, in the presence of UG, the formation of Fn-Fn aggregates was inhibited by UG in a concentration-dependent manner (FIG. 3E). To delineate whether there is any difference in the binding affinity of Fn for UG and that of Fn for itself, we performed binding experiments in which ^{125}I-Fn was incubated with unlabeled Fn, immobilized on multiwell plates together with varying concentrations of UG. In separate experiments, we also carried out binding studies of ^{125}I-Fn with unlabeled, immobilized Fn using various concentrations of unlabeled soluble Fn. The Scatchard analyses of the data from both of these binding experiments yielded straight lines with K_d values of 13 nM for the binding of UG with Fn and 176 nM for Fn binding to itself. Due to a relatively higher binding affinity of UG for Fn, these results suggest that UG may effectively counteract Fn self-aggregation. We also performed affinity cross-linking experiments in which ^{125}I-collagen I was incubated with unlabeled Fn in the absence or presence of UG, as described above for Fn. The results indicated that UG counteracts the formation of high-molecular-weight ^{125}I-collagen-Fn aggregates (FIG. 3F).

To test whether UG protects the renal glomeruli from Fn accumulation, we administered soluble human Fn (hFn) alone or hFn mixed with equimolar concentrations of UG intravenously to UG$^{+/+}$ and apparently healthy UG$^{-/-}$ littermates. The rationale for injecting hFn was to be able to discriminate between endogenous murine Fn and administered hFn. The method of intravenous administration and immunohistochemical detection of hFn in various tissues has been described previously.[19] After 24 h, histological sections of the kidneys were examined by immunofluorescence with a monospecific hFn antibody. Human Fn immunofluorescence in the glomeruli of wild-type mice injected either with a mixture of hFn and UG (1:1 molar ratio) or with hFn alone was similar (FIGS. 4A and 4B). However, although the UG$^{-/-}$ mice injected with a mixture of hFn + UG had little hFn-specific immunofluorescence in the glomeruli (FIG. 4C), those receiving Fn alone had higher intensity immunofluorescence (FIG. 4D). Administration of hFn + BSA had no protective effect (data not shown). To determine whether this protection by UG could be overcome by injecting larger quantities of Fn in UG$^{+/+}$ mice, we injected 1 mg of hFn per animal daily for three consecutive days. While intravenous administration of hFn to UG$^{+/+}$ mice at lower doses (500 µg/animal) caused no appreciable glomerular deposition, the administration of higher doses (3 mg/animal) led to a significant accumulation (data not shown). These results suggest that UG prevents glomerular Fn deposition, and UG$^{+/+}$ as opposed to UG$^{-/-}$ mice may have a higher threshold for the accumulation of soluble Fn.

Since Fn self-aggregation leads to abnormal tissue deposition and fibrillogenesis, we sought to determine whether UG prevents Fn fibrillogenesis and matrix assembly *in vitro*. Accordingly, we cultured mouse embryonic fibroblasts in medium containing either soluble hFn or a mixture of equimolar concentrations of hFn and UG. The results showed that the level of fibrillogenesis was much higher when hFN was used alone (FIG. 4E) compared with those cells that were treated with a mixture of hFn and UG (FIG. 4F). These results strongly suggested that UG plays a critical role in preventing Fn fibrillogenesis.

SUPPRESSION OF UG PRODUCTION WITHOUT DISRUPTING THE ENDOGENOUS UG GENE

Because UG has been shown by *in vitro* experiments to have many important biochemical and biological properties,[3,4] we rationalized that targeted disruption of this gene may result in an embryonically lethal phenotype. Thus, concomitant with the UG gene-targeting studies, we undertook the development of antisense-transgenic (UG-AS) mice. These mice expressed UG antisense RNA so that UG production was suppressed, but not completely stopped, without disruption of the endogenous UG gene.

Generation and Characterization of UG-AS Mice

For generating the UG-AS mice, we used a full-length murine UG cDNA[12] in its antisense orientation downstream of the MMTV-LTR promoter of the eukaryotic expression vector pMAM*neo* (FIG. 5a). Out of seven putative founders, four harbored a 262-bp DNA fragment (FIG. 5b) detected by PCR analysis of the genomic DNA. This fragment was lacking in the remaining three pups (FIG. 5b) and was considered negative for the transgene integration into their genome. Two out of the four

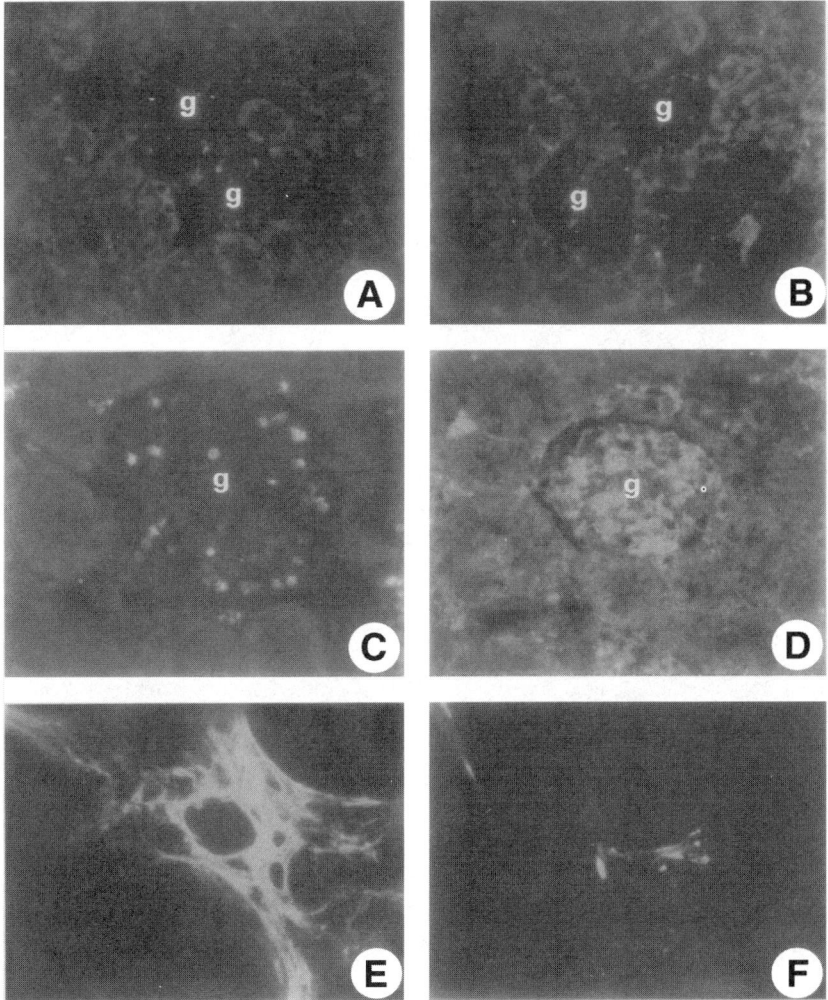

FIGURE 4. Inhibition of glomerular Fn deposition and Fn matrix assembly by UG: **(A)** kidney sections of a wild-type mouse that received equimolar concentrations of Fn + UG; **(B)** a $UG^{+/+}$ mouse that received the same dose of Fn as in panel A, but without UG; **(C)** an apparently healthy $UG^{-/-}$ mouse that received equimolar concentrations of Fn + UG; **(D)** a $UG^{-/-}$ mouse that received the same dose of Fn as in panel C, but without UG; **(E)** Fn fibrillogenesis by mouse embryonic fibroblasts cultured in medium containing soluble hFn alone; **(F)** an identical cell culture as in panel E that was maintained in medium containing hFN + UG. Magnification (A–F): ×145; g, glomerulus. Reprinted with permission from reference 6. [Figure reduced to 75%.]

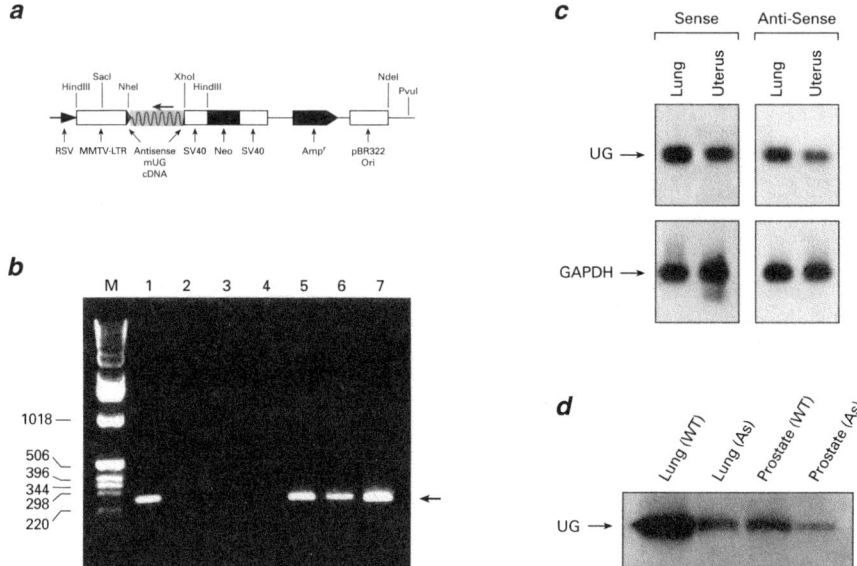

FIGURE 5. Generation and characterization of the UG-AS transgenic mice. (**a**) Diagrammatic representation of the UG-AS transgene construct. Full-length mouse UG cDNA was subcloned in the antisense (UG-AS) orientation between the Nhe I and Xho I sites of the eukaryotic expression vector pMAM*neo* (Clontech, CA). (**b**) Identification of UG-AS transgenic mice by PCR. Genotyping of UG-AS mice using genomic DNA isolated from tail biopsies from 7 pups (*lanes 1–7*) was subjected to PCR analyses using primers mUG-L and mUG-R as described in the text. M = DNA molecular weight markers. (**c**) RT-PCR analysis of total RNA from the lung and uterus; the RT-PCR products were further hybridized with a murine UG cDNA probe. (**d**) Western blot analysis of UG protein expression in the lung and prostate using monospecific polyclonal rabbit anti-mouse UG antibody. Note the reduction of intensity of the UG protein bands in UG-AS mice. Reprinted with permission from reference 7.

founders were successfully bred to the C57BL/6J strain of mice and two independent lines of UG-AS mice were established. The UG-AS mice were identified by both Southern blot and PCR analyses that confirmed the genomic integration and germline transmission of the antisense transgene to the progeny of both founders (data not shown).

To detect the antisense transgene expression, total RNA from the lungs and uteri was analyzed by RT-PCR analysis. The products of RT-PCR were further hybridized with murine UG cDNA probe. The results show that both UG sense and antisense RNAs are expressed in these organs (FIG. 5c). In order to delineate whether UG antisense RNA expression suppressed UG protein production, we performed Western blot analysis of the tissue homogenates using a monospecific murine UG antibody. The results clearly demonstrated that UG protein production in the lungs and prostate of UG-AS mice is drastically reduced versus wild-type littermates (FIG. 5d).

Renal Pathology of UG-AS Mice Is Virtually Identical to That of the UG-Knockout Mice

The results of the UG gene-targeting experiments uncovered that both homozygous ($UG^{-/-}$) and heterozygous ($UG^{+/-}$) mice manifest a phenotype characterized by abnormal glomerular deposition of Fn and collagen.[6] Thus, we sought to determine whether the UG-AS mice also had similar phenotypic abnormalities. We carried out the histopathological analyses using hematoxylin-eosin (H & E) staining. The renal glomeruli of the wild-type littermates showed no abnormalities (FIG. 6a), while abnormal depositions of an eosinophilic material were readily detectable in the glomeruli of UG-AS mice (FIG. 6b). We further characterized the eosinophilic glomerular deposits by Fn-specific immunofluorescence and Masson trichrome staining in order to determine whether they contain Fn and collagen, respectively. While Fn-specific immunofluorescence in the renal glomeruli of wild-type littermates was totally lacking (FIG. 6c), that in the glomeruli of UG-AS mice was abundant (FIG. 6d). Moreover, while Masson trichrome staining failed to detect collagen in the glomeruli of wild-type controls (FIG. 6e), characteristic collagen staining (blue) was found in the kidneys of UG-AS mice (FIG. 6f). These results strongly suggested abnormal accumulation of both Fn and collagen in the glomeruli of UG-AS mice, while being absent in controls. In a previous study[6] with heterozygous UG-knockout ($UG^{+/-}$) mice, we observed that these animals accumulate lower levels of Fn and collagen in the glomeruli compared with those found in $UG^{-/-}$ mice. We obtained similar results in the glomeruli of UG-AS mice, suggesting that the depositions of Fn and collagen in the kidneys of these mice are comparable to those of the $UG^{+/-}$ mice. Thus, using two entirely different experimental approaches, we generated two UG-deficient animal models: one in which the endogenous UG gene is disrupted by gene targeting (UG-knockout) and the other in which the UG protein production is suppressed by the expression of UG antisense RNA (i.e., UG-AS), without disrupting the endogenous UG gene. The fact that both of these models develop an identical disease phenotype clearly defines a nephroprotective physiological role of UG.

Manifestation of Glomerular Disease in UG-AS and UG-Knockout Mice Is Virtually Identical to Human IgA Nephropathy

Since abnormal deposition of Fn and collagen is often associated with the glomerular accumulation of immunocomplexes, we sought to determine whether the glomeruli of UG-AS mice manifest this abnormality. Accordingly, we performed immunohistological analyses of fresh frozen sections of kidneys from UG-AS mice and their wild-type littermates using antibodies against murine IgA, IgG, and IgM. While the glomeruli of wild-type mice were free of any glomerular IgA-specific immunofluorescence (FIG. 6g), those of the UG-AS mice manifested a distinct IgA-specific immunofluorescence (FIG. 6h). No IgG- or IgM-specific fluorescence was detectable in the glomeruli of either wild-type or UG-AS mice (data not shown). These data demonstrate that suppression of UG production in the UG-AS mice results in abnormal glomerular deposition of IgA in addition to Fn and collagen, strongly suggesting a role of UG in preventing abnormal glomerular deposition of not only Fn and collagen, but also IgA.

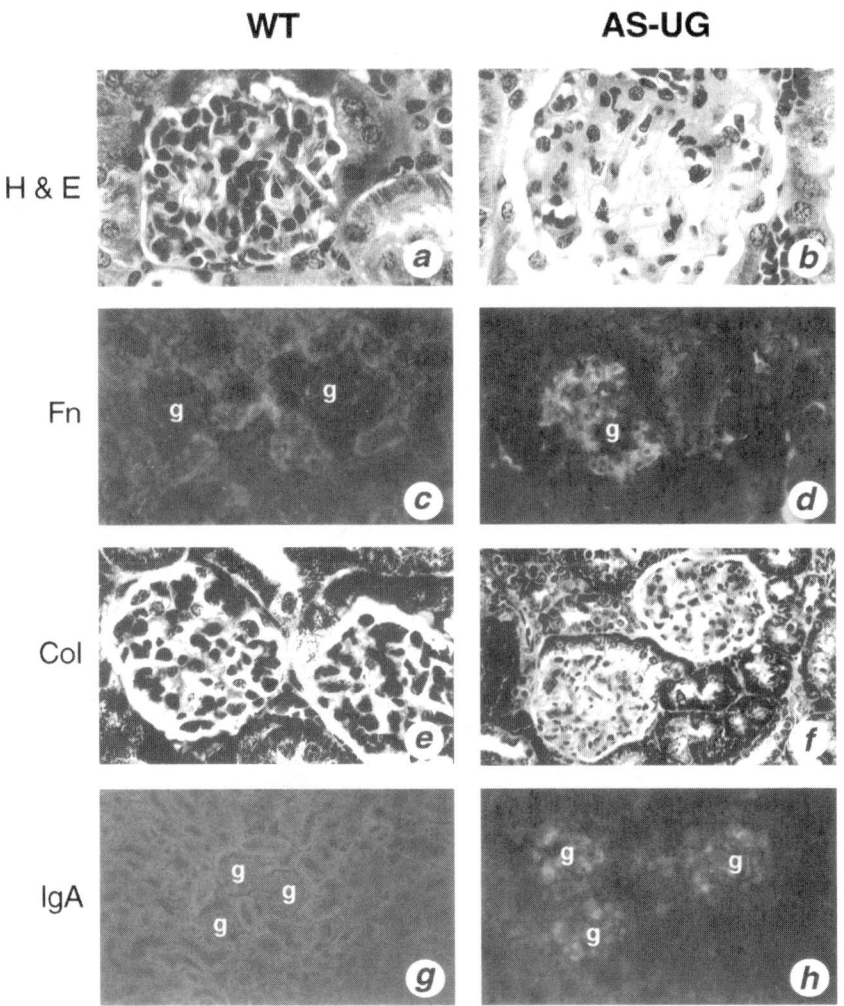

FIGURE 6. Histopathological and immunohistochemical analyses of the kidneys of the UG-AS and wild-type littermates. The kidney sections in the left panels are wild type and those in the right panels are from the UG-AS mice. (*a* & *b*) Hematoxylin and eosin staining. Note the deposition of eosinophilic material in the glomeruli of the UG-AS mouse (*b*). (*c* & *d*) Immunofluorescence using anti-mouse Fn antibody. Note the bright Fn-specific fluorescence localized primarily in the glomeruli of the UG-AS mouse (*d*) that is undetectable in wild-type animals (*c*). (*e* & *f*) Masson's trichrome staining for collagen. Note the positive (brighter) staining in the glomeruli of the UG-AS mice (*f*) that is absent from those of the wild type (*e*). (*g* & *h*) Immunofluorescence to detect IgA deposition. While bright IgA-specific immunofluorescence is readily detectable in UG-AS glomeruli (*h*), it is clearly undetectable in those of the wild type (*g*). Reprinted with permission from reference 7.

The discovery of abnormal IgA deposition in the glomeruli of UG-AS mice led us to reexamine the renal pathology of both $UG^{-/-}$ and $UG^{+/-}$ mice in further detail for glomerular deposition of IgA, in addition to the accumulation of Fn and collagen, reported previously.[6] Using anti-IgA, IgG, and IgM antibodies, we performed immunofluorescence analyses of frozen or partially protease-digested, paraffin-embedded sections of the kidneys from $UG^{-/-}$, $UG^{+/-}$, and $UG^{+/+}$ littermates. While the glomeruli of $UG^{-/-}$ (FIG. 7A) and $UG^{+/-}$ (FIG. 7B) mice accumulate heavy and moderate deposits of IgA, respectively, this immunofluorescence is undetectable in $UG^{+/+}$ littermates (FIG. 7C). Moreover, none of these animals appear to have detectable IgG or IgM immunocomplexes in the glomeruli (data not shown). We found that the use of frozen kidney sections or partial protease digestion of the paraffin-embedded tissue sections prior to the application of the first antibody (not used in our previous study[6]) facilitated the detection of glomerular IgA deposition in the present investigation.

Complement C3 Deposition in the Renal Glomeruli

Since a large percentage of patients with IgAN manifest glomerular accumulation of complement C3, we performed immunofluorescence studies using rabbit anti-complement C3 and FITC-conjugated swine anti-rabbit immunoglobulins. While glomerular deposition of complement C3 is readily detectable in the glomeruli of $UG^{-/-}$ (FIG. 7D) and $UG^{+/-}$ (FIG. 7E) mice, this was not detectable in $UG^{+/+}$ littermates (FIG. 7F). These results indicate that our mouse model manifests pathologic features similar to those of human IgAN.

Presence of Microhematuria

As stated earlier, the characteristic pathologic features of human IgAN include microhematuria, high levels of circulating IgA-Fn complex, and abnormal deposition of IgA, Fn, and collagen in the renal glomeruli. In order to determine whether the UG-knockout mice manifest the pathologic features of human IgAN, we first performed light microscopic examination of their urine. We found that the erythrocyte count in the urine of the $UG^{-/-}$ mice (>200/high-power field) is strikingly higher than that in the urine of the $UG^{+/+}$ littermates (0–5/high-power field), suggesting the presence of significant microhematuria in the UG-knockout mice. The abnormal accumulations of Fn and collagen in the glomeruli of UG-knockout mice,[6] together with the presence of massive IgA deposition and hematuria, make the pathologic findings of the UG-knockout mice consistent with those of the human IgAN.

Circulating IgA-Fn Complex

In addition to the abnormal deposition of IgA in the glomeruli, patients suffering from IgAN manifest high levels of circulating IgA-Fn complex.[20,21] Recently, Waga et al.[22] have reported that the 43-kDa fragment of plasma Fn interacts with IgA and that circulating IgA-Fn complex plays a major role in IgAN. In agreement with this finding, an elevated level of this complex formation has been one of the suggested

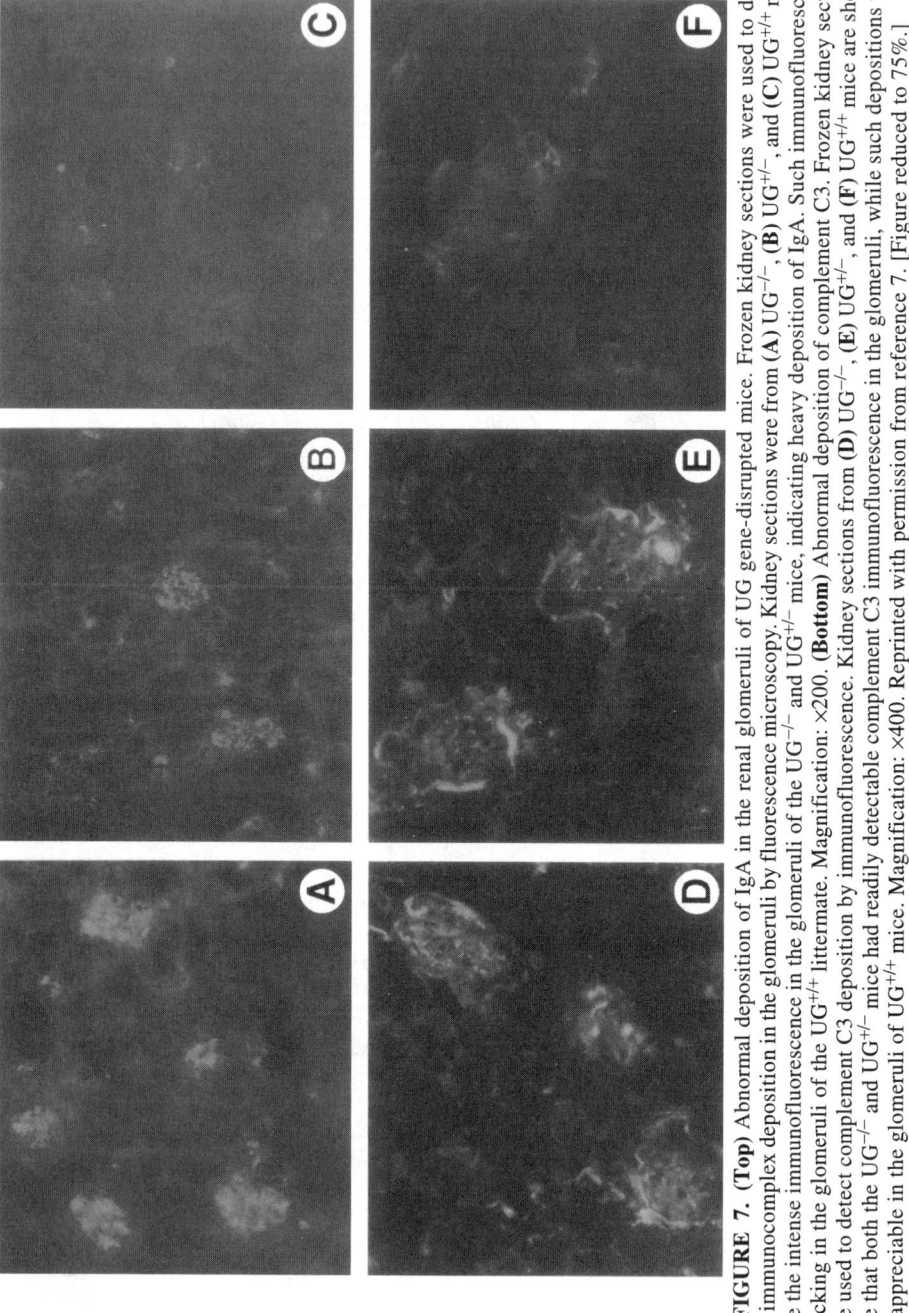

FIGURE 7. **(Top)** Abnormal deposition of IgA in the renal glomeruli of UG gene-disrupted mice. Frozen kidney sections were used to detect IgA immunocomplex deposition in the glomeruli by fluorescence microscopy. Kidney sections were from **(A)** UG$^{-/-}$, **(B)** UG$^{+/-}$, and **(C)** UG$^{+/+}$ mice. Note the intense immunofluorescence in the glomeruli of the UG$^{-/-}$ and UG$^{+/-}$ mice, indicating heavy deposition of IgA. Such immunofluorescence is lacking in the glomeruli of the UG$^{+/+}$ littermate. Magnification: ×200. **(Bottom)** Abnormal deposition of complement C3. Frozen kidney sections were used to detect complement C3 deposition by immunofluorescence. Kidney sections from **(D)** UG$^{-/-}$, **(E)** UG$^{+/-}$, and **(F)** UG$^{+/+}$ mice are shown. Note that both the UG$^{-/-}$ and UG$^{+/-}$ mice had readily detectable complement C3 immunofluorescence in the glomeruli, while such depositions were not appreciable in the glomeruli of UG$^{+/+}$ mice. Magnification: ×400. Reprinted with permission from reference 7. [Figure reduced to 75%.]

mechanisms of abnormal deposition of IgA and Fn in the glomeruli of human IgAN patients. Accordingly, we sought to determine whether a high level of IgA-Fn complex could be detected in the plasma of these mice by an enzyme-linked immunosorbent assay (ELISA). The results (OD at 405 nm) show the following: while the plasma of control mice contained a very low level (0.026 ± 0.016) of IgA-Fn complex, the levels of this complex in the plasma of UG-knockout mice (0.619 ± 0.141) were significantly ($p < 0.05$) higher. Compared with the wild-type littermates, no apparent elevation of the IgA level in the plasma of UG-knockout mice was observed (data not shown). Together with the findings that UG-knockout mice manifest heavy deposits of IgA and Fn in the glomeruli, and the presence of high levels of IgA-Fn complex in the plasma of both human IgAN patients and these mice, this suggests that the elevated level of this complex is a harbinger of abnormal glomerular IgA and Fn deposition. Furthermore, these results also indicate that the pathological findings in UG-knockout mice are very similar, if not identical, to those reported in human IgAN[23] and strongly suggest that this is a valid animal model for the human disease.

UG Disrupts IgA-Fn Complex Formation

In our previous report,[6] we demonstrated that one of the molecular mechanisms by which UG prevents Fn deposition in the renal glomeruli of normal mice is through its binding with Fn at high affinity ($K_d = 13$ nM) to form Fn-UG heteromers that compete with Fn self-aggregation, essential for abnormal tissue deposition.[24,25] We also demonstrated the presence of Fn-UG heteromers in the plasma of wild-type mice, but not in UG-knockout mice.[6] Thus, in the present study, we sought to determine if UG also interferes with the IgA-Fn interaction, thereby averting glomerular deposition of both IgA and Fn. Accordingly, we determined the effect of UG on IgA-Fn complex formation. The results show that UG prevents the formation of IgA-Fn complex (FIG. 8a), even though UG itself does not bind with IgA (data not shown). A nonspecific protein, myoglobin, showed no such effect (FIG. 8a). These results suggest that UG plays an essential role in suppressing IgA-Fn complex formation and, consequently, it is a potent inhibitor of IgA deposition in the glomeruli.

Glomerular Mesangial Cells Fail to Bind IgA-Fn Complex in Presence of UG

It has been reported that IgA binds with the renal mesangial cells,[26] monocytes, and neutrophils,[27] and such binding is enhanced in IgAN.[28] Moreover, IgA binding has been reported to activate the renal mesangial cells *in vitro*.[29] Thus, we sought to determine whether the IgA-Fn complex also binds with mesangial cells and whether UG interferes with this binding. Accordingly, we carried out binding studies using ^{125}I-IgA, ^{125}I-IgA-Fn complex, and cultured mesangial cells. While both ^{125}I-IgA and ^{125}I-IgA-Fn bind with these cells, the level of binding is much higher for IgA-Fn compared with IgA alone (FIG. 8b). More importantly, in the presence of UG, the level of ^{125}I-IgA-Fn binding was significantly ($p < 0.05$) reduced (FIG. 8b). These results may be interpreted in at least two different ways: (i) UG competes with the IgA-Fn complex for binding with the cells or (ii) UG disrupts the IgA-Fn complex formation by its high-affinity binding with Fn, forming Fn-UG heteromers, as previously reported.[6]

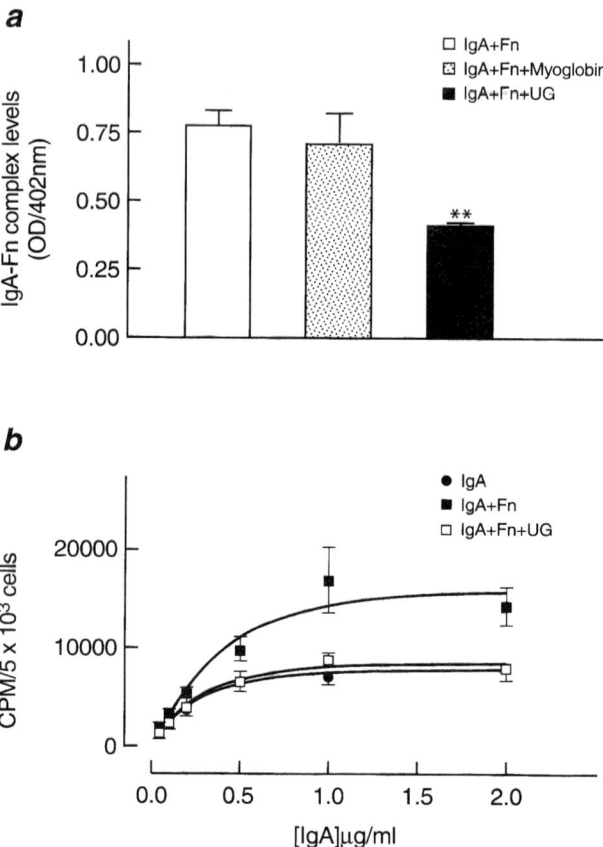

FIGURE 8. The effects of UG on IgA-Fn complex formation (**a**) and IgA-Fn complex binding (**b**) to cultured mesangial cells. (**a**) Equimolar concentrations of IgA and Fn were incubated at 37°C for 2 h in the absence and presence of UG. The IgA-Fn complex formation was quantitated by ELISA. Myoglobin was used as a nonspecific control for UG. (**b**) Effect of UG on the binding of ^{125}I-IgA and ^{125}I-IgA-Fn with mesangial cells. ^{125}I-IgA alone or ^{125}I-IgA in combination with equimolar concentrations of Fn was incubated at 4°C for 2 h as described. Note that the presence of UG caused a significant ($p < 0.05$) inhibition of IgA-Fn complex formation and binding with the mesangial cells. The results are expressed as the mean ($n = 4$) ± SD. Reprinted with permission from reference 7.

Glomerular Deposition of Exogenous IgA in UG Gene-Disrupted Mice Is Prevented by UG

To determine whether UG prevents the glomerular deposition of exogenous IgA *in vivo*, we performed experiments in which UG$^{-/-}$, UG$^{+/-}$, and UG$^{+/+}$ littermates were intra-arterially injected with fluorescein isothiocyanate (FITC)–conjugated IgA in the presence and absence of UG. The results showed a strong immunofluorescence in the glomeruli of the UG$^{-/-}$ mice that received FITC-IgA alone (FIG. 9A), but virtually no such accumulation was detected when a mixture of FITC-IgA and

UG was administered (FIG. 9B). Myoglobin, used as a nonspecific control, failed to provide any protection against FITC-IgA accumulation in the glomeruli of these mice (data not shown). Compared with the glomeruli of the $UG^{-/-}$ mice (FIG. 9A), those of the wild-type littermates manifested a total lack of immunofluorescence (data not shown). Interestingly, even when a higher dosage (300 µg) of FITC-IgA was injected into $UG^{+/+}$ mice, they showed no detectable glomerular accumulation (FIG. 9C), suggesting a higher threshold for abnormal glomerular IgA accumulation in these mice. To rule out the possibility that the observed glomerular fluorescence in $UG^{-/-}$ mice is not due to the accumulation of free FITC, we injected free FITC to $UG^{-/-}$ and $UG^{+/+}$ mice and found no detectable accumulation of fluorescence in the glomeruli of these animals (data not shown). These results clearly demonstrate that *in vivo* UG has a critical function in preventing abnormal glomerular accumulation of IgA.

Mesangial Cells from UG Gene-Disrupted Mice Show Elevated Fn and Collagen mRNA Levels

Since elevated expression of Fn and type IV collagen has been reported in the glomeruli of human IgAN patients, we isolated pure populations of glomeruli isolated from the kidneys of $UG^{+/+}$ and $UG^{-/-}$ mice by microdissection and evaluated the level of expression of Fn and α-chain-specific type IV collagen mRNAs. The Fn mRNA levels were determined by semiquantitative RT-PCR and the results show that these levels are significantly higher in the glomeruli of $UG^{-/-}$ mice compared with $UG^{+/+}$ littermates (FIG. 10a). Notably, while the higher Fn mRNA level was clearly detectable using total RNA from the isolated glomeruli, it was not detectable when total RNA from the whole kidneys was used in our previous investigation.[6] This discrepancy may be due to the fact that Fn mRNA is expressed locally in the glomeruli and, in the total RNA extracted from the whole kidneys, the concentration of Fn mRNA contributed by the glomeruli is too low to be detected. This observation raised the possibility that Fn produced locally in the glomeruli may also contribute to the glomerular deposition of IgA-Fn complexes. However, the level of Fn mRNA elevation observed in the UG-knockout mice alone may not explain the massive deposition of IgA and Fn in this organelle. Moreover, in experimental glomerulonephritis in the rat, animals expressing high levels of glomerular TGFβ and Fn *in situ* do not manifest heavy deposits of IgA.[30] Furthermore, Waga *et al.*[22] have recently suggested that a 43-kDa carboxy-terminal fragment of circulating Fn that binds IgA may cause abnormal renal glomerular deposition of this immunoglobulin in IgAN patients.

In addition to the accumulation of Fn and IgA, the glomeruli of IgAN patients as well as those of the UG-knockout mice show abnormal deposition of collagen. The accumulated collagen in the glomeruli has been reported to be of type IV. Thus, we determined the level of α-chain-specific type IV collagen mRNA from the glomeruli of $UG^{-/-}$ and $UG^{+/+}$ mice by quantitative RT-PCR. The results show that type IV collagen mRNA is expressed at a much higher level in the glomeruli of $UG^{-/-}$ mice compared with $UG^{+/+}$ littermates (FIGS. 10b and 10d). Since PDGF and TGFβ are known stimulators of Fn and collagen production[30] in glomerular cells, we also determined PDGF and TGFβ mRNA levels in isolated glomeruli from $UG^{-/-}$ and $UG^{+/+}$ littermates by competitive RT-PCR. The results show that PDGF mRNA is ex-

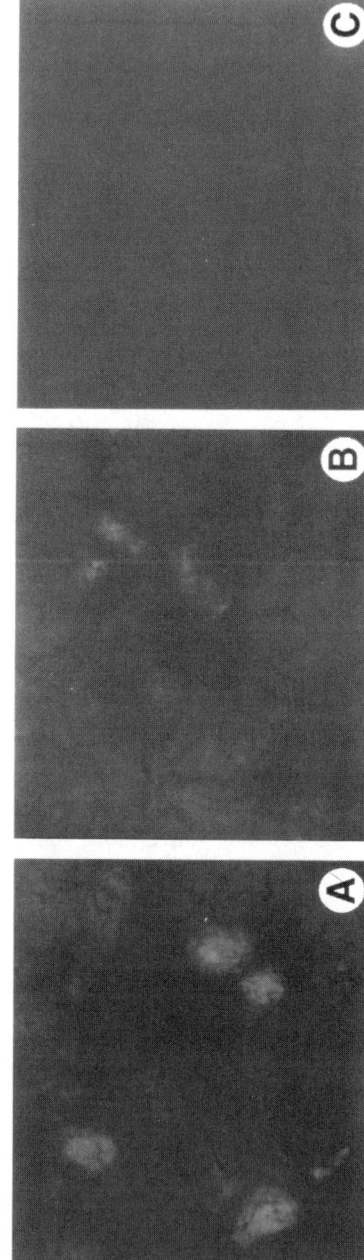

FIGURE 9. Inhibition of renal glomerular deposition of exogenous IgA by UG. UG−/− mice were injected with FITC-conjugated IgA in the presence and absence of UG. A nonspecific protein myoglobin served as a control. Glomerular deposition of FITC-IgA was determined by fluorescence microscopy. Kidney sections of a UG−/− mouse injected with 100 μg FITC-IgA alone (**A**) or 100 μg FITC-IgA + 100 μg of UG (**B**); kidney section of a UG+/+ mouse that received 300 μg of FITC-conjugated IgA alone (**C**). While the administration of only 100 μg of FITC-IgA caused dramatic glomerular accumulation in UG−/− mice (**A**), the presence of UG dramatically inhibited the glomerular deposition of FITC-IgA (**B**). In the wild-type (WT) littermates (UG+/+) having circulating UG, the injection of 300 μg of FITC-IgA caused no appreciable accumulation of fluorescence (**C**). Reprinted with permission from reference 7.

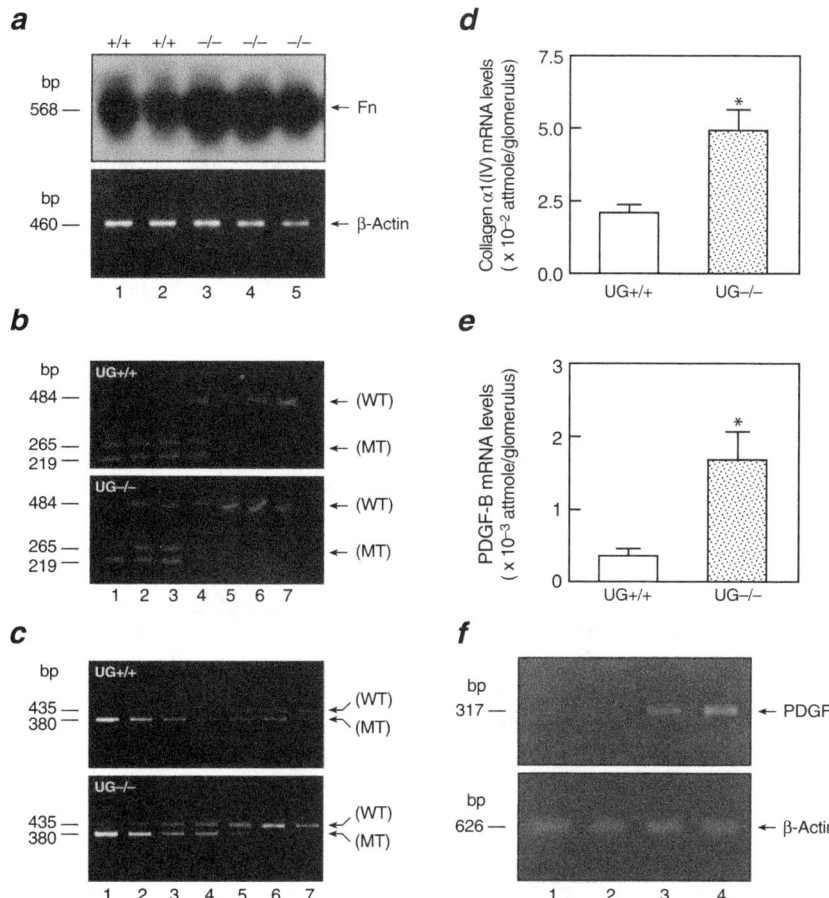

FIGURE 10. Expression of Fn, α1-chain of type IV collagen, and PDGF-B mRNAs in the glomeruli of UG-knockout mice. Homogeneous populations of glomeruli were isolated by microdissection as described (see reference 7). Total RNA from isolated glomeruli was purified and used in quantitative and semiquantitative RT-PCR to determine the mRNA levels. GAPDH mRNA was amplified as an internal control for the integrity of the RNA and the quantity loaded per lane. (a) *Upper panel*: Fn mRNA levels in the glomeruli of UG$^{-/-}$ and UG$^{+/+}$ mice; semiquantitative RT-PCR products were further hybridized with an Fn probe. *Lower panel*: β-Actin mRNA was used to standardize the Fn mRNA levels. Competitive RT-PCR was used to determine the levels of the α1-chain of type IV collagen (b) and PDGF-B mRNA (c). Glomerular cDNAs from wild-type (WT) (*upper*) and UG-knockout (*lower*) mice were competed with decreasing amounts of mutant (MT) cDNA. The MT cDNA ranged from 0.1 to 2 attomol (lanes 1–7) for the α1-chain of type IV collagen and from 0.0004 to 0.02 attomol for PDGF-B (lanes 1–7). Glomerular α-chain type IV collagen (d) and PDGF-B mRNA (e) levels were compared between UG$^{-/-}$ and UG$^{+/+}$ mice; *$p < 0.05$. (f) Induction of mesangial cell PDGF-B mRNA expression by IgA-Fn. Confluent mesangial cells in 6-well plates were incubated with DMEM + 0.1% BSA for 24 h. Some of the cultures were treated with IgA-Fn and the others were incubated with 100 μg/mL of Fn alone or with PBS that served as controls. Total RNA was isolated from cells at 4 h after

pressed at a significantly higher level in the glomeruli of $UG^{-/-}$ mice compared with $UG^{+/+}$ littermates (FIGS. 10c and 10e). The PDGF protein levels were qualitatively determined by immunohistochemical analyses. While the relative intensity of PDGF-specific staining was readily detectable (2+ to 3+) in the glomeruli of $UG^{-/-}$ mice ($n = 9$), such staining was virtually lacking (0 to 1+) in those of $UG^{+/+}$ mice ($n = 3$). The glomerular $TGF\beta_1$ mRNA levels, however, remained unchanged (data not shown). These findings in UG gene-disrupted mice are in agreement with those reported in human IgAN patients.[25]

Elevated Levels of PDGF mRNA and PDGF Protein Production by Mesangial Cells of UG-Deficient Mice

To determine if IgA or IgA-Fn complexes may stimulate PDGF mRNA production in the glomerular mesangial cells, we performed *in vitro* experiments in which cultured glomerular mesangial cells were stimulated with IgA, IgA + Fn, and Fn alone and measured the level of PDGF mRNA. The results show that the cells treated with IgA alone had a low level of PDGF mRNA expression, while treatment with IgA + Fn caused a highly elevated level of expression (FIG. 10f); Fn alone did not show such an effect (FIG. 10f). Thus, while IgA alone may cause a small increase in PDGF mRNA in mesangial cells, high levels of the IgA-Fn complex, found in the plasma of IgAN patients and in that of the UG-knockout mice, may stimulate even higher levels of PDGF mRNA. Consequently, in the glomeruli of IgAN patients, higher levels of Fn and collagen type IV expression would be expected. This may initiate a vicious cycle causing abnormal glomerular accumulation of IgA, Fn, and collagen, characteristically found in human IgAN patients. Since UG interferes with IgA-Fn complex formation, but does not bind IgA, it is reasonable to expect a small elevation of PDGF mRNA production by mesangial cells treated with IgA alone (FIG. 10f).

Possible Molecular Mechanism of UG Action

On the basis of our present experimental results and those from published reports (reviewed in references 3 and 4), we propose the following mechanism of UG action that prevents the development and progression of IgAN: under normal physiological conditions, there is a balance among the levels of circulating UG, IgA, and Fn. We have previously demonstrated that UG binds Fn with high affinity, forming Fn-UG heteromers.[6] This heteromer formation not only prevents Fn self-aggregation, essential for abnormal tissue deposition,[24,25] but also counteracts the formation of IgA-Fn

---

incubation. *Upper panel*: RT-PCR analyses of PDGF-B mRNA levels. *Lower panel*: β-Actin mRNA expression was used to determine the quality and amount of RNA loaded. Lane 1, control; lane 2, Fn only; lane 3, IgA only; lane 4, IgA-Fn. Note an apparent increase of intensity of the PDGF-B mRNA band in lane 4 (IgA-Fn) compared with that in lane 3 (IgA only). Reprinted with permission from reference 7.

complexes, found in the plasma of IgAN patients.[20,21] Moreover, lack of UG in mice has been reported to manifest increased sensitivity to hyperoxia and altered oxidant-induced pulmonary proinflammatory responses.[8] Similarly, in humans, significantly lower plasma levels of UG are found in patients with inflammatory lung disease such as asthma[31] and in upper respiratory viral infections.[32] In these scenarios, the circulating UG level may decrease and the balance among the levels of Fn, UG, and IgA may be disrupted, allowing free Fn to form IgA-Fn and Fn-Fn complexes more readily. In addition, our results in two different murine models of IgAN suggest that sustained suppression of UG production allows the IgA and IgA-Fn complexes to bind to the mesangial cells, stimulating PDGF production in the glomeruli *in situ*. The interaction of PDGF with its receptor on mesangial cells causes enhanced Fn and collagen production. Since Fn also binds collagen, the glomeruli of UG-deficient mice as well as those of the IgAN patients accumulate not only IgA and Fn, but also collagen. In agreement with our hypothesis, it has been reported that inflammation in the lung, a major source of plasma UG,[33] can lead to the development of IgAN (for review, see reference 34). Furthermore, the lack of an as yet uncharacterized "circulating factor" that prevents IgAN has been proposed as 60% of the IgAN patients that undergo renal transplantation have recurrence of the disease[35] in the transplanted kidney. Additionally, 48–68% of these patients also manifest high levels of plasma IgA-Fn complex.[20,21] Thus, it appears that high IgA-Fn levels in blood may contribute to the pathogenesis of IgAN and that the "circulating factor" may prevent this complex formation. It is tempting to speculate that this "circulating factor" may be UG.

Our proposed mechanism of pathogenesis of IgAN is consistent with the results of our present experiments as (i) both UG-AS and the UG-knockout mice manifest abnormal depositions of IgA, Fn, and collagen in the glomeruli that are also found in those of the human IgAN patients; (ii) high levels of IgA-Fn complex in the plasma of UG-knockout mice as well as in IgAN patients are readily detectable; (iii) UG inhibits Fn-Fn and IgA-Fn interactions despite the fact that it does not directly interact with IgA; (iv) UG inhibits the binding of IgA-Fn complex with the mesangial cells and consequently prevents the expression of PDGF and Fn mRNAs locally in the glomeruli; and finally (v) the administration of purified recombinant UG to UG-knockout mice prevents glomerular accumulation of FITC-IgA and averts glomerular Fn and collagen deposition as reported previously.[6]

Originally described by Berger and Hinglais[23] as idiopathic IgAN, it is currently recognized as the most common primary glomerulonephritis worldwide, especially in certain parts of Asia, Europe, and Australia.[23,36] It is also recognized that nearly 30–50% of the patients with IgAN develop end-stage renal disease requiring chronic hemodialysis and transplantation. Furthermore, in the majority of such IgAN patients, the disease recurs in the transplanted kidney.[35] The molecular mechanism(s) of IgAN is not yet clear. The results of our present study, for the first time, demonstrate that UG-knockout mice are a valid and unique animal model manifesting virtually all pathologic features of human IgAN.[23,36] Taken together, our results warrant further investigations to determine whether the mechanism(s) uncovered in the mouse models may underlie the human IgAN. A clear understanding of such a mechanism(s) may facilitate the development of rational therapeutic approaches to human IgAN, for which currently there is no effective treatment.

REFERENCES

1. KRISHNAN, R.S. & J.C. DANIEL, JR. 1967. "Blastokinin": inducer and regulator of blastocyst development in the rabbit uterus. Science **158:** 490–492.
2. BEIER, H.M. 1968. Uteroglobin: a hormone-sensitive endometrial protein involved in blastocyst development. Biochim. Biophys. Acta **160:** 289–291.
3. MUKHERJEE, A.B., G.C. KUNDU, A.K. MANDAL et al. 1998. Uteroglobin: physiological role in normal glomerular function uncovered by targeted disruption of the uteroglobin gene in mice. Am. J. Kidney Dis. **32:** 1106–1120.
4. MUKHERJEE, A.B., G.C. KUNDU, G. MANTILE-SELVAGGI et al. 1999. Uteroglobin: a novel cytokine? Cell. Mol. Life Sci. **55:** 771–787.
5. ZHANG, Z., D.B. ZIMONJIC, N.C. POPESCU et al. 1997. Human uteroglobin gene: structure, subchromosomal localization, and polymorphism. DNA Cell Biol. **16:** 73–83.
6. ZHANG, Z., G.C. KUNDU, C-J. YUAN et al. 1997. Severe fibronectin-deposit renal glomerular disease in mice lacking uteroglobin. Science **276:** 1408–1412.
7. ZHENG, F., G.C. KUNDU, Z. ZHANG et al. 1999. Uteroglobin is essential in preventing immunoglobulin A nephropathy in mice. Nat. Med. **5:** 1018–1025.
8. JOHNSTON, C.J., J.N. FINKELSTEIN, G. OBERDORSTER et al. 1999. Clara cell secretory protein–deficient mice differ from wild-type mice in inflammatory chemokine expression to oxygen and ozone, but not to endotoxin. Exp. Lung Res. **25:** 7–21.
9. KUNDU, G.C., G. MANTILE, L. MIELE et al. 1996. Suppression of cellular chemoinvasiveness by human uteroglobin via a novel high affinity binding site. Proc. Natl. Acad. Sci. U.S.A. **93:** 2915–2919.
10. KUNDU, G.C., A.K. MANDAL, Z. ZHANG et al. 1998. Uteroglobin (UG) suppresses extracellular matrix invasion by normal and cancer cells that express the high affinity UG-binding proteins. J. Biol. Chem. **273:** 22819–22824.
11. ZHANG, Z., G.C. KUNDU, D. PANDA et al. 1999. Loss of transformed phenotype in cancer cells by overexpression of the uteroglobin gene. Proc. Natl. Acad. Sci. U.S.A. **96:** 3963–3968.
12. RAY, M.K., S. MAGDALENO, B.W. O'MALLEY & F.J. DEMAYO. 1993. Cloning and characterization of mouse Clara cell specific 10 kDa protein gene: comparison of 5′-flanking region with the human, rat, and rabbit gene. Biochem. Biophys. Res. Commun. **197:** 163–171.
13. NAGY, A., R. ROSSANT, W. ABRAMOW-NEWERLY & J.C. RODER. 1993. Derivation of completely cell culture–derived mice from early-passage embryonic stem cells. Proc. Natl. Acad. Sci. U.S.A. **90:** 8424–8428.
14. WU, C., V.M. KEIVENS, T.E. O'TOOLE et al. 1995. Integrin activation and cytoskeletal interaction are essential for the assembly of fibronectin matrix. Cell **83:** 715–724.
15. LEVIN, S.W., J.D. BUTLER, K. SCHUMACHER et al. 1986. Uteroglobin inhibits phospholipase A2 activity. Life Sci. **38:** 1813–1819.
16. MIELE, L., E. CORDELLA-MIELE, A. FACCHIANO & A.B. MUKHERJEE. 1988. Novel anti-inflammatory peptides from the region of highest similarity between uteroglobin and lipocortin I. Nature **335:** 726–730.
17. FACCHIANO, A., L. MIELE, E. CORDELLA-MIELE & A.B. MUKHERJEE. 1991. Inhibition of pancreatic phospholipase A_2 activity by uteroglobin and antiflammin peptides: possible mechanism of action. Life Sci. **48:** 453–464.
18. BORDER, W.A. & E. RUOSLAHTI. 1992. Transforming growth factor-β: the dark side of tissue repair. J. Clin. Invest. **90:** 1–7.
19. OH, E., M. PIERSCHBACHER & E. RUOSLAHTI. 1981. Deposition of plasma fibronectin in tissues. Proc. Natl. Acad. Sci. U.S.A. **78:** 3218–3221.
20. CEDERHOLM, B., J. WIESLANDER, P. BYGREN & D. HEINEGARD. 1986. Patients with IgA nephropathy have circulating anti-basement membrane antibodies reacting with structures common to collagen I, II, and IV. Proc. Natl. Acad. Sci. U.S.A. **83:** 6151–6155.
21. BALDREE, B., R.J. WYATT, B.A. JULIAN et al. 1993. Immunoglobulin A–fibronectin aggregate levels in children and adults with immunoglobulin A nephropathy. Am. J. Kidney Dis. **22:** 1–8.
22. WAGA, S., K. SUGIMOTO, H. TANAKA et al. 1999. IgA interaction with carboxy-terminal 43-kD fragment of fibronectin in IgA nephropathy. J. Am. Soc. Nephrol. **10:** 256–263.

23. BERGER, J. & N. HINGLAIS. 1968. Intercapillary deposits of IgA-IgG. J. Urol. Nephrol. **74:** 694–695.
24. HYNES, R.O. 1986. Fibronectins. Sci. Am. **254:** 42–51.
25. NAITO, T., K. NITTA, H. OZU et al. 1997. Clinical assessment of the significance of platelet-derived growth factor in patients with immunoglobulin A nephropathy. J. Lab. Clin. Med. **130:** 63–68.
26. EMANCIPATOR, S.N., C.S. RAO, A. AMORE et al. 1992. Macromolecular properties that promote mesangial binding and mesangiopathic nephritis. J. Am. Soc. Nephrol. **2:** S149–S158.
27. LAUNAY, P., C. PATRY, A. LEHUEN et al. 1999. Alternative endocytic pathway for immunoglobulin A Fc receptors (CD89) depends on the lack of FcR-gamma association and protects against degradation of bound ligand. J. Biol. Chem. **274:** 7116–7125.
28. LAI, K.N., W.Y. TO, P.K. LI & J.C. LEUNG. 1996. Increased binding of polymeric lambda-IgA to cultured human mesangial cells in IgA nephropathy. Kidney Int. **49:** 839–845.
29. GOMEZ-GUERRERO, C., N. DUQUE & J. EGIDO. 1996. Stimulation of Fc (alpha) receptors induces tyrosine phosphorylation of phospholipase C–gamma(1), phosphatidylinositol phosphate hydrolysis, and Ca^{2+} mobilization in rat and human mesangial cells. J. Immunol. **156:** 4369–4376.
30. BORDER, W.A., S. OKUDA, L.R. LANGUINO et al. 1992. Transforming growth factor-beta in disease: the dark side of tissue repair. J. Clin. Invest. **90:** 1–7.
31. SHIJUBO, N., Y. ITOH, T. YAMAGUCHI et al. 1999. Serum levels of Clara cell 10 kDa protein are decreased in patients with asthma. Lung **177:** 45–52.
32. VOLVOVITZ, B., I. NATHANSON, G. DECASTRO et al. 1988. Relationship between leukotriene C_4 and a uteroglobin-like protein in nasal and tracheobronchial mucosa of children: implication in acute respiratory illness. Int. J. Allergy Appl. Immunol. **86:** 420–425.
33. KIKUKAWA, T. & A.B. MUKHERJEE. 1989. Detection of uteroglobin-like phospholipase A_2–inhibitory protein in the circulation of rabbits. Mol. Cell. Endocrinol. **62:** 177–187.
34. ENDO, Y. & H. KANBAYASHI. 1994. Etiology of IgA nephropathy syndrome. Pathol. Int. **44:** 1–13.
35. OHMACHT, C., V. KLIEM, M. BERG et al. 1997. Recurrent immunoglobulin A nephropathy after renal transplantation: a significant contributor to graft loss. Transplantation **64:** 1493–1496.
36. D'AMICO, G. 1998. Pathogenesis of immunoglobulin A nephropathy. Curr. Opin. Nephrol. Hypertens. **7:** 247–250.

Uteroglobin Binding Proteins: Regulation of Cellular Motility and Invasion in Normal and Cancer Cells

GOPAL C. KUNDU,[a,b] ZHONGJIAN ZHANG,[b] GIUDITTA MANTILE-SELVAGGI,[b,c] ASIM MANDAL,[b] CHIUN-JYE YUAN,[b,d] AND ANIL B. MUKHERJEE[b]

[b]*Section on Developmental Genetics, Heritable Disorders Branch, National Institute of Child Health and Human Development, National Institutes of Health, Bethesda, Maryland 20892-1830, USA*

ABSTRACT: Uteroglobin (UG) is a multifunctional, secreted protein with anti-inflammatory and antichemotactic properties. While its anti-inflammatory effects, in part, stem from the inhibition of soluble phospholipase A_2 (sPLA$_2$) activity, the mechanism(s) of its antichemotactic effects is not clearly understood. Although specific binding of UG on microsomal and plasma membranes has been reported recently, how this binding affects cellular function is not clear. Here, we report that recombinant human UG (hUG) binds to both normal and cancer cells with high affinity (20–35 nM, respectively) and specificity. Affinity cross-linking studies revealed that ^{125}I-hUG binds to the NIH 3T3 cell surface with two proteins of apparent molecular masses of 190 and 49 kDa, respectively. UG affinity chromatography yielded similar results. While both the 190- and 49-kDa proteins were expressed in the heart, liver, and spleen, the lung and trachea expressed only the 190-kDa protein. Some cancer cells (e.g., mastocytoma, sarcoma, and lymphoma) expressed both the 190- and 49-kDa proteins. Further, using functional assays, we found that UG dramatically suppressed the motility and extracellular matrix invasion of both NIH 3T3 and some cancer cells. In order to further characterize the anti-ECM-invasive properties of UG, we induced expression of hUG into cancer cell lines derived from organs that, under physiological circumstances, secrete UG at a high level. Interestingly, it has been reported that a high percentage of the adenocarcinomas arising from the same organs fail to express UG. Our results on induced hUG expression in these cells show that inhibition of motility and ECM invasion requires the expression of both UG and its binding proteins. Taken together, our data define receptor-mediated functions of UG in which this protein regulates vital cellular functions by both autocrine and paracrine pathways.

---

[a]Present address for correspondence: National Center for Cell Science, NCCS Complex, Pune 411007, India. Fax: 91-20-5672259.
gopalkundu@hotmail.com
[c]Present address: Hackensack University Medical Center, 1 Pavilion East, Room 1928, 30 Prospect Avenue, Hackensack, NJ 07601.
[d]Present address: Department of Biological Sciences and Technology, National Chiao Tung University, 70 Po-Ai Street, 300 Hsinchu, Taiwan.

INTRODUCTION

Cellular migration and extracellular matrix invasion are critical for embryonic development, wound healing, inflammation, and cancer cell metastasis. However, the exact molecular mechanisms that regulate these processes are not well understood. Blastokinin[1] or uteroglobin (UG)[2] is a steroid-dependent, multifunctional, secreted protein with anti-inflammatory and antichemotactic activities (reviewed in reference 3). Depending on the status of expression of this protein in different tissues or body fluids, UG has been identified by a number of names, including Clara cell 10-kDa (CC10) protein,[4,5] progesterone-binding protein,[6] urine protein-1,[7–10] retinol-binding protein,[11] and polychlorinated biphenyl–binding protein.[12]

Structurally, UG is a covalently linked, homodimeric protein in which the 70 amino acid subunits are connected to each other by two disulfide bonds in an antiparallel orientation. Each UG monomer forms four α-helices (α-helix 1–4) and a β-turn between α-helices 2 and 3, but there is no β-structure. Human UG (hUG)[13,14] is the counterpart of rabbit UG.[15,16] Both rabbit and human UGs are potent inhibitors of low-molecular-weight group I and group II extracellular phospholipase A_2 (E.C.3.1.1.4).[17–19] Several years ago, Robinson et al. described an active transport of UG into the blastocoele of preimplanted rabbit embryos by a carrier-mediated process.[20] Moreover, Nieto and coworkers have shown the specific binding of UG in microsomal and plasma membranes.[21]

In this paper, we summarize our findings on the characterization of the hUG-binding proteins demonstrating novel effects of recombinant, homodimeric hUG in suppressing the ability of NIH 3T3 cells to invade the extracellular matrix (ECM). This effect of hUG appears to be mediated via its high-affinity binding protein (putative receptor). This binding protein has a molecular mass of 190 kDa and is expressed on several cell types including the NIH 3T3 cells. Since UG exists as a homodimer with two interchain disulfide bonds, we sought to determine whether the reduced, monomeric form of UG, like its dimer, shows the same biological effect as well as interacts with the same 190-kDa protein. Interestingly, we discovered that monomeric UG not only interacts with the 190-kDa protein, but also binds with the 49-kDa protein in several normal (NIH 3T3) and cancer (mastocytoma, sarcoma, and lymphoma) cells with high affinity and specificity. Moreover, the reduced hUG, like dimeric hUG, also suppresses the ECM invasion in those normal and cancer cells that express the UG-binding proteins. The anti-ECM-invasive properties of hUG were further confirmed by inducing the expression of hUG into several cancer cells by stable transfection using hUG cDNA. These cell lines were derived from the adenocarcinomas of the lung, prostate, breast, and uterus, which all (under physiological conditions) secrete UG at a high level, but fail to do so when malignant transformation occurs. The data showed that only one of the cell lines was derived from an adenocarcinoma of the uterus on which the anti-invasive activity of UG could be demonstrated. It turned out that this was the only cell line, in which hUG expression was induced, that also expressed the UG-binding proteins. These results clearly showed that the anti-invasive effect of hUG is mediated via its putative receptor.

Data on tissue-specific expression indicated that both the 190- and 49-kDa proteins are detectable in the heart, liver, and spleen, whereas only the 190-kDa protein is present in the lung and trachea. Neither the 190- nor 49-kDa protein was present in the aorta. Pretreatment of the NIH 3T3 cells with IL-6 and LPS followed by ^{125}I-UG-binding showed a significant increase in expression of both the 190- and 49-kDa binding proteins. Purification of these proteins by UG affinity chromatography from bovine lungs and analysis by SDS-PAGE followed by silver staining identified two protein bands with molecular masses of 180 and 40 kDa, respectively. Taken together, these data demonstrate that UG plays critical roles in regulating the cellular motility and ECM invasiveness by interacting with these binding proteins.

MATERIALS AND METHODS

Materials

NIH 3T3, mouse mastocytoma, sarcoma, lymphoma, fibrosarcoma, and human adenocarcinoma cell lines HEC-1A (uterus), HTB-174 (lung), HTB-81 (prostate), and HTB-30 (breast) were obtained from the American Type Culture Collection (Rockville, MD). All cell culture–grade reagents were purchased from GIBCO-BRL. These cell lines were grown according to standard procedures. Recombinant hUG was expressed and purified as described.[15] BioCoat Matrigel invasion chambers were obtained from Collaborative Research. Disuccinimidyl suberate (DSS) was from Pierce.

In Vitro *Invasion Assay*

The *in vitro* invasion assay was performed as described earlier.[22] Briefly, semiconfluent cells were harvested with trypsin and EDTA, centrifuged, and washed with PBS containing 0.1% BSA. The cells were resuspended in either DMEM or McCoy's medium containing 0.1% BSA and seeded in the upper compartment of the prehydrated, Matrigel-coated invasion chamber. The cells were treated in the absence or presence of nonreduced or reduced hUG and then incubated at 37°C for 24–36 h in a humidified incubator. The lower compartment was filled with fibroblast conditioned medium, which acted as a chemoattractant. The noninvaded cells and the Matrigel were scraped and removed from the upper surface of the filter. The invaded cells were fixed with methanol, stained with Giemsa, and washed with PBS. The invaded cells were counted under an inverted microscope and the percentage of invaded cells was calculated.

Radioreceptor Assay

The radioiodination of hUG and the radioreceptor assay were performed as described earlier.[23,24] Briefly, the recombinant hUG was radioiodinated by using [^{125}I]-NaI (2 mCi; carrier-free) and IODO-BEADS. The ^{125}I-UG was purified by G-25 column chromatography and the specific activity of the purified, carrier-free ^{125}I-hUG was about 20–25 µCi/µg. The cells were grown to confluence in 12-well plates and washed with PBS. The cells were incubated with varying concentrations of nonreduced or reduced ^{125}I-hUG in the absence or presence of excess unlabeled hUG in 1 mL Hank's balanced salt solution (HBSS), pH 7.6, containing 0.1% bovine serum albumin (BSA) at 4°C for 2 h. In some experiments, the cells were incubated with nonreduced or reduced ^{125}I-hUG in the same HBSS, pH 7.6, in the absence or presence of increasing concentrations of unlabeled hUG under the same conditions as described above. The reaction was stopped by rapid removal of unbound ^{125}I-hUG. The cells were washed with PBS, pH 7.4, solubilized in 1 N NaOH followed by addition of 1 N HCl and the radioactivity was counted by a gamma-counter. The monomeric form of hUG was prepared by treating the dimeric hUG with 10 mM dithiothreitol at 37°C for 15 min. The specific binding was calculated by subtracting the nonspecific binding from the total binding, and the data were analyzed by Scatchard plot using the LIGAND computer program.[25]

Affinity Cross-linking Experiments

The cells were grown in 6-well plates and, when semiconfluent, they were incubated with nonreduced or reduced ^{125}I-hUG in the absence or presence of excess unlabeled hUG in 2 mL HBSS, pH 7.6, containing 0.1% BSA at 4°C for 2 h. The cells were washed with PBS, pH 7.4, and incubated further with 0.2 mM DSS in 2.0 mL HBSS, pH 7.6, for 20 min. The cells were collected by centrifugation (10,000g for 15 min) and lysed in 40 µL of lysis buffer (1% Triton X-100 containing 1 mM PMSF, 20 µg/mL leupeptin, and 2 mM EDTA). The supernatants were resuspended in sample buffer containing 5% β-mercaptoethanol, boiled, and resolved by electrophoresis on 4–20% gradient sodium dodecyl sulfate (SDS)–polyacrylamide gel (Bio-Rad). The gels were stained, dried, and autoradiographed using Kodak X-Omat AR X-ray film.

Tissue-Specific Expression

The membrane preparations from bovine spleen, trachea, heart, aorta, lung, and liver were prepared as described before.[23] Briefly, the membrane samples containing equal amounts of total proteins were incubated with reduced ^{125}I-hUG in HBSS, pH 7.6, containing 0.1% BSA in the absence or presence of unlabeled hUG for binding as described before. The samples were cross-linked with DSS, solubilized, electrophoresed, and autoradiographed. To check the regulation of hUG-receptor expression, the NIH 3T3 cells were treated with various cytokines and other agents. The treated cells were incubated with ^{125}I-hUG for binding, cross-linked with DSS, electrophoresed, and autoradiographed.

UG Receptor Purification

The hUG receptor was purified from bovine spleen by hUG affinity chromatography. The tissue was homogenized in 10 mM NaHCO$_3$ buffer, pH 8.0, and the homogenate was centrifuged at 600g for 10 min at 4°C. The supernatant was centrifuged at 24,000g for 60 min and the pellets were collected and solubilized in solubilization buffer [50 mM Tris-HCl buffer, pH 7.4, containing 1% Triton X-100, 0.4 mM phenylmethylsulfonyl fluoride (PMSF), 10 mg/mL leupeptin, and 2 mM EDTA] by stirring at 4°C for 6 h. The supernatant was collected by centrifugation and loaded to the CNBr-activated Sepharose 4B–coupled hUG affinity column. The hUG receptor protein was eluted by using 0.1 M glycine-HCl buffer, pH 3.0, containing 0.1% Triton X-100, 0.4 mM PMSF, 10 mg/mL leupeptin, and 2 mM EDTA. All fractions were neutralized with 2 M Tris-HCl, pH 8.0. The fraction containing receptor protein was analyzed by ^{125}I-hUG-binding and affinity cross-linking assays and the purity was checked by SDS-PAGE followed by silver staining.

Construction of pRC/RSV-hUG and Transfection

The full-length hUG cDNA, cloned in pGEM 4Z,[15] was excised by *EcoR*I digestion and subcloned into the TA vector. The orientation of the hUG cDNA construct in the TA vector was checked by DNA sequencing. The fragment was excised from the TA vector by using *Hind*III and *Xba*I and ligated further into the pRC/RSV expression vector (Invitrogen) to make the pRC/RSv-hUG construct. The cell lines derived from the adenocarcinomas of human uterus (HEC-1A), lung (HTB-174), prostate (HTB-81), and breast (HTB-30) were cultured as described before.[26] Both the pRC/RSV-hUG and pRC/RSV (mock) were individually transfected to each of the tumor cell lines by electroporation. G418 (400 µg/mL) was added to the cells after 24 h of the transfection. The stably transfected clones were isolated, expanded, and used for further experiments.

Detection of hUG mRNA by RT-PCR

In order to detect the level of hUG mRNA in different adenocarcinoma cells, total RNAs were isolated from the cells by using the RNAzol method according to the manufacturer's instructions (Tel Test, TX). Reverse transcription (RT) using equal amounts of total RNA and cDNA amplification were performed using DNA Thermal Cycler 480 (Perkin Elmer, Norwalk, CT). The sequences of hUG-specific primers were as follows—hUG-L: 5'-ATG AAA CTC GCT GTC ACC C-3'; hUG-R: 5'-TAC ACA GTG AGC TTT GGG C-3'. The sequences of GAPDH-specific primers were as follows—GAPDH-L: 5'-CCA TGG AGA AGG CTG GGG-3'; GAPDH-R: 5'-CAA AGT TGT CAT GGA TGA CC-3'. The Gene Amp Thermostable rTth Reverse Transcriptase PCR kit (Perkin Elmer) was used for the reaction according to the manufacturer's instructions. The PCR products were subjected to electrophoresis on agarose gel, transferred to nylon membranes, and UV cross-linked. The membranes were hybridized using hUG- and GAPDH-specific probes and autoradiographed. The sequences of the probes were as follows—hUG-P: 5'-TGA AGA AGC TGG TGG ACA CC-3'; GAPDH-P: 5'-TCC TGC ACC ACC AAC TGC TT-3'.

Detection of hUG by Immunoprecipitation and Western Blot Analysis

The confluent adenocarcinoma cells were washed with PBS, lysed in lysis buffer (50 mM Tris-HCl, pH 7.5, containing 150 mM NaCl, 1% Nonidet P-40, 15 µg/mL leupeptin, and 0.5 mM PMSF), and immunoprecipitated by using a kit according to the manufacturer's instructions (Boehringer Mannheim). Briefly, the cell lysates were centrifuged and the supernatants were incubated with rabbit hUG antibody for 1 h and incubated further with protein A-agarose at 4°C overnight. Bound complexes were collected by centrifugation, washed, and eluted by boiling in SDS-sample buffer. The samples were resolved on SDS-PAGE and transferred to the nitrocellulose membranes. In case of Western blot analysis, the membranes containing hUG protein were blocked, washed, and incubated with goat hUG antibody at room temperature for 1 h. The membranes were washed, incubated further with rabbit anti-goat horseradish peroxidase (HRP)–conjugated IgG, and detected by enhanced chemiluminescence (ECL) according to the manufacturer's instructions (Amersham Pharmacia).

Soft Agar Assay

The anchorage-independent growth of cells on soft agar was performed as described.[26] Briefly, wild-type, pRC/RSV-hUG-, and pRC/RSV (mock)–transfected adenocarcinoma cells were trypsinized. Usually, transfected cells with 3–4 passages were used for this assay. These single cell suspensions were stained with neutral red to check the cell viability. About 10,000 viable cells were suspended in 2.5 mL of McCoy's 5A medium containing 10% FBS and 0.3% Noble agar (Difco) and plated on 60-mm petri dishes. These cells were then layered over the agar-coated petri dishes. The medium used to prepare the top agar contained G418 (200 µg/mL) and was added for both the pRC/RSV-hUG- and pRC/RSV (mock)–transfected cells. These plates were incubated at 37°C for 3 weeks in an atmosphere of 5% CO_2 and 95% air. The colonies were stained with neutral red, counted, and photomicrographed.

RESULTS AND DISCUSSION

We have previously reported that recombinant, dimeric hUG suppresses ECM invasion of NIH 3T3 and human trophoblast cells by interacting with a specific hUG-binding protein (putative receptor).[22] Because hUG exists as a homodimer in nature with two interchain disulfide bonds, we sought to determine whether the reduced, monomeric form of hUG also suppresses the ECM invasion in normal and cancer cells. Accordingly, ECM invasion assays were performed by treating each of the cell lines (NIH 3T3, mastocytoma, sarcoma, lymphoma, and fibrosarcoma cells) with the monomeric form of hUG. The data showed that the monomeric hUG (1 µM) suppresses the ECM invasion in NIH 3T3 (82%), mouse mastocytoma (77%), sarcoma (79%), and lymphoma (75%), but there was no such effect when fibrosarcoma cells were used (FIG. 1a). Myoglobin, a nonspecific protein, was used as a negative control (FIG. 1a). In order to ascertain the possible roles of hUG in reversing the transformed phenotype of hUG-transfected adenocarcinoma cells, we tested them for ECM invasion and anchorage-independent growth on soft agar assays. We selected human cancer cell lines derived from the adenocarcinomas of the uterus, lung, prostate, and breast. All of these adenocarcinoma cell lines did not express hUG, although the organs in which these cancers arose constitutively expressed high levels of hUG. To determine the effects of endogenous hUG production in these cells, we induced its expression by stably transfecting the cells with an hUG cDNA construct, pRC/RSV-hUG. The results of ECM invasion using pRC/RSV-hUG-transfected HEC-1A cells showed a significant suppression of ECM invasion (88%) compared with nontransfected HEC-1A cells (FIG. 1b). The wild-type HEC-1A cells when treated with recombinant hUG yielded 79% suppression of ECM invasion (FIG. 1b). There were no differences in the monolayer growth properties among the hUG-transfected, mock-transfected, and nontransfected HEC-1A cells. Neither the wild-type nor the hUG-transfected adenocarcinoma-derived prostate cell lines (HTB-81) showed any suppression of ECM invasion (data not shown). We have also examined the levels of hUG mRNA and protein in hUG-transfected HEC-1A and HTB-81 cells

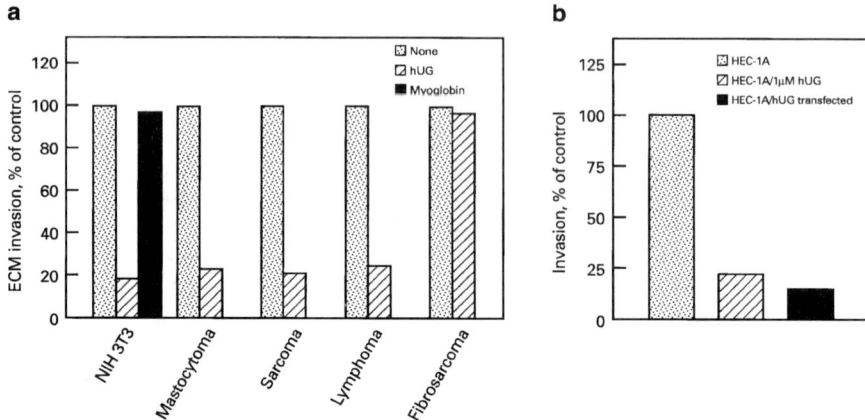

FIGURE 1. (a) Effect of recombinant hUG on extracellular matrix (ECM) invasion by NIH 3T3, mouse mastocytoma, sarcoma, lymphoma, and fibrosarcoma cells. The hUG-induced suppressions of ECM invasion were significant when NIH 3T3, mouse mastocytoma, sarcoma, and lymphoma cells were used, but there was no effect on fibrosarcoma cells. Myoglobin was used as a control. Data represent the average of three experiments. (b) Effect of hUG on ECM invasion by nontransfected and hUG-transfected HEC-1A cells. Note that there was significant suppression of ECM invasion when using either wild-type HEC-1A cells treated with 1 μM recombinant hUG or hUG-transfected HEC-1A cells. Data represent the average of three experiments. Reprinted with permission from reference 26.

as compared to nontransfected cells. Both RT-PCR and immunoprecipitation followed by Western blot analysis indicated that nontransfected (wild-type) cells neither express the hUG mRNA (FIG. 2a, upper left panel) nor the hUG protein (FIG. 2b), whereas the hUG-transfected HEC-1A and HTB-81 cells express both the hUG mRNA (FIG. 2a, upper middle and right panels) and the hUG protein (FIG. 2b), respectively. The expression of GAPDH, a housekeeping gene, is shown in the lower panel (FIG. 2a). The soft agar assay data demonstrated that the anchorage-independent growth of the pRC/RSV-hUG-transfected clones (HEC-1A) on soft agar was drastically reduced in size compared with the nontransfected or mock-transfected clones, although the numbers of cells seeded in each soft agar plate were equivalent (FIGS. 3a–3c). Both the nontransfected (FIG. 3a) and mock-transfected (FIG. 3b) cells grew a mixture of large and small colonies, whereas the cells transfected with pRC/RSV-hUG (FIG. 3c) showed very small and uniform-type colonies on soft agar plates after 4 weeks of incubation. The pRC/RSV-hUG-transfected colonies (HEC-1A) were transferred to the monolayer culture and these cells grew equally well as compared to nontransfected or mock-transfected cells, which indicates that the growth inhibition on soft agar was not because of nonviability of the cells, but because of the presence of hUG. Neither the nontransfected nor the hUG-transfected prostate adenocarcinoma cells (HTB-81) showed any effect on colony formation on soft agar plates (data not shown).

To delineate the mechanism(s) of suppression of cellular invasiveness, we first determined whether these normal and cancer cells express any functional receptor (binding protein) for hUG. Since hUG exists as a homodimer and earlier data have

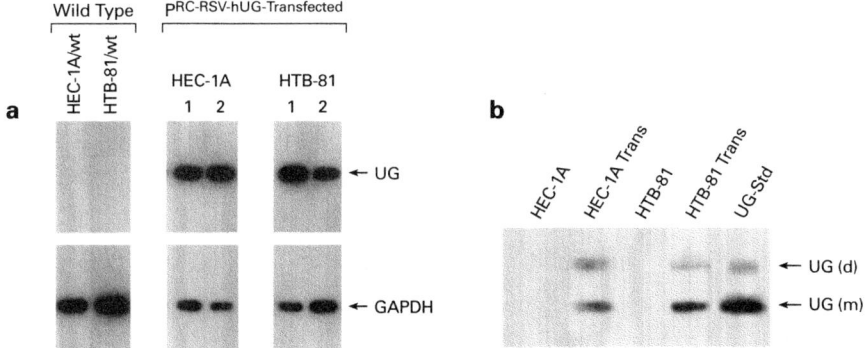

FIGURE 2. (a) Reverse transcription of total RNA isolated from wild-type and pRC/RSV-hUG-transfected HEC-1A (uterus) and HTB-81 (prostate) cells. PCR products were resolved on agarose gel, transferred to nylon membrane, and hybridized with hUG-specific oligonucleotide probe (hUG-P). GAPDH was used as a positive control in order to check the integrity and quality of the RNA. Left panels: wild-type HEC-1A and HTB-81 cells. Right panels: pRC/RSV-hUG-transfected HEC-1A and HTB-81 cells. Lanes 1 and 2 represent two independently derived pRC/RSV-hUG-transfected clones, each of HEC-1A (left) and HTB-81 (right), respectively. **(b)** Detection of hUG in nontransfected and hUG-transfected HEC-1A and HTB-81 cells by immunoprecipitation followed by Western blot analysis: hUG (d), hUG dimer; UG (m), hUG monomer. Reprinted with permission from reference 26.

shown that radiolabeled dimeric hUG interacts with the surface of NIH 3T3 and trophoblast cells with a molecular mass of 190 kDa, we sought to determine whether monomeric hUG also interacts with the same 190-kDa protein in these normal and cancer cells. Thus, the reduced form of ^{125}I-hUG was incubated with NIH 3T3 cells and Scatchard analysis of the binding data using this ^{125}I-hUG as a ligand showed a single class of specific binding in these cells with a dissociation constant (K_d) of 20 nM (FIG. 4a). The nonreduced and reduced forms of radiolabeled hUG were individually incubated with several cancer cells, such as mastocytoma, sarcoma, lymphoma, and fibrosarcoma, and the K_d values were in the range of 30–35 nM (nonreduced) and 20–25 nM (reduced), respectively (data not shown), except that no such binding was observed in the case of fibrosarcoma cells. In order to delineate the molecular size of the hUG-binding protein, we performed affinity cross-linking experiments by incubating the reduced form of ^{125}I-hUG with NIH 3T3 and other cancer cells (mastocytoma, sarcoma, lymphoma, and fibrosarcoma) in the absence or presence of excess unlabeled hUG and then cross-linked with DSS. A new 49-kDa radioactive protein band, in addition to the previously identified 190-kDa band, was detected when NIH 3T3,[22] mastocytoma, sarcoma, and lymphoma cells were used (FIG. 4b; lanes 2, 4, 6, and 8). However, both the 49- and 190-kDa bands were virtually undetectable when 1 μM hUG was added to the cells prior to ^{125}I-hUG-binding and affinity cross-linking (FIG. 4b; lanes 3, 5, 7, and 9). As expected, no protein bands were detected in the absence of DSS (FIG. 4b; lane 1). The 190-kDa protein band was also detected when the nonreduced form of ^{125}I-hUG was used as a ligand and incubated with each of the mastocytoma, sarcoma, and lymphoma cells (data not

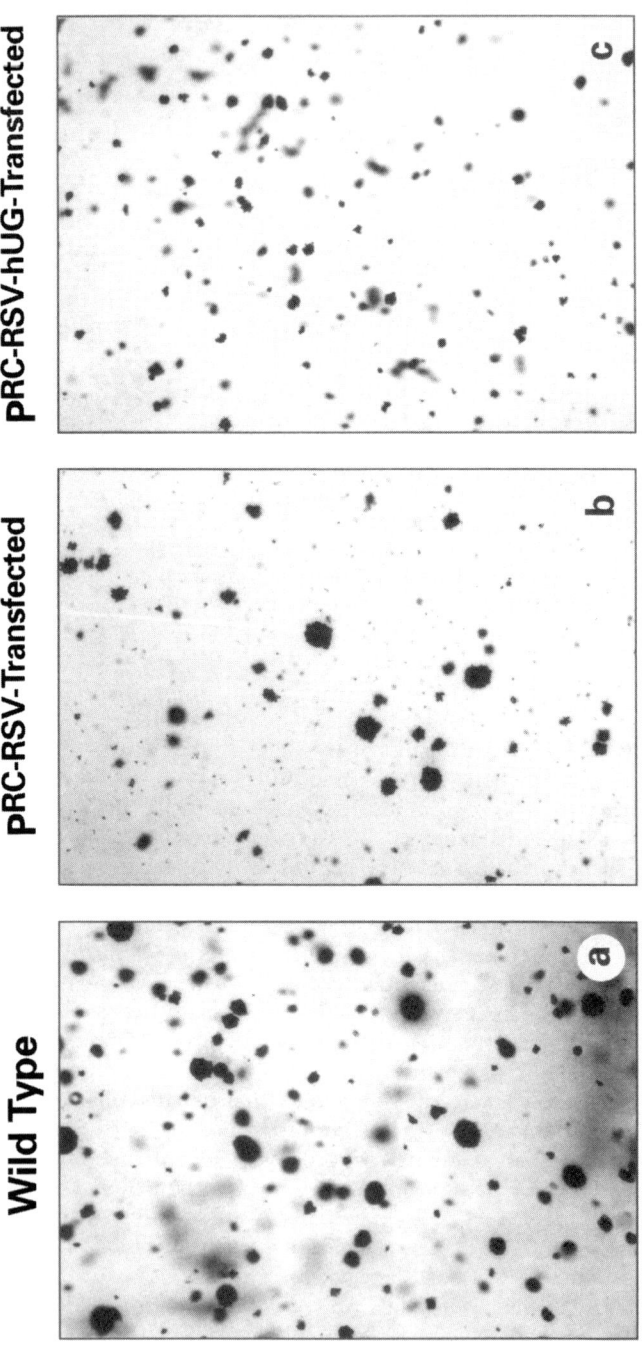

FIGURE 3. Anchorage-independent growth on soft agar: (**a**) nontransfected (wild-type), (**b**) pRC/RSV (mock)–transfected, and (**c**) pRC/RSV-hUG-transfected HEC-1A cells. Note that there was striking suppression of anchorage-independent growth of pRC/RSV-hUG-transfected HEC-1A cells. Magnification: ×20. Reprinted with permission from reference 26. [Figure reduced to 75%.]

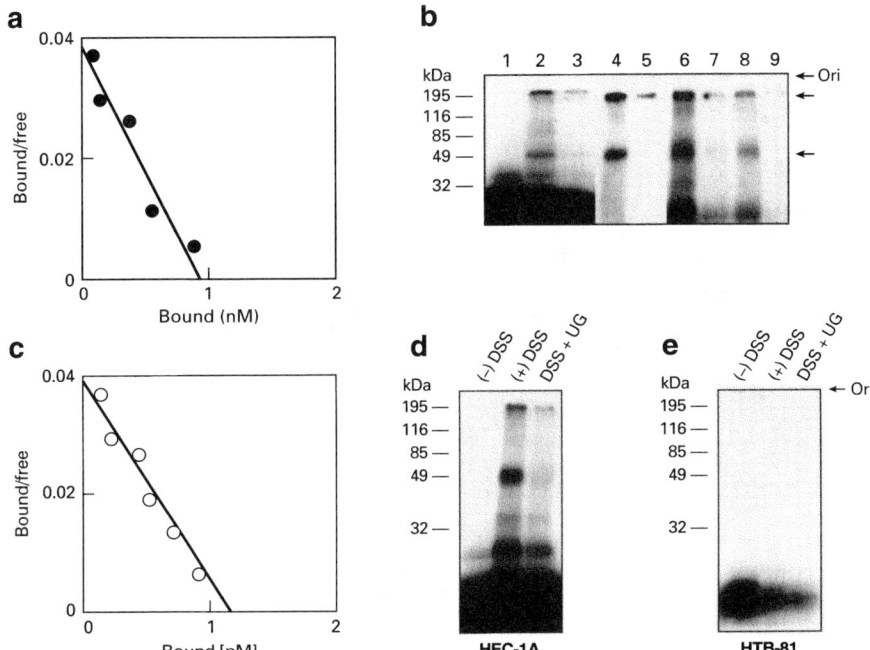

FIGURE 4. (a) Scatchard plot of specific binding of ^{125}I-hUG (reduced) on NIH 3T3 cells. The data are from three experiments and each point represents the mean of triplicate determinations. (b) Affinity cross-linking of hUG receptor(s) on NIH 3T3 (lanes 1–3), mastocytoma (lanes 4 & 5), sarcoma (lanes 6 & 7), and lymphoma (lanes 8 & 9) cells. Lane 1, −DSS; lane 2, +DSS; lane 3, +unlabeled hUG, +DSS; lane 4, +DSS; lane 5, +unlabeled hUG, +DSS; lane 6, +DSS; lane 7, +unlabeled hUG, +DSS; lane 8, +DSS; lane 9, +unlabeled hUG, +DSS. Note that both the 190- and 49-kDa protein bands are present and those bands are specifically displaced in the presence of excess unlabeled hUG. (c) Scatchard analysis of specific binding of reduced ^{125}I-hUG in nontransfected HEC-1A cells. The data are from three experiments and each point represents the mean of triplicate determinations. Affinity cross-linking of ^{125}I-hUG in (d) HEC-1A and (e) HTB-81 cells. Lane 1, −DSS; lane 2, +DSS; lane 3, +unlabeled hUG, +DSS. Note the absence of hUG-binding protein(s) in HTB-81 cells. Reprinted with permission from reference 26.

shown). However, the 49- and 190-kDa protein bands were not observed when either the nonreduced or reduced form of ^{125}I-hUG was incubated with fibrosarcoma cells (data not shown). To investigate the mechanism(s) of hUG-mediated suppression of ECM invasion and anchorage-independent growth in nontransfected and pRC/RSV-hUG-transfected HEC-1A cells, we first detected the expression of hUG-binding protein (receptor) in both cells by receptor binding and affinity cross-linking experiments. The results of the binding data using reduced ^{125}I-hUG as a ligand showed specific, high-affinity binding with a K_d value of 25 nM in nontransfected HEC-1A cells (FIG. 4c). Affinity cross-linking data indicated the presence of 190- and 49-kDa bands in nontransfected HEC-1A cells using reduced, radiolabeled hUG as a ligand and these bands were specifically displaced when excess unlabeled hUG was added

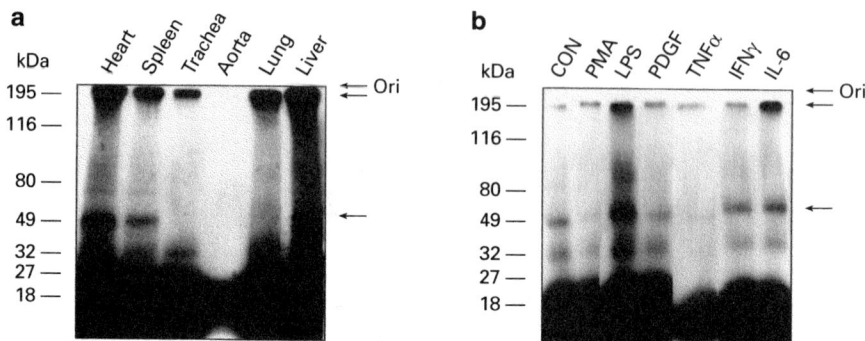

FIGURE 5. (a) Tissue-specific expression of hUG-binding protein(s) using different bovine tissues by affinity cross-linking technique. Note that both the 190- and 49-kDa binding proteins are detectable in bovine heart, spleen, and liver. The trachea and lung express only the 190-kDa protein, while none of these proteins are present in the aorta. (b) Effect of different cytokines and other agents on the expression of hUG-binding protein(s) by NIH 3T3 cells. The intensities of both the 190- and 49-kDa protein bands were considerably higher when the cells were treated with LPS and IL-6 compared to the control. However, there was no increase in intensities of these protein bands when other agents were used. Reprinted with permission from reference 23.

prior to the binding assays (FIG. 4d). Similar results were obtained when pRC/RSV-hUG-transfected HEC-1A cells were used for these studies (data not shown). In contrast, both the binding and affinity cross-linking data demonstrated that there was no such detectable specific, cross-linked protein band in nontransfected (FIG. 4e) and transfected (data not shown) prostate adenocarcinoma cells (HTB-81). These data led us to conclude that induced hUG expression or pretreatment of the adenocarcinoma cells with purified hUG inhibits ECM invasion of those cells that also express the hUG-binding protein(s). These results raise the strong possibility that hUG regulates the cellular invasiveness of uterine adenocarcinoma cells by both autocrine and paracrine pathways.

To determine the tissue-specific expression of hUG-binding proteins, the reduced, radiolabeled hUG was incubated with several bovine tissues (heart, spleen, trachea, aorta, lung, and liver) for binding and then cross-linked with DSS. The data indicated that the heart, spleen, and liver expressed both the 190- and 49-kDa binding proteins, whereas only the 190-kDa protein was present in the trachea and lung (FIG. 5a). Neither the 190- nor 49-kDa protein was present in the aorta (FIG. 5a). These results showed that the 190- and 49-kDa proteins are expressed in a tissue-specific manner. Since hUG is an anti-inflammatory protein, we sought to determine whether hUG-binding protein(s) is up- or downregulated by different inflammatory cytokines and other agents such as lipopolysaccharide (LPS). The NIH 3T3 cells were pretreated with cytokines and LPS alone and in combination followed by binding and affinity cross-linking with ^{125}I-hUG. The data showed that there was a substantial increase in the expressions of both the 190- and 49-kDa protein bands when the cells were pretreated with interleukin-6 (IL-6) and LPS as compared to nontreat-

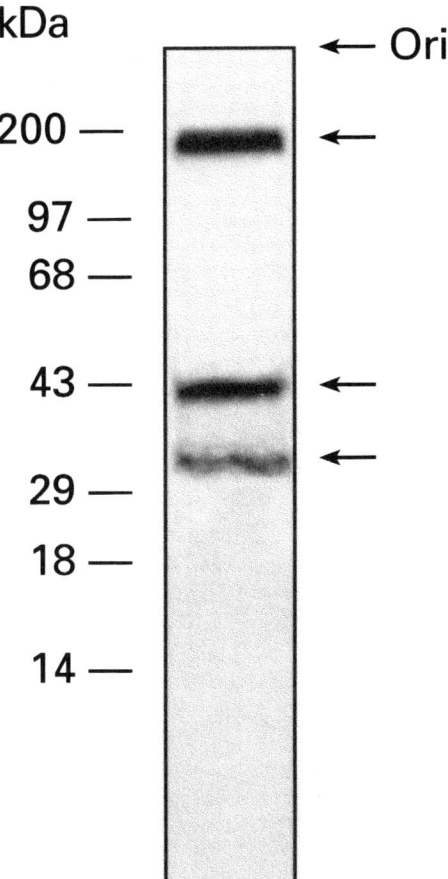

FIGURE 6. Purification of the hUG-binding protein(s) by hUG affinity chromatography. The affinity-purified samples were resolved by SDS-PAGE under denaturing and reducing conditions and then stained with silver staining. Note that two protein bands with apparent molecular masses of 180 and 40 kDa were visualized. In addition, a faint band with a molecular mass of 32 kDa was also seen. However, the 32-kDa band did not interact with the radiolabeled hUG, indicating that this may be either a degradation product of upper bands that lacks the hUG-binding epitope or an artifact. Reprinted with permission from reference 23.

ed cells (FIG. 5b). We found that both IL-6 and LPS mediated an apparent increase in the expression of hUG receptor(s) by these agents. These results suggest that the homeostatic mechanism to control the inflammatory response may involve not only hUG production, but also the dynamics of the interaction of hUG with its binding proteins that is regulated by proinflammatory agents.

In an attempt to purify the hUG-binding protein(s), the bovine spleen tissues were homogenized and the solubilized extract of the homogenate was passed through the

Sepharose 4B–linked hUG affinity column. The bound protein was eluted by lowering the pH of the elution buffer. The affinity-purified fraction was examined by binding and affinity cross-linking assays (data not shown). The purity was checked by SDS-PAGE followed by silver staining. Two major protein bands with apparent molecular masses of 180 and 40 kDa were detected (FIG. 6). It is not clear if these two protein bands are the subunits of the hUG-binding protein or if the lower band is the degradation product of the upper one with the putative receptor–binding epitope remaining the same. Molecular cloning and further characterization will answer which of the two alternatives is correct. In addition, a faint protein band with an apparent molecular mass of 30 kDa was also noticed that did not interact with radiolabeled hUG. Data suggested that this 30-kDa band might be a degradation product of a high-molecular-weight hUG-binding protein band that does not recognize the receptor-binding epitope or it might be an artifact.

Earlier, UG was thought to be a steroid-inducible secretory protein.[1,2] However, now it is reported that a nonsteroid hormone like prolactin enhances the progesterone-induced expression of the hUG gene.[27,28] Both progesterone and prolactin have been suggested to possess anti-inflammatory/immunomodulatory effects and it is possible that hUG may act as the effector molecule for both of these hormones for exerting these cellular effects. Although the biochemistry, molecular biology, and structural biology of hUG have been studied extensively, the physiological functions of this protein remained unsolved until recently. To understand the physiological function of hUG, we performed targeted disruption of the hUG gene and generated hUG-deficient mice. These hUG-deficient mice were found to develop multimeric fibronectin-deposited renal glomerular disease, distal tubular hyperplasia, and renal parenchymal fibrosis.[29] Recent studies also demonstrated that both UG knockout as well as UG antisense transgenic mice have developed immunoglobulin A (IgA) nephropathy.[30] The molecular mechanisms by which UG prevents abnormal fibronectin (Fn) deposition in the glomeruli at least in part is due to the formation of Fn-UG heteromers that compete with Fn homomerization, required for abnormal deposition of Fn.[31–33] As UG/hUG is an anti-inflammatory protein,[34] the development of glomerulonephritis, an inflammatory disease,[35] and renal parenchymal fibrosis, a sequela of the inflammatory process, is understandable. However, the mechanism(s) of distal tubular hyperplasia is not yet clear. In summary, we have demonstrated for the first time the receptor-mediated anti-ECM-invasive function of hUG in specific normal and cancer cells in an autocrine and paracrine manner. In addition, hUG also showed the inhibition of anchorage-independent growth of uterine-derived adenocarcinoma cells on soft agar and this is a characteristic property of most of the cancer cells. Therefore, it is likely that UG/hUG is a novel cytokine[3] that functions as a tumor suppressor via its interaction with the cell surface binding proteins.

ACKNOWLEDGMENTS

We thank I. Owens, J. Chou, J. D. Butler, and S. W. Levin for critical review of the manuscript and helpful suggestions, and Mr. Tiawari and S. Philip for editorial assistance. We also thank Rick Dreyfuss and Shauna Everett of Medical Arts and Photography, NIH, for their expert photomicrographic assistance.

REFERENCES

1. KRISHNAN, R.S. & J.C. DANIEL, JR. 1967. "Blastokinin": inducer and regulator of blastocyst development in the rabbit uterus. Science **158**: 490–492.
2. BEIER, H.M. 1968. Uteroglobin: a hormone-sensitive endometrial protein involved in blastocyst development. Biochim. Biophys. Acta **160**: 289–291.
3. MUKHERJEE, A.B., G.C. KUNDU, G. MANTILE-SELVAGGI et al. 1999. Uteroglobin: a novel cytokine? Cell. Mol. Life Sci. **55**: 771–787.
4. SINGH, G., J. SINGH, S.L. KATYAL et al. 1988. Identification, cellular localization, isolation, and characterization of human Clara cell–specific 10 kD protein. J. Histochem. Cytochem. **36**: 73–80.
5. SINGH, G. & S.L. KATYAL. 1997. Clara cells and Clara cell 10 kD protein (CC10). Am. J. Respir. Cell Mol. Biol. **17**: 141–143.
6. BEATO, M. 1976. Binding of steroids to uteroglobin. J. Steroid Biochem. **7**: 327–334.
7. JACKSON, P.J., R. TURNER, J.N. KEEN et al. 1988. Purification and partial amino acid sequence of human urine protein 1: evidence for homology to rabbit uteroglobin. J. Chromatogr. **452**: 359–367.
8. BERNARD, A.M., R. LAUWERYS, A. NOEL et al. 1989. Urine protein 1: a sex-dependent marker of tubular or glomerular dysfunction. Clin. Chem. **35**: 2141–2142.
9. BERNARD, A., R. LAUWERYS, A. NOEL et al. 1991. Determination by latex immunoassay of protein 1 in normal and pathological urine. Clin. Chim. Acta **201**: 231–245.
10. OKUTANI, R., Y. ITOH, H. HIRATA et al. 1992. Simple and high-yield purification of urine protein 1 using immunoaffinity chromatography: evidence for the identity of urine protein 1 and human Clara cell 10-kilodalton protein. J. Chromatogr. **577**: 25–35.
11. LOPEZ DE HARO, M.S., M. PEREZ MARTINEZ, C. GARCIA & A. NIETO. 1994. Binding of retinoids to uteroglobin. FEBS Lett. **349**: 249–251.
12. GILLNER, M., J. LUND, C. CAMBILLAU et al. 1988. The binding of methylsulfonyl-polychloro-biphenyls to uteroglobin. J. Steroid Biochem. **31**: 27–33.
13. DHANIREDDY, R., T. KIKUKAWA & A.B. MUKHERJEE. 1988. Detection of a rabbit uteroglobin-like protein in human neonatal tracheobronchial washings. Biochem. Biophys. Res. Commun. **152**: 1447–1454.
14. KIKUKAWA, T., B.D. COWAN, R.I. TEJADA & A.B. MUKHERJEE. 1988. Partial characterization of a uteroglobin-like protein in the human uterus and its temporal relationship to prostaglandin levels in this organ. J. Clin. Endocrinol. Metab. **67**: 315–321.
15. MANTILE, G., L. MIELE, E. CORDELLA-MIELE et al. 1993. Human Clara cell 10-kDa protein is the counterpart of rabbit uteroglobin. J. Biol. Chem. **268**: 20343–20351.
16. PERI, A., E. CORDELLA-MIELE, L. MIELE & A.B. MUKHERJEE. 1993. Tissue-specific expression of the gene coding for human Clara cell 10-kD protein, a phospholipase A_2–inhibitory protein. J. Clin. Invest. **92**: 2099–2109.
17. LEVIN, S.W., J.D. BUTLER, U.K. SCHUMACHER et al. Uteroglobin inhibits phospholipase A_2 activity. Life Sci. **38**: 1813–1819.
18. MIELE, L., E. CORDELLA-MIELE, A. FACCHIANO & A.B. MUKHERJEE. 1988. Novel antiinflammatory peptides from the region of highest similarity between uteroglobin and lipocortin I. Nature (London) **335**: 726–730.
19. MIELE, L., E. CORDELLA-MIELE & A.B. MUKHERJEE. 1990. High level bacterial expression of uteroglobin, a dimeric eukaryotic protein with interchain disulfide bridges in its natural quaternary structure. J. Biol. Chem. **265**: 6427–6437.
20. ROBINSON, D.H., K.L. KIRK & D.J. BENOS. 1989. Macromolecular transport in rabbit blastocyst: evidence for a specific uteroglobin transport system. Mol. Cell. Endocrinol. **63**: 227–234.
21. DIAZ GONZALEZ, K. & A. NIETO. 1995. Binding of uteroglobin to microsomes and plasmatic membranes. FEBS Lett. **361**: 255–258.
22. KUNDU, G.C., G. MANTILE, L. MIELE et al. 1996. Recombinant uteroglobin suppresses cellular invasiveness via novel class of high-affinity cell surface binding site. Proc. Natl. Acad. Sci. U.S.A. **93**: 2915–2919.
23. KUNDU, G.C., A.K. MANDAL, Z. ZHANG et al. 1998. Uteroglobin (UG) suppresses extracellular matrix–invasion by normal and cancer cells that express the high-affinity UG-binding proteins. J. Biol. Chem. **273**: 22819–22824.

24. KUNDU, G.C. & A.B. MUKHERJEE. 1997. Evidence that porcine pancreatic phospholipase A_2 via its high-affinity receptor can stimulate extracellular matrix invasion by normal and cancer cells. J. Biol. Chem. **272:** 2346–2353.
25. MUNSON, P.J. & D. RODBARD. 1980. Ligand: a versatile computerized approach for characterization of ligand binding systems. Anal. Biochem. **107:** 220–239.
26. ZHANG, Z., G.C. KUNDU, D. PANDA et al. 1999. Loss of transformed phenotype in cancer cells by overexpression of the uteroglobin gene. Proc. Natl. Acad. Sci. U.S.A. **96:** 3963–3968.
27. CHILTON, B.S., S.K. MANI & D.W. BULLOCK. 1988. Servomechanism of prolactin and progesterone in regulating uterine gene expression. Mol. Endocrinol. **2:** 1169–1175.
28. RANDALL, G.W., J.C. DANIEL, JR. & B.S. CHILTON. 1991. Prolactin enhances uteroglobin gene expression by uteri of immature rabbits. J. Reprod. Fertil. **91:** 249–257.
29. ZHANG, Z., G.C. KUNDU, C-J. YUAN et al. 1997. Severe fibronectin-deposit renal glomerular disease in mice lacking uteroglobin. Science **276:** 1408–1412.
30. ZHENG, F., G.C. KUNDU, Z. ZHANG et al. 1999. Essential role of uteroglobin in preventing IgA-nephropathy. Nat. Med. **5:** 1018–1025.
31. RUOSLAHTI, E. 1988. Fibronectin and its receptors. Annu. Rev. Biochem. **57:** 375–413.
32. ZHANG, Q. & D.F. MOSHER. 1996. Cross-linking of the NH_2-terminal region of fibronectin to molecules of large apparent molecular mass. J. Biol. Chem. **271:** 33284–33292.
33. HYNES, R.O. 1990. Fibronectins. Springer-Verlag. New York/Berlin.
34. MUKHERJEE, A.B., G.C. KUNDU, A.K. MANDAL et al. 1998. Uteroglobin (UG): its physiological role in normal glomerular function uncovered by targeted disruption of the UG gene in mice. Am. J. Kidney Dis. **32:** 1106–1120.
35. BORDER, W.A. & E. RUOSLAHTI. 1992. Transforming growth factor-beta in disease: the dark side of tissue repair. J. Clin. Invest. **90:** 1–7.

The Role of CC10 in Pulmonary Carcinogenesis: From a Marker to Tumor Suppression

R. ILONA LINNOILA,[a,b] EVA SZABO,[b] FRANCESCO DEMAYO,[c] HANSPETER WITSCHI,[d] CAROL SABOURIN,[e] AND AL MALKINSON[f]

[b]*National Cancer Institute, National Institutes of Health, Rockville, Maryland 20850, USA*

[c]*Baylor College of Medicine, Houston, Texas 77030, USA*

[d]*University of California, Davis, California 95616, USA*

[e]*Ohio State University, Columbus, Ohio 43210, USA*

[f]*University of Colorado, Denver, Colorado 80262, USA*

> ABSTRACT: CC10 is infrequently expressed in human non–small cell lung cancers (NSCLCs), despite being abundantly produced by progenitor cells for normal and neoplastic epithelium. Many abnormalities in the surrounding lung associated with field carcinogenesis, which reflect prolonged exposure to such carcinogens as tobacco smoke, also revealed altered expression of CC10. Exposure of hamsters and mice to the tobacco-specific carcinogen NNK led to reduced CC10 expression, which was partially reversible. Overexpression of CC10 in immortalized bronchial epithelial cells delayed the induction of anchorage-independent growth in response to NNK. The data suggest that downregulation of CC10 contributes to carcinogenesis because CC10 antagonizes the neoplastic phenotype.

INTRODUCTION

Lung cancer continues to be the leading cause of cancer-related deaths for both men and women in the United States. It is estimated that 164,100 new cases and 156,900 deaths will occur in the year 2000.[1] Approximately 20% will be small cell lung cancers (SCLCs), which are very aggressive neoplasms featuring neuroendocrine differentiation, while 75–80% will be non-SCLCs (NSCLCs). NSCLCs include three major histologic types: squamous cell carcinoma, adenocarcinoma, and large cell carcinoma.[2] For reasons not fully understood, adenocarcinoma has become the most common histologic form of lung cancer in recent years.[3] Moreover, in many hospital-based series, features of peripheral airway cell differentiation associated with adenocarcinomas appear to be much more common than previously appreciated.[4,5]

---

[a]Address for correspondence: Ilona Linnoila, Cell and Cancer Biology Department, Medicine Branch, Division of Clinical Sciences, National Cancer Institute, NIH, 9610 Medical Center Drive, Suite 300, Rockville, MD 20850. Voice: 301-402-3128, ext. 313; fax: 301-402-4422.
il17h@nih.gov

In contrast to the centrally located squamous cell carcinomas of the lung with known premalignant changes, the early events in the genesis of adenocarcinomas are poorly understood. Clinically, this contributes to a delay in diagnosis, with initial detection of the cancer occurring at an advanced stage that is associated with poor prognosis. The precursor cells of peripheral lung adenocarcinomas include metaplastic mucin-producing cells, type II cells, and Clara cells, whose major product is the Clara cell specific 10-kDa (CC10) protein.[6] CC10 comprises 7% of the total protein found in bronchioalveolar lavage fluid.[7] While a myriad of biological activities including the inhibition of phospholipase A_2 activity and binding fibronectin have been associated with CC10, its physiological role in the adult lung remains unknown.[8] The purpose of this article is to review existing data and present new observations suggesting that pulmonary CC10 contributes to tumor suppression. Our goal is to study CC10 in order to provide a better understanding of the early phases of lung carcinogenesis, which can then be applied to early detection and chemoprevention of pulmonary adenocarcinoma.

DECREASED EXPRESSION OF CC10 IN LUNG TUMORS

Ultrastructurally, Clara cells, which are nonciliated bronchiolar epithelial cells, are characterized by abundant, discrete electron-dense granules.[9] In the normal lung, they are intensely positive for CC10 protein by immunohistochemistry, while RNA-RNA *in situ* hybridization reveals strong labeling consistent with abundant CC10 mRNA in these cells (FIG. 1A; also FIG. 3 below). CC10-containing nonciliated secretory cells are most prominent in the terminal and respiratory bronchioli, but to varying degree can also be found in larger airways, including bronchi.[10] Clara cells are progenitors for themselves and ciliated cells, and they can rapidly repopulate damaged or denuded airway epithelium.[11,12]

In the earlier studies on lung cancer, Clara cell differentiation was recognized by the presence of the electron-dense granules, which led to the widely held assumption that the majority of human lung adenocarcinomas are of bronchiolar origin, despite the frequently recognized clinicopathological heterogeneity in these tumors.[13–15] Thus, it was surprising that most of the tumors were actually negative for CC10 in the subsequent immunohistochemical studies (see FIG. 1).[16,17] Many NSCLCs, regardless of their histological type, reveal scattered CC10 immunoreactivity in a small subpopulation of the cells. Furthermore, in most cases, the tumor immunoreactivity appears lighter than what is found in the bronchiolar cells of the adjacent normal lung. The pattern of staining is generally cytoplasmic, with occasional membranous distribution or nuclear reactivity. Most importantly, among the >500 tumors that we stained, only 10% demonstrate marked CC10 staining in substantial numbers of tumor cells (≥10% of tumor cells positive), which is in contrast with previous ultrastructural studies (FIG. 1B). The low incidence of CC10 expression was confirmed by studies using RNA-RNA *in situ* hybridization where only 1 squamous cell carcinoma out of 19 NSCLCs revealed the presence of mRNA.[10] Nevertheless, the fact that CC10 was often detected in squamous cell and large cell carcinomas is in accordance with the theory that all lung cancers may derive from common precursor cells. Clinically, patients with CC10-positive tumors tended to be significantly

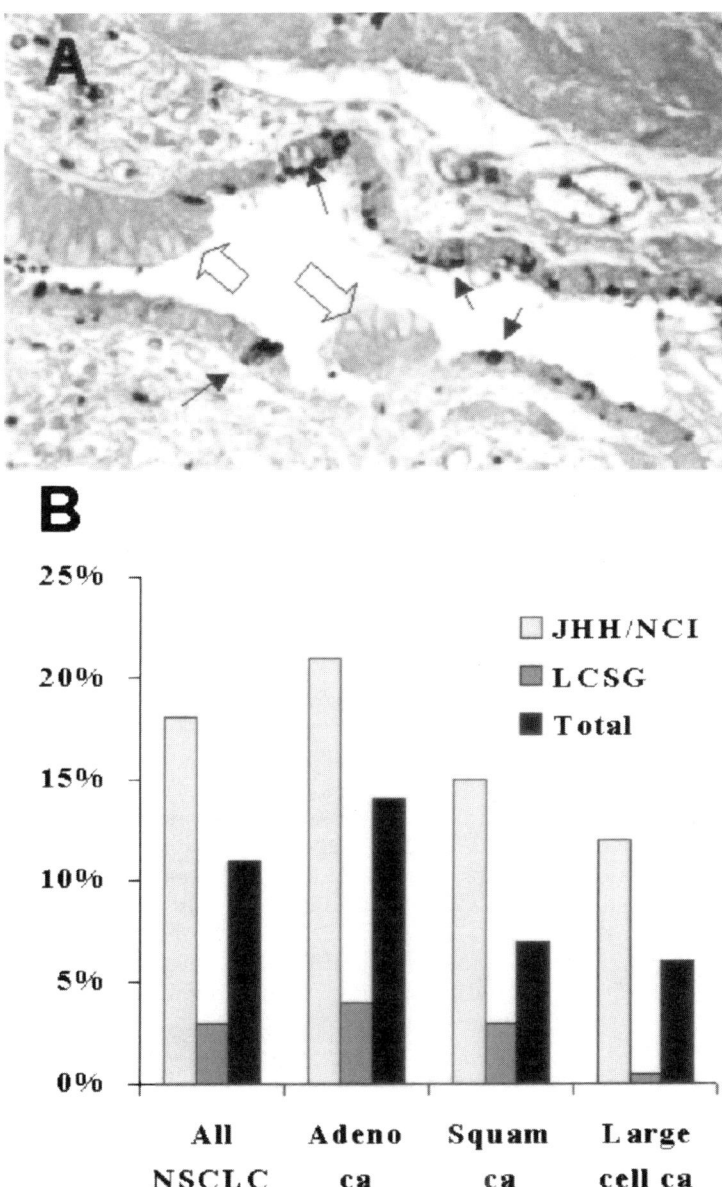

FIGURE 1. Expression of CC10 in human lung cancers. (**A**) A photomicrograph of a bronchiolus invaded by an adenocarcinoma (block arrows) that is negative for CC10, while the surrounding nonneoplastic epithelium contains dark, intensely positive CC10-containing cells (small arrows; immunoperoxidase staining). (**B**) A barograph illustrating the percentage of CC10-positive tumors in various histological types. The tumors were ob-

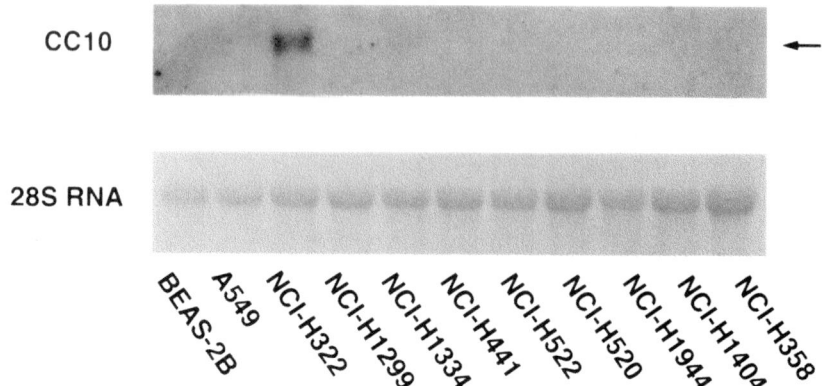

FIGURE 2. Expression of CC10 in lung-derived cell lines. Total cellular RNA was isolated from logarithmically growing NSCLC cell lines and the immortalized bronchial epithelial cell line BEAS-2B. After Northern transfer, hybridization was performed with ^{32}P-labeled CC10 cDNA. Ethidium bromide shadowing revealed equal RNA loading in all lanes. (From reference 18 with permission from the publisher.)

younger, and in women there was a correlation with lighter smoking.[16] No difference was detected between survival and the CC10 status.

The expression of CC10 by a small number of NSCLC cell lines is consistent with the data on primary tumors. As shown in FIGURE 2, CC10 mRNA was expressed in only 2 of 10 established, well-characterized NSCLC cell lines surveyed by Northern blot analysis.[18] Ultrastructural evidence for Clara cell differentiation was initially detected in 5 out of 32 NSCLC cell lines (16%).[19] Later on, we detected Clara cell granules in 11 out of 39 NSCLC cell lines (28%) when we included more cell lines derived from lung tumors exhibiting papillolepidic growth pattern, which is characteristic of peripheral adenocarcinomas. Interestingly, most of the cell lines also showed evidence of type II cell differentiation either biochemically or ultrastructurally.[20] Likewise, 12 out of 50 (24%) NSCLC cell lines revealed the expression of CC10 mRNA by *in situ* hybridization, while 10 out of these 12 cell lines also coexpressed SPA, a marker for type II cell differentiation. These studies have demonstrated that NSCLC cell lines rarely express exclusively CC10, suggesting a close relationship with the type II cell lineage. In general, NSCLC cell lines express only reduced levels of CC10, requiring sensitive detection methods including electron microscopy and RNA-RNA *in situ* hybridization.

---

tained from two Maryland institutions (light gray bars)—The Johns Hopkins Hospital (JHH) and the National Cancer Institute (NCI)[16]—as well as from a multi-institutional study of the Lung Cancer Study Group (LCSG; dark gray bars).[17] Black bars represent a summary of the two studies (total). Tumors were considered positive if ≥10% of tumor cells stained.

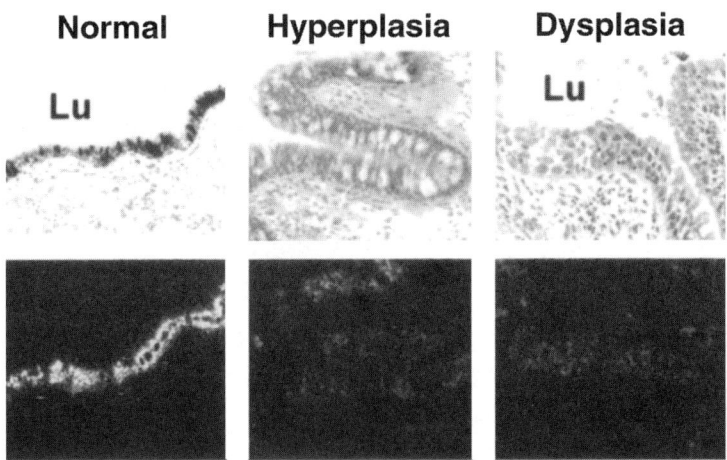

FIGURE 3. Photomicrographs of decreased expression of CC10 mRNA in histologically abnormal human bronchi. The panels on the left reveal intense signal for CC10 in the normal epithelium: bright field view on the top and dark field at the bottom. Middle panels show diffuse goblet cell hyperplasia resulting in only a few grains of labeling, indicating the presence of very low amounts of CC10 mRNA. Sharply reduced expression is seen also in the dysplastic squamous cell metaplasia on the right. Note the paucity of signal in the corresponding dark field picture (RNA-RNA *in situ* hybridization).

HISTOLOGIC ABNORMALITIES IN NONNEOPLASTIC LUNG CORRELATE WITH ALTERED CC10 EXPRESSION

According to the concept of field cancerization, initially described by Slaughter *et al.* in 1954, who observed a variety of histologic abnormalities in benign epithelium adjacent to oral cancers, the entire respiratory tract is exposed to carcinogens such as tobacco smoke.[21] Consequently, injured bronchopulmonary epithelium may accumulate genetic and morphologic changes that are multifocal and proceed towards cancer.[22] Interestingly, CC10 levels were significantly lower in serum and bronchioalveolar lavage specimens obtained from smokers and lung cancer patients compared with specimens from healthy nonsmokers.[23] This prompted us to examine the morphological basis of CC10 dysregulation in field carcinogenesis using surrounding nonneoplastic lungs obtained from patients undergoing resection for cancer.

In morphologically normal lungs, strong and moderate levels of CC10 mRNA were observed in bronchioli and bronchi, respectively, but the expression was rarely observed in the alveolar region.[10,24] CC10 hybridization signal decreased markedly in bronchi containing diffuse goblet cell hyperplasia or squamous metaplasia (FIG. 3). Variable results were seen in bronchi with dysplasia. Basal cell hyperplasia or focal goblet cell hyperplasia did not alter CC10 expression. While a significant decrease in Clara cell number in distal airways has been described, our results on CC10 expression in bronchioli were inconclusive.[25–27] This may be due to the fact

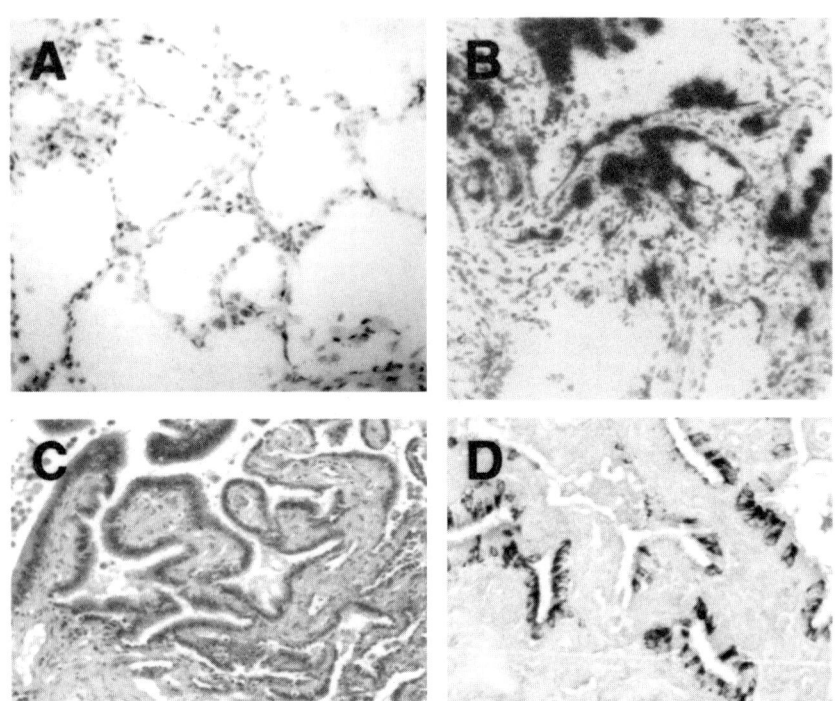

FIGURE 4. Photomicrographs of alveolar abnormalities in the surrounding lung. (**A**) Histologically normal alveolar septa that lack CC10 mRNA expression (RNA-RNA *in situ* hybridization). (**B**) Type II cell hyperplasia in fibrotic alveoli expressing high levels of CC10 mRNA (RNA-RNA *in situ* hybridization). (**C**) Bronchiolization of alveoli (BOA) where thickened alveolar walls are lined by bronchiolar-type epithelium (hematoxylin-eosin stain). (**D**) BOA typically composed of dark CC10-immunoreactive cells (immunoperoxidase stain).

that those histologic abnormalities were rarely detected in the bronchioli in our material. Moreover, bronchiolar CC10 mRNA levels remained unchanged in sections containing abnormalities elsewhere. Interestingly, in the alveolar compartment, there was frequently an increase in the amount of CC10 mRNA (FIG. 4). The distinct alveolar abnormalities, which were associated with increased CC10 expression, included type II cell hyperplasia, atypical alveolar hyperplasia (AAH), and bronchiolization of the alveoli. A comparison of mRNA expression and clinicopathological features demonstrated that the amount of histologic abnormalities increased with smoking history; however, no correlation between CC10 mRNA expression and sex, age, or smoking history was found.[24]

Approximately 20% of the benign, reactive type II cell hyperplasias revealed CC10-containing cells,[16] while AAH with rare CC10-containing cells is a potential precursor for bronchioalveolar carcinoma. However, the most striking increase in CC10 mRNA was detected in areas of fibrosis with BOA (FIG. 4). BOA is a histologically distinct lesion that is thought to arise from the "colonization" of alveolar walls with bronchiolar epithelium either via cell migration through alveolar pores or from the transformation of alveolar type II cells into bronchiolar-type epithelium.[28] It occurs in a variety of pathological conditions. We have detected BOA lesions in up to 12% of specimens derived from over 400 NSCLC resections. The BOAs exhibited a range of morphological alterations associated with a significant decrease in CC10 expression and focal expression of p53 and 3p abnormalities in high-grade lesions, while c-myc and cJun protein were observed in all grades.[29,30] BOA lesions were rarely present in control lungs obtained from patients without neoplasias. Our results indicated that BOA is a multifocal metaplasia of unknown significance in lung resection specimens that shares features often seen in premalignant lesions.

OVEREXPRESSION OF CC10 MODIFIES PULMONARY NEOPLASTIC POTENTIAL

Premalignant lesions and tumors frequently display disorderly cell differentiation, prompting the question of whether the loss of at least some differentiation markers by cancer cells is a necessary event during carcinogenesis.[31–33] Since only 5–25% of human lung cancers express CC10, despite the abundant CC10 production by peripheral airway progenitor cells, we hypothesized that the loss of CC10 may contribute to pulmonary carcinogenesis. Further evidence strengthening this hypothesis came from studies showing that naturally occurring CC10 and recombinant CC10 have potent anti-invasive effects on fibroblasts, human trophoblasts, and prostate cancer cells, potentially acting through specific high-affinity cell surface binding sites.[34,35]

To characterize CC10 expression during pulmonary carcinogenesis, CC10 was first examined in a panel of lung-derived cell lines by Northern blot analysis.[18] As shown in FIGURE 2, CC10 mRNA was expressed in only 2 of 10 established, well-characterized NSCLC cell lines, showing a distinct signal in NCI-H322, but being barely detectable in NCI-H1334. The nontumorigenic immortalized bronchial epithelial cell line BEAS-2B did not express CC10, whereas normal lung showed a very strong hybridization signal. We overexpressed CC10, under the control of a mouse metallothionin promoter, in the A549 NSCLC cell line and examined its effects on various aspects of the neoplastic phenotype. Three clones (A5-7-6, A5-8-7, and A5-8-8) that expressed the highest amounts of CC10 mRNA from two independent transfections were studied along with three control clones (FIG. 5A).

When the effect of CC10 on the growth of A549 cells was examined, monolayer growth under high-serum or low-serum conditions did not differ between CC10-overexpressing and control cell lines. Anchorage-independent growth in soft agar, on the other hand, was significantly reduced in CC10-expressing cell lines. On average, there was a 45% decrease in clonogenic growth (FIG. 6A).

Our data show that enforced CC10 expression does, indeed, decrease the neoplastic potential of these cells. CC10-expressing cells exhibited a marked reduction in invasiveness that was paralleled by diminished metalloproteinase activities by zymography (FIG. 6B). This was attributable to the near absence of the two corresponding metalloproteinases, MMP-2 and MMP-9, in the CC10-transfected lines, but not in the vector-transfected controls. Curiously, the CC10 transfectants also showed decreased adhesion to fibronectin. Given that the binding of fibronectin to

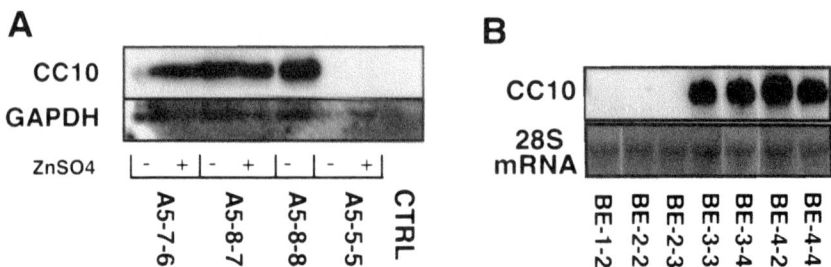

FIGURE 5. CC10 expression in transfected cell lines. **(A)** CC10 expression in the NSCLC cell line A549. A549 cells were transfected with CC10 cDNA under the control of a mouse metallothionin promoter (A5-7-6, A5-8-7, and A5-8-8) or a control plasmid (A5-5-5, shown; and A5-6-1, not shown) and clonal cell lines were isolated. Total cellular RNA was isolated after cell lines were grown in $ZnSO_4$ (to induce CC10 expression), RNA was reverse-transcribed, and the resulting cDNA was amplified by PCR for CC10 or the housekeeping gene GAPDH. After Southern transfer, nitrocellulose filters were probed with ^{32}P-labeled CC10 cDNA or a ^{32}P-labeled GAPDH oligomer. As shown, CC10 was expressed in the transfected cell lines even in the absence of Zn, while it was absent in the control transfectants. CTRL indicates a PCR reaction control, with substitution of water for RNA before PCR amplification. (From reference 18 with permission from the publisher.) **(B)** CC10 expression in the immortalized bronchial epithelial cell line BEAS-2B. BEAS-2B cells were transfected with CC10 cDNA in an episomally integrated vector, and clonal cell lines expressing CC10 (BE-3-3, BE-3-4, BE-4-2, and BE-4-4) or control plasmids (BE-1-2, BE-2-2, and BE-2-3) were isolated. Total cellular RNA was isolated, and Northern transfer followed by hybridization with ^{32}P-labeled CC10 cDNA was performed. Ethidium bromide shadowing revealed equal RNA loading in all lanes.

its integrin receptors initiates a signal transduction cascade that may result in stimulation of growth, metalloproteinase expression, or suppression of apoptosis, we hypothesize that the interruption of such signal transduction cascades by CC10 may in part be responsible for its anticarcinogenic effects.[36,37] In contrast to the observed interference with adhesion to fibronectin, cell-cell homotypic adhesion was increased in the CC10-overexpressing cell lines compared with that in controls. Thus, in summary, our studies confirmed that CC10 is downregulated in cancer cells and demonstrated that CC10 expression antagonizes the malignant phenotype at multiple levels through decreased expression of MMPs, altered adhesive properties, and decreased anchorage-independent growth.

To further assess if CC10 has any effect on cells prior to the establishment of the invasive malignant phenotype, we overexpressed CC10 cDNA in the immortalized bronchial epithelial cell line BEAS-2B and selected four clones that expressed CC10 as well as three vector-transfected controls for further study (FIG. 5B). Under standard growth conditions, BEAS-2B does not express CC10 (FIG. 2) and grows as a monolayer, but not in an anchorage-independent fashion. However, anchorage-independent growth with colony formation in soft agarose can be induced by transiently exposing these cells to 3 µg/mL of the tobacco-specific nitrosamine, 4-(methylnitrosamino)-1-(3-pyridyl)-1-butanone (NNK).[38] We exposed three vector-transfected control cell lines, three CC10-expressing transfectant cell lines, and one CC10-transfected cell line that had lost its CC10 expression to NNK and then

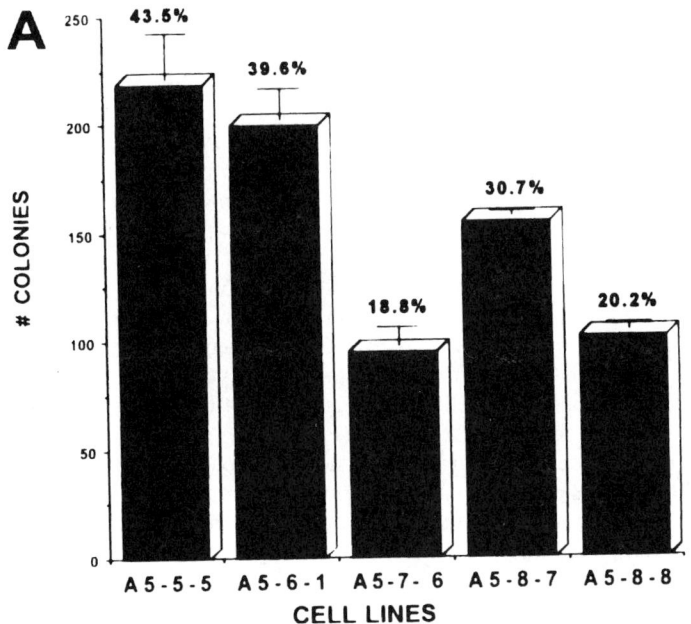
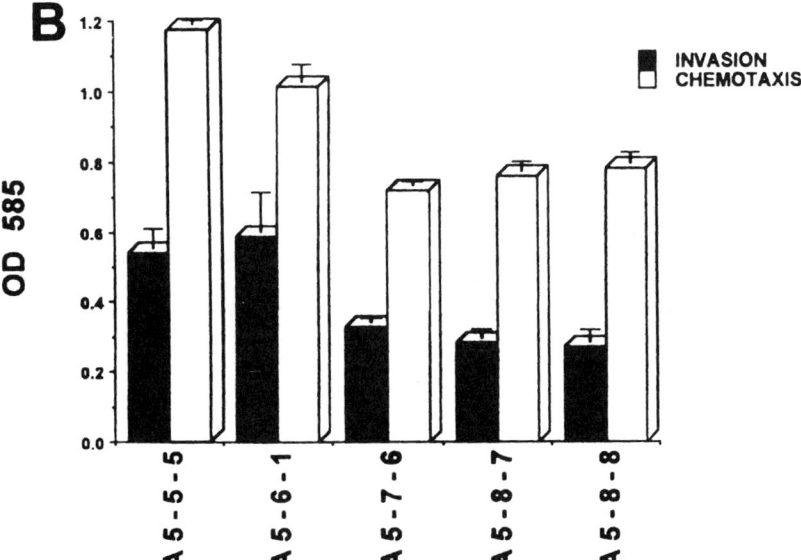

FIGURE 6. Reversion of the neoplastic phenotype in CC10-transfected A549 cells. (**A**) Anchorage-independent growth: Clonogenic growth in soft agar was assessed for CC10-transfected and control cell lines. The number of colonies was significantly reduced in the CC10-transfected cell lines. Error bars represent the standard deviation; the numbers above

assessed their capacity for colony formation over multiple passages. In all three CC10 clones, CC10 overexpression was associated with minimal colony formation until five passages following the NNK exposure (FIG. 7). However, at later passages (passages 7–9 post-NNK), one of the CC10-expressing clones also started to form colonies (FIG. 7C). In contrast, 3 out of 4 CC10-negative cell lines revealed increasing colony-forming efficiency, ranging from 1.5% to 24.7%, beginning much earlier after exposure to NNK. Curiously, the two cell lines with the greatest capacity for colony formation (the CC10-expressing cell line that began to clone at passages 5–7 and the transfectant that lost its CC10 expression) both expressed substantially higher levels of cyclin D1 compared with the other cell lines (FIG. 8). Since cyclin D1 is a key positive regulator of cell growth, and deregulation of the cyclin D1-p16-Rb pathway occurs frequently during lung carcinogenesis, acquired upregulation of cyclin D1 may have been responsible for the change in the growth characteristics.[39,40] The expression of p16, a negative regulator of cyclin D1 kinase activity, and cdk4, the catalytic subunit of cyclin D1 kinase, did not differ between the various transfectants before or after NNK exposure (FIG. 8).

Taken together, these results suggest that not only does reexpression of CC10 in overtly malignant cells antagonize the malignant phenotype, but CC10 overexpression may also counteract the acquisition of anchorage-independent growth, a hallmark of epithelial malignancy, very early in carcinogenesis.

FIGURE 7. Effect of CC10 overexpression on anchorage-independent growth of immortalized bronchial epithelial cells exposed to NNK. CC10-transfected BEAS-2B clonal cell lines were treated with the tobacco-specific nitrosamine NNK for 20 h and then propagated in growth media. BE-4-2, while transfected with CC10 and initially expressing CC10 (see FIG. 5), lost its expression prior to NNK exposure. Cloning in soft agarose to assess anchorage-independent growth was performed at various passages after NNK exposure. **(A)** Soft agarose cloning prior to and shortly after NNK exposure. Cloning efficiency was negligible prior to NNK treatment, but became measurable in 2 of 4 nonexpressing cell lines within two passages after NNK. **(B)** Expression of CC10 mRNA at the same time as clonogenic growth was assessed. **(C)** Soft agarose cloning at later passages after NNK exposure. With increasing passage number after NNK exposure, 3 of 4 CC10-nonexpressing cell lines and 1 of 3 CC10-expressing cell lines acquired the capacity for clonogenic growth. **(D)** Expression of CC10 mRNA at the same time as clonogenic growth was assessed.

the error bars represent cloning efficiency. (From reference 18 with permission from the publisher.) **(B)** Invasion: Invasion through reconstituted basement membrane (Matrigel) was measured using a Boyden chamber assay and fibroblast-conditioned medium as chemoattractant. Chemotaxis was assessed in the same way using type IV collagen instead of Matrigel. Cells that traversed the coated filters were stained with crystal violet, the dye was solubilized, and absorbance (as a measure of invasion) was determined. CC10-transfected cell lines exhibited significantly less invasion than control cell lines. (From reference 18 with permission from the publisher.)

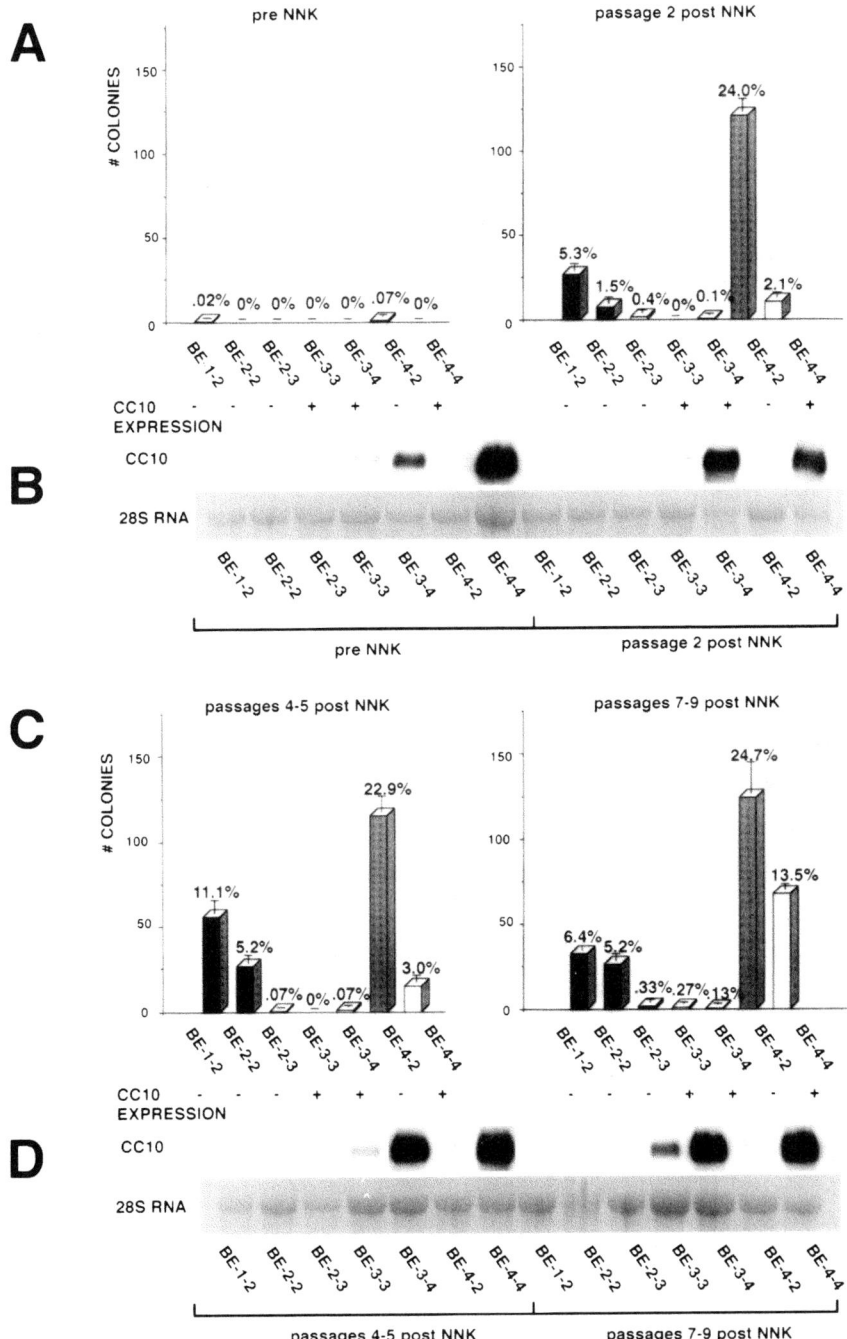

FIGURE 7. *See previous page for caption.*

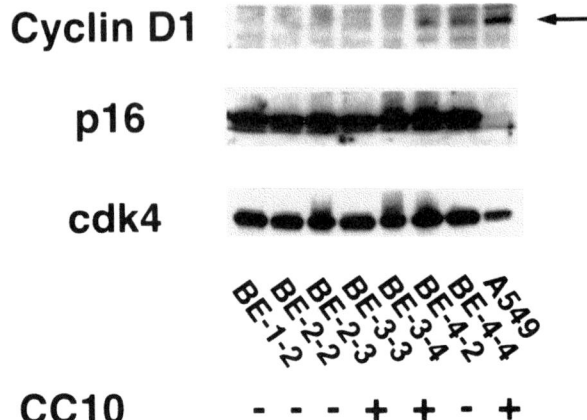

FIGURE 8. Expression of cell cycle regulatory proteins in CC10-transfected BEAS-2B cells exposed to NNK. Protein extracts were prepared from clonal BEAS-2B cell lines grown for 7–9 passages after NNK exposure, and Western analysis was performed using antibodies to the indicated cell cycle regulatory proteins. The NSCLC cell line A549, which is known to have high cyclin D1 and absent p16 expression, was used as a control. Cyclin D1 expression was detectable only in the two cell lines with the greatest capacity for clonogenic growth.

ANIMAL MODELS

While a variety of animal models have been developed to date on various aspects of lung carcinogenesis, relatively few studies have systematically assessed the histogenesis regarding the NSCLCs originating from the conducting airway epithelium, which includes the Clara cells or nonciliated secretory cells. The distinct advantages of animal models include the possibility of sequential tissue sampling, which in humans is very difficult. In hamsters, a number of features of multicentric, peripheral neoplastic lesions of the lung are similar to the morphological characteristics observed in humans, making this species an appealing candidate for the investigations.[41] Systemic exposure to nitrosamines induces adenomas, adenocarcinomas (including the bronchioalveolar subtype), and adenosquamous and squamous cell carcinomas in these animals in 3–4 months. Ultrastructural and immunohistochemical evidence has suggested Clara cell origin for these tumors.[42]

In two independent series, 50% of the tumors, which resulted from the exposure of Syrian golden hamsters to N-nitrosodiethylamine (DEN) or tobacco-specific nitrosamine (NNK), revealed CC10 immunoreactivity.[43,44] Moreover, another study by Rehm et al. suggested that bronchioalveolar carcinomas in hamsters are derived from bronchiolar secretory (Clara) cells growing along alveolar walls, differentiating into other bronchiolar cell types and entrapping resident alveolar type II cells.[45] In this study, the animals were exposed to N-nitrosomethyl-n-heptylamine. The authors postulated that the active migration of Clara cells contributing to tumor formation in the alveolar compartment was similar to that seen in bronchiolization of the alveoli or BOA[28,29] and might have been induced by a fibronectin-derived

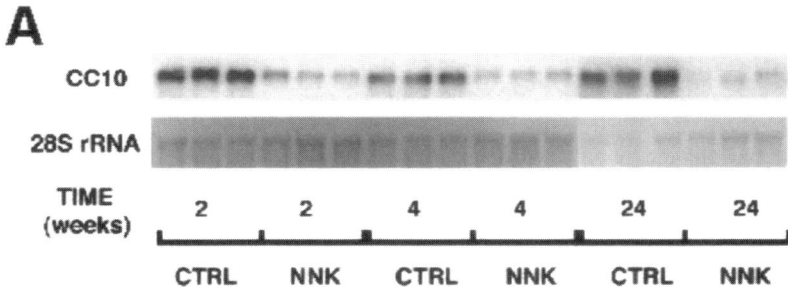
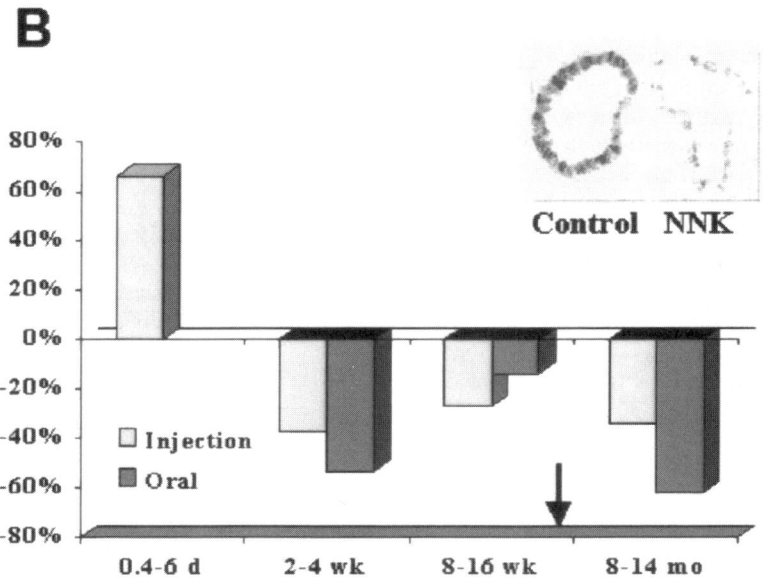

FIGURE 9. Alterations in pulmonary CC10 expression in hamsters and mice exposed to NNK. **(A)** Northern blot analysis of CC10 expression in hamster lungs. Total lung RNA was isolated from hamsters treated with biweekly injections of NNK for 2–24 weeks as indicated. After Northern transfer, hybridization was performed with ^{32}P-labeled CC10 cDNA. Ethidium bromide shadowing revealed equal RNA loading in all lanes. Lower levels of CC10 are seen in lanes from animals exposed to NNK. (From reference 18 with permission from the publisher.) **(B)** Morphometry of bronchiolar CC10-immunolabeled areas in mice exposed to NNK. Sections of lungs were immunostained for CC10 using supraoptimal (1:250,000) dilution, which revealed reduced staining in the airways of NNK-exposed mice (inset). Area of staining relative to the length of epithelium evaluated in five small airways from each animal was determined using computer-assisted image analysis. The results are expressed as the percent of change as compared with control animals in each experimental group at the same time point. Each time point is composed of data derived from 16–25 NNK-exposed mice. Tumors started to develop after 16 weeks of exposure (arrow). Injection = a single peritoneal injection of NNK at a dose of 10 μmol. Oral = each animal received 9.1 mg of NNK in drinking water over an 8-week period.

chemoattractant. Moreover, we exposed Syrian golden hamsters to biweekly injections of NNK and were able to show a marked reduction in the expression of CC10 mRNA expression in the experimental animals (FIG. 9A) by 2 weeks.[18] Immunohistochemically, this was associated with the depletion of the majority of CC10-containing cells prior to the emerging histologic atypias and tumor formation.[46,47] Epithelial changes were also associated with marked neuroendocrine cell hyperplasia. The decrease in CC10-containing cells was partially reversible as evidenced by the staining pattern in recovering animals. Most tumors appearing at 24 weeks were positive for CC10 immunoreactivity. In other words, the histologic spectrum of tumors, the premalignant changes, and the decrease in CC10 expression in hamsters are similar to the abnormalities seen in humans, while the tumors themselves may express CC10 at higher percentages. Most of the hamster lung tumors harbor ras mutations, while alterations in p53 are rare.[48]

In contrast, most peripheral lung tumors induced in mice tend to be adenomas or adenocarcinomas, which originate from type II cells, progenitors for the nonneoplastic and neoplastic epithelium in the alveolar compartment. Various mouse strains display marked differences in their sensitivity to develop tumors.[49] According to a commonly used regimen, we exposed the sensitive A/J mice to NNK intraperitoneally or in drinking water in order to study CC10 expression during carcinogenesis. Multiple lung tumors were observed starting at 24–32 weeks. As expected, these tumors contained ras mutations and were negative for CC10 in the immunohistochemical analysis.[50] However, as seen in humans and hamsters, the expression of CC10 along the epithelium of conducting airways was reduced up to 60%. To visualize this change, we applied a supraoptimal dilution technique with computer-assisted imaging (FIG. 9B). Curiously, the acute phase of the exposure of mice was associated with a transient increase in the CC10 content in the bronchioli, which is similar to the response seen in rats exposed to a domestic insecticide or firefighters exposed to smoke inhalation.[51,52] Yet another model to study bronchiolar cell carcinogenesis in this species might be mice exposed to *N*-nitrosobis-(2-chloroethyl) ureas.[53] In contrast to the typical mouse alveolar type II cell tumors, the exposure resulted in squamous cell carcinomas and adenocarcinomas, followed by adenocarcinomas. In several mice, hyperplasias and tumors were actually composed of cells expressing CC10.

In addition to the chemical approach, transgenic mice whose Clara cells were transformed with SV40 large T antigen (TAg) develop lung tumors expressing reduced levels of CC10 mRNA.[54–56] TAg has been shown to disrupt normal cell cycle control by binding and inhibiting the cell cycle checkpoint proteins, Rb and p53, in epithelial cells.[57] Removal of these proteins, concomitant with secondary genetic events, leads to the loss of cell cycle control, increased cellular proliferation, and eventual tumor development.[56] We were interested in the mechanisms regulating neuroendocrine differentiation in normal lung and lung cancer. Our studies suggested that ASH1 (also termed MASH1), a basic helix-loop-helix transcription factor conserved from the *Drosophila* achaete-scute complex, plays a critical role in regulating the lung neuroendocrine phenotype.[58] We therefore created a transgenic model in order to constitutively express ASH1 in Clara cells that normally lack this factor. Remarkably, we found that this neural transcription factor could promote airway epithelial proliferation and dramatically potentiate the tumorigenic impact of

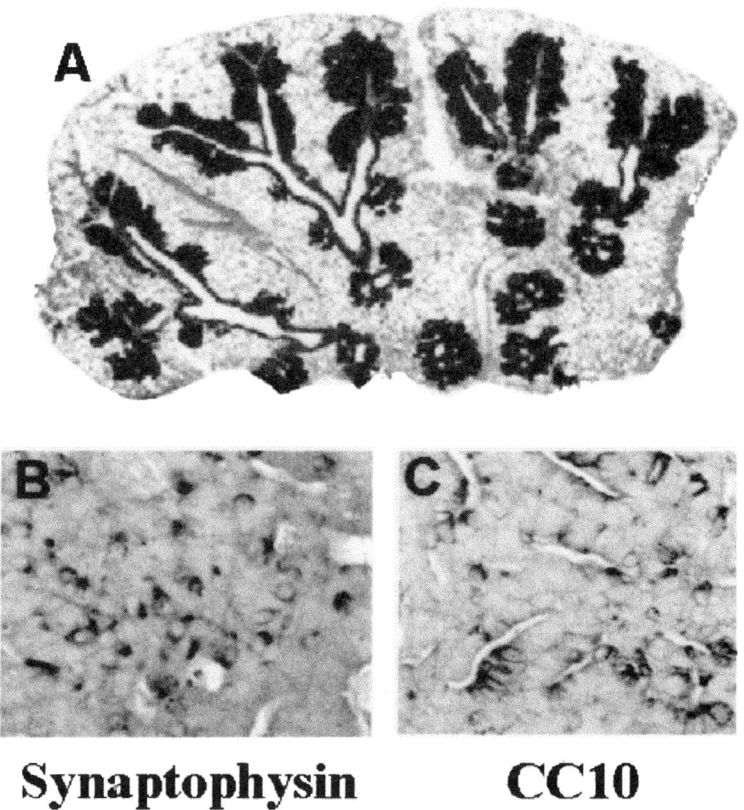

FIGURE 10. Enhanced tumorigenesis with multidirectional differentiation in CC10-Tag-hASH1 bitransgenic mice. (**A**) Microscopic cross section of a lung reveals large tumor masses in the parenchyma along peripheral airways of a 6-week-old bitransgenic mouse (hematoxylin-eosin stain with autoenhancement). (**B–C**) Photomicrographs of a representative tumor at 12 weeks displaying both neuroendocrine differentiation with clusters of synaptophysin-containing cells and peripheral airway cell differentiation with CC10-containing cells (immunoperoxidase staining).

TAg, when we crossed CC10-hASH1 mice with an existing strain of CC10-SV40 TAg.[59] By 2–4 months of life, doubly transgenic mice had extensive lung replacement by solid adenocarcinomas (FIG. 10A). These tumors were far more extensive than those seen in the CC10-SV40 TAg mice. Moreover, the resulting adenocarcinomas exhibited frequent neuroendocrine differentiation and CC10 immunoreactivity as well (FIGS. 10B and 10C). In the absence of TAg, hASH1 promoted airway epithelial cell hyperplasia and metaplasia, with bronchiolization of proximal alveolar spaces (BOA). We suspect that developmental plasticity of airway epithelial precursors utilizing the CC10 promoter may contribute to the proliferation of hASH1-

expressing cells in this model system. In summary, tumors emerging from the doubly transgenic mouse background bear striking resemblance to the NSCLCs with neuroendocrine features, which account for approximately 8% of human lung cancers.[2,60]

SUMMARY AND CONCLUSIONS

Our basic hypothesis is that genes such as CC10, which are involved in differentiation, are relevant to pulmonary carcinogenesis. We have shown that CC10 is infrequently expressed in human non–small cell lung cancers (NSCLCs) or tumor cell lines, despite being abundantly produced by progenitor cells for normal and neoplastic epithelium. Many abnormalities in the surrounding lung associated with field carcinogenesis, which reflect prolonged exposure to such carcinogens as tobacco smoke, also revealed altered expression of CC10. To determine whether the loss of this protein contributes to pulmonary carcinogenesis, we overexpressed CC10 cDNA in cancer cells and immortalized normal human bronchial epithelial cells. Enhanced expression in an NSCLC cell line resulted in decreased clonogenic survival, decreased invasiveness, and decreased metalloproteinase expression. Overexpression of CC10 in immortalized bronchial epithelial cells delayed the induction of anchorage-independent growth in response to NNK. In addition, exposure of hamsters and mice to the tobacco-specific carcinogen NNK led to reduced CC10 expression, which was partially reversible. An intriguing bitransgenic model for human lung neuroendocrine carcinomas was established via constitutive expression of a neural transcription factor under the CC10 promoter. Taken together, these data suggest that downregulation of CC10 contributes to carcinogenesis because CC10 antagonizes the neoplastic phenotype. The findings also confirm the pivotal role that CC10-containing cells play in pulmonary carcinogenesis.

REFERENCES

1. AMERICAN CANCER SOCIETY. 2000. Cancer Facts and Figures 2000. Am. Cancer Soc. Atlanta.
2. LINNOILA, R.I. & S.C. AISNER. 1994. Pathology of lung cancer: exercise in classification. *In* Lung Cancer: Current Clinical Oncology, pp. 73–95. Wiley. New York.
3. ALBERG, A.J. & J.M. SAMET 1998. Epidemiology of lung cancer. *In* Biology of Lung Cancer, pp. 11–51. Dekker. New York.
4. LINNOILA, R.I. 1990. Pathology of non–small cell lung cancer: new diagnostic approaches. Hematol. Oncol. Clin. North Am. **4:** 1027–1051.
5. BARSKY, S., R. CAMERON, K. OSANN *et al.* 1994. Rising incidence of bronchioloalveolar lung carcinoma and its unique clinicopathological features. Cancer **73:** 1163–1170.
6. GREENBERG, S.D. 1987. Carcinomas of the peripheral airways. *In* Lung Carcinomas, pp. 287–309. Churchill Livingstone. New York.
7. BERNARD, A., H. ROELS, R. LAUWERYS *et al.* 1992. Human urinary protein 1: evidence for identity with Clara cell protein and occurrence in respiratory tract and urogenital tract secretions. Clin. Chim. Acta **207:** 239–249.
8. MUKHERJEE, A.B., G.C. KUNDU, G. MANTILE-SELVAGGI *et al.* 1999. Uteroglobin: a novel cytokine? Cell. Mol. Life Sci. **55:** 771–787.
9. PLOPPER, C.G., L.H. HILL & A.T. MARIASSY. 1980. Ultrastructure of the nonciliated bronchiolar epithelial (Clara) cell of mammalian lung. III. A study of man with comparison of 15 mammalian species. Exp. Lung Res. **1:** 171–180.

10. BROERS, J.L.V., S.M. JENSEN, W.D. TRAVIS et al. 1992. Expression of surfactant associated protein-A and Clara cell 10 kilodalton mRNA in neoplastic and non-neoplastic human lung tissue as detected by in situ hybridization. Lab. Invest. **66:** 337–346.
11. EVANS, M.J., L.J. CABRAL-ANDERSON & G. FREEMAN. 1978. Role of the Clara cell in renewal of the bronchiolar epithelium. Lab. Invest. **38:** 648–655.
12. BRODY, A.R., G.E. HOOK, G.S. CAMERON et al. 1987. The differentiation capacity of Clara cells isolated from the lungs of rabbits. Lab. Invest. **57:** 219–229.
13. OGATA, T. & K. ENDO. 1984. Clara cell granules of peripheral lung cancers. Cancer **54:** 1635–1644.
14. CLAYTON, F. 1988. The spectrum and significance of bronchioloalveolar carcinomas. Pathol. Annu. **23**(part 2): 361–394.
15. SHIMOSATO, Y. 1989. Pulmonary neoplasms. In Diagnostic Surgical Pathology, pp. 785–800. Raven Press. New York.
16. LINNOILA, R.I., S.M. JENSEN, S.M. STEINBERG et al. 1992. Peripheral airway cell marker expression in non–small cell lung carcinoma. Am. J. Clin. Pathol. **97:** 233–243.
17. KURITA, Y., J.A. WHITSETT, G. SINGH et al. 1989. Frequent expression of bronchioloalveolar cell markers in pulmonary large cell and adenocarcinomas by immunohistochemistry [abstract]. Lab. Invest. **60:** 50A.
18. SZABO, E., A. GOHEER, H. WITSCHI et al. 1998. Overexpression of CC10 modifies neoplastic potential in lung cancer cells. Cell Growth Differ. **9:** 475–485.
19. GAZDAR, A.F., R.I. LINNOILA, Y. KURITA et al. 1990. Peripheral airway cell differentiation in human lung cancer cell lines. Cancer Res. **50:** 5481–5487.
20. JENSEN, S.M., Y. KURITA, H. OIE et al. 1991. Detection of peripheral airway cell differentiation in human lung cancer lines by RNA-RNA in situ hybridization with ultrastructural correlation [abstract]. Proc. Am. Assoc. Cancer Res. **32:** 33.
21. SLAUGHTER, D.P., H.W. SOUTHWICK & W. SMEJKAL. 1954. "Field cancerization" in oral stratified squamous epithelium: clinical implications of multicentric origin. Cancer **6:** 963–968.
22. SOZZI, G., M. MIOZZO, U. PASTORINO et al. 1995. Genetic evidence for independent origin of multiple preneoplastic and neoplastic lung lesions. Cancer Res. **55:** 135–140.
23. MARCHANDISE, B.A., F.X. DEPELCHIN, R. LAUWERYS et al. 1992. Clara cell protein in serum and bronchoalveolar lavage. Eur. Respir. J. **5:** 1231–1238.
24. JENSEN, S.M., H. PASS, J.E. JONES et al. 1994. Clara cell 10 kD protein mRNA in normal and atypical regions of human respiratory epithelium. Int. J. Cancer **58:** 629–637.
25. LUMSDEN, A.B., A. MCLEAN & D. LAMB. 1984. Goblet and Clara cells of human distal airways: evidence for smoking induced changes in their numbers. Thorax **39:** 844–849.
26. EBERT, R.V. & M.J. TERRACIO. 1975. The brochiolar epithelium in cigarette smokers. Am. Rev. Respir. Dis. **111:** 4–11.
27. SHIJUBO, N., Y. ITOH & T. YAMAGUCHI. 1997. Serum and BAL Clara cell 10 kDa protein (CC10) levels and CC10 positive bronchiolar cells are decreased in smokers. Eur. Respir. J. **10:** 1108–1114.
28. NETTESHEIM, P. & M.S. SZAKAL. 1972. Morphogenesis of alveolar bronchiolization. Lab. Invest. **26:** 210–219.
29. JENSEN, S.M. & R.I. LINNOILA. 1996. The pathology of bronchiolization of alveoli: progressive changes during human lung carcinogenesis. Lab. Invest. **74:** 157A.
30. LINNOILA, R.I., S.M. JENSEN, W. LASKIN et al. 1995. Frequent alterations of chromosome 3p in peripheral lung during pulmonary carcinogenesis in man [abstract]. Proc. Am. Assoc. Cancer Res. **36:** 130.
31. PIERCE, G.B. & W.C. SPEERS. 1988. Tumors as caricatures of the process of tissue renewal: prospects for therapy by directing differentiation. Cancer Res. **48:** 1996–2004.
32. LOTAN, R., X-C. XU, S.M. LIPPMAN et al. 1995. Suppression of retinoic acid receptor-β in premalignant oral lesions and its up-regulation by isotretinoin. N. Engl. J. Med. **332:** 1405–1410.
33. MAO, L., A.K. EL-NAGGAR, V. PAPADIMITRAKOPOULOU et al. 1998. Phenotype and genotype of advanced premalignant head and neck lesions after chemopreventative therapy. J. Natl. Cancer Inst. **90:** 1545–1551.
34. ZHANG, Z., G.C. KUNDU, D. PANDA et al. 1999. Loss of transformed phenotype in cancer cells by overexpression of the uteroglobin gene. Proc. Natl. Acad. Sci. U.S.A. **96:** 3963–3968.

35. LEYTON, J., L. MANYAK, A.B. MUKHERJEE et al. 1994. Recombinant human uteroglobin inhibits the in vitro invasiveness of human metastatic prostate tumor cells and the release of arachidonic acid stimulated by fibroblast-conditioned medium. Cancer Res. **54:** 3696–3699.
36. MORTARINI, R., A. GISMONDI, A. SANTONI et al. 1992. Role of the $\alpha_5\beta_1$ integrin receptor in the proliferative response of quiescent human melanoma cells to fibronectin. Cancer Res. **52:** 4499–4506.
37. SEFTOR, R.E.B., E.A. SEFTOR, W.G. STETLER-STEVENSON et al. 1993. The 72-kDa type IV gelatinase is modulated via differential expression of $\alpha_v\beta_3$ and $\alpha_5\beta_1$ integrins during human melanoma cell invasion. Cancer Res. **53:** 3411–3415.
38. LANGENFELD, J., F. LONARDO, H. KIYOKAWA et al. 1996. Inhibited transformation of immortalized human bronchial epithelial cells by retinoic acid is linked to cyclin E down-regulation. Oncogene **13:** 1983–1990.
39. SCHAUER, I.E., S. SIRIWARDANA, T.A. LANGAN et al. 1994. Cyclin D1 overexpression vs. retinoblastoma inactivation: implications for growth control evasion in non–small cell and small cell lung cancer. Proc. Natl. Acad. Sci. U.S.A. **91:** 7827–7831.
40. KRATZKE, R.A., T.M. GREATENS, J.B. RUBINS et al. 2000. Rb and p16^{INK4a} expression in resected non–small cell lung tumors. Cancer Res. **56:** 3415–3420.
41. MOHR, U., M. EMURA, D.L. DUNGWORTH et al. 1996. Tumors of the lower respiratory tract. In Pathology of Tumors in Laboratory Animals. Volume III: Tumors of the Hamster, pp. 189–222. IARC. Lyon.
42. SCHÜLLER, H.M. 1987. Experimental carcinogenesis in the peripheral lung. In Lung Carcinomas, pp. 243–254. Churchill Livingstone. New York.
43. REHM, S., M. TAKAHASHI, J.M. WARD et al. 1989. Immunohistochemical demonstration of Clara cell antigen in lung tumors of bronchiolar origin induced by N-nitrosodiethylamine in Syrian golden hamsters. Am. J. Pathol. **134:** 79–87.
44. SUNDAY, M.E., C.G. WILLET, S.A. GRAHAM et al. 1995. Histochemical characterization of non-neuroendocrine tumors and neuroendocrine cell hyperplasia induced in hamster lung by 4-(methylnitrosamino)-1-(3-pyridyl)-1-butanone with or without hyperoxia. Am. J. Pathol. **147:** 740–752.
45. REHM, S., W. LIJINSKY, B.J. THOMAS et al. 1993. Clara cell antigen in normal and migratory dysplastic Clara cells, and bronchioalveolar carcinoma of Syrian hamsters induced by N-nitrosomethyl-n-heptylamine. Virchows Arch. B Cell Pathol. **64:** 181–190.
46. BUNNAG, T., S. JENSEN, H. WITSCHI et al. 1996. Reduced expression of Clara cell secretory protein (CC10) during early lung carcinogenesis in a hamster model [abstract]. Proc. Am. Assoc. Cancer Res. **37:** 112
47. JENSEN, S.M., T. BUNNAG, C. CASTRO et al. 1998. Early changes during experimental lung carcinogenesis detected by laser capture microdissection and immunohistochemistry [abstract]. Proc. Am. Assoc. Cancer Res. **39:** 22
48. OREFFO, V.I.C., H.W. LIN, R. PADMANABHAN et al. 1993. K-ras and p-53 point mutations in 4-(methylnitrosamino)-1-(3-pyridyl)-1-butanone-induced hamster lung tumors. Carcinogenesis **14:** 451–455.
49. RITTINGHAUSEN, S., D.L. DUNGWORTH, H. ERNST et al. 1996. Naturally occurring pulmonary tumors in rodents. In Respiratory System, pp. 183–206. Springer-Verlag. Berlin/New York.
50. LINNOILA, R.I., A. GOHEER, C. CASTRO et al. 2000. Premalignant changes involving neuroendocrine (NE) cells and Clara cell secretory protein (CC10) during tobacco-specific nitrosamine (NNK) induced lung carcinogenesis in mice [abstract]. Proc. Am. Assoc. Cancer Res. **41:** 839–840.
51. BERNARD, A., C. HERMANS & G. VAN HOUTE. 1997. Transient increase of serum Clara cell protein (CC16) after exposure to smoke. Occup. Environ. Med. **54:** 63–65.
52. ELIA, J., A. AOKI & C.A. MALDONADO. 2000. Response of bronchiolar Clara cells induced by a domestic insecticide: analysis of CC10 kDa protein content. Histochem. Cell Biol. **113:** 125–133.
53. REHM, S., W. LIJINSKY, G. SINGH et al. 1991. Mouse bronchiolar cell carcinogenesis: histologic characterization and expression of Clara cell antigen in lesions induced by N-nitrosobis-(2-chloroethyl) ureas. Am. J. Pathol. **139:** 413–422.

54. SANDMOLLER, A., R. HALTER, G. SUSKE et al. 1995. A transgenic mouse model for lung adenocarcinoma. Cell Growth Differ. **6:** 97–103.
55. DEMAYO, F.J., M.J. FINEGOLD, T.N. HANSEN et al. 1991. Expression of SV40 T antigen under control of rabbit uteroglobin promoter in transgenic mice. Am. J. Physiol. **261**(2, part 1): L70–L76.
56. MAGDALENO, S.M., G. WANG, V. MIRELES et al. 1997. Cyclin-dependent kinase inhibitor expression in pulmonary Clara cells transformed with SV40 large T antigen in transgenic mice. Cell Growth Differ. **8:** 145–155.
57. KAO, C., J. HUANG, S.Q. WU et al. 1993. Role of SV40 T antigen binding to pRb and p53 in multistep transformation *in vitro* of human uroepithelial cells. Carcinogenesis **14:** 2297–2302.
58. BORGES, M., R.I. LINNOILA, H.J.K. VAN DE VELDE et al. 1997. An achaete-scute homologue essential for neuroendocrine differentiation in the lung. Nature **386:** 852–855.
59. LINNOILA, R.I., B. ZHAO, J.L. DEMAYO et al. 2000. Constitutive achaete-scute homolog-1 (hASH1) promotes airway dysplasia and lung neuroendocrine tumors in transgenic mice. Cancer Res. **60:** 4005–4009.
60. LINNOILA, R.I., S. PIANTADOSI & J.C. RUCKDESCHEL. 1994. Impact of neuroendocrine differentiation in non–small cell lung cancer: the LCSG experience. Chest **106:** 367S–371S.

Development of an Enzyme-Linked Immunosorbent Assay for Clara Cell 10-kDa Protein: In Pursuit of Clinical Significance of Sera in Patients with Asthma and Sarcoidosis

NORIHARU SHIJUBO,[a,b] YOSHIHISA ITOH,[c] TETSUJI YAMAGUCHI,[c] AND SHOSAKU ABE[b]

[b]*Third Department of Internal Medicine, Sapporo Medical University School of Medicine, Sapporo, Japan*

[c]*Department of Clinical Pathology, Jichi Medical School, Tochigi-ken, Japan*

ABSTRACT: We have produced nine monoclonal antibodies to human CC10/protein-1 and analyzed their characterization. TY-5, TY-7, and TY-8 recognized restricted possible hydrophobic epitopes and their binding to CC10 prevented the other clones from CC10 binding, suggesting that these antibodies induce strong conformational change. TY-1, TY-2, TY-3, TY-6, and 6D4 recognize amino acid residues 61–68 and the presence of disulfide bonds might be essential for epitope expression of these five clones. The best combination was TY-1 and TY-2 in developing an enzyme-linked immunosorbent assay (ELISA), whereas TY-5 was most suitable for immunohistochemistry and immunoblotting. We found significantly lower serum CC10 levels in asthmatic subjects and higher serum CC10 levels in sarcoidosis subjects than in controls. Data of CC10 levels in BAL fluids of sarcoidosis subjects were similar to those in the circulation. CC10-positive epithelial cells were significantly lower in small airways of asthmatic subjects than in controls, and CC10-positive epithelial cells were inversely correlated with T cell and mast cell accumulation in the airways of asthmatic subjects. CC10 may be a downregulator in both Th1- and Th2-mediated chronic inflammatory diseases. The use of these MoAbs and recombinant CC10 is a powerful tool to investigate the clinical roles of CC10/P1 and the structure and function of CC10/P1.

Clara cell 10-kilodalton protein (CC10) is the predominant product from nonciliated bronchiolar epithelial cells (Clara cells) in the lung.[1] Human CC10 has been proven identical with human protein-1 (P1)[2] and uteroglobin (UG).[3] CC10/UG possesses varied biochemical and biological properties, including phospholipase A_2 (PLA_2)[4] and phospholipase C (PLC) inhibitory activity.[5] CC10 is also a potent inhibitor of IFN-γ, tumor necrosis factor-α (TNF-α), and IL-1β.[6,7] CC10 deficiency results in

---

[a]Address for correspondence: N. Shijubo, Third Department of Internal Medicine, Sapporo Medical University School of Medicine, South-1, West-16, Chuo-ku, Sapporo, 060-8543 Japan. Voice: 81-11-611-2111, ext. 3239; fax: 81-11-613-1543.
shijubo@sapmed.ac.jp

increased sensitivity to hyperoxia-induced lung injury by increasing proinflammatory cytokine expression.[8] Evidence has led to speculation that CC10 may function as a downregulator of inflammation. In addition, CC10 mRNA was upregulated by IFN-γ in an animal model.[9] We have produced nine monoclonal antibodies (MoAb) against CC10/P1 and analyzed their characteristics.[10] We measured CC10 levels in sera of patients with sarcoidosis (Th1-dominant disease) and with asthma (Th2-dominant disease). Local expression of CC10 was also analyzed in these diseases. The CC10 expression was compared with clinical parameters.

CHARACTERIZATION OF MONOCLONAL ANTIBODIES AGAINST CC10/P1

CC10 is structurally a homodimer of a single polypeptide of 70 amino acid residues, in which two subunits are covalently bound in an antiparallel manner by two disulfide bonds between cysteine (Cys) 3 and Cys 69.[11] Nine MoAbs against CC10/P1 have been produced (clone names: 6D4, TY-1, TY-2, TY-3, TY-4, TY-5, TY-6, TY-7, and TY-8).[10]

We analyzed epitopes of the nine clones using synthetic peptides on pins examined by an ELISA as described previously.[10] Clone TY-5 recognized residues 7–16 of CC10, clone TY-7 recognized residues 19–28, and clone TY-8 recognized residues 39–46 (FIG. 1), all of which comprise the hydrophobic cavity of CC10, possibly associated with chemical binding function. There were no reactivities of TY-1, TY-2, TY-3, TY-4, TY-6, or 6D4 to synthetic peptides.

To evaluate reactivities of the clones, we made truncated forms of human CC10, such that $CC10_{1-60}$, $CC10_{1-68}$, $CC10_{4-60}$, $CC10_{4-68}$, and $CC10_{4-70}$.[10] The clones 6D4,

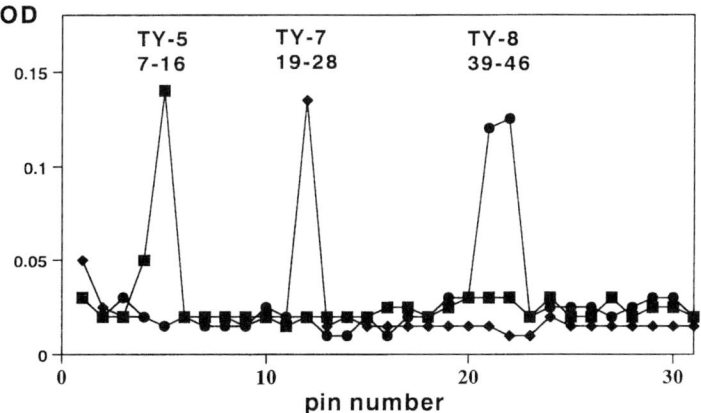

FIGURE 1. Epitope analysis of monoclonal antibodies (MoAbs) by a multipin ELISA. The reactivities of MoAbs against synthetic peptides on pins were examined by the ELISA. Each pin bore 10 residues of CC10 overlapping with 8 residues in the next pin. Three MoAbs (TY-5, TY-7, and TY-8) reacted with distinct synthetic CC10 peptides. The other six MoAbs yielded no reaction. Terms: 7–16, amino acid (aa) residues 7–16; 19–28, aa residues 19–28; 39–46, aa residues 39–46.

TABLE 1. Reactivities of MoAbs against truncated recombinant proteins

MoAb	$CC10_{1-68}$	$CC10_{4-68}$	$CC10_{4-70}$	$CC10_{1-60}$	$CC10_{4-60}$
TY-1	0.20	0.06	0.29	0.05	0.06
TY-2	0.21	0.06	0.20	0.05	0.06
TY-3	0.24	0.06	0.29	0.06	0.06
TY-4	0.24	0.34	0.40	0.22	0.19
TY-5	0.27	0.31	0.34	0.31	0.32
TY-6	0.22	0.05	0.24	0.05	0.05
TY-7	0.22	0.29	0.33	0.16	0.23
TY-8	0.23	0.24	0.33	0.25	0.29
6D4	0.19	0.06	0.18	0.05	0.06

NOTE: Each truncated CC10 was detected by MoAbs. Absorbance at 450 nm is shown.

TABLE 2. ELISA for CC10/P1 using several combinations of MoAbs

Capture MoAb	Biotinylated MoAb								
	TY-1	TY-2	TY-3	TY-4	TY-5	TY-6	TY-7	TY-8	6D4
TY-1	0.12	0.49	0.46	0.36	0.05	0.38	0.09	0.05	0.12
TY-2	0.77	0.05	0.05	0.05	0.05	0.05	0.05	0.05	0.63
TY-3	0.77	0.06	0.06	0.06	0.05	0.06	0.07	0.06	0.68
TY-4	0.53	0.06	0.05	0.05	0.06	0.05	0.05	0.05	0.51
TY-5	0.06	0.05	0.05	0.05	0.05	0.05	0.05	0.05	0.05
TY-6	0.86	0.06	0.06	0.06	0.05	0.06	0.06	0.06	0.76
TY-7	0.06	0.05	0.05	0.05	0.05	0.05	0.05	0.05	0.06
TY-8	0.05	0.06	0.05	0.05	0.05	0.06	0.05	0.05	0.05
6D4	0.12	0.63	0.68	0.51	0.06	0.76	0.06	0.05	0.06

NOTE: Native CC10/P1 (100 ng/mL) was detected by ELISA using several combinations of MoAbs. Absorbance at 450 nm is shown.

TY-1, TY-2, TY-3, and TY-6 recognized $CC10_{1-68}$ and $CC10_{4-70}$, but not $CC10_{1-60}$, $CC10_{4-60}$, or $CC10_{4-68}$ (TABLE 1). These clones seemed to recognize the epitope containing at least a part of residues 61–68, but the presence of a disulfide bond(s) might be essential for the epitope expression of the five clones. The disulfide bonds (Cys 3–Cys 69′, Cys 3–Cys 3′, and Cys 69–Cys 69′) might induce a conformational change from a single peptide of these molecules. The epitope of clone TY-4 remained undetermined, although the epitope recognized by TY-4 included amino acid residues 4–60.

To assess the best combination of MoAbs from these clones, we examined reactivities in developing ELISAs (TABLE 2). Native CC10/P1 is blocked in a sandwich ELISA between TY-1 and 6D4 (TY-1 pattern) and among TY-2, TY-3, TY-4, and TY-6 (TY-2 pattern). We were able to measure native CC10/P1 in sandwich ELISAs

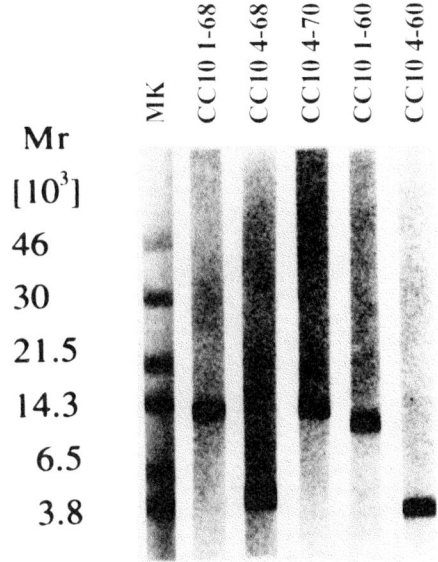

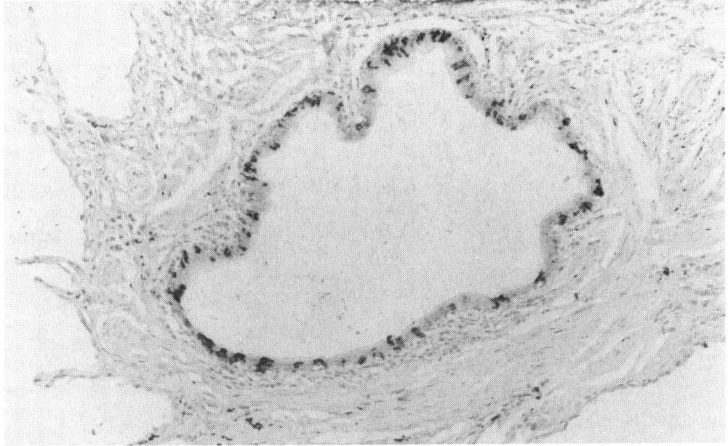

FIGURE 2. Immunoblotting with MoAb TY-5 of truncated CC10 (top) and immunohistochemistry of CC10 with MoAb TY-5 in a lung section (bottom). Even for recombinant CC10 deficient in one of the Cys residues ($CC10_{1-60}$, $CC10_{1-68}$, and $CC10_{4-70}$), positive bands were observed at a molecular weight of approximately 14 kDa. TY-5 reacted with nonciliated epithelial cells in a bronchiole.

employing TY-1 and TY-2 pattern clones. The epitopes recognized by TY-1 pattern clones or TY-2 pattern clones seemed to be very similar. The binding of each TY-5, TY-7, and TY-8 to CC10 prevents the other clones from binding to CC10, suggesting that the binding of these three clones to CC10 induces a strong conformational change (TABLE 2). The sensitivity level (reproducibly distinguishable from 0 level)

TABLE 3. CC10/P1 values in various body fluids

Sample	n	Mean (range, ng/mL)
Serum (20–50 years old)[a]		
Men	38	13 (6–25)
Women	40	11 (5–21)
Urine (20–50 years old)[a]		
Men	24	35 (4–89)
Women	25	3 (0.2–11)
BAL fluid[a]	15	760 (280–1900)
Seminal fluid[a]	46	1030 (150–8600)
Synovial fluid[b]	10	9 (3–16)
Pleural fluid[b]	20	11 (0.7–33)
CSF[b]	25	0.5 (0.02–6)

[a]Samples were obtained from healthy subjects. Serum and BAL fluid samples were obtained from lifelong nonsmokers.
[b]Samples were obtained under disease condition. CSF: cerebrospinal fluid.

of a combination of TY-1 and TY-2 (5 ng/mL) was lower than that of other combinations (TY-1 and TY-3, 10 ng/mL; TY-1 and TY-4, 40 ng/mL; TY-1 and polyclonal antibody, 100 ng/mL; and 6D4 and polyclonal antibody, 50 ng/mL).[12] We concluded that capture TY-1 and biotinylated TY-2 were the best combination in developing ELISAs. TABLE 3 shows CC10/P1 values in various body fluids when employing capture TY-1 and biotinylated TY-2.

In immunohistochemical analysis, all clones were able to be applicable in formalin-fixed and paraffin-embedded sections. However, TY-1, TY-2, TY-3, TY-6, and 6D4 were unstable as compared with TY-4, TY-5, TY-7, and TY-8.[10] TY-5 was most suitable for immunohistochemistry. In immunoblotting analysis, TY-5 showed high affinity of $CC10_{1-60}$, $CC10_{1-68}$, $CC10_{4-60}$, $CC10_{4-68}$, and $CC10_{4-70}$ (FIG. 2) and the reactivity of TY-5 by immunoblotting was good in reducing conditions as well as nonreducing conditions.[10] The characterization of MoAbs against CC10/P1 is summarized in TABLE 4.

CIRCULATING CC10 IN ASTHMA AND SARCOIDOSIS

Tobacco smoking has been associated with an approximately 30% decrease in serum CC10.[13,14] Therefore, lifelong nonsmokers (63 asthmatic and 31 sarcoidosis patients and 64 healthy subjects) were selected to compare serum CC10 values. As shown in FIGURE 3, serum CC10 levels were significantly lower in asthmatic patients than in healthy controls ($p < 0.0001$),[15] whereas they were significantly higher in sarcoidosis patients than in healthy controls ($p < 0.0001$).[16]

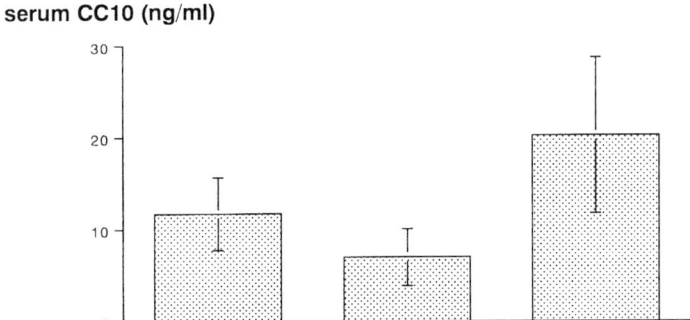
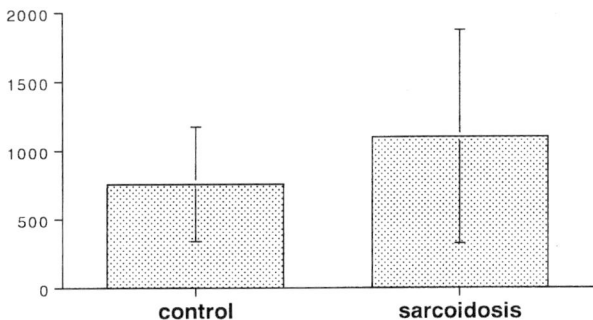

FIGURE 3. CC10 levels in sera (top) and bronchoalveolar lavage fluids (bottom) of patients with asthma (top) and sarcoidosis (top/bottom) and healthy controls (top/bottom). Serum CC10 levels were significantly decreased in patients with asthma ($p < 0.0001$) and increased in patients with sarcoidosis ($p < 0.0001$) as compared with healthy controls. There was no significant difference of BAL fluid CC10 levels between sarcoidosis patients and healthy controls. The Mann-Whitney U test was used to compare paired sets of data.

LOCAL CC10 IN ASTHMA AND SARCOIDOSIS

In bronchoalveolar lavage (BAL) examinations using three 50-mL aliquots, the mean CC10 in the first aliquots was 5.0 and 5.6 times, respectively, of that in the second and third lavages (FIG. 4), indicating that the first lavage includes predominantly bronchiolar lining fluids. The BAL examinations were also influenced by the BAL methods.[17]

Tobacco smoking has been associated with an approximately 50% decrease in BAL fluid CC10.[13,14] To compare CC10 levels in BAL fluids, we selected lifelong nonsmokers (31 sarcoidosis patients and 15 healthy controls). There was no significant difference of BAL fluid CC10 levels between sarcoidosis patients and healthy controls (FIG. 3).[16]

TABLE 4. Characterization of MoAbs against CC10/P1 including epitope analysis

Clone	Synthesized peptide	Reactivity with recombinant proteins				Supposed epitope	Sandwich ELISA	IHC
		$CC10_{1-68}$	$CC10_{4-68}$	$CC10_{4-70}$	$CC10_{1-60}$	$CC10_{4-60}$		
TY-1	NR	+	−	+	−	−	61–68	TY-1 pattern	±
TY-2	NR	+	−	+	−	−	61–68	TY-2 pattern	±
TY-3	NR	+	−	+	−	−	61–68	TY-2 pattern	±
TY-4	NR	+	+	+	+	+	ND	TY-2 pattern	+
TY-5	7–16	+	+	+	+	+	7–16	NA	++
TY-6	NR	+	−	+	−	−	61–68	TY-2 pattern	±
TY-7	19–28	+	+	+	+	+	19–28	NA	+
TY-8	39–46	+	+	+	+	+	39–46	NA	+
6D4	NR	+	−	+	−	−	61–68	TY-1 pattern	±

NOTE: IHC (immunohistochemistry) was assessed by staining for Clara cells; ± means unstable staining. NR: not reacted; ND: not determined; NA: not applicable.

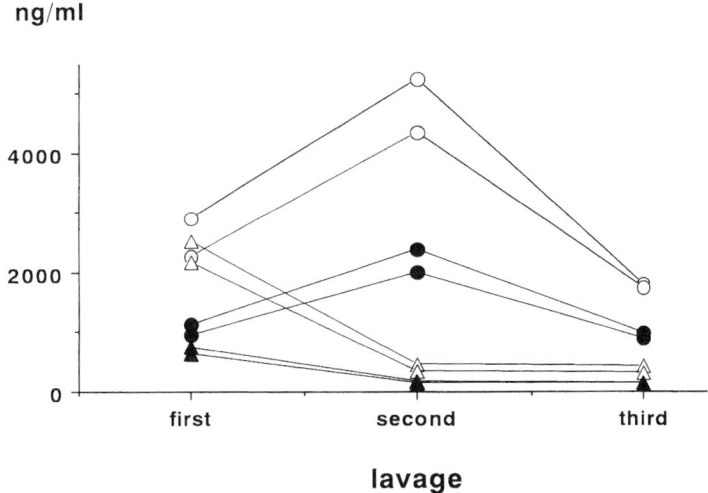

FIGURE 4. Surfactant protein A (SP-A) and CC10 levels in each lavage on BAL examinations using three 50-mL aliquots. Unlike SP-A, the mean CC10 in the first aliquots was 5.0 and 5.6 times, respectively, of that in the second and third lavages, indicating that the first lavage includes predominantly bronchiolar lining fluids. Circles, SP-A; triangles, CC10; open symbols, healthy nonsmokers; closed symbols, healthy smokers.

We have reported that CC10-positive bronchiolar epithelial cell proportions were significantly lower in smokers (16 ± 5%) than in all lifelong nonsmokers (27 ± 7%). To compare immunohistochemical CC10 expression in small airways, we selected lifelong nonsmokers (7 asthmatic patients and 8 controls) from whom lung specimens were obtained undergoing resection for a small lung carcinoma.

Airways with perimeter less than 6 mm were assessed as small airways. Cases where goblet cell metaplasia was found in more than 5% of the total epithelial cells were excluded in the present study. The CC10-positive epithelial cell proportions were significantly decreased in the small airways of asthmatic patients compared with controls (FIG. 5).[18]

RELATIONSHIP OF CIRCULATING CC10 LEVELS TO CLINICAL DATA

There was no significant difference of serum CC10 levels between atopic and nonatopic asthmatic patients. Asthmatic patients with a long duration of the disease (≥10 years) had significantly lower CC10 levels than those with a short duration of the disease (<10 years).[15]

There were no significant differences of serum CC10 levels according to radiologic stages. In sarcoidosis patients, serum CC10 levels were significantly higher in the regressive disease group (regressed within 3 years) than in the progressive disease group (progressed or did not change for 3 years or more).[16]

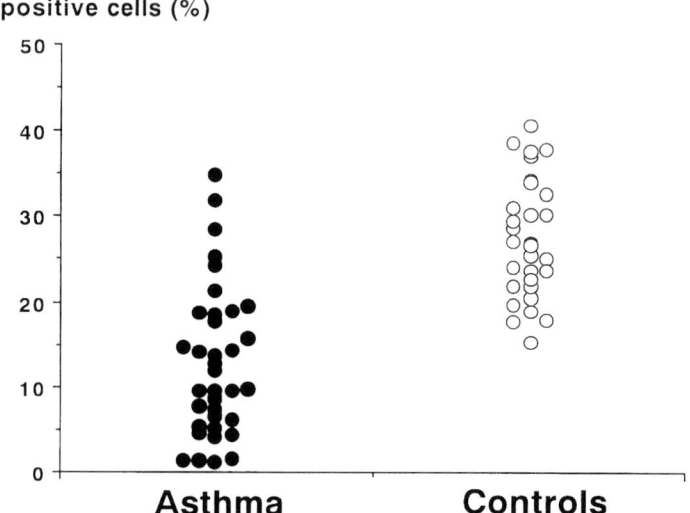

FIGURE 5. CC10-positive epithelial cells in small airways of patients with asthma and controls. CC10-positive epithelial cells in the small airways of patients with asthma were markedly decreased as compared with controls. See reference 18.

RELATIONSHIP OF LOCAL CC10 TO CLINICAL DATA

There were no significant differences of BAL fluid CC10 levels according to radiologic stages. In sarcoidosis patients, BAL fluid CC10 levels were significantly higher in the regressive disease group than in the progressive disease group.[15]

We also evaluated immunoreactive cells (T cells, activated eosinophils, and mast cells) and CC10-positive epithelial cells in the small airways using serous sections. The relationship was analyzed between CC10-positive epithelial cell proportions and inflammatory cells in the small airways of asthmatic patients.[18] There were significant negative correlations of CC10-positive epithelial cells with numbers of T cells (CD3; $r = -0.60, p < 0.0001$) and mast cells (tryptase; $r = -0.61, p < 0.0001$) in the small airways of asthmatic patients, whereas there was no significant correlation between CC10-positive epithelial proportions and activated eosinophils (EG2; $r = 0.18$). CC10 may thus play a role in controlling allergic inflammation of the lung.

DISCUSSION

In native CC10, two subunits are covalently bound in an antiparallel fashion by two disulfide bonds between Cys 3 and Cys 69. The monomer is composed of four α-helices: α1-helices from proline (Pro) 4 to leucine (Leu) 14, α2-helices from Pro 18 to Leu 27, α3-helices from glutamine (Gln) 32 to aspartic acid (Asp) 46, and α4-helices from Gln 50 to serine (Ser) 66. The first two are joined by short loops containing β-turns.[11] Of the clones examined, only TY-5, TY-7, and TY-8 reacted with distinct synthetic peptides on a multipin ELISA. These MoAbs were immunohistochemically demonstrated by the ELISA to react with CC10 in nonciliated bronchiolar epithelial cells, but not with CC10/P1 in body fluids, in which the MoAbs were used either as a capture antibody or a biotinylated antibody. TY-5 reacted with the monomeric form most strongly, as shown by immunoblotting. Epitopes of TY-5, TY-7, and TY-8 were shown to be located in the α1-, α2-, and α3-helices, respectively. These epitopes could be exposed by conformational changes in CC10/P1 induced by mechanical and chemical modification on immunoblotting, solid-phase ELISA,[12] or immunohistochemical study on formalin-fixed, paraffin-embedded sections, resulting in detection by these clones.

No reaction was obtained with any of the synthetic peptides by the rest of the clones. Therefore, several truncated CC10 molecules were generated and utilized in epitope analysis. Except for TY-4, the clones were found to recognize the epitope containing at least a part of residues 61–68 in the α4-helices, even if recombinant CC10 was lacking either of its Cys residues. However, the presence of at least one disulfide bond might be essential for the epitope expression of the five clones. The disulfide bonds (Cys 3–Cys 69′, Cys 3–Cys 3′, and Cys 69–Cys 69′) might induce a conformational change from a single peptide of these molecules. Since native CC10/P1 (homodimer) is blocked in sandwich ELISA between TY-1 and 6D4, the epitope recognized seemed to be similar between the two. However, these reactions were not completely inhibited. This may be explained as two epitopes expressed on the homodimer being shared by clones. TY-2, TY-3, and TY-6, which recognized distinct epitopes from the above two clones, were functionally the same since immunochemical results were completely identical among them. Furthermore, the immunoreactivity was lost with a combination of two clones from among them by the ELISA. The epitope of clone TY-4 remained undetermined, although the epitope recognized by TY-4 included amino acid residues 4–60. However, in its immunochemical behavior in ELISA, this clone seemed to be closely related to TY-2, TY-3, and TY-6. We found that capture TY-1 and biotinylated TY-2 were the best combination in developing ELISAs to detect native CC10/P1 (soluble form) in body fluid samples.

Sarcoidosis is a systemic granulomatous disorder of unknown etiology and is reported to be a Th1-mediated chronic inflammatory disease.[19] In active sarcoid lesions, elevated production of Th1 cytokines (IFN-γ and IL-2) and IL-12 has been found. Asthma is a Th2-mediated inflammatory disease. Elevated production of Th2 cytokines (IL-4, IL-5, and IL-13) has been reported in asthma and it is considered that IFN-γ production is markedly suppressed in asthma. We measured serum CC10 levels in patients with asthma and sarcoidosis. Circulating CC10 levels were increased in patients with sarcoidosis and decreased in patients with asthma as compared with healthy controls. Local production of CC10 also showed similar results. CC10 mRNA was upregulated by IFN-γ in an animal model, suggesting that IFN-γ upregulates CC10 production. In addition, CC10/UG possesses varied biochemical and biological properties, including phospholipase A_2 $(PLA_2)^4$ and phospholipase C (PLC) inhibitory activity.[5] CC10 was also shown to be a potent inhibitor of IFN-γ, TNF-α, and IL-1β.[6,7] We showed that CC10 levels in patients with sarcoidosis were significantly higher in the regressive disease group than in the progressive disease group.[15] CC10 partly inhibited IFN-γ production from LPS-stimulated BAL fluid cells obtained from patients with sarcoidosis.[16] TY-5 recovered IFN-γ production by blocking CC10 function.[16] Highly local production of CC10 may downregulate Th1 chronic inflammation. Distinct CC10 production in response to IFN-γ may be a clue to explain the clinical outcome in patients with sarcoidosis. CC10 expression in small airways of patients with asthma was reduced and was inversely correlated with T lymphocyte and mast cell accumulation. Evidence suggests that CC10 may influence Th2 chronic inflammation. CC10 may control both Th1 and Th2 inflammation in the lower respiratory tract.

Use of these MoAbs and recombinant CC10 is a powerful tool to investigate the clinical roles of CC10/P1 and the structure and function of CC10/P1.

ACKNOWLEDGMENTS

This work was supported in part by Grants-in-Aid for Scientific Research from the Ministry of Education, Science, and Culture [No. 10670556 to N. Shijubo and No. 08672648 to Y. Itoh].

REFERENCES

1. SINGH, G., S.L. KATYAL & S.A. GOTTRON. 1985. Antigenic, molecular, and functional heterogeneity of Clara cell secretory proteins in the rat. Biochim. Biophys. Acta **829:** 156–159.
2. OKUTANI, R., Y. ITOH, H. HIRATA et al. 1992. Simple and high-yield purification of urine protein 1 using immunoaffinity chromatography: evidence for the identity of urine protein 1 and human Clara cell 10-kilodalton protein. J. Chromatogr. **577:** 25–35.
3. MANTILE, G., L. MIELE, E. CORDELLA-MIELE et al. 1993. Human Clara cell 10 kDa protein is the counterpart of rabbit uteroglobin. J. Biol. Chem. **267:** 20343–20351.
4. LEVIN, S.W., J.D. BUTLER, U.K. SCHUMACHER et al. 1986. Uteroglobin inhibits phospholipase A2 activity. Life Sci. **38:** 1813–1819.
5. OKUTANI, R., Y. ITOH, T. YAMAGUCHI et al. 1996. Preparation and characterization of human recombinant protein 1/Clara cell 10 kDa protein. Eur. J. Clin. Chem. Clin. Biochem. **34:** 691–696.
6. DIERYNCK, I., A. BERNARD, H. ROELS et al. 1995. Potent inhibition of both human interferon-γ production and biological activity by the Clara cell protein CC16. Am. J. Respir. Cell Mol. Biol. **12:** 205–210.
7. DIERYNCK, I., A. BERNARD, H. ROELS et al. 1996. The human Clara cell protein: biochemical and biological characterization of a natural immunosuppressor. Multiple Sclerosis **1:** 385–387.
8. JOHNSTON, C.J., G.W. MANGO, J.N. FINKELSTEIN et al. 1997. Altered pulmonary response to hyperoxia in Clara cell secretory protein deficient mice. Am. J. Respir. Cell Mol. Biol. **17:** 147–155.
9. MAGDALENO, S.M., G. WANG, K.J. JACKSON et al. 1997. Interferon-γ regulation of Clara cell gene expression: *in vivo* and *in vitro*. Am. J. Physiol. **272:** L1142–L1151.
10. YAMAGUCHI, T., T. YAMADA, R. OKUTANI et al. 1999. Characterization of monoclonal antibodies to human protein 1/Clara cell 10 kilodalton protein. Clin. Chem. Lab. Med. **37:** 631–637.
11. UMLAND, T.C., S. SWAMINATHAN, G. SINGH et al. 1994. Structure of human Clara cell phospholipid-binding protein-ligand complex at 1.9 Å resolution. Nat. Struct. Biol. **1:** 538–545.
12. YAMAGUCHI, T., Y. ITOH, T. YAMADA et al. 1999. Development of a sensitive enzyme immunoassay for protein 1/Clara cell 10 kDa protein using monoclonal antibodies. Rinsho Byori **47:** 467–472 (in Japanese).

13. SHIJUBO, N., Y. ITOH, T. YAMAGUCHI et al. 1997. Serum and BAL Clara cell protein (CC10) levels and CC10-positive bronchiolar cells are decreased in smokers. Eur. Respir. J. **10:** 1108–1114.
14. HERMANS, C. & A. BERNARD. 1999. Lung epithelium-specific proteins: characteristics and potential applications as markers. Am. J. Respir. Crit. Care Med. **159:** 646–678.
15. SHIJUBO, N., Y. ITOH, T. YAMAGUCHI et al. 1999. Serum levels of Clara cell 10-kDa protein are decreased in patients with asthma. Lung **177:** 45–52.
16. SHIJUBO, N., Y. ITOH, K. SHIGEHARA et al. 2000. Association of Clara cell 10-kDa protein, spontaneous regression, and sarcoidosis. Eur. Respir. J. **16:** 414–419.
17. SHIJUBO, N., Y. HONDA, Y. ITOH et al. 1998. BAL surfactant protein A and Clara cell 10-kDa protein levels in healthy subjects. Lung **176:** 257–265.
18. SHIJUBO, N., Y. ITOH, T. YAMAGUCHI et al. 1999. Clara cell protein–positive epithelial cells are reduced in small airways of asthmatics. Am. J. Respir. Crit. Care Med. **160:** 930–933.
19. HUNNINGHAKE, G.W., U. COSTABEL, M. ANDO et al. 1999. ATS/ERS/WASOG statement on sarcoidosis. Sarcoidosis Vasc. Diffuse Lung Dis. **16:** 149–173.

Rationale for the Development of Recombinant Human CC10 as a Therapeutic for Inflammatory and Fibrotic Disease

APRILE L. PILON[a]

Claragen, Incorporated, College Park, Maryland 20742, USA

ABSTRACT: CC10/uteroglobin is a remarkable protein whose physiological roles have only recently been explored *in vivo*. Both transgenic mice that have been rendered deficient and humans that have been characterized as deficient in this protein exhibit tendencies toward inflammatory, fibrotic, and oncologic disease, demonstrating the potential of the protein as a therapeutic agent. The protein itself is an excellent candidate for clinical development because of its inherent physical properties. It is relatively small, resistant to proteases, stable to extremes of heat and pH, and can be produced by recombinant methods. The physiological roles of this multifunctional protein continue to be uncovered as research progresses *in vitro*, in animals, and eventually in humans. The pathways through which CC10 mediates its effects, its receptors, and other family members will be a rich source of exciting research, as well as potential diagnostic and therapeutic agents. This paper is an introductory, noncomprehensive review of some of the scientific and medical rationale in support of CC10-based therapies in selected clinical applications.

INTRODUCTION

Clara cell "10-kDa" protein (CC10) or uteroglobin (UG) is a small, homodimeric secretory protein produced by several mucosal epithelia and other organs of epithelial origin.[1] CC10 consists of two identical subunits of 70 amino acid residues, each with the "four helical bundle" secondary structure motif, joined in antiparallel orientation by two disulfide bonds between Cys 3 and 69′ and Cys 3′ and 69.[2,3] The fully oxidized homodimer appears to be its primary active form. CC10 is very compact and hardy, being resistant to denaturation by heat or acid and to protease digestion.[2] Both the secondary and tertiary structures of CC10, as well as the gene structure, have been highly conserved through evolution.[4,5]

In humans, the lung is the main site of CC10 production, while several other organs synthesize smaller amounts of mRNA encoding this protein.[6] Nonciliated Clara cells, the precursors of mature tracheobronchial epithelial cells, produce CC10 in the lung.[7] In female mammals, the uterine endometrium secretes CC10 under progesterone stimulation via the progesterone receptor,[8] while in males the prostate produces it constitutively.[6] CC10 also circulates in the blood and is excreted in urine.[9,10] Urinary CC10 has also been called "urine protein-1" or "P1".[11]

---

[a]Address for correspondence: Claragen, Incorporated, 387 Technology Drive, College Park, MD 20742. Voice: 301-405-8593; fax: 301-405-4989.
apilon@claragen.com

There have been many observations about the biological properties of CC10 *in vitro*. CC10 inhibits secretory phospholipase A_2 (PLA_2)[12,13] and is a substrate of transglutaminase.[14] CC10 inhibits FMLP-induced neutrophil and monocyte chemotaxis *in vitro*[15] as well as FCM-induced fibroblast chemotaxis *in vitro*.[16] CC10 also suppresses the IL-2-stimulated release of TNF-α, IL-1β, and IFN-γ in human peripheral blood lymphocytes.[17] CC10 inhibits the release of arachidonic acid from cells in culture, which is correlated with decreased chemotactic migration[16] and decreased extracellular matrix invasion.[18] CC10 also inhibits growth of tumor cells that express a CC10 receptor in culture.[19–21]

More recently, two transgenic mouse strains, both deficient in endogenous CC10, provide *in vivo* evidence for the role of CC10 in controlling inflammation and related processes. One strain of CC10/UG knockout mice develops a severe fibronectin-deposit glomerulopathy, a lethal phenotype.[22] In the same strain, cachexia, hypocalcemia, pancreatic inflammation, elevated circulating levels of PLA_2 activity, and elevated levels of lysophosphatidic acid (LPA, a potent proinflammatory lipid signaling molecule) are also observed.[23] The renal phenotype of this animal led to the discovery of a fundamental interaction between CC10 and fibronectin, an extracellular matrix protein that circulates in the blood.[24] Further work led to the discovery that CC10 inhibits the formation of a proinflammatory complex between fibronectin and IgA.[25] This strain also develops solid tumors with 100% incidence as it ages.[21] Unfortunately, the severe systemic inflammation seen in this mouse strain, as well as its limited viability, has curtailed efforts thus far for further experimentation such as adoptive transfer of tumor cells and pulmonary challenge testing.

However, another CC10/CCSP-deficient transgenic mouse strain exhibits limited renal proteinuria,[26] no systemic inflammation, and is sufficiently viable to perform pulmonary challenge experiments.[27] Exposure of these mice to 100% oxygen results in significantly accelerated mortality correlated with the earlier appearance of inflammatory cytokines and inflammatory cells, and more pronounced pulmonary edema.[28] Other pulmonary insults, such as ozone exposure,[29] virus infection,[30] and bacterial infection,[31] also result in significantly more severe inflammation compared to the parental mouse strain. It is clear from these data that CC10 has a physiologically important protective effect in the lungs against inflammation brought about by a variety of causes. This was recently directly visualized by electron microscopy of airway epithelia in these mice when treated with ozone compared to nonknockout controls.[32] These data provide powerful incentive to evaluate CC10 as an anti-inflammatory agent in animal models of pulmonary disease.

The combination of *in vivo* observations in CC10 knockout mice augments the considerable literature pertaining to the activities of CC10 derived from various species *in vitro*. The level of endogenous CC10 has been studied in a number of human disease conditions and will be discussed in the context of CC10 replacement therapy or other CC10-based interventions. The basic scientific, mechanistic, and medical rationale will be summarized for applications of CC10-based therapeutic intervention in pulmonary, renal, and oncologic disease. There are also potential applications of CC10-based therapies in reproductive medicine and immunomodulation that are not included due to space limitations. CC10-based intervention is not limited to the application of recombinant human CC10, but also extends to the use of gene-based therapies and therapies that target as yet undiscovered physiologic pathways affected by CC10.

PROSPECTS FOR CC10 THERAPY IN LUNG DISEASE

The potential for CC10-based therapy in acute and chronic pulmonary disease is enormous, based on available animal and human data. The principal animal data supporting the anti-inflammatory role of CC10 in the lungs have been generated using the transgenic CC10/CCSP-deficient mouse.[27] Although these animals appear healthy, are perfectly viable, and do not exhibit any pathological effects of CC10 deficiency, they are very susceptible to pulmonary challenge with a variety of insults. When exposed to inhaled 100% oxygen[28] or inhaled ozone,[29] these animals exhibit significantly higher inflammatory markers in lung fluids, pulmonary edema, and shorter survival times than control animals. When exposed to intratracheal adenovirus infection[30] or intratracheal *Pseudomonas* infection,[31] these animals exhibit significantly greater local neutrophil counts and higher levels of inflammatory markers such as IL-6, TNF-α, and IL-1β in lung homogenates. In the absence of CC10, the epithelial linings of the trachea and bronchi in these mice are severely damaged by treatment with ozone, as shown by electron microscopy of the tissue surfaces.[32] Furthermore, the development of abnormal alveoli and ARDS-like symptoms has recently been reported in these animals.[33] All of these data argue strongly for a protective anti-inflammatory role for CC10 in the upper airways and lungs.

In addition to the powerful data generated in the CC10/CCSP-deficient mouse, there is considerable indirect evidence for the role of CC10 in the lung's response to inflammatory stimuli. Pulmonary CC10 production is increased in response to lung inflammation[34] and to irritants such as asbestos,[35] as well as in response to elevations in TNF-α[36] and IFN-γ.[37] CC10 has been reported to bind small nonpolar compounds, including mutagenic xenobiotics such as polychlorinated biphenyls (PCBs),[27,38] which may play a role in the accumulation of such substances in Clara cells. Cigarette smoking causes a decrease in CC10 production and serum level,[39–41] which may worsen the chronic inflammation of the bronchial mucosa induced by smoke by-products. There are currently no known human populations that are genetically deficient in CC10. Although different alleles have been characterized in asthma populations, all DNAs to date have had intact exons, coding for a normal protein. However, the region to which the CC10 gene maps on chromosome 11q13[4] has been associated with atopy and bronchial hypersensitivity through genetic studies.[42,43] There are a number of clinical conditions in which either the local CC10 concentration in lung fluids or the systemic CC10 concentration in serum or plasma is abnormal compared to healthy controls. These include neonatal respiratory distress syndrome (nRDS),[44] ARDS,[45] ARDS due to cardiopulmonary bypass surgery,[46,47] idiopathic pulmonary fibrosis and bleomycin lung,[16] asthma,[48,49] lung cancer, sarcoidosis, chronic obstructive pulmonary disease (COPD),[41] and bacterial pneumonia.[50] Abnormalities in the pulmonary epithelial-blood barrier have been proposed as a mechanism for these deviations in CC10 levels.[47] Perturbation of Clara cell physiology could also account for skewed CC10 levels.[51] Undoubtedly, a combination of several processes contributes to alterations in CC10 levels and its dynamic role(s) in normal and pathological pulmonary physiology. These clinical conditions are characterized by pulmonary inflammation, elevated PLA$_2$ activity, and/or fibronectin deposition, and involve processes such as neutrophil and fibroblast infiltration of the lungs, all of which are consistent with a lack of local CC10 function.

NEONATAL RESPIRATORY DISTRESS SYNDROME AND BRONCHOPULMONARY DYSPLASIA

Very low concentrations of CC10 have been found in the BALF and TAF of ventilated premature babies as compared to adult levels. The difference in CC10 concentration between the TAF of premature babies born at 28–32 weeks of gestation[52] compared to the sputum of healthy adults[10] is 2–4 orders of magnitude, depending on how the data from each study are normalized. It is not surprising to find low concentrations of CC10 in premature lungs versus mature lungs because the architecture of the undeveloped lung is very different from the mature lung.[51,53] Mature Clara cells are not present in the pulmonary epithelium even in newborn mice.[53] However, the CC10 concentration in TAF increases steadily in the weeks after birth[54] as the lungs mature and the requirement for supplemental oxygen decreases. The concentration of CC10 in BALF or TAF of normal healthy full-term babies is not known, so it is difficult to gauge the magnitude of the deficiency in premature babies. Fetal lungs as early as 12 weeks exhibit CC10 immunoreactivity in tissue sections.[55]

CC10 appears in normal amniotic fluid at about 15 weeks of gestation and increases to a plateau at about 30 weeks at a concentration of 2–3 mg/mL.[56] This raises the possibility that CC10 could be a developmental factor in which the fetal tissues, particularly the respiratory and digestive tracts, are bathed. Human CC10 is known to bind to retinoids[57] and to progesterone[58] *in vitro* and could play a role in regulating the effects of these molecules on the reproductive tissues and developing fetus. It has been suggested that the level of CC10 in amniotic fluid could be used as a marker of fetal lung maturation.[56] However, it is not clear whether the source of this CC10 is primarily the fetal lung, the placenta, or maternal tissue; nor is the turnover rate of CC10 in amniotic fluid known. In the fetal rabbit lung, CC10 and surfactant protein synthesis are transcriptionally activated just prior to birth.[59] In rabbits,[60] rats,[61] and probably humans,[4,5] transcription of CC10 is induced in the lungs by corticosteroids. Likewise, transcription of surfactant proteins is also induced by corticosteroids, which are sometimes given to premature babies with nRDS to speed the maturation of the lungs, reduce time spent on ventilation, and decrease the severity of inflammation in nRDS.[62]

Surfactant synthesis in human fetal lungs is also dramatically upregulated at about 34 weeks of gestational age[62,63] and it is possible that CC10 synthesis is also upregulated at this point. Babies born after this point have a much lower incidence of respiratory distress. This observation contributed to the development of surfactant replacement therapy for the treatment of nRDS, which became the standard of care in the early 1990s. However, despite advances in treating the nRDS with surfactant and improved ventilator strategies, a significant proportion of these babies (40–60%) develop abnormal, poorly functioning lungs characterized by progressive pulmonary fibrosis termed bronchopulmonary dysplasia (BPD) and this proportion has not improved significantly with the use of surfactant replacement therapy. There is a clear relationship between the level of treatment required to keep the baby alive and the severity of pulmonary inflammation and the subsequent development of BPD or chronic lung disease (CLD). Treatment parameters such as volume and rate of ventilation, percent of supplemental oxygen, response to surfactant therapy, and duration of ventilation provide an indication of patient outcome. Barotrauma to the lungs, a wounding phenomenon, is associated with ventilation parameters, while inflammation can be attributed mainly to oxygen toxicity (summarized in reference 62).

Relevant to CC10 mechanisms of action, a statistically significant correlation between elevated levels of both soluble and insoluble fibronectin in BALF in babies treated for nRDS and the subsequent development of BPD has been shown in four studies.[64–67] Furthermore, approximately 10% of the nRDS babies do not respond to surfactant therapy and this is thought to be due to inactivation of the artificial surfactant by secretory PLA_2 ($sPLA_2$).[68] Pharmaceutical preparations of artificial surfactant are primarily composed of phospholipids, and artificial surfactant has been shown to be digested and inactivated by $sPLA_2$.[69] Therefore, in addition to the known deficiency of CC10 in premature babies with nRDS and the protective effects of CC10 against oxygen toxicity inferred from studies in the CC10/CCSP knockout mice, there are three other specific rationales based on the known biological activities of CC10 for the application of rhCC10 in this clinical indication. The first is that the inhibition of $sPLA_2$ activity by CC10 may decrease the severity or duration of pulmonary inflammation and prevent the inactivation of surfactant. It should be noted that no differences were seen in surfactant metabolism in the CC10/CCSP knockout mouse compared to normal mice, suggesting that CC10 has no role in surfactant turnover during viral infection.[70] However, the inhibition of $sPLA_2$ activity by CC10 is still important to consider in nRDS because elevated eicosanoids and platelet-activating factor, which are downstream products of PLA_2 activity, have been correlated with the development of BPD in TAF of premature babies with nRDS.[71] An inverse relationship between CC10 and eicosanoid concentrations has been shown in nasal lavages of children with upper respiratory tract infections.[72] The second is that CC10 may prevent the inappropriate deposition of fibronectin in lung tissue, which is a wound healing response and not a normal component of the extracellular matrix in the lungs.[73] The third is that rhCC10 may inhibit the migration of neutrophils, monocytes, and fibroblasts into the lungs of these babies, decreasing the severity of pulmonary inflammation. The combined rationale for the use of rhCC10 to treat nRDS and prevent neonatal BPD/CLD is quite powerful. Claragen has conducted preclinical studies in three animal models of inflammatory lung disease and results have been very encouraging, as reported at this meeting. [Enhanced blood-gas response to exogenous surfactant was observed in a small number of preterm lambs treated with a single intratracheal dose of rhCC10. Pulmonary function tests were normalized to control values in newborn piglets ventilated for 48 h with 100% oxygen by a single intratracheal dose of rhCC10. Also, airway resistance was normalized to control values by rhCC10 in an *ex vivo* rat lung model of LPS-induced pulmonary inflammation. The latter two results were statistically significant (data not shown).] A phase I safety study of intratracheal rhCC10 in premature babies will commence shortly.

ACUTE RESPIRATORY DISTRESS SYNDROME

Acute respiratory distress syndrome (ARDS) is similar in many respects to nRDS. It is characterized by pulmonary inflammation that is sometimes followed by fibrosis and is treated with ventilation and supplemental oxygen. However, ARDS occurs as a result of an insult to the mature, functional lung as opposed to the undeveloped premature lung. There are many different causes of ARDS, including pneumonia, smoke inhalation, aspiration, trauma, shock, multiple organ failure (MOF), sepsis, surgery, or severe inflammation of another organ or tissue such as the pancreas, intestines/colon, or liver. ARDS is associated with very high levels of circulating PLA_2 activity,[74] which may inactivate endogenous surfactant in the lungs. ARDS can be effectively treated with artificial surfactant, but the amounts required to reduce mortality are prohibitive[75] and the supply cannot meet the demand for this patient population. Effective PLA_2 inhibitor pharmaceuticals continue to be sought for application in this clinical indication and a few are in clinical development.[76]

CC10 is a good candidate therapy for ARDS not only because it is a PLA_2 inhibitor, but also because one study has shown that CC10 is deficient in ARDS patients and a second study demonstrated that CC10 is rapidly cleared from the circulation due to an elevated glomerular filtration rate during ARDS. In the first study, a statistically significant lower level of endogenous CC10 (2.7-fold lower, $p = 0.03$) was found in the BALF of 14 patients who developed ARDS subsequent to cardiopulmonary bypass surgery and died ($n = 8$) compared to those who developed ARDS and survived ($n = 6$).[46] In the second study, the turnover of CC10 was found to be very rapid (~18 min) and correlated with the glomerular filtration rate, which is increased in ARDS.[45] There was also a weak relationship between plasma CC10 concentration and paO_2 (partial pressure of oxygen in blood) and a weak inverse relationship between plasma CC10 concentration and $paCO_2$ (partial pressure of carbon dioxide in blood). In a similar study, the CC10 concentration in serum from 117 patients hospitalized with bacterial pneumonia was measured.[50] The majority of these patients were treated with oxygen and/or ventilation. The authors of this study reported that CC10 plasma levels were significantly lower in the pneumonia patients compared to healthy controls ($p < 0.0001$). They further observed that the plasma concentration of CC10 decreased between the onset of ARDS and death ($p < 0.01$), while CC10 increased in patients who recovered ($p < 0.001$). All of these studies strongly suggest that the loss of local and systemic CC10 correlates with mortality in ARDS and pneumonia. Moreover, there was a recent report of an ARDS-like syndrome in the CC10/CCSP-deficient mouse strain.[33] Therefore, rhCC10 therapy should be considered for the treatment or prevention of ARDS and pneumonia.

ASTHMA

Asthma is a bronchoconstrictive disease with a major inflammatory component that is typically treated with inhaled or systemic corticosteroids. There are many causes of asthma, including allergy, exercise, upper respiratory tract infections, emotional stress, and even cold temperatures, so it is not clear whether it is actually one disease or many. Some of the common features of asthma include a local elevation in eicosanoids such as leukotriene B4, infiltration of eosinophils, and acidic airway fluids. Eicosanoids are products of the arachidonic acid cascade, which is initiated by the release of arachidonic acid from intracellular phospholipid stores by PLA_2.[74] Corticosteroids are known to increase transcription of CC10 in the lungs,[60,61] so one of the anti-inflammatory mechanisms of corticosteroids may be to increase the production of pulmonary CC10. In long-term asthma sufferers, the surface of the airways undergoes remodeling and becomes less compliant with time. Because long-term use of corticosteroids often results in tachyphylaxis and is associated with a number of detrimental side effects, therapeutic alternatives continue to be pursued.

Asthma is the only pulmonary disease in which a search for CC10-based genetic causes has been done. There are four published reports to date that evaluate differences in the CC10 gene of asthma patients and two that examine CC10 concentrations in asthma patients. There are three short exons and two introns in the human CC10 gene, covering 4.1 kb of DNA.[4,5] In the first report, two types of polymorphisms in the CC10 gene were identified, but these did not correlate with any pulmonary condition.[43] The first type of polymorphism involved microsatellite DNA in intron #1 consisting of short (4–5 bp) repetitive elements, of which 5 alleles were found in a total of 43 individuals. The second type of polymorphism was the insertion of a rare human-specific Alu repeat in intron #2 of the CC10 gene in 3% of human samples. The CC10 gene was mapped to a region of chromosome 11 (11q11-q13) that had previously been linked to atopy and bronchial hyperresponsiveness (BHR).[77] In the second report, the chromosomal 11q11-q13 localization was confirmed and no mutations were found in either the three CC10 exons or in the 5′ promoter region (to base −265) of 52 asthma patients compared to 52 nonasthmatics from 26 families characterized with hereditary atopic asthma.[4] However, a new allele with a single base (A38G) substitution was found in the 5′ untranslated portion of exon #1 of the CC10 gene.[4] In the third report, the AA, AG, and GG alleles at position 38 of exon #1 were again reported.[78] In this study, 43.6% were homozygous (GG) for the allele and 46.2% were heterozygous (AG), suggesting that the GG allele is, in fact, the predominant allele in the Australian population. Further, the A38A genotype correlated in a statistically significant manner with a higher incidence of the asthmatic phenotype than did G38G or A38G in 226 patients genotyped. In the fourth report, the slight prevalence of the G38A over the A38G allele was confirmed; however, no association of the A38A genotype was found with asthma in 275 British and 300 Japanese individuals.[79] To summarize, there are at least three types of polymorphisms in the CC10 gene, but these are not mutations in the coding region of mature CC10 and would not produce an aberrant protein product. Rather, these alleles are in untranslated exonic sequences and intronic sequences that could, in the future, be associated with a lower CC10 output in asthmatics.

There are two reports that have evaluated CC10 protein concentrations in asthmatics. In the first,[48] the concentrations of CC10 in BALF showed a statistically significant ($p = 0.015$) decrease in 24 asthmatics evaluated versus 24 nonasthmatic nonsmokers. The asthmatic BALF contained about 70% of the actual CC10 concentration found in nonasthmatics. However, this deficiency was not statistically significant when normalized to albumin concentration in BALF. It is difficult to find a BALF parameter that provides an unbiased basis for normalization of the inherently variable BALF dilutions from patient to patient. Even albumin in extracellular lung fluids is known to vary with differences in vascular permeability;[80] therefore, it is not possible to dismiss the asthmatics' deficiency in CC10 concentration found in this study based on normalization to albumin. In the second, a statistically significant decrease in mean CC10 concentration in the serum of 63 asthmatics versus 64 nonsmoking nonasthmatics was found ($p < 0.0001$), which was more pronounced in asthmatics with >10-year duration of disease.[49] Comparison of CC10 levels in atopic versus nonatopic sera did not reveal any statistically significant difference. Asthmatic sera contained about 60% of the CC10 concentration found in nonasthmatics, which is comparable to the amount found in BALF in the previous study.

The discrepancy between genetic and protein findings is not necessarily inconsistent. There need not be an association between any genetic allele in the CC10 gene, resulting in a deficiency of CC10 protein due to decreased synthesis, and a predisposition towards asthma. A second type of deficiency, a deficiency of *functional* CC10 protein, could also be responsible for BHR in asthma. The natural form of CC10 is thought to be the fully oxidized homodimer, which is the form that mediates many of the known *in vitro* biological activities. Therefore, the oxidized homodimer is presumed to be the active form *in vivo*. [Claragen is developing the fully oxidized homodimer as a therapeutic agent and our data in experimental animals, presented at this meeting, indicate that this form is active *in vivo* (data not shown).] Thus, the CC10 peptides synthesized from mRNA must be correctly processed by the secretory machinery of the endoplasmic reticulum, assembled into a homodimer, and then fully oxidized in order to function as a secreted protein in the airways. Therefore, any aberrant step in this process could effectively reduce the amount of functional CC10 protein secreted into the airway fluids. Likewise, the chemistry of the airway fluids may also be important in altering the structure or function of correctly synthesized CC10. Excessive acidity in the sputum of asthma patients has recently been identified.[81] Abnormal pH could adversely affect the balance between oxidation and reduction in airway fluids, possibly lowering the amount of fully oxidized, active CC10. In fact, one method of characterizing an asthmatic versus a nonasthmatic is to determine the concentration of inhaled citric acid required to induce BHR.[82]

At least four isoforms of CC10 have been observed in mammalian body fluids. In the initial purification and identification of CC10 in rat lung lavage fluid, there were three isoforms of CC10 dimers.[7] These were all immunoreactive with anti-CC10 antibody and all migrated at the same apparent molecular weight by SDS-PAGE, but they eluted as different peaks from an ion exchange column. Similarly, Aoki et al.[83] observed three peaks of immunoreactive CC10 eluting from hydrophobic interaction HPLC, believed to be dimers. This protein was purified from human hemodialysate. Most recently, Lindhal[84] presented a poster at a conference on PLA_2 in which nasal lavage fluids from patients with respiratory symptoms resulting from occupational exposures to industrial chemicals were separated by 2D electrophoresis. A reducing, denaturing SDS-PAGE gel was followed by isoelectric focusing. The protein spots were blotted and CC10 detected by antibody. Four isoforms of CC10 monomer (5 kDa) were seen. The major isoform of reduced CC10 had an isoelectric point of 4.8, with other forms at 4.6, 4.9, and 5.2. The idea that environmental exposures and abnormal airway chemistry (aberrant pH in the case of asthma or BHR) can affect the function of anti-inflammatory airway proteins such as CC10 merits further investigation.

In summary, there is no solid genetic evidence as yet to confirm that a defect in the CC10 gene or protein could be responsible for the BHR in asthmatics. However, the application of fully oxidized recombinant human CC10 in a variety of animal models of pulmonary inflammation suggests that CC10 is an important factor in normalizing smooth muscle function and vascular permeability *in vivo* (data not shown). No studies have yet been done in traditional animal models of BHR to identify or confirm the role of CC10 in asthma. Nevertheless, it seems reasonable to hypothesize that some subset of asthma patients have decreased CC10 synthesis or function that makes a significant contribution to the pathophysiology of the asthmatic airways.

CYSTIC FIBROSIS (CF)

Cystic fibrosis is an inherited disease in which a defective chloride ion transport protein (CFTR) affects the composition and function of the mucosal fluids lining the respiratory and digestive systems (reviewed in reference 85). The CFTR protein mediates the transport of chloride ions from the cytosol of epithelial cells into the extracellular fluid lining the lung and gut. The lack of functional CFTR results in an abnormally viscous mucus in the lungs and pancreas that interferes with the normal functioning of these organs. In the pancreas, this mucus clogs the secretion pathways leading to the pancreatic duct, effectively blocking secretion of enzymes needed for digestion, including PLA_2 type 1b. These digestive enzymes may then destroy the surrounding pancreatic tissue, resulting in the formation of a cyst through the body's attempt to contain and heal the damage. In the lungs, the abnormal chemistry and composition of the mucosal fluid layer may render the fluid layer partially dysfunctional by altering protein secretion by epithelial cells, disrupting normal chemosensing pathways and water balance, and perhaps acting as a barrier to the normal transmission of the molecular signals. The thick mucus also clogs the smaller passages and facilitates infections by normally nonpathogenic microorganisms, particularly *Staphylococcus aureus* and *Pseudomonas aeruginosa*. These infections can be fatal and are often the primary cause of death in older CF patients. Patients with CF also have mild to severe lung inflammation. Many of the typical markers of inflammation are present in CF lungs. These include elevated protease activity, proinflammatory cytokines, neutrophil counts, and eicosanoid levels.

There is no information available on the characterization of the CC10 gene or the concentration of CC10 in the airway fluids of CF patients. However, the ongoing pulmonary inflammation and tracheobronchial rigidity characteristic of CF could, in part, be the result of a deficiency of functional CC10. A recent report indicated that CF sputum (not BALF) is a strongly reducing environment due to dramatically elevated levels of reduced glutathione (GSH), which is thought to be a direct result of the defect in chloride ion transport.[86] In fact, nebulized GSH strongly induces airway narrowing, cough, and breathlessness in mild asthmatics.[87] Thus, over time, chronic elevations in GSH in the proximal airways may produce CF-like airway rigidity. There is an equilibrium maintained in airway fluid between reduced and oxidized glutathione and this is important in maintaining the redox potential of all extracellular biological fluids, including the blood. GSH is a very effective reducing agent for rhCC10, requiring a 50-fold lower concentration compared to dithiothreitol (DTT) to achieve the same degree of protein reduction *in vitro* (data not shown). Because the redox potential of airway fluids is skewed in CF patients, a significant proportion of the endogenous CC10 in the airways of CF patients may be the reduced homodimer, which lacks the anti-inflammatory activity of the fully oxidized homodimer (data not shown).

There are some interesting parallels between the pattern of nonpulmonary disease manifestation in the organs of CF patients and the pattern of affected organs in the

NIH CC10/UG knockout mouse. In CF, these include malabsorption of nutrients from the gut, pancreatic inflammation, and renal impairment.[85] The CC10/UG knockout mouse suffers from cachexia or malabsorption, hypocalcemia, necrotic foci of the pancreas, spleen, and thymus, as well as lethal renal fibrosis.[22] No pulmonary challenge experiments have yet been performed in these mice. However, the CC10/CCSP-deficient mouse[27] does not suffer from these extreme phenotypes and has been subjected to a variety of pulmonary insults. This mouse strain is much more susceptible to pulmonary neutrophil infiltration in response to intratracheal infection with *Pseudomonas aeruginosa*.[30] In addition to the possible role of a strongly reducing airway environment on CC10 is the indication that the level of CC10 immunoreactivity is dramatically reduced during acute pulmonary infection in humans.[50] Thus, not only is the oxidation state of endogenous CC10 uncertain in CF airways, but there also may be a deficiency in the amount of CC10 present in the infected airways of CF patients. These observations provide a compelling rationale for the treatment of infectious pulmonary exacerbations in CF patients with rhCC10.

With respect to the expression of CC10 in human lungs, there is a correlation between the affected organs in CF and the primary nonreproductive organs in which CC10 is most highly produced and secreted on a constitutive basis, namely the trachea, proximal lungs, and pancreas. Furthermore, the pattern of fibrosis in CF in the proximal airways, rather than the distal airways, also matches the pattern of Clara cell distribution and CC10 expression, as well as the apparent proximal to distal concentration gradient of CC10 in the pulmonary mucosal fluids.[10] All of these correlations, together with the airway normalization effects of rhCC10 in two animal models of pulmonary inflammation reported at this meeting, strongly suggest that a deficiency of functional, oxidized, homodimeric CC10 exists in the proximal airways of CF patients.

OTHER PULMONARY DISEASES

Several human lung disease conditions have been shown to have low CC10 concentrations in BALF in small numbers of patients, as measured by immunoassay. As previously mentioned, these include COPD, emphysema, idiopathic pulmonary fibrosis, sarcoidosis, bleomycin lung, and bacterial pneumonia. Some of these conditions exhibit high plasma CC10 concentrations, suggesting that lung damage allows the leakage of lung-specific proteins across the pleural barrier into the blood. Although the possibility of pulmonary secretion of CC10 directly into the blood has not yet been ruled out, excessive CC10 in plasma or serum is gaining acceptance as a general marker of lung damage.[88]

The most extensively characterized human population known to have a chronic deficiency of CC10 is cigarette smokers. Several studies have verified, by immunoassay, both local pulmonary and systemic CC10 deficiencies on the order of 2- to 10-fold lower than healthy nonsmokers.[39–41] Clara cells are sensitive to toxic chemicals in cigarette smoke, particularly PCBs.[52,89,90] One group has even shown that PCBs bind specifically to CC10[91] and could, therefore, alter or diminish its activities. Thus, cigarette smoking may result in a primary deficiency of CC10 at the level of Clara cell protein production by killing off these specialized cells. Smoking may also result in a deficiency of functional protein in which the anti-inflammatory properties of the CC10 that is produced are altered or diminished by interaction of CC10 with PCBs in the cigarette smoke. It should be noted that insertion of PCBs into the hydrophobic cavity of the CC10 homodimer does not appear to be possible after the cysteine bridges are formed. Umland[92] pointed out that these bridges act like a gate at the mouth of the cavity and create severe size limitations on the molecules that can get in. Therefore, a determination of whether the disulfide bridges are formed inside the cell during dimer assembly or in the extracellular milieu after secretion becomes an important point in interpreting existing observations with regard to the effects of pollutants on CC10 deficiency and function.

The health risks associated with smoking are well known and the magnitude of the risk correlates directly with the amount of smoking. The predisposition of smokers to upper respiratory infection, asthma, BHR, inflammatory disease such as inflammatory bowel disease, and many types of cancer, especially lung cancer, is highly reversible with the cessation of smoking. This may be because Clara cells have a high proliferation potential and may quickly repopulate the tracheobronchial epithelium.[93] However, over time, the epithelia of the airways are progressively destroyed and the risk becomes less reversible. Based on current information, namely that chemicals in smoke inactivate CC10, CC10 protein replacement therapies for the treatment of diseases associated with cigarette smoking would not be very useful in individuals who continue to smoke. However, other potential CC10-related therapies targeting physiologic pathways affected by CC10 may be possible as research continues.

PROSPECTS FOR CC10 THERAPY IN RENAL DISEASE

The potential of CC10-based therapy in renal disease was first identified by Zhang *et al.*[22] and stems from their work in the CC10/UG knockout mouse. Lethal glomerulonephritis/renal fibrosis is a prominent phenotype in this mouse, among others. The renal disease in these mice was shown by immunohistochemistry to be due to the inappropriate and excessive deposition of fibronectin in the kidney glomeruli. The resulting inflammation and fibrosis of the kidneys result in death as early as 5 weeks after birth. The investigators postulated that a direct interaction between CC10 and fibronectin was responsible for preventing this type of disease in normal animals. This hypothesis was confirmed by directly demonstrating this interaction *in vitro* and by showing that CC10 prevented fibronectin-dependent fibrillogenesis mediated by fibroblasts in culture. The relevance of this mechanism *in vivo* was demonstrated by showing that recombinant human CC10 could prevent the renal deposition of human fibronectin injected into these mice. This work was followed by the same investigators with the further characterization of the involvement of IgA in the renal disease of these mice, as well as two additional transgenic mouse strains that expressed lower levels of endogenous CC10 due to the engineered expression of murine CC10 antisense RNA *in vivo*.[25] All of these animals exhibit not only fibronectin and collagen renal deposits, but also IgA deposits. The renal disease in these three strains of mice was also characterized not only by proteinuria, but also by hematuria as well, which is fairly unique among animal models of renal disease in its resemblance to human renal nephropathy. A more detailed review of the mechanisms involving CC10/UG in chronic renal disease (CRD) can be found in Mukherjee et al.[94]

CHRONIC RENAL DISEASE (CRD)

Examination of the CC10 gene in a family with hereditary fibronectin-deposit glomerulonephropathy did not reveal a mutation in the coding sequence of CC10.[95] However, an exhaustive search for mutations in noncoding sequences that could lead to lowered protein production was not performed in these patients. An analysis of known alleles in introns #1 and #2, for example, has not been done. More studies on both the concentration of CC10 in patients with renal disease, in particular IgA nephropathy, as well as a detailed genetic analysis of the CC10 gene in these patients are needed to further our understanding of this disease pathway in humans.

The prevention and treatment of renal disease with CC10-based therapies is quite promising. There are about 700,000 new patients each year in the United States who are diagnosed with chronic renal disease (CRD), and there are multiple causes of such disease. In the United States, most of these patients develop CRD subsequent to diabetes (40%) or chronic hypertension (30%), while the remainder are idiopathic (30%). In contrast to the United States, IgA nephropathy (also known as Berger's disease) represents over half of the CRD in Japan, Europe, and Australia. Therefore, identification of patients at risk and likely to benefit from CC10-based therapy will be important in developing clinical development strategies in CRD patients. For example, a significant correlation was found in children between severe upper respiratory infection immediately preceding presentation with hematuria and subsequent biopsy diagnosis of IgA nephropathy.[96] Patients with chronic bronchitis have lower levels of CC10.[94] Another study of patients hospitalized with pneumonia documented a significant decrease in pulmonary CC10 during acute infection.[50] Thus, there could be a link between a decrease in the pulmonary supply of CC10 during a respiratory infection concomitant with an increased output of IgA in response to the infection, creating the conditions for initiation of renal disease similar to that seen in the CC10 knockout mouse. Another clue to the link between pulmonary CC10 output and renal health is the recently identified correlation between cigarette smoking and renal fibrosis.[97] Direct studies must be done to determine the validity of any link between pulmonary inflammation and subsequent renal disease through CC10, and point the way towards appropriate patient populations for clinical testing and ultimate treatment with CC10-based therapies.

RENAL DIALYSIS

CRD patients with end-stage renal disease (ESRD) are placed on dialysis when their residual renal capacity reaches about 15% in order to remove small metabolic waste molecules such as urea. Dialysis patients suffer a number of complications, including accelerated loss of residual renal capacity, accelerated atherosclerosis and heart disease, loss of vascular access sites through localized inflammation and fibrosis, tendencies towards thrombosis (despite anticoagulant therapy), and inflammation and fibrosis of the vein returning the dialyzed blood.[98] The localization of inflammatory responses at the vascular access site and in the vein returning the dialyzed blood has often been interpreted as infection and may be treated with antibiotics. The localization of fibrosis around the vascular access site and in the vein returning the dialyzed blood has been interpreted as a response to the abnormal pressure to which these tissues are exposed as part of pumping the blood into and out of the body's circulation. In rare instances, an ARDS-like side effect, called Goodpasture syndrome, is seen, in which patients undergoing dialysis, or shortly after a dialysis session, develop respiratory insufficiency requiring ventilation and oxygen supplementation.[98]

CC10 passes through the glomerulus in the normal kidney and is present in urine. There are two studies that address the possibility that endogenous CC10 is removed from the blood by the hemodialysis process. The first study examined plasma samples from hemodialysis patients to determine whether CC10 is dialyzed out during a single hemodialysis session.[99] Contrary to the expected result, the mean serum CC10 concentration in 112 patients remained unchanged, or even increased slightly, between samples taken immediately before the dialysis session and immediately after the dialysis session. In the second study, CC10 was purified from human hemodialysate.[83] A significant quantity of CC10 was recovered after an extensive purification procedure from 40 liters of dialysis buffer. This recovered protein is difficult to reconcile with the other result that implied that CC10 was apparently not lost from the circulation of hemodialysis patients. An alternative explanation that accommodates both of the findings is that a significant quantity of endogenous CC10 was indeed dialyzed out of the patients in the first study, but that the loss was compensated by rapid replenishment. This would imply that circulating CC10 is maintained at a patient's inate set point by some homeostatic mechanism. Further detailed studies are required in these patients in order to determine whether circulating CC10 concentration is maintained by a homeostatic mechanism, whether pulmonary disease impairs the homeostatic mechanism in replenishing circulating CC10 lost during dialysis, and ultimately whether a precipitous drop in circulating CC10 is a factor in the development of Goodpasture syndrome.

ACUTE RENAL FAILURE

Because of the strong link between acute renal failure (ARF) and activation of the arachidonic acid cascade coupled with the presence of eicosanoids,[100] ARF must be considered a potential application for rhCC10. Like ARDS, there are many causes for ARF; therefore, there may only be a subset of ARF patients that can respond to any specific therapy, including rhCC10. The most likely subset that may respond to rhCC10 therapy are the patients that develop ARF as part of multiple or remote organ failure, including ARDS, in which the involvement of circulating $sPLA_2$ has been determined. Exogenous rhCC10 may prove useful in halting or preventing ARF in this subset because it is known to inhibit $sPLA_2$ (types 1b and 2a *in vitro*[12,13]). This is supported by the elevation in circulating $sPLA_2$ activity, as well as the increased plasma concentration of LPA, a downstream product of PLA_2 activity, in the CC10/UG knockout mouse.[94] No data are yet available on the concentration of endogenous CC10 in the serum, BALF, or urine of ARF patients before, during, or after ARF.

PROSPECTS FOR CC10 THERAPY IN CANCER

There is a growing body of literature that strongly suggests a role for CC10 deficiency in the development and/or progression of some types of human cancers. The experimental evidence relating CC10 and cancer can be categorized into three types of studies: (1) cell culture, (2) animal models, and (3) samples from human subjects.

In the cell culture studies, reported in five publications, a total of 13 different tumor cell lines were used to determine how the presence or absence of CC10, and its cell surface receptor(s), influences the neoplastic phenotype. Neoplastic cells are able to invade an artificial extracellular matrix (Matrigel) and grow in soft agar, and these properties were assayed in order to evaluate the effects of CC10 on the transformed phenotype. When the rhCC10 protein is preincubated with the highly invasive prostate cancer cell line, DU-145, it inhibits ECM invasion by these cells.[18] The invasiveness of a variety of certain transformed cell lines, of murine and human origin, was also inhibited by addition of rhCC10 protein, but this inhibition was linked to the presence of a CC10-specific receptor.[19,20] In two cell lines lacking this receptor, no inhibition of invasion was observed. Szabo and colleagues showed that a human lung cancer cell line transfected with a CC10 expression construct could not invade ECM or grow in soft agar as well as a comparable mock-transfected cell line.[101] These authors also reported two new observations in the CC10-transfected cell line, namely that of isotypic adhesion, a property of normal cells, and a decreased ability to adhere to fibronectin. However, the presence or absence of a CC10 receptor was not examined in this study. Zhang and colleagues were later able to link the suppression of tumor cell growth directly to the presence of the CC10 receptor by demonstrating that exogenous rhCC10 could inhibit ECM invasion and suppress tumor cell growth in soft agar only in receptor-positive cell lines.[21]

Another observation relevant to the role of CC10 in controlling cancer was made by Dierynck et al.[17] Here, the effect of CC10 on the cytotoxicity of human NK cells, freshly isolated from human peripheral blood lymphocytes, towards a tumor cell line, K562, was characterized. A dose-dependent increase in the ability of the NK cells to kill K562 was mediated by native human CC10 isolated from urine. This suggests another way in which CC10 could prevent cancer *in vivo*, that is, by playing a protective role in guarding against cancer by stimulating the immune system to destroy cancer cells.

The next set of data relating CC10 to cancer comes from two animal studies. In the first study, a tobacco-specific nitrosamine, NNK [4-(methylnitrosamino)-1-(3-pyridyl)-1-butanone], was given to golden Syrian hamsters for 2–24 weeks.[102] Animals were sacrificed at various intervals during the treatment period, CC10 concentration was measured, and tumor growth was monitored. CC10 levels decreased within 2 weeks of initiation of NNK treatment and tumors became detectable after 6 weeks of treatment. By 24 weeks, all of the animals had large tumors. This result is highly significant because it supports the hypothesis that CC10 deficiency correlates with tumor growth. It confirms an inference drawn from the tissue culture data that the neoplastic phenotype is not suppressed in the absence of CC10. Furthermore, it is significant that the CC10 level was decreased prior to the appearance of the tumors, indicating that the CC10 deficiency was not caused by tumor formation.

The second animal study involves observations made in an aging study of the CC10/UG knockout mice.[21] Relatively few of these animals survived for more than

10 months, but a total of 16 homozygous knockout mice survived for longer than 1 year. All 16 had solid organ tumors at the time of death, compared to 0/25 age-matched wild-type littermates. The parental strain of mice, C57/Bl6, is not predisposed to developing tumors, so this finding is significant.

The last set of data relating CC10 to cancer involves the analysis of samples and cancers from human subjects. The correlation between cigarette smoking and CC10 deficiency was reported in three references,[39–41] and the binding of CC10 to chemicals found in cigarette smoke[91] and the specific targeting of Clara cells by smoke by-products[90] have been discussed. The effect of tobacco-specific NNK on CC10 levels and tumor formation in hamsters reinforces the correlation between CC10 deficiency and cancer incidence in humans. In an analysis of CC10 expressed by various lung tumors, particularly NSCLC, the presence of CC10 was associated with younger age and less intense smoking.[103] These authors regarded CC10 as a differentiation marker for NSCLC. Another study examined CC10 expression in 19 pairs of nonneoplastic and neoplastic lung tissue.[89] Comparing both bronchiolar and bronchial epithelium to lung carcinomas, high levels of expression were found in the normal tissue, but only 1 out of 19 carcinomas expressed any detectable CC10. More recently, two studies have correlated a decrease or loss of CC10 expression with advancing tumor grade. The first study focused on lung cancer and correlated decreases in CC10 protein with increasing grade of bronchiolization of alveoli (BOA) lesions of the lung, irrespective of lung tumor type.[104] The second study focused on prostate cancer. Like the lungs, the prostate normally produces large amounts of CC10 constitutively.[6] CC10 was found at high levels in normal prostate tissue, benign prostatic hyperplasia, and prostatic atrophy, but it decreased significantly in intraepithelial neoplasia and in Gleason's pattern where grade was less than or equal to 2.[105] Higher grade tumors displayed no detectable CC10.

Taken together the experimental evidence strongly implicates the loss of CC10 and/or its receptor as factors accompanying, facilitating, or causing the cancer process. The antitumor effects of CC10 appear to be dependent upon the expression of a high-affinity CC10 receptor by the tumor cells.[19–21] Therefore, the utility of CC10-based cancer therapies will depend upon an analysis of the receptor in the candidate tumor. To date, tumor cell lines derived from the prostate, uterine endometrium, and lungs, as well as a lymphoma, a sarcoma, and a mastocytoma, have been receptor-positive. Other prostate cell lines, a choriocarcinoma, a cervical carcinoma, and a fibrosarcoma were receptor-negative. Clearly, not all tumors have the receptor or will be responsive to the pathway(s) affected by CC10. A prescreening of tumor cells for the presence of the receptor may be necessary prior to initiating CC10-based therapy, analogous to screening of breast cancers for expression of the estrogen receptor prior to initiation of hormonal therapy. Further characterization of the CC10 receptor is required in order to advance CC10-based cancer therapy into clinical development. A direct demonstration of the effectiveness of rhCC10 in preventing or slowing the growth or metastasis of tumor cells in an established tumor model is now in progress.

SUMMARY

There are multiple rationales for the application of rhCC10 or potential CC10-based therapies in several pulmonary, renal, and oncologic indications. For some of these clinical indications, rhCC10 is currently under clinical development, and much will be learned about the physiological role and metabolism of CC10 in humans through this process. CC10-based therapies also have applications in a variety of other inflammatory diseases, immunologic diseases, and organ transplantations. However, the supporting rationale and preclinical work are not as advanced as those presented here. Major leaps forward in understanding the physiological function(s) of CC10 have been made through the use of transgenic knockout technology. It is now becoming apparent how important CC10, its family members, and receptors might be in human health, development, and reproduction. Our expanding knowledge of these proteins, their functions, and the pathways through which they work will certainly provide new or better diagnostic and therapeutic tools in the future.

ACKNOWLEDGMENTS

I wish to thank the organizers of, and contributors to, the recent international meeting on the CC10/UG family of proteins, from which this volume is drawn.

REFERENCES

1. MUKHERJEE, A.B., G.C. KUNDU, G. MANTILE-SELVAGGI et al. 1999. Uteroglobin: a novel cytokine? Cell. Mol. Life Sci. **55:** 771–787.
2. MATTHEWS, J.H., N. PATTABIRAMAN, K.B. WARD et al. 1994. Crystallization and characterization of the recombinant human Clara cell 10-kDa protein. Proteins **20:** 191–196.
3. MORIZE, I. et al. 1987. Refinement of the C222(1) crystal form of oxidized uteroglobin at 1.34 Å resolution. J. Mol. Biol. **159:** 353–358.
4. ZHANG, Z. et al. 1997. Human uteroglobin gene: structure, subchromosomal localization, and polymorphism. DNA Cell Biol. **16:** 73–83.
5. WOLF, M. et al. 1982. Human CC10, the homologue of rabbit uteroglobin: genomic cloning, chromosomal localization, and expression in endometrial cell lines. Hum. Mol. Genet. **1:** 371–378.
6. PERI, A., E. CORDELLA-MIELE, L. MIELE & A.B. MUKHERJEE. 1993. Tissue-specific expression of the gene coding for human Clara cell 10-kD protein, a phospholipase A_2–inhibitory protein. J. Clin. Invest. **92:** 2099–2109.
7. SINGH, G. et al. 1987. Isolation and amino acid composition of the isotypes of a rat Clara cell specific protein. Exp. Lung Res. **13:** 299–309.
8. RAUCH, M., H. LOOSFELT, D. PHILIBERT & E. MILGROM. 1985. Mechanism of action of an antiprogesterone, RU486, in the rabbit endometrium: effects of RU486 on the progesterone receptor and on the expression of the uteroglobin gene. Eur. J. Biochem. **148:** 213–218.
9. KIKUKAWA, T. & A.B. MUKHERJEE. 1989. Detection of a uteroglobin-like phospholipase A2 inhibitory protein in the circulation of rabbits. Mol. Cell. Endocrinol. **62:** 177–187.
10. BERNARD, A. et al. 1992. Human urinary protein 1: evidence for identity with the Clara cell protein and occurrence in respiratory tract and urogenital secretions. Clin. Chim. Acta **207:** 239–249.
11. JACKSON, P.J. & R. TURNER. 1989. Purification and partial amino acid sequence of human urine protein 1: evidence for homology with rabbit uteroglobin. J. Chromatog. **452:** 359–367.
12. LEVIN, S.W. et al. 1986. Uteroglobin inhibits phospholipase A2 activity. Life Sci. **38:** 1813–1819.
13. MANTILE, G., L. MIELE, E. CORDELLA-MIELE et al. 1993. Human Clara cell 10-kDa protein is the counterpart of rabbit uteroglobin. J. Biol. Chem. **268:** 20343–20351.
14. MANJUNATH, R., S.I. CHUNG & A.B. MUKHERJEE. 1984. Crosslinking of uteroglobin by transglutaminase. Biochem. Biophys. Res. Commun. **121:** 400–407.

15. VASANTHAKUMAR, G. *et al.* 1987. Inhibition of phagocyte chemotaxis by potent phospholipase A2 inhibitory protein, uteroglobin. Biochem. Pharmacol. **37:** 389–394.
16. LESUR, O. *et al.* 1995. Clara cell protein (CC-16) induces a phospholipase A_2-mediated inhibition of fibroblast migration *in vitro*. Am. J. Respir. Crit. Care Med. **152:** 290–297.
17. DIERYNCK, I.A. *et al.* 1996. The human Clara cell protein: biochemical and biological characterization of a natural immunosuppressor. Multiple Sclerosis **1:** 385–387.
18. LEYTON, J. *et al.* 1994. Recombinant human uteroglobin inhibits the *in vitro* invasiveness of human metastatic prostate tumor cells and the release of arachidonic acid stimulated by fibroblast-conditioned medium. Cancer Res. **54:** 3696–3699.
19. KUNDU, G.C., G. MANTILE, L. MIELE *et al.* 1996. Recombinant human uteroglobin suppresses cellular invasiveness via a novel class of high-affinity cell surface binding site. Proc. Natl. Acad. Sci. U.S.A. **93:** 2915–2919.
20. KUNDU, G.C., A.K. MANDAL, Z. ZHANG *et al.* 1998. Uteroglobin (UG) suppresses extracellular matrix invasion by normal and cancer cells that express the high affinity UG-binding proteins. J. Biol. Chem. **273:** 22819–22824.
21. ZHANG, Z. *et al.* 1999. Loss of transformed phenotype in cancer cells by overexpression of the uteroglobin gene. Proc. Natl. Acad. Sci. U.S.A. **96:** 3963–3968.
22. ZHANG, Z. *et al.* 1997. Severe fibronectin-deposit renal glomerular disease in mice lacking uteroglobin. Science **276:** 1408–1412.
23. MIELE, L., E. CORDELLA-MIELE, G. MANTILE *et al.* 1994. Uteroglobin and uteroglobin-like proteins: the uteroglobin family of proteins. J. Endocrinol. Invest. **17:** 679–692.
24. RUOSLAHTI, E. 1988. Fibronectin and its receptors. Annu. Rev. Biochem. **57:** 375–413.
25. ZHENG, F. *et al.* 1999. Uteroglobin is essential in preventing immunoglobulin A nephropathy in mice. Nat. Med. **5:** 1018–1025.
26. REYNOLDS, S.J. *et al.* 1999. Normal function and lack of fibronectin accumulation in kidneys of Clara cell secretory protein/uteroglobin deficient mice. Am. J. Kidney Dis. **33:** 541–551.
27. STRIPP, B.R. *et al.* 1996. Clara cell secretory protein: a determinant of PCB bioaccumulation in mammals. Am. J. Physiol. **271:** L656–L664.
28. JOHNSON, C.J. *et al.* 1997. Altered pulmonary response to hyperoxia in Clara cell secretory protein deficient mice. Am. J. Respir. Cell Mol. Biol. **17:** 147–155.
29. MANGO, G.W. *et al.* 1998. Clara cell secretory protein deficiency increases oxidant stress response in conducting airways. Am. J. Physiol. **275:** L348–L356.
30. HARROD, K.S. *et al.* 1998. Clara cell secretory protein decreases lung inflammation after acute virus infection. Am. J. Physiol. **275:** L924–L930.
31. HAYASHIDA, K.S. *et al.* 2000. Role of Clara cell secretory protein in the endotoxin-induced neutrophil migration into the lungs. *In* Abstracts of the American Thoracic Society Meeting: A87, p. 454.
32. PLOPPER, C. *et al.* 2000. Ablation of the Clara cell secretory protein gene accelerates and alters the pattern of early acute tracheobronchial injury from ozone exposure. *In* Abstracts of the American Thoracic Society Meeting: G19, p. 374.
33. REYNOLDS, S.D. *et al.* 2000. Chronic Clara cell depletion leads to alveolar alterations and the development of ARDS-like symptoms. *In* Abstracts of the American Thoracic Society Meeting: H3, p. 199.
34. LESUR, O. *et al.* 1996. Clara cell protein (CC-16) and surfactant-associated protein A (SP-A) in asbestos-exposed workers. Chest **109:** 467–474.
35. ANDERSSON, O. *et al.* 1991. Purification and level of expression in bronchoalveolar lavage of a human polychlorinated biphenyl (PCB)–binding protein: evidence for a structural and functional kinship to the multihormonally regulated protein uteroglobin. Am. J. Respir. Cell Mol. Biol. **5:** 6–12.
36. YAO, X. *et al.* 1998. Tumor necrosis factor-alpha stimulates human Clara cell secretory protein production by human airway epithelial cells. Am. J. Respir. Cell Mol. Biol. **19:** 629–635.
37. MAGDALENO, S.M. *et al.* 1997. Interferon-gamma regulation of Clara cell gene expression: *in vivo* and *in vitro*. Am. J. Physiol. **272:** L1142–L1151.
38. ANDERSSON, O. *et al.* 1994. Heterologous expression of human uteroglobin/polychlorinated biphenyl–binding protein. J. Biol. Chem. **269:** 19081–19087.

39. BERNARD, A.M., H.A. ROELS, J.P. BUCHET & R.R. LAUWERYS. 1994. Serum Clara cell protein: an indicator of bronchial cell dysfunction caused by tobacco smoking. Environ. Res. **66:** 96–104.
40. SHIJUBO, N. *et al.* 1997. Serum and BAL Clara cell 10 kDa protein (CC10) levels and CC10-positive bronchiolar cells are decreased in smokers. Eur. Respir. J. **10:** 1108–1114.
41. BERNARD, A. *et al.* 1992. Clara cell protein in serum and bronchoalveolar lavage. Eur. Respir. J. **5:** 1231–1238.
42. DOULL, I.J. *et al.* 1996. Allelic association of gene markers on chromosomes 5q and 11q with atopy and bronchial hyperresponsiveness. Am. J. Respir. Crit. Care Med. **153:** 1280–1284.
43. HAY, J.G. *et al.* 1995. Human CC10 gene expression in airway epithelium and subchromosomal locus suggest linkage to airway disease. Am. J. Physiol. **268:** L565–L575.
44. DHANIREDDY, R., T. KIKUKAWA & A.B. MUKHERJEE. 1988. Detection of a rabbit uteroglobin-like protein in human neonatal tracheobronchial washings. BBRC **152:** 1447–1454.
45. DOYLE, I.R. *et al.* 1996. Clearance of Clara cell secretory protein 16 (CC16) and surfactant proteins A and B from blood in acute respiratory failure. Am. J. Respir. Crit. Care Med. **158:** 1528–1535.
46. JORENS, P. *et al.* 1995. Potential role of Clara cell protein, an endogenous phospholipase A_2 inhibitor, in acute lung injury. Eur. Respir. J. **8:** 1647–1653.
47. HERMANS, C. & A. BERNARD. 1998. Pneumoproteinaemia: a new perspective in the assessment of lung disorders. Eur. Respir. J. **11:** 801–803.
48. VAN VYVE, T. *et al.* 1995. Protein content in bronchoalveolar lavage fluid of patients with asthma and control subjects. J. Allergy Clin. Immunol. **95:** 60–68.
49. SHIJUBO, N. *et al.* 1999. Serum levels of Clara cell 10-kDa protein are decreased in patients with asthma. Lung **177:** 45–52.
50. NOMORI, H. *et al.* 1995. Protein 1 (Clara cell protein) serum levels in healthy subjects and patients with bacterial pneumonia. Am. J. Respir. Crit. Care Med. **152:** 746–750.
51. SINGH, G. & S.L. KATYAL. 1997. Clara cells and Clara cell 10 kDa protein (CC10). Am. J. Respir. Cell Mol. Biol. **17:** 141–143.
52. DHANIREDDY, R. *et al.* 1988. Uteroglobin-like protein in premature infants: effect of gestational age. Pediatr. Res. **23:** 463A.
53. WUENSCHELL, C.W. *et al.* 1996. Embryonic mouse lung epithelial progenitor cells coexpress immunohistochemical markers of diverse mature cell lineages. J. Histochem. Cytochem. **44:** 113–123.
54. DHANIREDDY, R. *et al.* 1993. Uteroglobin-like protein levels in premature infants on long term ventilator support. Pediatr. Res. **33:** 323A.
55. KHOOR, A., M.E. GRAY, G. SINGH & M.T. STAHLMAN. 1996. Ontogeny of Clara cell–specific protein and its mRNA: their association with neuroepithelial bodies in human fetal lung and in bronchopulmonary dysplasia. J. Histochem. Cytochem. **44:** 1429–1438.
56. BERNARD, A. *et al.* 1994. Clara cell protein in human amniotic fluid: a potential marker of fetal lung growth. Pediatr. Res. **36:** 771–775.
57. LOPEZ DE HARO, M.S. *et al.* 1994. Binding of retinoids to uteroglobin. FEBS Lett. **349:** 249–251.
58. SINGH, G. *et al.* 1990. Clara cell 10 kDa protein (CC10): comparison of structure and function to uteroglobin. Biochim. Biophys. Acta **1039:** 348–355.
59. PERI, A. *et al.* 1995. Uteroglobin gene expression in the rabbit uterus throughout gestation and in the fetal lung: relationship between uteroglobin and eicosanoid levels in the developing fetal lung. J. Clin. Invest. **96:** 343–353.
60. LOPEZ DE HARO, M.S. & A. NIETO. 1985. Glucocorticoids induce the expression of the uteroglobin gene in rabbit fetal lung explants cultured *in vitro*. Biochem. J. **225:** 255–258.
61. ARSALANE, K. *et al.* 2000. Clara cell specific protein (CC16) expression after acute lung inflammation induced by intratracheal lipopolysaccharide administration. Am. J. Respir. Crit. Care Med. **161:** 1624–1630.
62. DAVIS, J.M. & W.N. ROSENFELD. 1994. Chronic lung disease. *In* Neonatology: Pathophysiology and Management of the Newborn, pp. 453–477.

63. WHITSETT, J.A. et al. 1994. Acute respiratory disorders. In Neonatology: Pathophysiology and Management of the Newborn, pp. 429–452.
64. GERDES, J.S. et al. 1988. Tracheal lavage and plasma fibronectin: relationship to respiratory distress syndrome and development of bronchopulmonary dysplasia. J. Pediatr. **113:** 727–731.
65. GERDES, J.S. et al. 1988. Effects of dexamethasone and indomethacin on elastase, alpha 1–proteinase inhibitor, and fibronectin in bronchoalveolar lavage fluid from neonates. J. Pediatr. **113:** 727–731.
66. WATTS, C.L., A.A. FANAROFF & M.C. BRUCE. 1992. Elevation of fibronectin levels in lung secretions of infants with respiratory distress syndrome and development of bronchopulmonary dysplasia. J. Pediatr. **120:** 614–620.
67. WATTS, C.L. & M.C. BRUCE. 1992. Effect of dexamethasone therapy on fibronectin and albumin levels in lung secretions of infants with bronchopulmonary dysplasia. J. Pediatr. **121:** 597–607.
68. FUNG, T. & H. TAEUSCH. 1992. Surfactant: coming of age for the treatment of RDS. J. Respir. Dis. **13:** 609–964.
69. DUNCAN, J.E., G.M. HATCH & J. BELIK. 1996. Susceptibility of exogenous surfactant to phospholipase A2 degradation. Can. J. Physiol. Pharmacol. **74**(8): 957–963.
70. IKEGAMI, M., K.S. HARROD, J.A. WHITSETT & A.H. JOBE. 1999. CCSP deficiency does not alter surfactant homeostasis during adenoviral infection. Am. J. Physiol. **277**(5, part 1): L983–L987.
71. STENMARK, K. et al. 1987. Potential role of eicosanoids and PAF in the pathophysiology of bronchopulmonary dysplasia. Am. Rev. Respir. Dis. **136:** 770–772.
72. VOLOVITZ, B. et al. 1988. Relationship between leukotriene C4 and a uteroglobin-like protein in nasal and tracheobronchial mucosa of children: implication in acute respiratory illnesses. Int. Arch. Allergy Appl. Immunol. **86:** 420–425.
73. HYNES, R.O. 1990. Fibronectins. Springer-Verlag. New York/Berlin.
74. MUKHERJEE, A.B., E. CORDELLA-MIELE & L. MIELE. 1992. Regulation of extracellular phospholipase A2 activity: implications for inflammatory diseases. DNA Cell Biol. **11:** 233–243.
75. GREGORY, T.J. et al. 1997. Bovine surfactant therapy for patients with acute respiratory distress syndrome. Am. J. Respir. Crit. Care Med. **155:** 1309–1315.
76. GLASER, K.B. 1995. Regulation of phospholipase A_2 enzymes: selective inhibitors and their pharmacological potential. Adv. Pharmacol. **32:** 31–66.
77. COOKSON, W.O. et al. 1989. Linkage between immunoglobulin E responses and underlying asthma and rhinitis and chromosome 11q. Lancet **1:** 1292–1295.
78. LAING, I.A. et al. 1998. A polymorphism of the CC16 gene is associated with an increased risk of asthma. J. Med. Genet. **35:** 463–467.
79. GAO, P.S. et al. 1998. Negative association between asthma and variants of CC16 (CC10) on chromosome 11q13 in British and Japanese populations. Hum. Genet. **103:** 57–59.
80. JONES, K. et al. 1990. A comparison of albumin and urea as reference markers in bronchoalveolar lavage fluid from patients with interstitial lung disease. Eur. Respir. J. **3:** 152–156.
81. HUNT, J.F. et al. 2000. Condensed breath exhalate falls during experimental rhinovirus 16 infection in asthmatic subjects. In Abstracts of the American Thoracic Society Meeting, p. A104.
82. WONG, C.H., R. MATAI & A.H. MORICE. 1999. Cough induced by low pH. Respir. Med. **93:** 58–61.
83. AOKI, A. et al. 1996. Isolation of human uteroglobin from blood filtrate. Mol. Hum. Reprod. **2:** 489–497.
84. LINDHAL, M., J. SVARTZ & C. TAGESSON. 1999. Demonstration of different forms of lipocortin-1 and Clara cell protein-16 in human nasal and bronchioalveolar lavage fluids. Electrophoresis **20:** 881–890.
85. HUDSON, V.L. & M.F. GULL. 1998. New developments in cystic fibrosis. Pediatr. Ann. **27:** 515–520.
86. DAULETBAEV, N. et al. 2000. Glutathione and other antioxidants are abundant in cystic fibrosis sputum and effectively inhibit damaging effects of hydrogen peroxide in vitro. In Abstracts of the American Thoracic Society Meeting, p. A289.

87. MARRADES, R.M. et al. 1997. Nebulized glutathione induces bronchoconstriction in patients with mild asthma. Am. J. Respir. Crit. Care Med. **156:** 425–430.
88. HERMANS, C. et al. 1999. Clara cell protein as a marker of Clara cell damage and bronchoalveolar blood barrier permeability. Eur. Respir. J. **13:** 1014–1021.
89. BROERS, J.L. et al. 1992. Expression of surfactant associated protein-A and Clara cell 10 kilodalton mRNA in neoplastic and non-neoplastic human lung tissue as detected by *in situ* hybridization. Lab. Invest. **66:** 337–346.
90. BARON, J. et al. 1988. Sites for xenobiotic activation and detoxification within the respiratory tract: implications for chemically induced toxicity. Toxicol. Appl. Pharmacol. **93:** 493–505.
91. HARD, T. et al. 1995. Solution structure of a mammalian PCB-binding protein in complex with a PCB. Nat. Struct. Biol. **2:** 983–989.
92. UMLAND, T.C. & M. SAX. 1995. Twixt form and function. Nat. Struct. Biol. **2:** 919–922.
93. EVANS, M.J., L.J. CABRAL-ANDERSON & G. FREEMAN. 1978. Role of the Clara cell in renewal of the bronchiolar epithelium. Lab. Invest. **38:** 648–655.
94. MUKHERJEE, A.B. et al. 1998. Uteroglobin: physiological role in normal glomerular function uncovered by targeted disruption of the uteroglobin gene in mice. Am. J. Kidney Dis. **32:** 1106–1120.
95. VOLLMER, M., R. FRAPF & F. HILDEBRANDT. 1998. Exclusion of the uteroglobin gene as a candidate for fibronectin glomerulopathy. Nephrol. Dial. Transplant. **13:** 2417–2418.
96. HOGG, R.J. 1988. IgA nephropathy: clinical features and natural history—a pediatric perspective. Am. J. Kidney Dis. **12:** 358–361.
97. GAMBARO, G. et al. 1998. Renal impairment in chronic cigarette smokers. J. Am. Soc. Nephrol. **9:** 562–567.
98. NISSENSON, A.R., R.N. FINE & D.E. GENTILE. 1995. Clinical Dialysis. Third edition. Appleton & Lange. Norwalk, Connecticut.
99. KABANDA, A. et al. 1994. Determinants of the serum concentrations of low molecular weight proteins in patients on maintenance hemodialysis. Kidney Int. **45:** 1689–1696.
100. KLAHR, S. 1994. Role of arachidonic acid metabolites in acute renal failure and sepsis. Nephrol. Dial. Transplant. **9:** 52–56.
101. SZABO, E. et al. 1998. Overexpression of CC10 modifies neoplastic potential in lung cancer cells. Cell Growth Differ. **9:** 475–485.
102. SUNDAY, M.E. et al. 1995. Histochemical characterization of non-neuroendocrine tumors and neuroendocrine cell hyperplasia induced in hamster lung by 4-(methylnitrosoamino)-1-(3-pyridyl)-1-butanone with or without hyperoxia. Am. J. Pathol. **147:** 740–752.
103. LINNOILA, R.I. et al. 1992. Peripheral airway cell marker expression in non–small cell lung carcinoma: association with distinct clinicopathologic features. Am. J. Clin. Pathol. **97:** 233–243.
104. JENSEN-TAUBMAN, S.M., S.M. STEINBERG & R.I. LINNOILA. 1998. Bronchiolization of the avcoli in lung cancer: pathology, patterns of differentiation, and oncogene expression. Int. J. Cancer **75:** 489–496.
105. WEERARATNA, A.T. et al. 1997. Loss of uteroglobin expression in prostate cancer: relationship to advancing grade. Clin. Cancer Res. **3:** 2295–2300.

C/EBPα and TTF-1 Synergistically Transactivate the Clara Cell Secretory Protein Gene

TOBIAS N. CASSEL,[a,b] GUNTRAM SUSKE,[c] AND MAGNUS NORD[b]

[b]*Department of Medical Nutrition, Karolinska Institute, Novum, S-141 86 Huddinge, Sweden*

[c]*Institut für Molekularbiologie und Tumorforschung, Philipps-Universität Marburg, Marburg, Germany*

In the lung, the Clara cell secretory protein/uteroglobin (CCSP/UG) is specifically expressed in the Clara cells of the small airway epithelium in a differentiation-dependent manner. Although the physiological function of CCSP is unknown, the differentiation-dependent expression of this protein makes it a good model to study the mechanisms behind pulmonary epithelial gene expression during lung development and differentiation. In the rat developing lung, low levels of CCSP expression are first detected around embryonic day 16. As development proceeds, CCSP expression starts increasing at day 18–19 and further increases until birth.[1] Previously, the homeodomain thyroid transcription factor-1 (TTF-1) and the forkhead hepatocyte nuclear factor-3 (HNF-3) have been identified as major regulators of pulmonary CCSP gene expression.[2-4] Both transcription factors are expressed in the bronchiolar epithelium and in Clara cells. However, during lung development, the expression of HNF-3 factors and TTF-1 does not temporally reflect the expression pattern of CCSP. Therefore, the increase in CCSP expression levels seen late in lung development cannot be explained by the action of HNF-3 factors or TTF-1 alone. The CCAAT/enhancer binding protein alpha (C/EBPα) is expressed in the lung epithelium and is first detected in the rat lung at embryonic day 18–19,[5] correlating to the increase in CCSP expression. Recently, we have characterized a compound C/EBP-response element in the proximal CCSP promoter.[6] C/EBPα as well as C/EBPδ, another C/EBP family member that is expressed in the bronchiolar epithelium, transactivate the CCSP promoter through interaction with the two C/EBP-binding sites that form the compound element.[6] The C/EBP-binding sites are situated in close proximity to the previously characterized TTF-1 and HNF-3 binding sites. Together, this prompted us to investigate putative cooperative interactions between these transcription factors in the regulation of CCSP.

To address the question of whether C/EBP, TTF-1, and HNF-3 factors interact in the regulation of CCSP, we used *Drosophila* SL-2 cells. These cells lack many mammalian transcription factors and are thus well suited for studies of cooperative inter-

---

[a]Address for correspondence: T. N. Cassel, Department of Medical Nutrition, Karolinska Institute, F60 Novum, S-141 86 Huddinge, Sweden. Voice: +46-8-5858 3725; fax: +46-8-711 6659.
tobias.cassel@mednut.ki.se

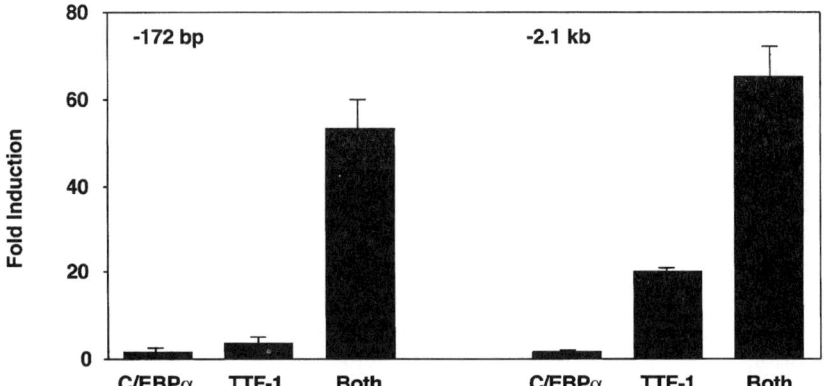

FIGURE 1. Cotransfection studies were performed with reporter plasmids containing the mouse CCSP promoter sequences −172 bp to +7 bp (left) and −2.1 kb to +7 bp (right) in front of a luciferase reporter gene. Error bars represent SD ($n = 3$).

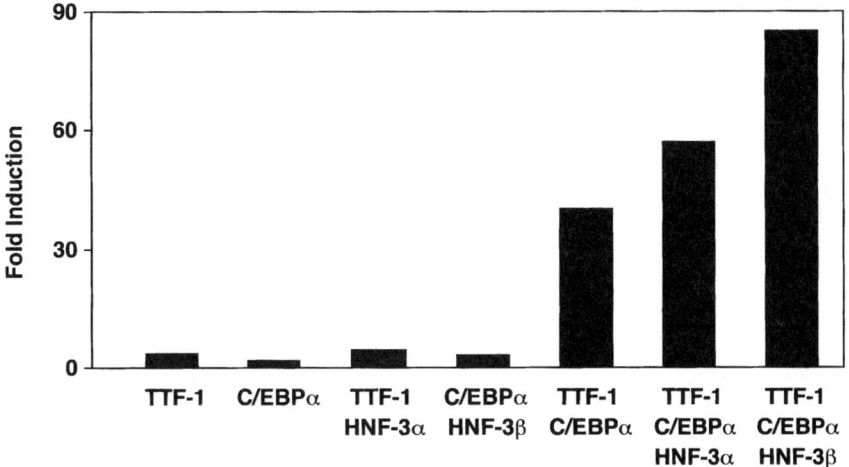

FIGURE 2. Cotransfection studies, using the respective insect expression vectors together with a reporter plasmid containing the mouse proximal CCSP promoter sequence. Assays were carried out in duplicate.

actions between transcription factors. In transient transfection studies with a 172-bp CCSP promoter fragment, together with C/EBPα or TTF-1 expression plasmid, the maximal induction of the reporter gene was 1.5- and 3.5-fold, respectively. However, when TTF-1 and C/EBPα expression plasmids were transfected together, a strong synergistic activity resulting in a 53-fold induction of the reporter gene was detected (FIG. 1). Previously, important TTF-1 sites in the CCSP promoter have been shown to reside more distally in the promoter. To investigate whether further synergism could be detected in a longer fragment of the CCSP promoter, transient transfections were performed with a 2.1-kb promoter fragment. Although the induction by TTF-1 alone was higher, no further synergistic activity was observed (FIG. 1). This suggests that the synergistic activity between C/EBPα and TTF-1 occurs in the proximal promoter. The strong synergistic activity that was detected between C/EBPα and TTF-1 was not observed between C/EBPα and HNF-3 or between TTF-1 and HNF-3. However, the synergistic effect of C/EBPα and TTF-1 on the CCSP promoter was further increased when HNF-3 expression plasmid was added in the transfection assay (FIG. 2).

In summary, this study suggests that the onset of C/EBPα expression in the developing lung is imperative for the high-level expression of CCSP. Furthermore, C/EBPα is acting in concert with TTF-1 and HNF-3 in the regulation of CCSP. This study supports a critical role for C/EBP factors in the control of lung epithelial differentiation and suggests that the appearance of C/EBPα in the bronchiolar epithelium is a critical event in the differentiation of this part of the lung.

REFERENCES

1. NORD, M., O. ANDERSSON, M. BRÖNNEGÅRD & J. LUND. 1992. Rat lung polychlorinated biphenyl–binding protein: effect of glucocorticoids on the expression of the Clara cell–specific protein during fetal development. Arch. Biochem. Biophys. **296:** 302–307.
2. SAWAYA, P.L., B.R. STRIPP, J.A. WHITSETT & D.S. LUSE. 1993. The lung-specific CC10 gene is regulated by transcription factors from the AP-1, octamer, and hepatocyte nuclear factor-3 families. Mol. Cell. Biol. **13:** 3860–3871.
3. BINGLE, C.D. & J.D. GITLIN. 1993. Identification of hepatocyte nuclear factor-3 binding sites in the Clara cell secretory protein gene. Biochem. J. **295:** 227–232.
4. RAY, M.K., C.Y. CHEN, R.J. SCHWARTZ & F.J. DEMAYO. 1996. Transcriptional regulation of a mouse Clara cell–specific protein (mCC10) gene by the NKx transcription factor family members thyroid transcription factor 1 and cardiac muscle–specific homeobox protein (CSX). Mol. Cell. Biol. **16:** 2056–2064.
5. LI, F., E. ROSENBERG, C.I. SMITH et al. 1995. Correlation of expression of transcription factor C/EBP alpha and surfactant protein genes in lung cells. Am. J. Physiol. **269:** L241–L247.
6. CASSEL, T.N., L. NORDLUND-MÖLLER, O. ANDERSSON et al. 2000. C/EBP alpha and C/EBP delta activate the Clara cell secretory protein gene through interaction with two adjacent C/EBP-binding sites. Am. J. Respir. Cell Mol. Biol. **22:** 469–480.

Human Uteroglobin Gene Polymorphisms and Genetic Susceptibility to Asthma

MOONSUK CHOI,[a] ZHONGJIAN ZHANG,[a] LEO P. TEN KATE,[b]
J. MARGRIET COLLEE,[b] J. GERRITSEN,[c] AND ANIL B. MUKHERJEE[a]

[a]*Section on Developmental Genetics, Heritable Disorders Branch, National Institute of Child Health and Human Development, National Institutes of Health, Bethesda, Maryland 20892-1830, USA*

[b]*Department of Clinical and Human Genetics, Vrije Universiteit Medical Center, Amsterdam, the Netherlands*

[c]*Department of Pediatric Pulmonology, Beatrix Children's Hospital, Groningen, the Netherlands*

Asthma is an inflammatory disease of the respiratory system. Compelling evidence suggests the involvement of genetic factors in predisposing individuals to this disease. To date, several chromosomal loci harboring putative asthma susceptibility genes have been mapped. Notable among these are the ones mapped on chromosome 5q31-33 and 11q13. The gene encoding human uteroglobin (hUG), a steroid-inducible, secretory protein with anti-inflammatory/immunomodulatory properties, is mapped to chromosome 11q12.2-13.1. It is constitutively expressed at a high level by the mucosal epithelial cells of the respiratory tract. We previously reported a polymorphism (38 A→G) in the noncoding region of the hUG exon 1. In the present study, we sought to determine if there is an association between polymorphisms in the hUG gene and susceptibility to asthma. Accordingly, we analyzed the hUG gene for polymorphisms in individuals with asthma and in controls without the disease. Our results show that there is a significant increase in the +38G allele frequency of A38G polymorphism in the asthma patients versus the control group. Moreover, we identified a novel polymorphism, (GTTT)m STR, adjacent to the (ATTT)n STR region at around −3100 bp from the transcription start point of the hUG gene. We found a significant difference ($p = 0.007$) in the haplotype frequencies of the polymorphic regions between the asthmatic and control groups. This may be primarily due to a significant difference of the allele frequencies within the (GTTT)m region.

Uteroglobin (UG) is a multifunctional secreted protein with anti-inflammatory and immunomodulatory properties (for review, see reference 1). This protein is constitutively expressed at a high level in the epithelial lining of the respiratory tract and in nonciliated Clara cells. It has been demonstrated in the rabbit that the pulmonary UG is the major source of its counterpart in the blood.[2] Recently, it has been reported that, compared to normal controls, plasma hUG levels are reduced in patients with asthma, an inflammatory disease of the airways. During the past decade, several laboratories have reported the existence of candidate asthma susceptibility genes on human chromosomes. Most investigations have focused on identifying candidate genes on chromosome 5q31-33[3,4] and 11q13[5–7] either by linkage analysis or by association studies. Human chromosome 11q13 is also the region in which the hUG gene is mapped.[8,9] Because of its anti-inflammatory/immunomodulatory properties, we sought to determine if a genetic association exists between hUG gene polymorphism(s) and asthma susceptibility in 41 asthmatic children from 22 different families and 34 nonasthmatic controls. Our results appear to indicate that there is an association between A38G and (GTTT)m STR polymorphisms in the asthmatic patients compared with nonasthmatic controls.

MATERIALS AND METHODS

Sample Composition

Genomic DNA samples from 41 well-characterized patients with asthma from the Netherlands[10] were included in the study. Genomic DNAs from 34 unrelated individuals in the DNA Polymorphism Discovery Resource from Coriell Cell Repositories (Camden, NJ) served as controls.

Detection and Genotyping of Single Nucleotide Polymorphism (SNP)

The promoter region (−97 to +54) and the 5′-flanking region (−265 to −97) of the hUG gene were amplified by PCR. Electrophoresis of the amplified products was carried out on 0.5× MDE gel for 5 h at 2 W in a cold room for SSCP analysis. The amplified products were also sequenced using Thermo Sequenase Radiolabeled Terminator Cycle Sequencing Kit (USB Corporation, OH) following the manufacturer's manual. The genotypes of amplified products were detected by digestion with *Sau* 96I restriction endonuclease enzyme for 2 h at 37°C and the electrophoresis was carried out on 2% agarose gel for 2 h at 100 V.

Detection and Genotyping of Short Tandem Repeats (STRs) Region

The region at around −3100 bp from the transcription start point of the hUG gene was amplified. The amplified products were sequenced using Thermo Sequenase Radiolabeled Terminator Cycle Sequencing Kit (USB Corporation, OH). The (GTTT)m region and (ATTT)n region were amplified separately. About 0.5 μL of amplified products was loaded on 6% or 10% polyacrylamide gel with 5× loading buffer at constant 150–200 V for 15–17 h. The bands were detected by silver staining.

Statistical Analysis

The allele and genotype frequencies of the asthma group and controls were calculated. The allele frequencies of each group were compared by means of the chi-square test. Values of $p < 0.05$ are considered significant.

RESULTS

Allele Frequencies of SNP

We detected a single nucleotide polymorphism (SNP) at the +38 site in the noncoding region of exon 1 by PCR/*Sau* 96I restriction endonuclease digestion. The amplified product harboring the +38G substitution could be digested by *Sau* 96I restriction endonuclease, showing 149-bp and 41-bp fragments. If there is no +38G, the amplified product could not be digested by *Sau* 96I, showing a 190-bp fragment.

Allele Frequencies of Two Tetranucleotide STR Polymorphisms (STRPs)

It has been reported that there was a polymorphic (ATTT)n repeated sequence at around −3100 bp from the transcription start point of the hUG gene.[11] During the sequencing of the (ATTT)n region, in addition to this polymorphism, we identified a novel (GTTT)m repeated sequence adjacent to the (ATTT)n region. Genotyping of these two short tandem repeat (STR) sites was carried out by the amplified fragment length polymorphism (Amp-FLP) and sequencing. There were 5 alleles representing the 4–9 repeated numbers of the (GTTT)m site. The common allele was the 4 repeated number, showing more than that of 0.5 in frequency. There were 6 alleles representing the 2–9 repeated numbers of the (ATTT)n site. From sequencing and genotyping data, we detected the haplotypes of the (GTTT)m(ATTT)n polymorphic site. There were 9 haplotypes in the asthma group and 11 haplotypes in the controls.

DISCUSSION

Human uteroglobin, also known as Clara cell 10 kDa, is a low-molecular-weight, homodimeric, secreted protein.[1,8,10] The hUG gene comprises 3 exons and 2 introns.[8,10] We have previously reported a 38 A→G polymorphism in the noncoding region of exon 1 of the hUG gene.[8] In the present study, we found that there is a significant ($p = 0.029$) increase in the frequency of the G allele of the 38 A→G polymorphism in the asthma group versus the control group. We identified an additional polymorphism, (GTTT)m, in the 5′-flanking region of the hUG gene. When we analyzed this as well as the (ATTT)n polymorphic region, we uncovered a significant difference ($p = 0.047$) in the allele frequencies of the (GTTT)n polymorphism, whereas there was no significant difference ($p = 0.983$) in the allele frequency of the (ATTT)n polymorphism. Moreover, we found a significant difference ($p = 0.007$) in the haplotype frequency of the asthma group versus controls. This is most likely due to a significant difference in the polymorphic (GTTT)m region. These results raise the possibility that hUG gene polymorphisms [38 A→G and (GTTT)m] may be a predisposing factor for airway inflammation that leads to a susceptibility to asthma.

Recently, Szelestei et al.[12] have reported that there is an association between the 38 A→G polymorphism in the hUG gene and disease progression in immunoglobulin A (IgA)–nephropathy (IgAN) patients. We have reported that, in mice, UG deficiency caused by either targeted disruption of the UG gene or expression of UG antisense-RNA leads to the development of IgAN.[13,14] It has also been reported that inflammation in the pulmonary system, the major source of circulating UG, caused abnormal deposition of IgA in the glomeruli.[15] These data suggest a link between the levels of circulating UG, pulmonary inflammation, and IgA deposition in the renal glomeruli. Moreover, Shijubo et al.[16] have recently reported that serum UG levels are lower in asthma patients compared with nonasthmatic controls. Whether the UG polymorphisms and the lower levels of UG in the serum of asthmatic patients have a cause and effect relationship is not yet clear. In contrast to our results, it should be noted that Liang et al.[17] have suggested that the risk of asthma is increased

6.9-fold in individuals with the 38AA allele rather than 38GG. The reason for this discrepancy is not clear at this time. Thus, our results underscore the necessity of further studies with a larger asthma patient population to determine whether 38 A→G and (GTTT)m polymorphisms correlate with disease susceptibility and/or severity and whether the A or G allele positively correlates with susceptibility to asthma.

REFERENCES

1. MUKHERJEE, A.B., G.C. KUNDU, G. MANTILE-SELVAGGI et al. 1999. Uteroglobin: a novel cytokine? Cell. Mol. Life Sci. **55:** 771–787.
2. KIKUKAWA, T. & A.B. MUKHERJEE. 1989. Partial characterization of a uteroglobin-like phospholipase A2 inhibitory protein in the circulation of rabbits. Mol. Cell. Endocrinol. **62:** 177–187.
3. MARSH, D.G., J.D. NEELY, D.R. BREAZEALE et al. 1994. Linkage analysis of IL-4 and other chromosome 5q31.1 markers and total serum immunoglobin E concentrations. Science **264:** 1162–1166.
4. POSTMA, D.S., E.R. BLEECKER, P.J. AMELUNG et al. 1995. Genetic susceptibility to asthma-bronchial hyperresponsiveness coinherited with a major gene for atopy. N. Engl. J. Med. **333:** 894–900.
5. COOKSON, W.O., P.A. SHARP, J.A. FAUX & J.M. HOPKIN. 1989. Linkage between immunoglobulin E responses underlying asthma and rhinitis and chromosome 11q. Lancet **1(8650):** 1292–1295.
6. SANFORD, A.J., T. SHIRAKAWA, M.F. MOFFATT et al. 1993. Location of atopy and β subunit of high-affinity IgE receptor (FceRI) on chromosome 11q. Lancet **341:** 332–334.
7. STAFFORD, A.N., S.H. RIDER, J.M. HOPKIN et al. 1994. A 2.8 Mb YAC contig in 11q12-q13 localizes candidate genes for atopy: FceRI-β and CD20. Hum. Mol. Genet. **3:** 770–785.
8. ZHANG, Z., D.B. ZIMONJIC, C.P. NICHOLAS et al. 1997. Human uteroglobin gene: structure, subchromosomal localization, and polymorphism. DNA Cell Biol. **16(1):** 73–83.
9. COLLEE, J.M., L.P. TEN KATE, H.G. DEVRIES et al. 1993. Allele sharing on chromosome 11q13 in sibs with asthma and atopy. Lancet **242:** 936.
10. HAY, J.G., C. DANEL, C.S. CHU & T.G. CRYSTAL. 1995. Human CC10 gene expression in airway epithelium and subchromosomal locus suggest linkage to airway disease. Am. J. Physiol. **268:** L565–L575.
11. STÖHR, H. & B.H.F. WEBER. 1994. (ATTT)n-tetranucleotide repeat polymorphism in the 5′-flanking region of the UGB gene. Hum. Mol. Genet. **3:** 2086.
12. SZELESTEI, T., S. BAHRING, T. KOVACS et al. 2000. Association of uteroglobin polymorphism with rate of progression in patients with IgA nephropathy. Am. J. Kidney Dis. **36:** 468–473.
13. ZHANG, Z., G.C. KUNDU, C.Y. YUAN et al. 1997. Severe fibronectin-deposit renal glomerular disease in mice lacking uteroglobin. Science **276:** 1408–1412.
14. ZHENG, F., G.C. KUNDU, Z. ZHANG et al. 1999. An essential role of uteroglobin in preventing IgA-nephropathy. Nat. Med. **5:** 1018–1025.
15. WALDHERR, R., M. RAMBAUSEK, W.D. DUNCKER & E. RITZ. 1989. Frequency of mesangial IgA deposits in a non-selected autopsy series. Nephrol. Dial. Transplant. **4(11):** 943–946.
16. SHIJUBO, N., Y. ITOH, T. YAMAGUCHI et al. 1999. Serum levels of Clara cell 10-kDa protein are decreased in patients with asthma. Lung **177:** 45–52.
17. LIANG, I.A., J. GOLDBLATT, E. EBER et al. 1998. A polymorphism of the CC16 gene is associated with an increased risk of asthma. J. Med. Genet. **35(6):** 463–467.

Amino Acid Residues in α-Helix-3 of Human Uteroglobin Are Critical for Its Phospholipase A_2 Inhibitory Activity

BHABADEB CHOWDHURY,[a] GIUDITTA MANTILE-SELVAGGI,[a,b] GOPAL C. KUNDU,[c] LUCIO MIELE,[d] ELEONORA CORDELLA-MIELE,[e] ZHONGJIAN ZHANG, AND ANIL B. MUKHERJEE

Section on Developmental Genetics, Human Genetics Branch, National Institute of Child Health and Human Development, National Institutes of Health, Bethesda, Maryland 20892-1830, USA

Rabbit blastokinin[1] or uteroglobin (UG)[2] is a steroid-dependent, multifunctional, secreted protein.[3,4] This protein was first discovered in the rabbit uterus during early pregnancy and subsequently found in many other nonreproductive organs. Interestingly, the highest level of UG expression occurs in the rabbit[1,2] and human endometrium[5] during the progesterone-dominated phase of the ovarian menstrual cycle, when implantation of the embryo in the uterus takes place. This protein is also expressed in many extrauterine tissues, including the thymus, pituitary gland, lungs, gastrointestinal tract, pancreas, mammary gland, prostate, and seminal vesicle.[6] UG is also present in the blood[7,8] and urine,[9] although it is not synthesized in the kidneys. Currently, this protein is known by several names, which are primarily derived from the organ or body fluid in which it is detectable or from the type of xenobiotics with which it interacts. Thus, it is called progesterone-binding protein,[10] Clara cell 10-kDa protein,[11] urine protein-1,[9] polychlorinated biphenyl–binding protein,[12] and retinol-binding protein.[13] A unified nomenclature is being developed (see Nomenclature Committee Report in this volume). Structurally, UG is a homodimer in which the two identical 70-amino-acid subunits are covalently linked in an antiparallel orientation by two interchain disulfide bonds.[14–20] Each monomer consists of four α-helices and one β-turn between α-helix-2 and -3, but there is no β-structure. Recently, we and others have identified a high-affinity binding site (putative receptor) of UG on several cell types.[21–23] Through this pathway, UG appears to inhibit cellular motility and invasion of the extracellular matrix,[21] suggesting an antichemokine-like property of this protein. Interestingly, human UG (hUG) is encoded by a single copy gene located on chromosome 11q12.3-13.1,[24] a region in which a number of candidate disease genes have been mapped by linkage analyses.

---

[a]Equal contributions from these two main authors.
[b]Present address: Hackensack University Medical Center, 30 Prospect Avenue, Hackensack, NJ 07601.
[c]Present address: National Center for Cell Science, NCCS Complex, Pune 411007, India.
[d]Present address: Cardinal Bernadin Cancer Center and Department of Pathology, Loyola University Medical Center, 2160 South First Avenue, Maywood, IL 60153.
[e]Present address: Department of Anesthesiology, Hindsdale Hospital, Hindsdale, IL 60521.

Several years ago, we reported the presence of a UG-like protein in the human uterus, lungs, and prostate[3] and, subsequently, the human UG cDNA[25] and the gene[24,26] were characterized. Additionally, cloning of the cDNAs and genes encoding UG in the mouse,[27,28] rat,[29] and hamster[30] has demonstrated a remarkable similarity in the UG gene structure of various species. Moreover, the results of immunohistochemical analyses suggest that a protein immunoreactive to UG antibody is expressed in the mucosal epithelial cells of virtually all vertebrate classes,[24] raising the possibility that this protein plays important physiological roles. Both natural and recombinant human UG have an anomalous electrophoretic mobility in SDS-PAGE that is consistent with that of a 10-kDa protein, while their calculated molecular mass is 17.8 kDa. The availability of substantial quantities of recombinant rabbit[31] and human[32] UGs allowed the comparison of their structural features as determined by X-ray crystallography.[20] The results of this and other studies[33] showed that rabbit and human UGs are indistinguishable proteins both structurally and functionally,[32] although there is only 68% protein sequence similarity between them.

We have previously demonstrated that both natural and recombinant UG are potent inhibitors of extracellular (soluble) phospholipase A_2 (sPLA_2) activity.[34,35] The PLA_2 inhibitory and anti-inflammatory activities appear to reside in a nonapeptide (amino acid residues 39–47) region of the α-helix-3 of UG.[35,36] The present investigation was carried out in order to understand whether site-specific mutagenesis of the α-helix-3 of UG has any effect on its PLA_2 inhibitory activity. Our results show that site-specific mutation of Lys 43 → Asp, Asp 46 → Lys, or both mutated in the same protein can cause a virtual lack of its PLA_2 inhibitory activity.

MATERIALS AND METHODS

Reagents and Chemicals

Porcine pancreatic PLA_2 and arachidonic acid were purchased from Boehringer. Both [^{14}C]-phosphatidylcholine and [^{14}C]-arachidonic acid were purchased from Amersham Pharmacia Biotech. Silica Gel–prechanneled thin-layer chromatography (TLC) plates, 250 μm, were from Analtech. All other reagents were of analytical grade from Sigma (St. Louis, MO), Fisher (Pittsburgh, PA), or Amersham (Piscataway, NJ).

Site-Specific Mutagenesis of hUG

Site-specific mutagenesis was carried out by replacement of segments of hUG-cDNA with synthetic, double-stranded oligonucleotides. The detailed methodology for mutagenesis will be reported elsewhere. The mutant cDNAs, thus generated, code for the following amino acid replacements compared to wild-type CC10: Lys 43 → Asp (MH1); Asp 46 → Lys (MH2); and the double replacement of Lys 43 → Asp and Asp 46 → Lys in the same construct (MH3). The nucleotide sequences of the inserts of mutant plasmids were determined to rule out any additional mutations or other cloning. Mutant hCC10 proteins were purified as previously described for wild-type rhCC10.[32]

Purification of Wild-Type and Mutated Proteins

Wild-type and mutated proteins were purified as previously described for wild-type hUG.[32] Protein was estimated by using the Bradford assay.[37]

PLA_2 Assay

PLA_2 assays were performed according to the method of Levin et al.[34]

RESULTS AND DISCUSSION

The purification of human recombinant mutated hUG from E. coli transfected with pGEL101 was accomplished following the method previously reported for wild-type recombinant hUG.[32] As expected, the chromatographic behavior of these proteins in the ionic exchange step was slightly different from that of the wild-type human recombinant CC10. In our experience, it appeared critical that the protein be extensively dialyzed against distilled water before the ion exchange step is carried out to ensure reproducible binding of proteins to the Econopak-S matrix. These proteins were soluble and were quantitatively recovered from the supernatant after bacterial lysis. After the last gel filtration step, the mutant proteins appear homogeneous, as judged by the appearance of a homogeneous protein band in overloaded, silver-stained SDS-polyacrylamide gels. The purified material reacted with hUG antibodies and was always accompanied by a small amount of undissociated dimer migrating around 10 kDa, which can be detected by Western blotting. The yield of the purified mutant protein was virtually identical to that found for the wild-type recombinant hUG.[32]

One of the important biochemical properties of UG is its ability to inhibit the activity of PLA_2.[31,32,34,35] Thus, in the present study, we investigated the possible PLA_2 inhibitory activity of recombinant mutated hUG using a previously established, phosphatidylcholine/deoxycholate mixed micellar assay.[32,34] We used these purified mutated proteins to determine the PLA_2 inhibitory activity. We evaluated the concentration and time dependence of enzyme activity in addition to the effects of detergent concentrations. In our assay conditions, we found that the optimum deoxycholate concentration is 1.5 mM. Under the conditions used, a maximum inhibitory activity of about 61% was observed above 2 nM of wild-type UG, whereas for the mutated proteins only 8–10% inhibition was achieved. Nonspecific proteins (lysozyme or BSA), used at concentrations identical to that of recombinant hUG (1 to 10 nM), failed to cause any detectable inhibition of PLA_2 activity. Our results show that both Lys 43 → Asp and Asp 46 → Lys mutations abrogate the PLA_2 inhibitory activity of hUG. We have demonstrated that recombinant hUG, like rabbit UG, is a potent inhibitor of sPLA$_2$ activity in vitro. This inhibition of PLA_2 activity is specific and dose-dependent. Furthermore, our demonstration of specific residues in the α-helix-3 of hUG abolishing its PLA_2 inhibitory activity establishes our previous observation that this region of hUG is at least partially responsible for this activity of UG.[35] We conclude that the Lys 43 and Asp 46, corresponding to an epitope in α-helix-3 (which is exposed to the solvent and has potential for interaction with other molecules), are critical residues responsible for the PLA_2 inhibitory activity in intact UG.

REFERENCES

1. KRISHNAN, R.S. & J.C. DANIEL, JR. 1967. Blastokinin: inducer and regulator of blastocyst development in the rabbit uterus. Science **158**: 490–492.
2. BEIER, H.M. 1968. Uteroglobin: a hormone sensitive endometrial protein involved in blastocyst development. Biochim. Biophys. Acta **160**: 289–291.
3. MUKHERJEE, A.B., G.C. KUNDU, G. MANTILE-SELVAGGI et al. 1999. Uteroglobin: a novel cytokine? Cell. Mol. Life Sci. **55**: 771–787.
4. MIELE, L., E. CORDELLA-MIELE, G. MANTILE et al. 1994. Uteroglobin and uteroglobin-like proteins: the uteroglobin family of proteins. J. Endocrinol. Invest. **17**: 679–692.
5. PERI, A., B.D. COWAN, D. BHARTIYA et al. 1994. Expression of Clara cell 10-kDa gene in the human endometrium and its relationship to ovarian menstrual cycle. DNA Cell Biol. **13**: 495–503.
6. PERI, A., E. CORDELLA-MIELE, L. MIELE et al. 1993. Tissue-specific expression of the gene coding for human Clara cell 10-kD protein, a phospholipase A2–inhibitory protein. J. Clin. Invest. **92**: 2099–2109.
7. KIKUKAWA, T. & A.B. MUKHERJEE. 1989. Detection of a uteroglobin-like phospholipase A2 inhibitory protein in the circulation of rabbits. Mol. Cell. Endocrinol. **62**: 177–187.
8. AOKI, A., H.A. PASOLI, M. RAIDA et al. 1996. Isolation of human uteroglobin from blood filtrate. Mol. Hum. Reprod. **2**: 489–497.
9. JACKSON, P.J., R. TURNER, J.N. KEEN et al. 1988. Purification and partial amino acid sequence of human urine protein 1: evidence for homology with rabbit uteroglobin. J. Chromatogr. **452**: 359–367.
10. BEATO, M. 1976. Binding of steroids to uteroglobin. J. Steroid Biochem. **7**: 327–334.
11. SINGH, G., J. SINGH, S.L. KATYAL et al. 1988. Identification, cellular localization, isolation, and characterization of human Clara cell-specific 10 kd protein. J. Histochem. Cytochem. **36**: 73–80.
12. GILLNER, M., J. LUND, C. CAMBILLAU et al. 1988. The binding of methylsulfonyl-polychloro-biphenyls to uteroglobin. J. Steroid Biochem. **31**: 27–33.
13. LOPEZ DE HARO, M.S., M. PEREZ MARTINEZ, C. GARCIA et al. 1994. Binding of retinoids to uteroglobin. FEBS Lett. **349**: 249–251.
14. MORNON, J.P., R. BALLY, F. FRIDLANSKY et al. 1979. Characterization of two new crystal forms of uteroglobin. J. Mol. Biol. **127**: 237–239.
15. BUEHNER, M. & M. BEATO. 1978. Crystallization and preliminary crystallographic data of rabbit uteroglobin. J. Mol. Biol. **120**: 337–341.
16. NIETO, A., H. PONSTINGL & M. BEATO. 1977. Purification and quaternary structure of the hormonally induced protein uteroglobin. Arch. Biochem. Biophys. **180**: 82–92.
17. ATGER, M., J.C. MERCIER, G. HAZE et al. 1979. N-terminal sequences of uteroglobin and its precursor. Biochem. J. **177**: 985–988.
18. BUEHNER, M., A. LIFCHITZ, R. BALLY et al. 1982. Use of molecular replacement in the structure determination of the P21212 and the P21 (pseudo P21212) crystal forms of oxidized uteroglobin. J. Mol. Biol. **159**: 353–358.
19. MORIZE, I., E. SURCOUF, M.C. VANEY et al. 1987. Refinement of the C222 (1) crystal form of oxidized uteroglobin at 1.34 Å resolution. J. Mol. Biol. **194**: 725–739.
20. MATTHEWS, J.H., N. PATTABIRAMAN, K.B. WARD et al. 1994. Crystallization and characterization of the recombinant human Clara cell 10-kDa protein. Proteins **20**: 191–196.
21. KUNDU, G.C., G. MANTILE, L. MIELE et al. 1996. Recombinant human uteroglobin suppresses cellular invasiveness via a novel class of high-affinity cell surface binding site. Proc. Natl. Acad. Sci. U.S.A. **93**: 2915–2919.
22. KUNDU, G.C., A.K. MANDAL, Z. ZHANG et al. 1998. Uteroglobin (UG) suppresses extracellular matrix invasion by normal and cancer cells that express the high-affinity cell UG-binding proteins. J. Biol. Chem. **273**: 22819–22824.
23. GONZALEZ, K.D. & A. NIETO. 1995. Binding of uteroglobin to microsomes and plasmatic membranes. FEBS Lett. **361**: 255–258.
24. ZHANG, Z., D.B. ZIMONJIC, N.C. POPESCU et al. 1997. Human uteroglobin gene: structure, subchromosomal localization, and polymorphism. DNA Cell Biol. **16**: 73–83.

25. SINGH, G., S.L. KATYAL, W.E. BROWN et al. 1988. Amino-acid and cDNA nucleotide sequences of human Clara cell 10-kDa protein. Biochim. Biophys. Acta **950:** 329–337.
26. HAY, J.G., C. DANEL, C.S. CHU et al. 1995. Human CC10 gene expression in airway epithelium and subchromosomal locus suggest linkage to airway disease. Am. J. Physiol. **268:** L565–L575.
27. RAY, M.K., S. MAGDALENO, B.W. O'MALLEY et al. 1993. Cloning and characterization of the mouse Clara cell specific 10-kDa protein gene: comparison of the 5′-flanking region with the human, rat, and rabbit gene. Biochem. Biophys. Res. Commun. **197:** 163–171.
28. SINGH, G., S.L. KATYAL, W.E. BROWN et al. 1993. Mouse Clara cell 10-kDa (CC10) protein: cDNA nucleotide sequence and molecular basis for the variation in progesterone binding of CC10 from different species. Exp. Lung Res. **19:** 67–75.
29. HAGEN, G., M. WOLF, S.L. KATYAL et al. 1990. Tissue-specific expression, hormonal regulation, and 5′-flanking gene region of the rat Clara cell 10-kDa protein: comparison to rabbit uteroglobin. Nucleic Acids Res. **18:** 2939–2946.
30. DOMINGUEZ, P. 1995. Cloning of a Syrian hamster cDNA related to sexual dimorphism: establishment of a new family of proteins. FEBS Lett. **376:** 257–261.
31. MIELE, L., E. CORDELLA-MIELE & A.B. MUKHERJEE. 1990. High level bacterial expression of uteroglobin, a dimeric eukaryotic protein with two interchain disulfide bridges, in its natural quaternary structure. J. Biol. Chem. **265:** 6427–6435.
32. MANTILE, G., L. MIELE, E. CORDELLA-MIELE et al. 1993. Human Clara cell 10-kDa protein is the counterpart of rabbit uteroglobin. J. Biol. Chem. **268:** 20343–20351.
33. WOLF, M., J. KLUG, R. HACKENBERG et al. 1992. Human CC10, the homologue of rabbit uteroglobin: genomic cloning, chromosomal localization, and expression in endometrial cell lines. Hum. Mol. Genet. **1:** 371–378.
34. LEVIN, S.W., J.D. BUTLER, U.K. SCHUMACHER et al. 1986. Uteroglobin inhibits phospholipase A2 activity. Life Sci. **38:** 1813–1819.
35. MIELE, L., E. CORDELLA-MIELE, A. FACCHIANO et al. 1988. Novel anti-inflammatory peptides from the region of highest similarity between uteroglobin and lipocortin I. Nature **335:** 726–730.
36. CAMUSSI, G., C. TETTA, F. BUSSOLINO et al. 1990. Anti-inflammatory peptides (antiflammins) inhibit synthesis of platelet-activating factor, neutrophil aggregation and chemotaxis, and an intradermal inflammatory reaction. J. Exp. Med. **171:** 913–927.
37. BRADFORD, M.M. 1976. A rapid and sensitive method for the quantitation of microgram quantities of protein utilizing the principle of protein-dye binding. Anal. Biochem. **72:** 248–254.

Mammaglobin Complexes with BU101 in Breast Tissue

T. L. COLPITTS,[a,b] P. BILLING,[b] E. GRANADOS,[b] S. HODGES,[b] N. MENHART,[c] J. RUSSELL,[b] AND S. STROUPE[b]

[b]*Breast Cancer Venture, Abbott Park, Illinois 60064-6015, USA*

[c]*Department of Chemistry, Loyola University, Chicago, Illinois 60626, USA*

We queried the Lifeseq EST database (Incyte Genomics) with the known human uteroglobin sequences to investigate their tissue specificity. The database comprises over 1100 libraries prepared from well-characterized tissues, both diseased and normal. The libraries have significant depth, averaging 4000 ESTs of greater than 200 bases each. The database provides an electronic link between the tissue, its pathology, and the profile of expressed sequences generated from that tissue. One can query the database with a sequence and retrieve the tissues in which the sequence was found. The human sequences queried included mammaglobin,[1] mammaglobin B,[2] BU101,[3] ESSBF I,[4] and lipophilin A.[5] TABLE 1 presents the abundance (number of ESTs) found in the tissue that encodes the sequence of interest.

As other investigators have observed,[1,5] mammaglobin is very specific for mammary tissue (included under exocrine tissue). Mammaglobin B does not present the same profile. It is more specific for tissues of the female reproductive system, specifically the uterus. BU101 is present in both tissue types, exocrine (breast) and reproductive (uterus), whereas ESSBF I is present in only the reproductive tissues. Lipophilin A was not found in the database, which may indicate its lower expression level or a difficulty in generating ESTs from the message. For comparison, CC10 (Clara cell 10-kDa protein) was present only in respiratory tissue.

This tissue distribution profile of the uteroglobin member sequences provides a platform for proposing potential partners in protein complex formation. In breast tissue, Mam and BU101 are expressed at measurable levels, unlike MamB, which is barely detected. The presence of MamB in breast tissue is suspect until abundance levels rise to include multiple specimens of the same tissue type. Based on known uteroglobin protein structures, the potential is there for Mam and BU101 to be involved in complex formation in breast tissue. In uterine tissue, BU101 may partner with MamB and/or ESSBF I.

The rat prostatein complex provides an interesting model[6] for these human sequences. BU101, ESSBF I, and lipophilin A are closely related to the C1 and C2 chains, which appear only once in the rat heterotetramer. Mam and MamB, however, are more closely related to the C3 chain, which appears twice in the heterotetramer.

[a]Address for correspondence: Tracey Colpitts, Breast Cancer Venture, Department 90M, AP20, 100 Abbott Park Road, Abbott Park, IL 60064-6015. Voice: 847-935-0139; fax: 847-938-7996.

tracey.colpitts@add.ssw.abbott.com

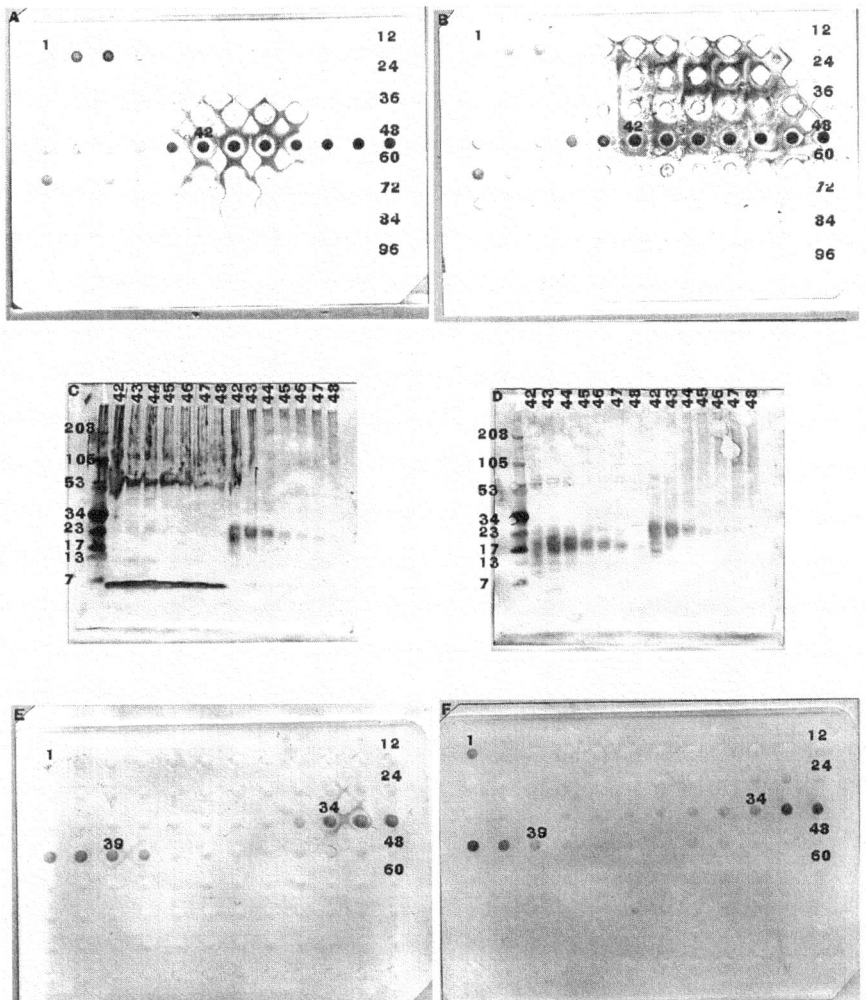

FIGURE 1. Purification of breast tumor extract. Breast tumor tissue extract was purified by ion-exchange chromatography as described in the text. The elution was accomplished using a salt gradient, from 0 mM NaCl, at fraction 30, to 1000 mM, at fraction 80. Samples from the fractions were blotted onto membranes and detected by dot blot analysis using anti-BU101 (**A**) or anti-Mam (**B**) antibodies. Positive reactions for each antibody were observed to coelute in fractions 40–48, approximately 200 mM. To further confirm the identities of the proteins that caused these positive reactions, SDS-PAGE was performed on fractions 42–48, followed again by anti-BU101 (**C**) or anti-Mam (**D**) Western blot analysis. This was done under reducing (lanes 2–8) and nonreducing (lanes 9–15) conditions. In the fully reduced lanes, bands of appropriate molecular weight were observed in each case: approximately 7 kDa for the anti-BU101-specific blot and approximately 18 kDa for Mam. For the nonreduced samples, a band of approximately 24 kDa was observed, indicating that a disulfide-linked complex is involved. To further demonstrate that these proteins formed a complex, an additional molecular sieve purification step was performed. Again, proteins detected by anti-BU101 (**E**) or anti-Mam (**F**) antibodies copurified, eluting in fractions 34–39, with approximate molecular mass of 45 kDa.

TABLE 1. Human uteroglobins: names and tissue distribution

	Mam SBP C2	MamB ESSBF III Lipophilin C	ESSBF II Lipophilin B SPB C1	ESSBF I	Lipophilin A	Institution Wash U HGS UCLA Incyte
	Mammaglobin	EU250	BU101	TU104		Abbott
Tissue type						
Exocrine	167	1	40	0	0	
Genit. Male	0	1	3	0	0	
Genit. Female	0	8	7	4	0	
Respiratory	0	1	0	0	0	
Parotid	0	0	0	0	0	

NOTE: Number of ESTs found in the indicated tissue source in the Lifeseq database. Tissue sources with only one or two hits in the same library for only one marker are not included. Generally, multiple libraries are shown for those markers with >1 EST in a tissue.

By analogy, one could envision a complex comprising BU101-Mam, specific for breast tissue, and BU101-MamB and/or ESSBF I–MamB, specific for uterine tissue.

Experimentally, the presence of Mam and BU101 was confirmed in breast tissue and their expression levels correlated at both the mRNA and protein levels. Ribonuclease protection assays were carried out on 10 breast tissues for both sequences. Mam and BU101 mRNA levels in the 5 normal tissues ranged between 0.2 and 20 molecules/cell. Two of the cancer specimens had undetectable levels of MAM and BU101 mRNA, and 2 cancer specimens were upregulated. Western blots of 9 breast tissues probed for each sequence clearly showed Mam and BU101 protein expression at similar levels. BU101 has a predicted MW of 7.7 kDa and the Western blot shows the corresponding species observed in tissue. Mam is similarly small, with a predicted MW of 8.4 kDa, but it has two potential N-linked glycosylation sites, which undergo processing and result in a heterogeneous pattern. Mam has an observed molecular weight spanning ~9 to 18 kDa.

The presence of a complex comprising BU101 and Mam could best be shown by isolating the species from human tissue. Breast tissue was homogenized and fractionated by ion-exchange chromatography, specifically a Mono Q column run with 20 mM piperazine buffer, pH 6.0. These conditions were selected based upon the calculated isoelectric points of the individual sequences. The mature sequence of Mam has a pI of 3.8 and a net charge of −10 at pH 6.0. The mature sequence of BU101 has a pI of 8.4 and a net charge of +2 at pH 6.0. Under these conditions, the sequences would be expected to separate during chromatography and would be detected when probed with specific antisera. Each fraction from the chromatography run was analyzed by dot blot with either anti-BU101 polyclonal antisera (FIG. 1A) or anti-Mam polyclonal antisera (FIG. 1B). Fractions 42–48 were positive for BU101 and Mam. These fractions were then analyzed by Western blot under reducing and nonreducing conditions. Using the anti-BU101 polyclonal antisera (FIG. 1C), BU101 is clearly visible under reducing conditions at approximately 7 kDa, the predicted molecular weight. Under nonreducing conditions, a higher molecular weight species is observed at approximately 24 kDa and the 7-kDa band is gone. Using the anti-Mam polyclonal antisera (FIG. 1D), Mam at approximately 18 kDa (under reducing conditions) is observed in the same fractions. Under nonreducing conditions, the 18-kDa band is lost and the 24-kDa band is detected. This simple experiment gives strong support to the model of a disulfide-linked heterodimer consisting of Mam and BU101 present in human breast tissue.

The material in these fractions was pooled, dialyzed, concentrated 50-fold, and applied to a gel filtration column, specifically Superose 12 (16 × 30) run with phosphate-buffered saline. The 400-µL fractions were analyzed by Dot blot, using 200 µL each for probing with anti-BU101 (FIG. 1E) and anti-Mam (FIG. 1F) polyclonal antisera. The resultant blots show that fractions 34–39 are positive for both BU101 and Mam. Separation experiments based on molecular weight confirmed the results observed with separation experiments based on charge, namely that BU101 and Mam coelute. The apparent molecular weight of the species observed from the Superose 12 column was calculated to be approximately 45 kDa.

REFERENCES

1. WATSON, M.A. & T.P. FLEMING. 1996. Mammaglobin, a mammary-specific member of the uteroglobin gene family, is overexpressed in human breast cancer. Cancer Res. **56:** 860–865.
2. BECKER, R.M., C. DARROW, M.A. WATSON & T.P. FLEMING. 1998. Identification of mammaglobin B, a novel member of the uteroglobin gene family. Genomics **54:** 70–78.
3. BILLING-MEDEL, P.A., M. COHEN, T. COLPITTS et al. 1998. Reagents and methods useful for detecting diseases of the breast. International Publication #WO9807857A1.
4. HUMAN GENOME SCIENCES, INC. 1997. Human endometrial specific steroid-binding factor I, II, and III—used to treat inflammation, asthma, rhinitis, cystic fibrosis, airway disease, neoplasia, atopy, etc. WPI 1997-480206.
5. ZHAO, C., T. NGUYEN, T. YUSIFOV et al. 1999. Lipophilins: human peptides homologous to rat prostatein. Biochem. Biophys. Res. Commun. **256:** 147–155.
6. MOUS, J.M., B.L. PEETERS, W.J. HEYNS & W.A. ROMBAUTS. 1982. Assembly, glycosylation, and secretion of the oligomeric rat prostatic binding protein in *Xenopus* oocytes. J. Biol. Chem. **257:** 11822.

Uteroglobin *In Situ* Hybridization: Novel Monitoring of Epithelial Differentiation in the Rabbit Endometrium

C. A. KRUSCHE,[a] A. HERRLER, AND H. M. BEIER

Department of Anatomy and Reproductive Biology, RWTH University of Aachen, 52057 Aachen, Germany

INTRODUCTION

Endometrial differentiation is a complex programmed sequence of molecular processes under endocrine and paracrine regulation. In particular, the endometrial epithelium has to fulfill different functional properties. First, preimplantation embryos have to be supported in development by proteins and factors secreted by the endometrial epithelium. Second, the endometrial epithelium has to develop towards a status, which allows embryo adhesion and implantation. To fulfill these needs, the endometrial epithelium of many mammals differentiates into two distinct cell populations: surface and glandular cells. To understand differentiation and regulation of these two endometrial cell types, marker molecules are needed, which discriminate between these cell populations. Such markers are helpful to monitor the effects of hormones, therapeutic treatment schedules, paracrine factors, and (last, but not least) embryonic effects on endometrial cell differentiation.

MATERIALS AND METHODS

We used sexually mature nulliparous New Zealand white rabbits with a body weight of 3.2–4.0 kg. Pseudopregnancy was induced by an iv injection of 75 IU hCG. The day of hCG injection is designated as day 0 of pseudopregnancy (d0 p-hCG). Experiment 1: Endometrial uteroglobin (Ugl) mRNA expression was analyzed in estrous as well as pseudopregnant animals on d1, d2, d3, and d5 p-hCG ($n = 5$/day). Further, 18 animals were treated with the progesterone receptor antagonist onapristone (ONA, Schering AG, Berlin, Germany; 20 mg/kg body weight) on d0 p-hCG, and Ugl mRNA expression was assessed on d1, d2, d3, and d5 p-hCG. Another 10 animals received ONA on d2 p-hCG, and Ugl expression was analyzed on d3 and d5 p-hCG. Experiment 2: Superovulation was induced in 10 animals by treatment with the gonadotropin preparation FSH-P® prior to ovulation induction. Five animals were treated with ONA on d2 p-hCG and 5 received the vehicle only. Pseudopregnant animals served as controls. Endometrial Ugl mRNA expression was

---

[a]Address for correspondence: Department of Anatomy and Reproductive Biology, RWTH University of Aachen, Wendlingweg 2, 52057 Aachen, Germany. Voice: +49-241-8088926; fax: +49-241-8888508.
ckrusche@post.klinikum.rwth-aachen.de

A Endometrial Ugl mRNA distribution

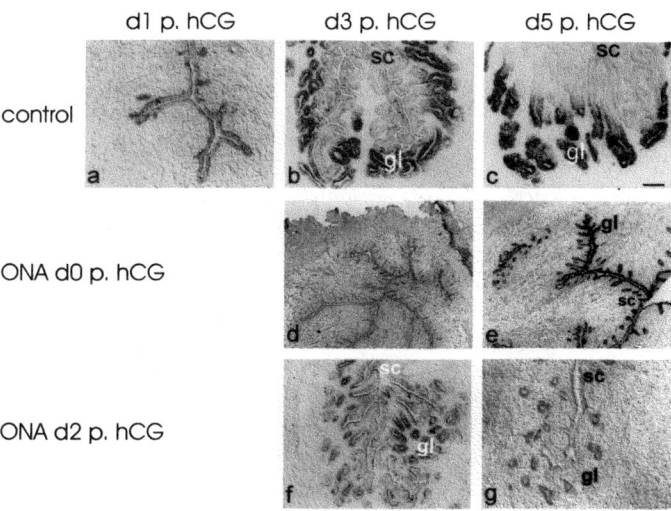

B Quantitative Ugl mRNA expression

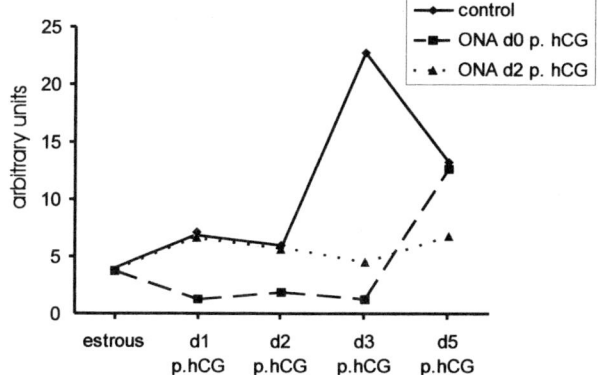

FIGURE 1. (**A**) Uteroglobin *in situ* hybridization: panels a, b, and c show the Ugl mRNA expression in pseudopregnant control animals on d1, d3, and d5 p-hCG, respectively; panels d and e show a representative animal on d3 and d5 p-hCG, respectively, which was treated with 20 mg ONA/kg body weight on d0 p-hCG; panels f and g demonstrate animals on d3 and d5 p-hCG, respectively, after administration of 20 mg ONA/kg body weight on d2 p-hCG. Magnification in all pictures: 100×. The bar (panel c) represents 100 μm; sc: surface cells; gl: glandular cells. (**B**) Quantitative assessment of the endometrial Ugl mRNA expression in experiment 1. The Ugl mRNA expression was corrected against the 18S rRNA expression to adjust RNA load (arbitrary units = $OD_{Ugl}/OD_{18S\ rRNA}$). [Figure reduced to 72%.]

analyzed on d5 p-hCG. Northern blot analysis and *in situ* hybridization experiments were performed as already described.[1,2]

RESULTS AND DISCUSSION

Endometrial Ugl mRNA expression is under control of estrogen and progesterone (P). During estrus, Ugl mRNA expression is low (FIG. 1B). All endometrial epithelial cells evenly express Ugl mRNA like animals on d1 p-hCG (FIG. 1A, a). On d3 and d4 p-hCG, Ugl mRNA is most strongly expressed; thereafter, Ugl mRNA expression declines (FIG. 1B). On d3 p-hCG, Ugl mRNA is more strongly expressed in the cells of endometrial glands (gl) than in endometrial surface cells (sc) lining the uterus cavum (FIG. 1A, b and c). We interpret this difference of Ugl mRNA expression as the beginning of the specific gland and surface cell differentiation. This is paralleled by an increase of P receptor expression in the designated gland cells.[3]

It is generally accepted that P drives the differentiation of the endometrial epithelium.[4,5] Thus, we investigated the effects of the preovulatory P serum peak within the first 10 hours after the hCG injection and the effects of the following transient rise in the P serum concentration on d2 p-hCG.[6] The endometrial P effects were studied by the specific inhibition of P action by a single application of the P receptor antagonist ONA. Ugl mRNA expression is suppressed below estrous levels on d1, d2, and d3 p-hCG after ONA treatment on d0 p-hCG (FIGS. 1A, d; and 1B), although the endometrium is under P influence on these days as ONA is already metabolized. On d5 p-hCG, Ugl mRNA is expressed as strong as in d5 control animals. However, the equal expression of Ugl mRNA in endometrial gland and surface cells (FIG. 1A, e) indicates that endometrial differentiation is heavily retarded. We conclude that the preovulatory P serum peak is the initial starting signal to induce endometrial epithelial cell differentiation in the rabbit. The quantitative Ugl mRNA expression and distribution demonstrate that the blockade of the P serum peak postpones endometrial development. Application of ONA on d2 p-hCG after the initial induction of endometrial differentiation on d0 p-hCG also leads to an inhibition of endometrial differentiation. On d3 and d5 p-hCG, Ugl mRNA expression remains low; however, expression is as high as in estrous animals (FIG. 1B). Thus, inhibition of endometrial differentiation is not as strong as after ONA application on d0 p-hCG. The distribution of Ugl mRNA expression indicates that differentiation of glandular and surface cells is inhibited (FIG. 1A, d and e). We conclude that continuous P influence is needed to support differentiation of the two specific endometrial cell populations, which is initiated by the preovulatory P serum peak.

The second experiment demonstrates that Ugl *in situ* hybridization is as well a suitable method to assess the effects of combined hormonal treatments on the differentiation of the endometrial epithelium. Superovulation induced an enhancement of endometrial development (FIGS. 2a and 2b). Ugl mRNA expression in the surface epithelium of superovulated animals on d5 p-hCG is as low as that in pseudopregnant control animals on d6/7 p-hCG (data not shown) and is lower than that in pseudopregnant control animals on d5 p-hCG (FIG. 2a). However, by the application of ONA on d2 p-hCG, the enhancement of epithelial cell differentiation can be delayed. The Ugl mRNA distribution as well as the gland morphology of super-

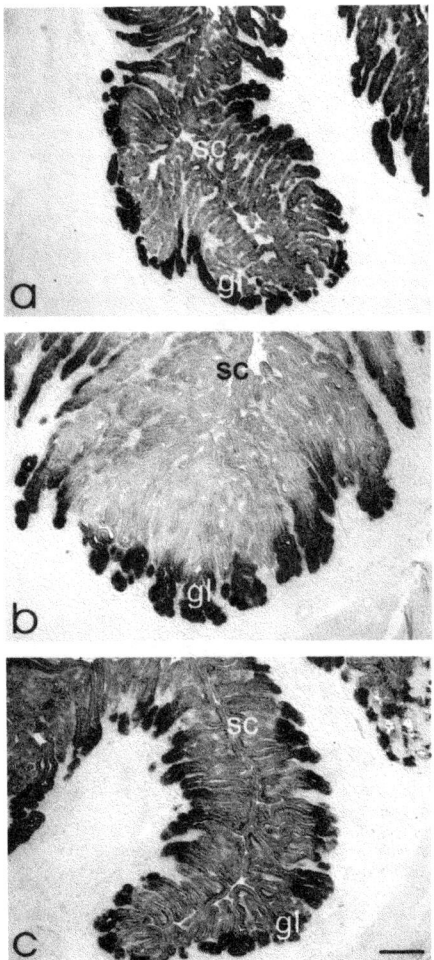

FIGURE 2. Assessment of the effects of superovulation and progesterone receptor antagonist treatment on the Ugl mRNA expression and morphology (experiment 2). Endometrium of a pseudopregnant control animal on d5 p-hCG (**a**); a superovulated, pseudopregnant animal on d5 p-hCG (**b**); and a superovulated, pseudopregnant animal on d5 p-hCG after treatment with the progesterone receptor antagonist onapristone on d2 p-hCG (**c**). The bar represents 100 µm; sc: surface cells; gl: gland cells.

ovulated, ONA-treated animals are similar to the Ugl mRNA distribution and gland size of control animals (FIGS. 2a and 2c).

Taken together, the presented results show that endometrial Ugl mRNA distribution is a novel marker to analyze physiological epithelial cell differentiation as well as the influence of various hormones on the differentiation of the endometrial epithelium in the rabbit.

REFERENCES

1. KRUSCHE, C.A. & H.M. BEIER. 1994. Localization of uteroglobin mRNA by nonradioactive *in situ* hybridization in the pregnant rabbit endometrium. Ann. Anat. **176:** 23–31.
2. KRUSCHE, C.A. *et al.* 2000. Modulation of endometrial transformation in gonadotropin-stimulated and unstimulated pseudopregnant rabbits: studies with the progesterone receptor antagonist onapristone. Mol. Hum. Reprod. **6:** 726–734.
3. HEGELE-HARTUNG, C. *et al.* 1992. Distribution of estrogen and progesterone receptors in the uterus: an immunohistochemical study in immature and adult pseudopregnant rabbits. Histochemistry **97:** 39–50.
4. ROTELLO, R.J. *et al.* 1992. Characterization of uterine epithelium apoptotic cell death kinetics and regulation by progesterone and RU 486. Am. J. Pathol. **140:** 449–456.
5. HEGELE-HARTUNG, C. *et al.* 1992. Luteal control of endometrial receptivity and its modification by progesterone antagonists. Endocrinology **131:** 2446–2460.
6. HILLIARD, J. & L.W.M. EATON, JR. 1971. Estradiol 17β, progesterone, and 20α-hydroxypregn-4-en-3-one in rabbit ovarian venous plasma. II. From mating through implantation. Endocrinology **89:** 522–527.

Prolactin Augments Progesterone-Dependent Expression of a Nuclear P-Type ATPase that Associates with the RING Domain of RUSH Transcription Factors in the Endometrium

MALINI MANSHARAMANI AND BEVERLY S. CHILTON[a]

Department of Cell Biology and Biochemistry, Texas Tech University Health Sciences Center, Lubbock, Texas 79430, USA

Links between hormone action, gene structure and function, and evolution comprise some of the most compelling themes of molecular endocrinology. Multiple sequence alignments have played a key role in the identification of functionally important regions in gene families. The RING-finger motif is an example of a sequence-structure family[1] that is purported to be involved in the formation of large protein complexes.[2] Originally identified in only 7 proteins in the OWL database,[3] the RING motif is the common feature of a superfamily of nearly 200 otherwise unrelated proteins, many of which are transcription factors. In an effort to understand the function of the RING domain, the search for RING-finger binding partners has become quite intense.

We recently cloned and characterized a RING-Finger Binding Protein (RFBP; Genbank™ accession number AF236061) from a rabbit uterine epithelial (HRE-H9) cell line. RFBP binds the RING domain of RUSH nuclear phosphoproteins that are potential mediators[4] of uteroglobin gene transcription. The coding sequence for the RFBP has all of the structural features that are diagnostic of a Type IV P-type ATPase,[5] including seven of eight core segments and nine transmembrane domains potentially involved in substrate binding and translocation.

P-type ATPases are classified into five main groups based on substrate specificity and sequence identity.[5] These groups include Type I (heavy metal pumps), Type II (Ca^{2+}, Na^+/K^+, and H^+/K^+ pumps), Type III (H^+ and Mg^{2+} pumps), Type IV (phospholipid pumps), and Type V (no assigned substrate specificity). A search of sequence databases using the BLAST program at the National Center for Biotechnology Information (http://www.ncbi.nlm.nih.gov/) revealed that the predicted RFBP protein and Type IV P-type ATPases from *Bos taurus* (protein ID AAD03352), *Mus musculus* (protein ID AAB18627), and *Saccharomyces cerevisiae* (protein ID L01795) are 50% similar and 34% identical. The three Type IV P-type ATPases are putative aminophospholipid transporters.

However, the newly identified RFBP differs from these authentic Type IV P-type ATPases in two important ways. The most remarkable feature is that it has an odd number of transmembrane domains. All known P-type ATPases have an even num-

---

[a]Author for correspondence: B. S. Chilton, Department of Cell Biology and Biochemistry, Texas Tech University Health Sciences Center, 3601 4th Street, Lubbock, TX 79430. Voice: 806-743-2709; fax: 806-743-2990.
beverly.chilton@ttmc.ttuhsc.edu

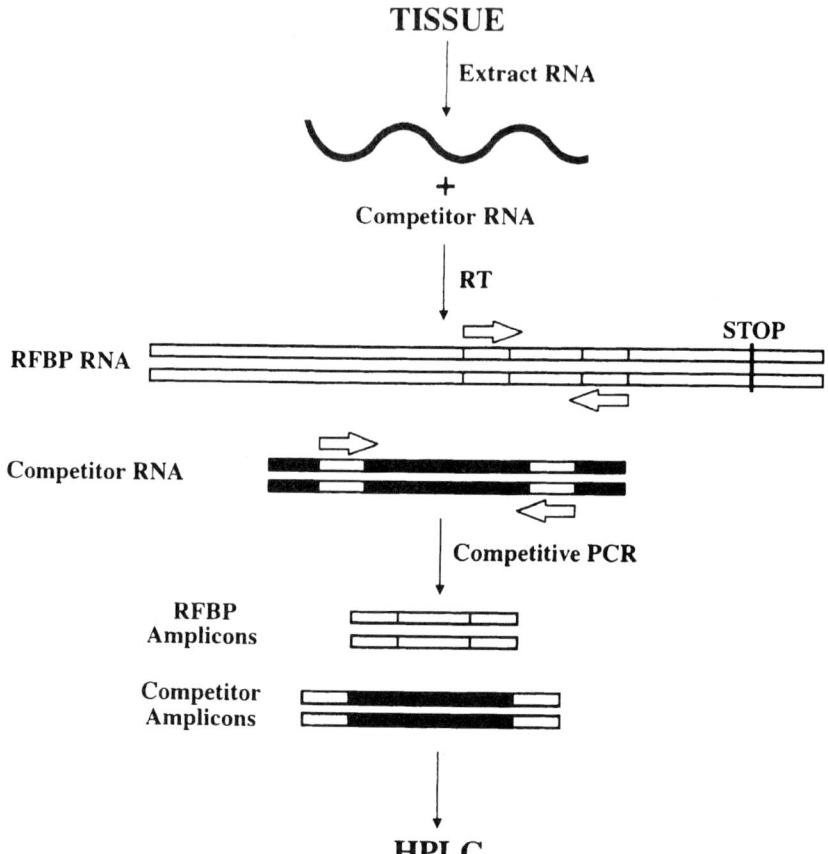

FIGURE 1. A schematic representation of a competitive RT-PCR reaction. Quantitative RT-PCR was performed on total RNA samples extracted from the endometrium of hormonally manipulated rabbits. Heterologous competitor RNA was developed from the MIMIC system (Clontech). Gene-specific PCR primers were synthesized by Midland Certified Reagent Company (Midland, TX). The relative abundance of RFBP was determined by ion-pair reversed-phase HPLC analysis of the RT-PCR reaction products or amplicons.

ber of such domains. The deletion of Core Region D results in the elimination of a transmembrane domain that is essential for energy transduction. The resultant protein that is altered topologically is unlikely to participate in aminophospholipid transport. The second equally remarkable feature of RFBP is its presence in the inner nuclear membrane. Physical contact between the conformationally flexible loop that extends between transmembrane domains 3 and 4 and the euchromatin was confirmed with immunoelectron microscopy. Thus, we hypothesize that RFBP interacts with transcription factors with RING domains and is involved in transcriptional regulation in the nucleus.

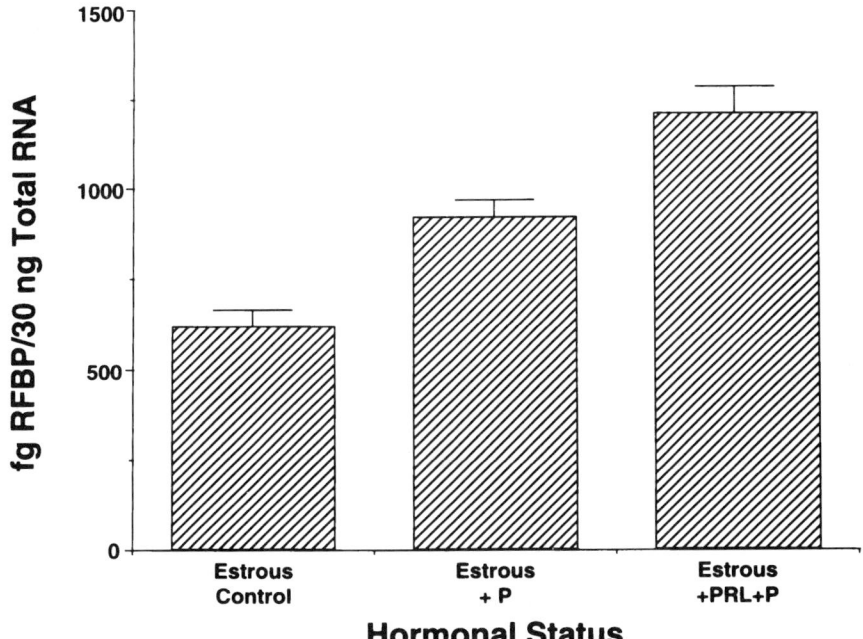

FIGURE 2. The amount of RFBP message in the endometrium changes in response to hormones. The relative abundance of RFBP was determined by HPLC analysis of native and competitor products from competitive RT-PCR. Estrous animals ($n = 12$) were used as follows: endometria from 6 animals were pooled and used as controls, whereas the remaining animals were divided equally into two hormone treatment groups designated +P (progesterone, 3 mg/kg/day for 5 days) and +PRL+P (prolactin, 2 mg/day for 5 days, followed by progesterone for 5 days). Animals were killed 24 hours after the last injection. Each assay was performed in triplicate and values are expressed as the mean ± SEM. Data were analyzed by ANOVA and Student-Newman-Keul's multiple range test ($p < 0.05$). Each value differs significantly from every other value.

Since RFBP interacts with RUSH, which is hormonally regulated,[6] we wanted to determine whether its expression was also affected by the hormonal milieu. Currently, there is a paucity of information regarding hormonal regulation of message levels for P-type ATPases. What little is known relates to changes in protein activity or protein availability. For example, the incubation of cultured cortical neurons with 17β-estradiol and pregnenolone sulfate increased the activity of the calcium ATPase in the neuronal plasma membranes.[7] Menkes and Wilson ATPases have been implicated in genetic disorders of copper homeostasis. In the case of Menkes P-type ATPase, treatment of the human breast carcinoma cells, PMC42, with lactational hormones increased perinuclear (Golgi) and punctate (endosome) protein, as measured by indirect immunofluorescence,[8] suggesting a hormone-dependent increase in the amount of protein.

We employed quantitative RT-PCR and denaturing HPLC to measure changes in the amount of RFBP mRNA in the endometrium. Briefly, Clontech's MIMIC system was used to develop a heterologous RNA that was used in competitive RT-PCR reactions with RNA from the endometrium of hormonally manipulated rabbits (FIG. 1). Native and mutant products were identified and quantified by denaturing HPLC, an analytical technique that preserves the accuracy of competitive RT-PCR beyond the log-linear phase of the reaction. A computer model developed for competitive PCR was used for data analysis. This model is available at http://www.grad.ttuhsc.edu/archive/index.html.

Quantification of competitive RT-PCR reaction products by ion-pair reversed-phase HPLC showed that RFBP message expression in rabbit endometrium is increased ($p < 0.05$) in response to treatment with progesterone (FIG. 2). Prolactin plus progesterone further increased ($p < 0.05$) the amount of message over the value for progesterone alone. Prolactin alone had no effect. The amount of RFBP message corresponded to similar changes in the expression of RUSH under identical hormonal conditions. This suggests there is tight regulation between the expression of the RING-finger containing RUSH proteins and the binding partner, RFBP.

ACKNOWLEDGMENTS

This work was supported by NIH Grant No. HD29457 to B. S. Chilton.

REFERENCES

1. BORDEN, K.L. & P.S. FREEMONT. 1996. The RING finger domain: a recent example of a sequence-structure family. Curr. Opin. Struct. Biol. **6:** 395–401.
2. BORDEN, K.L. 2000. RING domains: master builders of molecular scaffolds? J. Mol. Biol. **295:** 1103–1112.
3. FREEMONT, P.S., I.M. HANSON & J. TROWSDALE. 1991. A novel cysteine-rich sequence motif. Cell **64:** 483–484.
4. HEWETSON, A. & B.S. CHILTON, 1997. Novel elements in the uteroglobin promoter are a functional target for prolactin signaling. Mol. Cell. Endocrinol. **136:** 1–6.
5. AXELSEN, K.B. & M.G. PALMGREN. 1998. Evolution of substrate specificities in the P-type ATPase superfamily. J. Mol. Evol. **46:** 84–101.
6. HAYWARD-LESTER, A., A. HEWETSON, E.G. BEALE *et al.* 1996. Cloning, characterization, and steroid-dependent posttranscriptional processing of RUSH-1α and β, two uteroglobin promoter binding proteins. Mol. Endocrinol. **10:** 1335–1349.
7. ZYLINSKA, L. & B. LEGUTKO. 1998. Neuroactive steroids modulate *in vitro* the $Mg^{(2+)}$-dependent $Ca^{(2+)}$ ATPase activity in cultured rat neurons. Gen. Pharmacol. **30:** 533–536.
8. ACKLAND, M.L., E.J. CORNISH, J.A. PAYNTER *et al.* 1997. Expression of Menkes disease gene in mammary carcinoma cells. Biochem. J. **328:** 237–243.

A Novel *In Situ* Method of SV40 Transfection for the Establishment of Immortal Pulmonary Alveolar Type II Cell Lines

KANAKO MOMEDA,[a,b,c] ZHONGJIAN ZHANG,[b] ANIL B. MUKHERJEE,[b] AND R. DHANIREDDY[a,d]

[a]*Division of Neonatology, Department of Pediatrics, Georgetown University School of Medicine, Washington, District of Columbia, USA*

[b]*Section on Developmental Genetics, Heritable Disorders Branch, National Institute of Child Health and Human Development, National Institutes of Health, Bethesda, Maryland, USA*

Despite considerable progress in our understanding of mammalian lung development and respiratory physiology, there is a paucity of knowledge of alveolar type II cell differentiation and function. A major problem in studying the mechanism(s) of differentiation of these cells is the lack of a long-term *in vitro* model that can simulate *in vivo* characteristics of type II cells. This is due to the fact that these cells either fail to proliferate *in vitro* or cease to express differentiated function in long-term culture conditions. An additional problem often encountered in establishing cell lines from highly specialized organs, like the lung, is the lack of biochemical markers to characterize the cell types originally intended to be studied. The type II cells of rats and rabbits possess well-characterized biochemical markers—for example, the synthesis and secretion of surfactant, a phospholipid-protein complex,[1,2] deficiency of which causes neonatal respiratory distress syndrome (RDS).[3] In addition, these cells also synthesize and secrete a potent anti-inflammatory protein, uteroglobin (UG) (for review, see reference 4). It is also known as Clara cell 10-kDa protein (CC10).[5,6] UG is a steroid-dependent, low-molecular-weight (15.8 kDa), secretory protein, first discovered in the uterus of the rabbit during early pregnancy.[7,8] It has several important biological properties, including phospholipase A_2 (PLA_2)–inhibitory, anti-inflammatory (immunomodulatory), antichemotactic, and antithrombotic activities (reviewed in reference 4).

During the past decade, techniques have been developed by which cells with differentiated, tissue-specific function(s) could be immortalized. These techniques involve the transformation of normal differentiated cells with temperature-sensitive (ts) mutants of RNA and DNA tumor viruses that are temperature-sensitive for the maintenance of transformation (for review, see reference 9). When judged by a variety of criteria, such cells at the permissive temperature (33°C) manifest the transformed phenotype, but mimic the characteristics of nontransformed, differentiated cells at the nonpermissive temperature (39°C). It has been suggested that much in-

---

[c]Present address: Department of Anesthesiology, JR Tokyo General Hospital, Tokyo, Japan.

[d]Author for correspondence: R. Dhanireddy, Division of Neonatology, Department of Pediatrics, LSU Medical Center, Shreveport, LA 71105.
Rdhani@LSUHSC.edu

sight could be gained from studying gene regulation in these cells at nonpermissive temperature, which may be directly applicable to the mechanism(s) regulating normal cells. One of the major problems associated with this technique is that the resulting cell lines fail to produce the differentiated products at a level similar to that of the normal cells. One of the reasons for the low level of expression of differentiated products may be due to the fact that the cells are generally transformed after establishment of a primary culture on plastic tissue culture flasks without the extracellular matrix (i.e., basement membrane) upon which these cells are attached in an intact organ. Thus, if the type II cells could be transformed in their natural environment (i.e., within an intact lung), we rationalized that they may maintain their differentiated function at a higher level *in vitro*.

Here, we report the biochemical and biological characterization of immortal type II cell lines derived from the rat and rabbit fetal lungs using a novel method of transfection of these cells by a ts mutant of SV40 virus (A_{250}) introduced directly into the isolated lungs in organ culture. These cells are temperature-sensitive for differentiation and synthesize and secrete near-normal levels of both surfactant and UG *in vitro*. At the nonpermissive temperature (39°C), these cells synthesize and secrete surfactant, transcribe the UG gene, and secrete this protein in culture media upon stimulation with corticosteroid. Most interestingly, when grown on artificial basement membrane (Matrigel™), they differentiate to form complex, multilayer structures reminiscent of the architecture of mammalian lung. To our knowledge, this is the first report of establishment of temperature-sensitive, immortal type II cell lines from rodent pulmonary alveoli that differentiate *in vitro*, forming complex, organized structures on Matrigel and expressing near-normal levels of two well-characterized, steroid-responsive, biochemical markers that characterize these cells, surfactant and UG.

MATERIALS AND METHODS

Animals

Timed-pregnant New Zealand rabbits (3–3.7 kg) were obtained from Charles River Laboratories after approval of a protocol by the Institutional Animal Care and Use Committee. One animal was housed per cage and the animals received food and water *ad libitum*. A 12-h light and 12-h dark cycle was maintained in the animal room.

Transformation of Type II Cells by **In Situ** *Transfection*

On the 29th day of gestation, the animals were subjected to deep coma by administration of intravenous sodium pentobarbital (150 mg). Immediately thereafter, a longitudinal incision was made in the uterine horn and fetuses were removed and put on a sterile petri dish on ice. All subsequent procedures were carried out by following sterile techniques. The thorax of each fetus was opened and the trachea clamped to prevent entry of blood into the lungs. Residual blood was removed from the pulmonary circulation by perfusion with 0.15 M NaCl. After the trachea catheter was inserted, saline perfusion was continued until the lungs were excised. The lungs were removed and lavaged with 2 mL of cold 0.15 M NaCl three times. The lavage was retained for isolation of alveolar macrophage. The lungs were inflated with 1 mL of virus stock of the ts A mutants (A_{255}),[9] giving a multiplicity of infection of 5–20 PFU/fetus. The lungs inflated with virus stock were placed in 100-mm bacteriologic plastic dishes (Falcon, Cockeysville, MD) and incubated at 33°C for 3 hours. The lungs were kept inflated throughout the incubation. After this incubation period, the lungs were placed on a clean glass plate and the parenchymal tissue was scraped free of obvious blood vessels and bronchi. The parenchymal fragments were cut into small pieces of about 2 to 3 mm. These tissue fragments were collected in 25-cm^2 flasks and incubated in Alpha modified minimal essential medium (Alpha-MEM) containing penicillin (100 U/mL), streptomycin (100 µg/mL), gentamicin (100 µg/mL), and 4% fetal calf serum. These tissue cultures were then incubated at 33°C in an atmosphere of 5% CO_2 and 95% air. The medium was first changed after 1 week and subsequently twice a week. After 7–8 weeks at 33°C, transformed clones were visible as distinct islands. These transformed clones were picked up by using a stainless-steel cloning cylinder placed directly over the island of cells. The bottom of the cylinder was sealed by sterile silicone grease to isolate the clone from the rest of the monolayer of cells. The clone within the cylinder was removed enzymatically with 0.5% trypsin-EDTA and grown into separate 25-cm^2 flasks (Corning Inc., Corning, NY) until confluent. The identification and characterization of transformants is then performed as previously reported.[9]

RESULTS

Morphological Analysis of Transformed Type II Alveolar Cells

The transformed cells grown at the permissive temperature revealed spindle-shaped morphology (FIG. 1A) like fibroblasts. However, at the nonpermissive temperature, the size is larger and the appearance is flatter and with more granular cytoplasm (FIG. 1B). Ultrastructure analysis by transmission electron microscopy of the

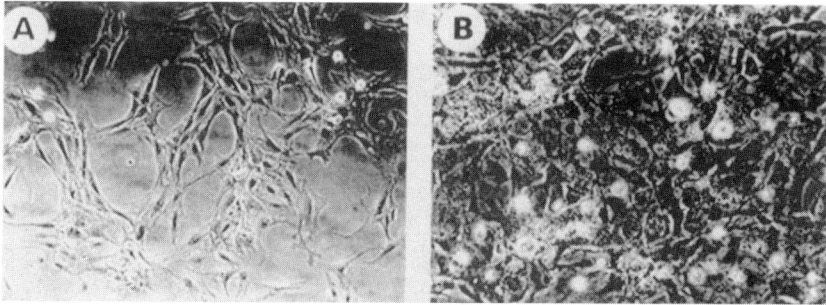

FIGURE 1. Light microscopic analysis of transformed alveolar type II cells: (**A**) cells incubated at 33°C; (**B**) cells incubated at 39°C. Note that the cells maintained at permissive temperature (33°C) reveal a spindle-shaped morphology, whereas those maintained at nonpermissive temperature (39°C) are more epithelial-like and with granular cytoplasm.

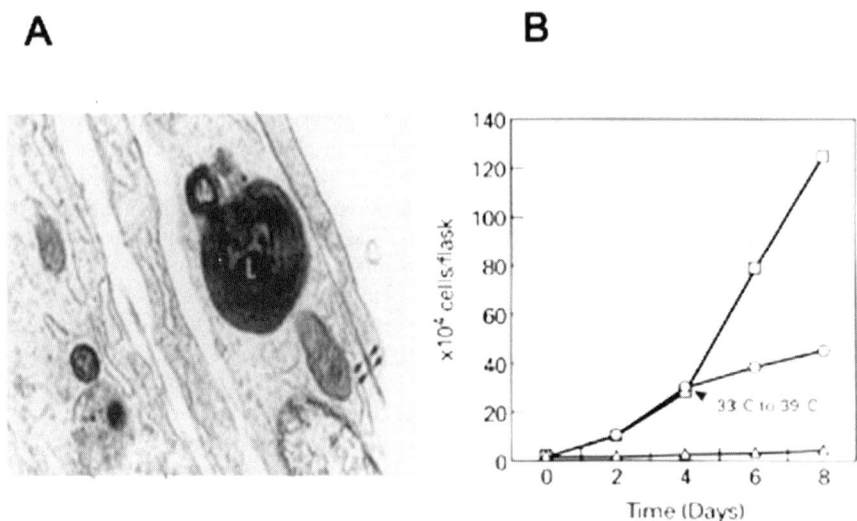

FIGURE 2. (A) Ultrastructural analysis of transformed alveolar type II cells by transmission electron microscopy. Note the presence of a lamellar body (L), characteristic of these cells. Magnification: ×115,000. [Figure reduced to 50%.] **(B)** Temperature-dependent growth of transformed alveolar type II cells. Cultures were plated at 33°C and shifted to 39°C, causing a shift in the growth curve.

cells grown at 39°C revealed the presence of lamellar bodies, a characteristic feature of alveolar type II cells (FIG. 2A).

Temperature-Dependent Growth of Transformed Alveolar Type II Cells

The growth characterization of these cells varies with temperature as shown in FIGURE 2B. These cells grow exceptionally well at 33°C, which is the permissive temperature for transformed cells; they grow rapidly to high cell densities. At the nonpermissive temperature (39°C), however, these cells grow more slowly and to lower saturation densities than cells grown at 33°C. This indicates that these cells are temperature-sensitive for growth. The growth inhibition is observed even after the cultures are grown at 33°C for 4 days and shifted to 39°C.

Effect of TRH and Dexamethasone on [^{3}H]Choline Incorporation at 33°C and 39°C

In order to determine whether the transformed type II cells were still hormonally responsive as *in vivo*, the effect of dexamethasone (DEX) alone and of TRH and DEX in combination on choline incorporation into saturated phosphatidylcholine was measured. Data from studies using transformed type II cells are shown in FIGURES 3A and 3B. Exposure of the transformed type II cells to DEX only had a modest stimulatory or light suppressive effect on the incorporation of choline into saturated

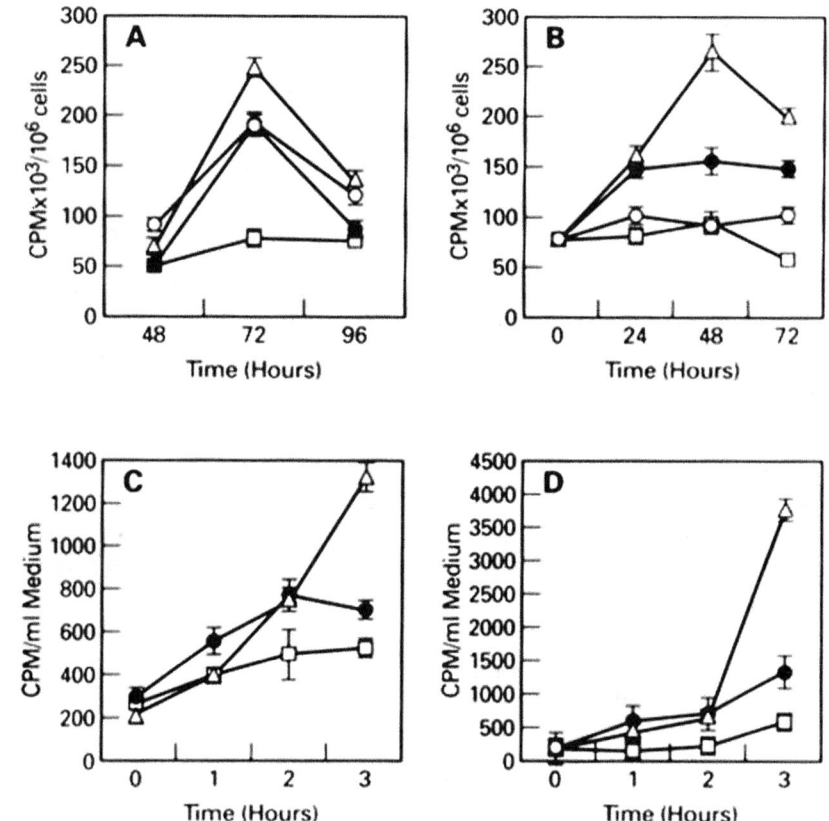

FIGURE 3. Effect of TRH and DEX on the stimulation of ^{3}H-choline incorporation (**A, B**) and secretion of saturated phosphatidylcholine (**C, D**) by transformed alveolar type II cells: permissive temperature (A, C); permissive temperature (B, D).

phosphatidylcholine (data not shown). As it has been demonstrated previously that both TRH and DEX increase the incorporation rate into saturated phosphatidylcholine,[10] we added both DEX and TRH into the medium. At both permissive (33°C) and nonpermissive (39°C) temperature, the stimulative effect of TRH and DEX on choline incorporation into saturated phosphatidylcholine was time-dependent and dose-dependent. The stimulation effect was the strongest at a DEX concentration of 10^{-7} M at both temperatures. The time courses for stimulation of choline incorporation into phosphatidylcholine were similar at permissive and nonpermissive temperatures. The response to TRH plus DEX in this study was maximal by 72 hours in permissive temperature (33°C) and 48 hours in nonpermissive temperature (39°C).

Secretion of Saturated Phosphatidylcholine

The alveolar type II cells secrete phosphatidylcholine. In FIGURES 3C and 3D, phosphatidylcholine secretion profiles of the transformed type II cells are shown. It

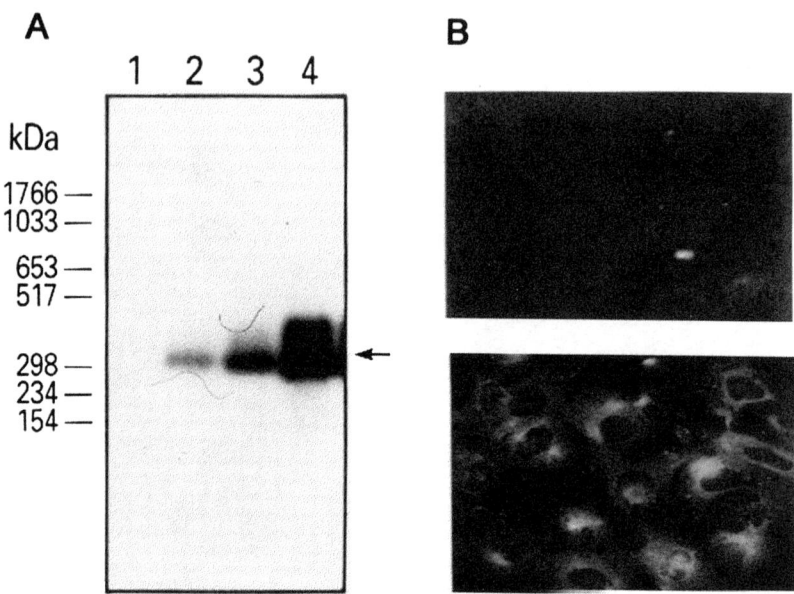

FIGURE 4. Production of UG by transformed type II cells. The results of **(A)** RT-PCR and **(B)** immunofluorescence. Upper panel in part B: control without first antibody; lower panel in part B: with antibody against UG, confirming that these cells are capable of producing UG. Note the cytoplasmic fluorescence.

is clear that these cells release saturated phosphatidylcholine into the medium and that this release is stimulated by isoproterenol in a time- and dose-dependent manner. As seen, the secretion is maximum at an isoproterenol concentration of 10^{-5} M at both permissive and nonpermissive temperatures, although at the nonpermissive temperature these cells secrete a much greater quantity of phosphatidylcholine.

UG Production by Transformed Alveolar Type II Cells

By using RT-PCR, Western blotting, and immunofluorescence, we determined that the transformed type II cells produce UG. The results of RT-PCR (FIG. 4A) and immunofluorescence (FIG. 4B) showed that UG mRNA and protein are produced by these cells. The results of Western blotting of both cell lysate and conditioned medium showed that the UG is secreted into the medium (data not shown). In fact, our results confirmed the previous finding that UG is secreted at a rapid rate immediately after synthesis by transformed type II cells.[11] It has been reported that prolactin acts directly on the fetal rat lung by stimulating the synthesis of surfactant phospholipids and that prolactin enhances the effect of DEX.[12] Prolactin receptor is regulated by progesterone, and PRL augments the progesterone-dependent increase in UG mRNA.[13] UG is also a model for the dynamic interaction between prolactin and progesterone in the regulation of gene expression.[14,15] Therefore, in order to see the correlation between UG gene expression and prolactin and DEX treatment, we performed RT-PCR using poly(A)$^+$ RNA extracted from cells grown at nonpermissive temperature (39°C) in the presence or absence of prolactin and/or DEX. The results show that UG mRNA was expressed in these transformed alveolar type II cells. These results were further confirmed by Western blotting for UG protein expression. A major drawback of this method is that the cells immortalized by this technique fail to produce UG after revival from cryopreservation. Efforts are now being directed to understand why these cells do not undergo cryopreservation without losing their differentiated characteristics.

In summary, our results indicate that the *in situ* transfection of type II alveolar cells results in the maintenance of normal differentiated functions for a longer period of time in culture. This technique may allow detailed investigations on the regulation of surfactant production using these cells as *in vitro* model systems.

REFERENCES

1. MENDELSON, C.R. 2000. Annu. Rev. Physiol. **62:** 875–915.
2. JOBE, A.H. & M. IKEGAMI. 2000. Annu. Rev. Physiol. **62:** 825–846.
3. BATEMAN, S.T. & J.H. ARNOLD. 2000. Curr. Opin. Pediatr. **12:** 233–237.
4. MUKHERJEE, A.B., G.C. KUNDU, G. MANTILE-SELVAGGI *et al.* 1999. Cell. Mol. Life Sci. **55:** 771–787.
5. KRISHNAN, R.S. & J.C. DANIEL, JR. 1967. Science **158:** 490–492.
6. BEIER, H.M. 1968. Biochim. Biophys. Acta **160:** 289–291.
7. SINGH, G., J. SINGH, S.L. KATYAL *et al.* 1988. J. Histochem. Cytochem. **36:** 73–80.
8. SINGH, G. & S.L. KATYAL. 1997. Am. J. Respir. Cell Mol. Biol. **17:** 141–143.
9. CHOU, J.Y. 1985. Methods Enzymol. **109:** 385–396.
10. PHILIP, L., M. BALLARED, H. LESLIE & L. GONZALAS. 1984. J. Clin. Invest. **74:** 898–905.
11. GUY, J., R. DHANIREDDY & A.B. MUKHERJEE. 1992. Biochem. Biophys. Res. Commun. **189:** 662–669.
12. MULLON, D.K., Y.F. SMITH, L.L. RICHARDSON & P. HAMOSH. 1983. Biochim. Biophys. Acta **751:** 166–174.
13. CHILTON, B.S., S.K. MANI & D.W. BULLOCK. 1984. Mol. Endocrinol. **2:** 1169–1175.
14. DANIEL, J.C., A.E. JETTON & B.S. CHILTON. 1984. J. Reprod. Fertil. **72:** 443–452.
15. RANDALL, G.W., J.C. DANIEL & B.S. CHILTON. 1991. J. Reprod. Fertil. **91:** 249–257.

Uteroglobin Expression and Release in the Human Endometrium

F. MÜLLER-SCHÖTTLE,[a,b] I. CLASSEN-LINKE,[b] K. BEIER-HELLWIG,[b] K. STERZIK,[c] AND H. M. BEIER[b]

[b]*Department of Anatomy and Reproductive Biology, RWTH University of Aachen, 52057 Aachen, Germany*

[c]*Institute for Reproductive Medicine, Ulm, Germany*

INTRODUCTION

Initially, uteroglobin (UGL) was identified as the major protein component of rabbit uterine secretion and blastocyst fluid during the time before and at implantation.[1,2] Recently, its progesterone-dependent expression was shown in the equine endometrium.[3] Our research efforts have focused on the biological significance of UGL in human reproduction, particularly in the endometrium. An essential prerequisite to start with those functional studies is the exact cellular localization and studies on the expression of this protein throughout the menstrual cycle.[4,5] To study the endocrine/paracrine regulation of UGL expression, we use a cell culture system of isolated human endometrial cells.[6] By means of this primary cell culture system, we hope to elucidate the cell-cell interactions between epithelial and mesenchymal cells, the cytokine signaling on UGL expression, and its ligand binding dynamics since these studies are not permitted *in vivo* because of ethical reservations.

METHODS

RT-PCR

Reverse transcription of total RNA endometrial samples was performed using 0.8 µg RNA from each sample. The amplification of the cDNA was performed with a specific primer pair for the human UGL gene. To obtain semiquantitative results, the housekeeping gene, cytochrome oxidase subunit I (CYT), was amplified with the same cDNA samples.

Immunohistochemistry

Endometrial tissue samples were fixed in formalin and embedded in paraffin. Paraffin sections were incubated with a polyclonal antihuman UGL antibody (Klug and Suske, Marburg, Germany).

---

[a]Address for correspondence: Department of Anatomy and Reproductive Biology, RWTH University of Aachen, Wendlingweg 2, 52057 Aachen, Germany. Voice: +49-241-8089 110; fax: +49-241-8888 508.

fmueller-schoettle@post.klinikum.rwth-aachen.de

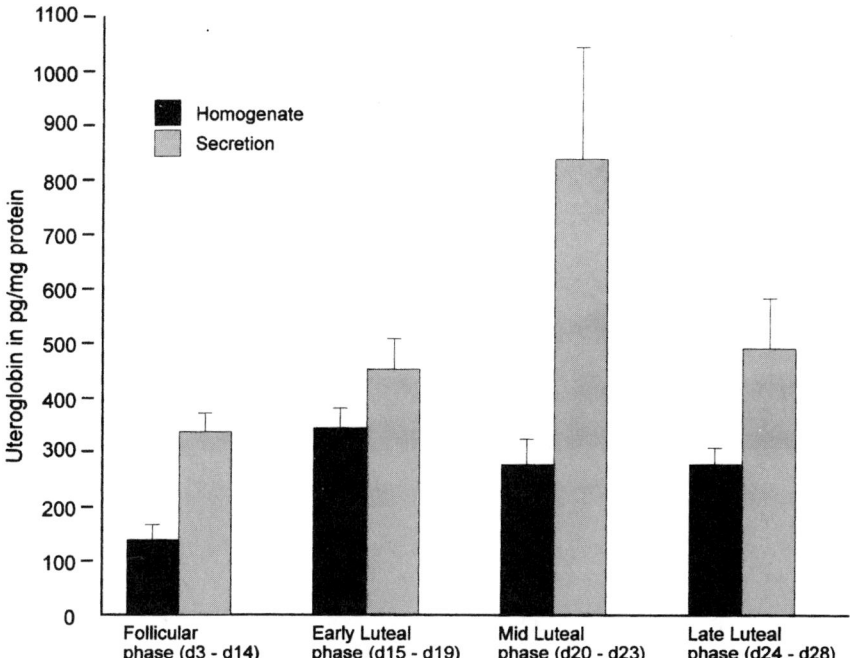

FIGURE 1. UGL expression *in vivo*. UGL concentrations assessed by ELISA in endometrial tissue homogenates and uterine secretion samples throughout the menstrual cycle. The UGL concentrations are significantly different in secretions versus tissue homogenates ($p = 0.003$, mid-luteal phase). In endometrial tissue homogenates, the differences are significant comparing the follicular versus early luteal phase ($p = 0.04$). In the uterine secretion samples, the differences are significant comparing the mid-luteal phase versus follicular phase ($p = 0.019$), early luteal phase ($p = 0.007$), and late luteal phase ($p = 0.006$).

UGL ELISA

UGL ELISA was performed with secretion samples and tissue homogenates from human endometrium and serum from the same volunteers. For UGL detection in the cell culture, we used the correspondent cell culture medium. The ELISA (CC16 ELISA, Bosmans, Eurogenetics, Belgium) recognizes only human UGL using a monoclonal antibody from the mouse.

Cell Culture

Endometrial tissue of the proliferative phase was separated in epithelia and fibroblasts. After adhesion of cells and a priming phase with 10% FCS and 10^{-8} M estradiol (E_2), the cells were further cultured in a serum-free medium with addition of 10^{-8} M E_2 and 10^{-6} M medroxyprogesterone acetate (MPA) for 2 and 4 days. Basal and apical media were measured in CC16 ELISA. After finishing the experiment, the RNA of the cells was isolated for UGL RT-PCR analysis.

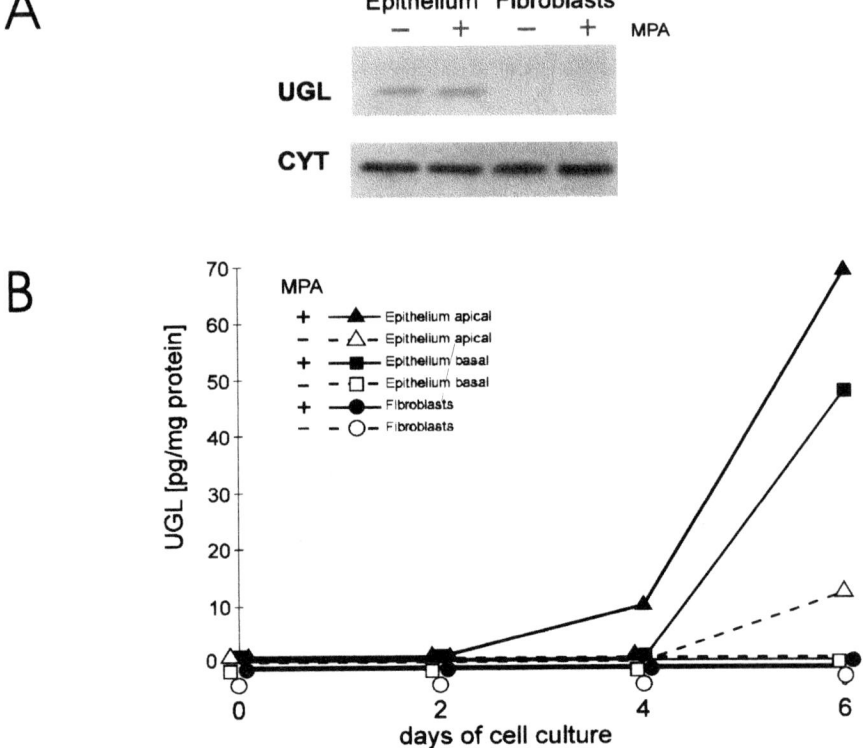

FIGURE 2. UGL expression *in vitro*. In an endometrial cell culture, UGL mRNA is expressed by epithelial cells, but not by fibroblasts. UGL release into the cell culture medium was measured by ELISA. After substitution with MPA for 4 days, UGL could be detected in the apical and basal culture medium of epithelial cells and not in fibroblast medium.

RESULTS

UGL expression, localization, and secretion in the human endometrium are presented in a consistent and conclusive synopsis.[4] Our studies on 115 volunteers combine RT-PCR, immunohistochemistry, and quantitative assessment by an ELISA for UGL (FIG. 1). Secretory UGL is found in endometrial tissue homogenates with the highest levels of expression during the early luteal phase (d15–d19: 340 pg/mg total protein). In turn, UGL is released into the uterine lumen in peak amounts during the receptive phase of the menstrual cycle (mid-luteal phase d20–d23: secretion level of 833.4 pg/mg total protein). Our immunohistochemical studies match with these results as UGL is located during the early and mid-luteal phase in the apical compartments of endometrial gland cells. However, RT-PCR analysis throughout the menstrual cycle showed the highest expression of the UGL gene between days 18 and 24.

In addition to these *in vivo* studies, we studied the expression of UGL *in vitro* in a cell culture system of isolated human endometrial fibroblasts and epithelial cells (FIG. 2). UGL expression was detected by RT-PCR in the epithelial cells and not in the fibroblasts. UGL release into the cell culture medium was measured by ELISA. After substitution for 4 days with MPA, a synthetic progesterone, UGL could be detected in the apical and basal cell culture medium of epithelial cells. In fibroblast culture medium, no UGL was detectable after MPA treatment at any time. Basic supplementation of E_2 was not changed during the course of this experiment.

DISCUSSION

UGL is expressed in the human endometrium, although at a lower level compared to the human lung. The highest level of UGL expression in endometrium was found between days 18 and 24 throughout the menstrual cycle. These *in vivo* observations strongly suggest an involvement of UGL in endometrial preparations for implantation. Furthermore, as is well known, mammalian mucosal epithelia provide a protective barrier in such organs like the lung or uterus. While these studies of UGL suggest a significance of this molecule as an immunomodulatory or immunosuppressive cytokine,[7] our focus of interest has also been directed towards a reproductive significance during the luteal phase of the menstrual cycle. In an attempt to clarify the endocrine and paracrine regulation of UGL, we used a primary cell culture system and demonstrated that UGL expression and release from epithelial endometrial cells are stimulated by progesterone as the culture time proceeds. Our observation indicates a 4-day epithelial differentiation to achieve UGL release. Further investigations, using particularly this primary cell culture system, are in progress to elucidate some of the key functions of this protein at the time of progesterone-prepared epithelial differentiation.

REFERENCES

1. BEIER, H.M. 1966. Das Proteinmilieu im Blutserum, Uterus und in den Blastocysten des Kaninchens vor der Nidation. Coll. Biochim. Morphogenese, DFG, Bonn, Germany.
2. BEIER H.M. 1968. Uteroglobin: a hormone-sensitive endometrial protein involved in blastocyst development. Biochim. Biophys. Acta **160:** 289–291.
3. BEIER-HELLWIG, K. *et al.* 1995. Partial sequencing and identification of three proteins from equine uterine secretion regulated by progesterone. Reprod. Domest. Anim. **30:** 295–298.
4. PERI, A. *et al.* 1994. Expression of Clara cell 10 kD gene in the human endometrium and its relationship to ovarian menstrual cycle. DNA Cell Biol. **13:** 495–503.
5. MÜLLER-SCHÖTTLE, F. *et al.* 1999. Expression of uteroglobin in the human endometrium. Mol. Hum. Reprod. **5:** 1155–1161.
6. CLASSEN-LINKE, I. *et al.* 1997. Establishment of a human endometrial cell culture system and characterization of its polarized hormone responsive epithelial cells. Cell Tissue Res. **287:** 539–549.
7. DIERYNCK, I. *et al.* 1995. Potent inhibition of both human interferon γ production and biologic activity by the Clara cell protein CC16. Am. J. Respir. Cell Mol. Biol. **12:** 205–210.

Expression of Inflammatory Cytokines in a Mouse Transformed Clara Cell Line by Tumor Necrosis Factor-α

MOON S. PARK,[a] BIHONG ZHAO,[b] PATRICIA L. RAMSAY,[a,b]
ALBERT S. Y. CHANG,[c] MICHAEL J. REARDON,[c] AND FRANCESCO J. DEMAYO[b]

*Department of [a]Pediatrics, [b]Molecular and Cellular Biology, and [c]Surgery,
Baylor College of Medicine, Houston, Texas 77030*

Airway epithelium plays an important role in providing a barrier against external stimuli, but also plays a critical role in regulating airway inflammation by modulating the release of a variety of pro- and anti-inflammatory mediators. One of the pulmonary epithelial cell types involved in this regulation is the nonciliated secretory cell of the bronchiolar epithelium, the Clara cell. The major secretory product of the Clara cell is a 16-kDa protein, Clara cell secretory protein (CCSP), which protects the respiratory tract from oxidative stress,[1] viral infection,[2] and other potentially deleterious inflammatory states. Recent studies have demonstrated that various inflammatory mediators, including tumor necrosis factor-α (TNF-α), interferon-γ, and phospholipase A_2, may alter the production of CCSP.[3,4]

Cytokines are important soluble signaling molecules that dictate and coordinate inflammatory and immune responses. TNF-α, an early-response proinflammatory cytokine, plays a major role in commanding inflammation. This 17-kDa polypeptide is released from alveolar macrophages and other inflammatory cells, initiating the production of numerous cytokines and chemokines from pulmonary epithelial cells.[5] Furthermore, there is evidence that TNF-α is also important in the development of various inflammatory lung diseases such as chronic bronchitis, cystic fibrosis, asthma, and ARDS.[6]

We have generated a mouse transformed Clara cell line, mtCC, which expresses endogenous markers characteristic of Clara cells, including CCSP, surfactant protein A, and a repertoire of transcriptional factors expressed in the Clara cells *in vivo*.[7] The goal of this study is to test the hypothesis that exposure of mtCC cells to TNF-α will lead to the expression of pro- and anti-inflammatory molecules.

The mtCC cells were treated with 0 ng/mL, 5 ng/mL, or 20 ng/mL TNF-α and subsequently harvested after 24 hours of exposure. Total RNA, isolated in Trizol (Life-Technology, MD), was assayed for the expression of mRNAs for specific cytokines and chemokines in the mtCC cell line by ribonuclease protection assays (RPA II kit, Ambion, TX). The probes were generated for the cytokines and chemokines using four separate cytokine multiprobe templates (PharMingen, CA).

Our data show that the mtCC cell produces multiple cytokines and chemokines, including IL-1RA, LT-β, TNF-α, IFN-γ, TGF-β1, TGF-β2, MIF, RANTES, MIP-1β, MIP-1α, MCP-1, and TCA-3. The proinflammatory cytokine TNF-α, IFN-γ, as well as the anti-inflammatory cytokine IL-1RA were upregulated by exposure to TNF-α. Additionally, many chemokines such as RANTES, MIP-1β, MIP-1α, and MCP-1 were also upregulated.

These findings demonstrate the ability of the mtCC cells to respond to a proinflammatory stimulus. Further, they support the hypothesis that the Clara cell may be a key component for the modulation of inflammatory events.

REFERENCES

1. MANGO, G.W., C.J. JOHNSTON, S.D. REYNOLDS *et al.* 1998. Clara cell secretory protein deficiency increases oxidant stress response in conducting airways. Am. J. Physiol. **275**(2, part 1): L348–L356.
2. IKEGAMI, M., K.S. HARROD, J.A. WHITSETT & A.H. JOBE. 1999. CCSP deficiency does not alter surfactant homeostasis during adenoviral infection. Am. J. Physiol. **277**(5, part 1): L983–L987.
3. YAO, X.L., S.J. LEVINE, M.J. COWAN *et al.* 1998. Tumor necrosis factor-alpha stimulates human Clara cell secretory protein production by human airway epithelial cells. Am. J. Respir. Cell Mol. Biol. **19**(4): 629–635.
4. YAO, X.L., T. IKEZONO, M. COWAN *et al.* 1998. Interferon-gamma stimulates human Clara cell secretory protein production by human airway epithelial cells. Am. J. Physiol. **274**(5, part 1): L864–L869.
5. FUJISAWA, T., Y. KATO, J. ATSUTA *et al.* 2000. Chemokine production by the BEAS-2B human bronchial epithelial cells: differential regulation of eotaxin, IL-8, and RANTES by TH2- and TH1-derived cytokines. J. Allergy Clin. Immunol. **105**(1, part 1): 126–133.
6. LUSTER, M.I., P.P. SIMEONOVA, R. GALLUCCI & J. MATHESON. 1999. Tumor necrosis factor alpha and toxicology. Crit. Rev. Toxicol. **29**(5): 491–511.
7. DEMAYO, F.J., M.J. FINEGOLD, T.N. HANSEN *et al.* 1991. Expression of SV40 T antigen under control of rabbit uteroglobin promoter in transgenic mice. Am. J. Physiol. **261**(2, part 1): L70–L76.

Binding of rhCC10 to Fibronectin and Its Effect on Cellular Adhesion

JEFFREY FARROW, JAMES MELBY, LAURA WIESE, JERRY LOHNAS, RICHARD WELCH, AND APRILE L. PILON[a]

Claragen, Incorporated, College Park, Maryland 20742, USA

INTRODUCTION

The discovery that CC10 binds to human fibronectin in solution by Zhang *et al.*[1] has profound implications. Their further observations of the ability of CC10 to inhibit aggregation of fibronectin *in vitro*, to inhibit fibronectin-mediated fibrillogenesis in cell culture, and to inhibit renal fibronectin deposition *in vivo* demonstrates an important physiological role of endogenous CC10 in all mammals. Fibronectin is one of the most well-characterized mediators of cellular adhesion and is involved in several physiologic processes including wound healing, fibrosis, inflammatory cell and fibroblast adhesion, tumor metastases, and extracellular basement membrane formation.[2] Fibronectin in solution has a disk-like, roughly globular conformation that is distinct from its insoluble form in an unfolded and "stretched out" conformation across solid surfaces.[3] When fibronectin undergoes this conformational switch in tertiary structure, it becomes capable of polymerizing. Zhang *et al.*[1] demonstrated that CC10 inhibits the conversion of fibronectin, indicating that CC10 may regulate processes dependent upon this conversion *in vivo*. The inflammatory, fibrotic, and oncogenic phenotypes of a uteroglobin/CC10 knockout mouse illustrate the significance of preventing the initiation of these processes *in vivo* (reviewed in reference 4). In a CCSP/CC10-deficient mouse, the threshold for the initiation of these processes is significantly lowered, resulting in the exacerbation of inflammation in response to pulmonary insults.[5]

When fibronectin converts to its insoluble form, it changes conformation and deposits on the surfaces of cells and in the extracellular matrix (ECM), where it may act as an anchor for neutrophils, macrophages, monocytes, and fibroblasts during an inflammatory episode. The "receptors" for fibronectin on cell surfaces are numerous and fall into different classes of molecules, including cell adhesion molecules and integrin complexes. Interactions between deposited fibronectin and these types of molecules not only mediate cellular adhesion, but also provide a means through which the cell senses, and reacts to, its environment. Therefore, we evaluated the potential binding of CC10 to insoluble, plated fibronectin, which is thought to represent a functional form of fibronectin that mediates cellular adhesion *in vitro* and to simulate deposited fibronectin on cell surfaces and/or in the ECM *in vivo*. We found that CC10 does indeed bind to the plated fibronectin conformation. We further explored the potential of this interaction to effect the biological process of cellular

---

[a]Author for correspondence: A. L. Pilon, Claragen, Inc., 387 Technology Drive, College Park, MD 20742. Voice: 301-405-8593; fax: 301-405-4989.
  apilon@claragen.com

adhesion to fibronectin and found that CC10 does act as an inhibitor of cellular adhesion to fibronectin *in vitro* for certain cell types.

MATERIALS AND METHODS

Purified human plasma fibronectin was purchased from Sigma, resuspended in PBS to 1–2 mg/mL, aliquoted, and stored at −80°C until use. Frozen fibronectin was thawed slowly on ice and then allowed to equilibrate at 37°C for 1 hour. The fibronectin was then diluted in serum-free Dulbecco's Modified Eagle's Medium (DMEM) and approximately 300 µg was pipetted to each well of a 96-well microtiter plate (Nunc Polysorb™). Plates were then placed in a Ziploc bag and kept overnight at room temperature. The fibronectin solution was then aspirated and the plates were washed three times with sterile 150 mM NaCl. Plates were stored in Ziploc bags at 4°C with 250 µL of serum-free DMEM per well until use. Prior to use, the medium was aspirated and the plates were blocked for 2 hours at room temperature using 5% BSA (Fisher). The blocking solution was then aspirated and the plate washed three times with 150 mM NaCl. Then, varying amounts of recombinant human CC10 (rhCC10) in 150 mM NaCl mixed with a fixed amount of an rhCC10-HRP conjugate was added to the wells and incubated at room temperature for 1 hour. The rhCC10 was then aspirated off and the wells were washed three times with 150 mM NaCl. The amount of bound rhCC10-HRP was then quantitated by absorbance at 450 nm and the degree of competition from the unlabeled rhCC10 could be estimated. All rhCC10 concentrations were run in triplicate and a standard curve with varying amounts of the rhCC10-HRP conjugate, in duplicate, is run with each experiment.

Cellular adhesion assays were performed essentially as described by Retta, Ternullo, and Tarone.[6] Briefly, a 96-well fibronectin-coated Biocoat™ plate (Becton-Dickinson) is blocked to prevent nonspecific binding of cells or proteins to the plastic with 1× Pierce Blocker™, containing 5% BSA, at 37°C in an incubator for at least 1 hour. The blocking reagent is then removed and the wells washed three times with PBS (phosphate-buffered saline). The prepared plate was stored in the incubator while preparing the cells. NIH 3T3 were used in the assay and were grown with standard tissue culture techniques in DMEM containing 10% fetal calf serum. The cells were removed from culture flasks using 5 mM EDTA and spun down gently. The cells were then washed three times by alternately resuspending in serum-free medium and repeat centrifugation. The final cell pellet was resuspended to an estimated density of $2.5–7.5 \times 10^4$ cells per 200 mL. Cells were partitioned into aliquots to receive either CC10, a monoclonal antibody control, myoglobin (dog heart) control, or the PBS-only control. Cells were then added to each well of the fibronectin-coated plate as quickly as possible. Then, the plates were incubated at 37°C in a CO_2 incubator for 1 hour. While the plated cells were incubating, live cells in the pipetted suspension were counted by trypan blue dye exclusion using a hemacytometer. At 1 hour, the wells of the plate were aspirated and washed one time with PBS to remove nonadhered cells. The adhered cells were then fixed to the plate with 3.7% paraformaldehyde in PBS for 10 min at room temperature. The plates were then washed once with PBS and the adherent cells were stained by adding 200 mL of

TABLE 1. Effect of rhCC10 on the adhesion of NIH 3T3 cells to human fibronectin

CC10 (50 ng/well)	α-Fibronectin monoclonal Ab (10 µg/well)	Myoglobin (10 µg/well)
54% ± 21%	60% ± 21%	21% ± 43%

NOTE: Data are expressed as the percent of control in which no protein (PBS only) was used. Numbers represent the mean and standard deviation of three independent experiments, each run in triplicate.

0.25% R250 Coomassie blue to each well and leaving at room temperature for 1 hour. The stain solution was then aspirated and the wells washed three times with PBS. Adherent cells were then quantitated by reading the optical density at 540 nm in a microtiter plate reader. Tabulated results are reported as the percent inhibition of cellular adhesion. Percent inhibition is calculated as 100% minus the ratio of the mean OD of the test protein over the mean OD of the PBS-alone (no protein) control. All protein groups were run in triplicate in each experiment. All numbers represent the mean percent inhibition for three separate experiments.

RESULTS

Fibronectin Binding Assay

FIGURE 1 shows results from a typical assay. Panel A shows a standard curve for direct binding of the rhCC10-HRP conjugate to purified human plasma fibronectin. Panel B shows a representative competition curve when unlabeled rhCC10 competes for binding sites on the plated fibronectin molecule with the HRP conjugate. As the rhCC10 binding sites on the fibronectin are used up, less of the rhCC10-HRP conjugate can bind and the signal goes down. The classic shape of each of these curves indicates a real, dose-dependent binding of rhCC10 to plated fibronectin. This result is consistent with that already reported by Zhang et al.[1] in a solution phase format, but has further implications for the role of rhCC10 in regulating cellular adherence to fibronectin in inflammation and fibrosis. This assay is also used to characterize the relative activities of different preparations of rhCC10 with respect to fibronectin binding.

Cellular Adhesion Assay

TABLE 1 shows the effects of rhCC10 on a cellular adhesion assay. NIH 3T3 cells, which are mouse lung fibroblasts, were used. A mouse monoclonal antibody specific for the "RGD" region of human fibronectin and known to block cellular adhesion to fibronectin, primarily via integrin-dependent pathways, was used as a positive control. Myoglobin, a serum protein of roughly the same size and charge as CC10, was used as a negative control. Data are expressed as the percent of control and represent the mean of three independent experiments, all run in triplicate. Despite rather large standard deviations, the effect of rhCC10 is obvious. A relatively small quantity of rhCC10, just 50 ng/well, was found to significantly inhibit adhesion of NIH 3T3 to plated human fibronectin. A comparable level of inhibition was seen only with a large amount (10 mg/well) of the monoclonal antibody. Although myoglobin at 10 μg/well did inhibit adhesion (presumably nonspecifically) to some degree, the inhibition observed with a 200-fold lower amount of rhCC10 was at least twice that.

Panel A: Direct binding

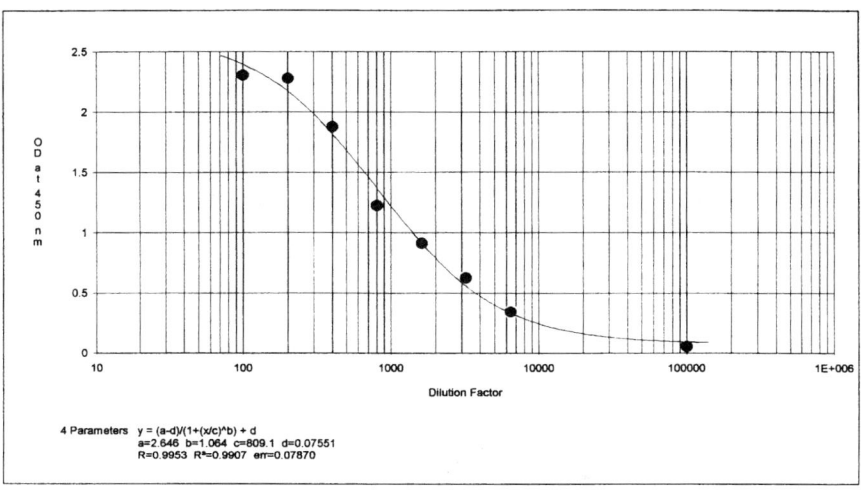

Panel B: Competition

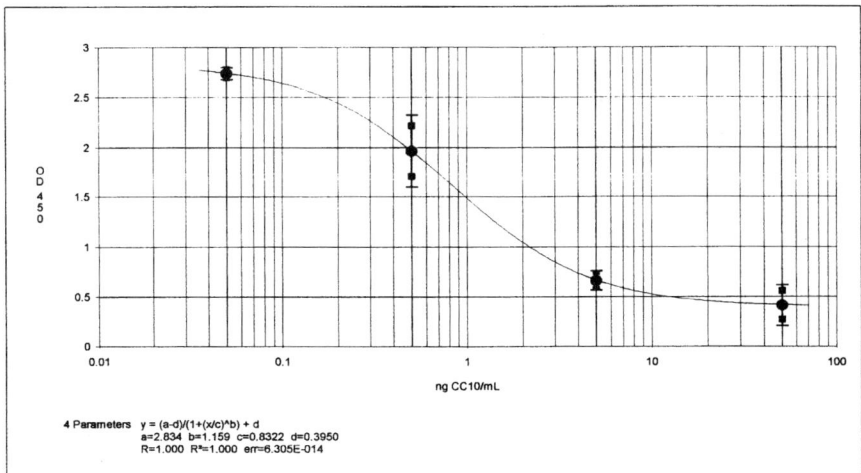

FIGURE 1. Binding and competition of rhCC10 to fibronectin-coated microtiter plates. (**A**) Titration of rhCC10-HRP conjugate on human plasma fibronectin-coated microtiter plates. Human plasma fibronectin was coated onto the wells of a microtiter plate at a concentration of 0.25 mg/mL. (**B**) Displacement of rhCC10-HRP by unlabeled CC10 binding to fibronectin-coated plates.

These observations suggest that rhCC10 can inhibit cellular adhesion to fibronectin and this may be due to its ability to bind to plated fibronectin and block access to the "RGD" sequence. However, since we added the rhCC10 and the cells to the plates at the same time, the possibility remains that the CC10 acted on the cells to reduce their ability to bind to fibronectin. In this scenario, the CC10 may have interrupted the integrin-mediated pathway responsible for this adhesion pathway. Further work must be done to clarify these possibilities.

CONCLUSIONS

These observations confirm the initial observation of Zhang *et al.*[1] that CC10 may form a complex with human fibronectin, with at least one binding site exposed in both the soluble and plated conformations of fibronectin. The data also indicate that CC10 mediates inhibition of cellular adhesion to fibronectin, although the mechanism of this inhibition is not yet clear. Both of these activities are consistent with the proposed inhibitory role for CC10 in the processes of inflammation, fibrosis, and tumor metastases.

REFERENCES

1. ZHANG, Z. *et al.* 1997. Severe fibronectin-deposit renal glomerular disease in mice lacking uteroglobin. Science **276:** 1408–1412.
2. HYNES, R.O. 1990. Fibronectins. Springer-Verlag. New York/Berlin.
3. ERICKSEN, H.P. & N.A. CARRELL. 1983. Fibronectin in extended and compact conformations: electron microscopy and sedimentation analysis. J. Biol. Chem. **258:** 14539–14544.
4. MUKHERJEE, A.B. 1999. Uteroglobin: a novel cytokine? Cell. Mol. Life Sci. **55:** 771–787.
5. JOHNSON, C.J. *et al.* 1997. Altered pulmonary response to hyperoxia in Clara cell secretory protein deficient mice. Am. J. Respir. Cell Mol. Biol. **17:** 147–155.
6. RETTA, S.F., M. TERNULLO & G. TARONE. 1999. Adhesion to matrix proteins. *In* Methods in Molecular Biology. Volume 96, pp. 125–130. Humana Press. Totowa, NJ.

Mammaglobin as a Marker for the Detection of Tumor Cells in the Peripheral Blood of Breast Cancer Patients

OTTO ZACH,[a,b] HEDWIG KASPARU,[b] HELGA WAGNER,[c] OTTO KRIEGER,[b] AND DIETER LUTZ[b]

[b]*Elisabethinen Hospital, A-4010 Linz, Austria*

[c]*Institute for Applied Statistics, University of Linz, Linz, Austria*

The detection of circulating tumor cells in the peripheral blood (pB) of breast cancer (BC) patients might become an important factor for the prognosis of BC patients. Sensitive molecular techniques, primarily based upon the reverse-transcriptase polymerase chain reaction (RT-PCR), have been developed using the expression of tissue- and/or tumor-specific genes as a marker for the presence of tumor cells.

In 1996, the cDNA of a novel gene termed human mammaglobin (hMAM) was described.[1] As far as known, the expression of hMAM is restricted to the adult mammary gland and to breast carcinoma cells; therefore, it might be a specific marker for BC cells.

We developed a nested RT-PCR assay for the detection of hMAM mRNA molecules in the pB of BC patients as a marker for the presence of tumor cells.[2] The assay is highly sensitive (1 tumor cell detectable in 10^6–10^7 white blood cells) and specific for breast carcinoma cells (no hMAM expression in pB of healthy volunteers).

Peripheral blood samples from 286 BC patients (27–94 years, mean of 58 years) were classified into four defined clinical subgroups: before (pre) and after (post) surgery (without metastases and before any further adjuvant treatment), no evidence of disease (NED, stages I–III or relapsed, and after chemotherapy-induced remission), and metastatic disease (MD, stage IV and relapses of earlier stages).

RNA was isolated out of 2× 5 mL pB per sample and patient, cDNA was synthesized, and nested PCR was performed with two PCR setups per cDNA, resulting in four PCR setups per sample.[2] In all assays, freshly prepared cDNAs from dilutions of 10 cells of the mammary carcinoma cell line SKBR5 in 5 mL pB and from pB of healthy volunteers were used as positive and negative controls, respectively.

Results of pB samples from BC patients tested for hMAM mRNA expression via a nested RT-PCR assay were as follows: 2/46 (4%) pre, 2/24 (8%) post, 4/135 (3%) NED, and 35/81 (43%) MD (TABLE 1). Significantly more MD patients were hMAM-positive than patients of the other subgroups ($p < 0.0001$).

The detection of tumor cells in pB of BC patients is limited to tumors that do express hMAM. It has been shown that about 80% of mammary carcinomas are strongly immunopositive for hMAM protein.[3] In addition, cells derived from breast

---

[a]Address for correspondence: Elisabethinen Hospital, I. Interne Abteilung, Fadingerstrasse 1, A-4010 Linz, Austria. Voice: 732 7676 4410; fax: 732 7676 4436.

otto.zach@elisabethinen.or.at

TABLE 1. Results of hMAM nested RT-PCR assay with pB samples from BC patients

	No. of patients	No. hMAM-positive	% hMAM-positive
pre	46	2	4%
post	24	2	8%
NED	135	4	3%
MD	81	35	43%

TABLE 2. Characteristics of relapsed BC patients

Patient	Stage	+LN	ER	Localization	Months	hMAM
\multicolumn{7}{c}{Early relapse (≤24 months)}							
GW	II	21	negative	visceral/bone	8	positive
AZ	III	12	negative	visceral	10	positive
AM	III	3	positive	local	11	positive
IP	III	19	positive	local	16	positive
AP	II	17	positive	local	21	negative
KA	II	19	negative	visceral/contralat.	24	positive
LM	II	2	positive	local	24	negative
\multicolumn{7}{c}{Late relapse (>24 months)}							
RW	II	9	positive	bone	25	positive
GM	II	15	n.a.	visceral	27	positive
CM	II	8	negative	cerebral	32	negative
LL	II	n.a.	positive	local	36	negative
AH	III	13	positive	visceral/bone	39	negative
RE	n.a.	1	negative	visceral/bone	40	negative
HB	II	24	negative	local	42	positive
RK	II	8	negative	visceral	47	negative

tumors overexpressing hMAM may be detected easier in pB compared to other markers. In summary, we cannot exclude the presence of tumor cells in pB of BC patients having a negative hMAM nested RT-PCR result, but hMAM positivity indicates a high probability of circulating BC cells.

Sixty-eight BC patients with NED and repeatedly negative for hMAM mRNA in their pB were tested for at least 6 up to 36 months. Fifteen of these patients relapsed. Eight of them (53%) were hMAM-positive: 5 at the time of relapse, 1 patient at 13 months before, and 2 patients at 10 and 17 months after relapses were diagnosed. In pB samples of 1 patient, hMAM mRNA transcripts were detected without clinical relapse of the patient during the following 16 months. Negative results from relapsed BC patients might be due either to the absence of tumor cells in pB at relapse or to a loss of hMAM expression of the tumor cells.

Seven of 15 BC patients relapsed within 24 months, with 5 of them being hMAM-positive, versus only 3 of 8 patients with relapses later than 24 months. No differences were seen between patients with early and late relapses with regard to stage of disease, number of infiltrated lymph nodes (LN), estrogen-receptor (ER) status at diagnosis, or localization of metastases at relapse (TABLE 2).

We conclude that the hMAM nested RT-PCR assay is a sensitive and specific method for the detection of tumor cells in pB of BC patients. The clinical relevance of hMAM-based tumor cell detection should be evaluated in prospective studies.

REFERENCES

1. WATSON, M.A. *et al.* 1996. Mammaglobin, a mammary-specific member of the uteroglobin gene family, is overexpressed in human breast cancer. Cancer Res. **56:** 860–865.
2. ZACH, O. *et al.* 1999. Detection of circulating mammary carcinoma cells in the peripheral blood of breast cancer patients via a nested reverse transcriptase polymerase chain reaction assay for mammaglobin mRNA. J. Clin. Oncol. **17:** 2015–2019.
3. WATSON, M.A. *et al.* 1999. Mammaglobin expression in primary, metastatic, and occult breast cancer. Cancer Res. **59:** 3028–3031.

In Vivo and *In Vitro* Analysis of Hyperoxia-Induced Gene Expression in Mouse Lung and Mouse Transformed Clara Cells

BIHONG ZHAO,[a] PATRICIA L. RAMSAY,[a,b] MOON S. PARK,[b] STEPHEN E. WELTY,[b] AND FRANCESCO J. DEMAYO[a]

Department of [a]Molecular and Cellular Biology and [b]Pediatrics, Baylor College of Medicine, Houston, Texas 77030, USA

Clara cells are the nonciliated secretory cells characterized by large dome-shaped apical projections and abundant secretory granules and endoplasmic reticulum. Clara cells are localized throughout the respiratory tree and are the most actively dividing cells in the lower airways during prenatal as well as postnatal development. Clara cells abundantly produce the Clara cell secretory protein (CCSP), which is also called uteroglobin.[1–3] CCSP is an anti-inflammatory secretory product of the Clara cells, which has been shown to protect the lung from hyperoxic insults.[4] A deficiency in CCSP has been associated with decreased survival in hyperoxia exposure in mice,[4] as well as with poor neonatal outcomes in prematurely born infants.[5]

These observations have focused our attention on defining the alterations in gene expression induced by hyperoxia exposure in the Clara cell. Therefore, these studies were designed to test the hypothesis that hyperoxia exposure would result in alterations of specific genes *in vivo* and *in vitro*.

The *in vivo* analyses utilized male FVB mice at 6 weeks of age. The mice were exposed to >95% oxygen for 0, 24, 48, 72, or 96 h. Mice were anesthetized by intraperitoneal injections of 0.5 mL Avertin. Lung injury was assessed by measuring the total lung weight per total animal weight and by hematoxylin and eosin (H&E) staining for histological changes. To determine the changes in the expression of CCSP mRNA, mouse lungs were excised and total RNA was extracted using Trizol reagent following the instructions provided by the manufacturer (GIBCO BRL). RNase protection assay (Ambion) was employed to detect the CCSP mRNA level. Total lung homogenates were assessed for protein concentration by the Bradford method. Each Western blot was loaded with equal amounts of total protein, electrophoresed on 10% SDS-polyacrylamide denaturing gels, and transferred to PVDF membranes, and protein was detected using a primary antibody against mCCSP. Immunohistochemistry was performed also to detect *in situ* CCSP expression. These analyses provide the baseline time course for the comparison of additional gene changes to be identified by our *in vitro* investigations.

The *in vitro* analyses utilized total RNA extracted from mouse transformed Clara cells (mtCC), previously generated in our laboratory by clonal expansion of tumor cells derived from transgenic mice, secondary to the overexpression of SV40 large T antigen under the control of the CCSP promoter. Following 24 h of exposure to either 21% or 95% oxygen, total RNA was extracted and subsequently underwent polyA$^+$ enrichment for analysis of differential gene expression by Atlas cDNA Expression Arrays (Clontech).

We observed a time-dependent decrease in lung CCSP mRNA and protein expression in the FVB mice that preceded the onset of lung injury after hyperoxia exposure. The morphological alterations seen on H&E staining show disrupted alveoli, vascular congestion, and an inflammatory cell infiltration consistent with the observed increase in lung to body weight ratios. Immunohistochemistry demonstrated a reduction in CCSP immunoreactive positive staining cells in the airways in spite of the relatively intact epithelium. Moreover, hyperoxia induced specific alterations in gene expression in the mtCC cells, suggesting that the hyperoxia-induced lung damage could involve multiple mechanisms. These observations support the utility of mtCC cells for the analysis of hyperoxia-induced gene expression.

Our findings support a role of the mtCC cells for investigation of the physiological responses of the Clara cell. Also, hyperoxia exposure results in an increase or decrease of specific gene expressions, which may be important to our understanding of the pathophysiology of hyperoxia-induced lung injury.

REFERENCES

1. SINGH, G., S.L. KATYAL, W.E. BROWN et al. 1990. Clara cell 10 kDa protein (CC10): comparison of structure and function to uteroglobin. Biochim. Biophys. Acta **1039**: 348–355.
2. WAREMBOURG, M., O. TRANCHANT, M. ATGER & E. MILGROM. 1986. Uteroglobin messenger ribonucleic acid: localization in rabbit uterus and lung by *in situ* hybridization. Endocrinology **119**(4): 1632–1640.
3. PERI, A., N.H. DUBIN, R. DHANIREDDY & A.B. MUKHERJEE. 1995. Uteroglobin gene expression in the rabbit uterus throughout gestation and in the fetal lung: relationship between uteroglobin and eicosanoid levels in the developing fetal lung. J. Clin. Invest. **96**(1): 343–353.
4. JOHNSON, C.J., G.W. MANGO, J.N. FINKELSTEIN & B.R. STRIPP. 1997. Altered pulmonary response to hyperoxia in Clara cell secretory protein deficient mice. Am. J. Respir. Cell Mol. Biol. **17**(2): 147–155.
5. RAMSAY, P.L., S.E. HEGEMIER, M.E. WEARDEN & S.E. WELTY. 1999. Attenuation in the postnatal expression of the Clara cell secretory protein is associated with the development of bronchopulmonary dysplasia. Pediatr. Res. **45**: 316A.

Uteroglobin/Clara Cell 10-kDa Family of Proteins: Nomenclature Committee Report

COMMITTEE CHAIR: J. KLUG[a,b]
COMMITTEE MEMBERS: H. M. BEIER,[c] A. BERNARD,[d] B. S. CHILTON,[e]
T. P. FLEMING,[f] R. I. LEHRER,[g] L. MIELE,[h] N. PATTABIRAMAN,[i] AND G. SINGH[j]

[b]*Institut für Molekularbiologie und Tumorforschung (IMT), Philipps-Universität Marburg, D-35033 Marburg, Germany*

[c]*Department of Anatomy and Reproductive Biology, RWTH University of Aachen, D-52057 Aachen, Germany*

[d]*Unit of Industrial Toxicology and Occupational Medicine, Faculty of Medicine, Catholic University of Louvain, B-1200 Bruxelles, Belgium*

[e]*Department of Cell Biology and Biochemistry, Texas Tech University Health Sciences Center, Lubbock, Texas 79430, USA*

[f]*Department of Surgery, Washington University School of Medicine, St. Louis, Missouri 63110, USA*

[g]*Department of Medicine and Molecular Biology Institute, University of California (Los Angeles) School of Medicine, Los Angeles, California 90095-1690, USA*

[h]*Cardinal Bernardin Cancer Center and Department of Pathology, Loyola University Medical Center, Maywood, Illinois 60153, USA*

[i]*Advanced Biomedical Computing Center, SAIC-NCI/FCRDC, Frederick, Maryland 21702-1201, USA*

[j]*Department of Pathology, University of Pittsburgh School of Medicine and VA Medical Center, Pittsburgh, Pennsylvania 15240, USA*

During the meeting on the "Uteroglobin/Clara Cell 10-kDa Family of Proteins", a nomenclature conference was held. J. Klug was nominated as the chair of the session and the following, listed alphabetically, were nominated as cochairs: H. M. Beier, A. Bernard, I. Callebaut, B. S. Chilton, J. Daniel, F. DeMayo, T. P. Fleming, R. I. Lehrer, J. Lund, A. B. Mukherjee, M. Nord, J. H. Shelhamer, and G. Singh. Many conference participants joined this nomenclature session, indicating that a discussion on this topic was much needed. Investigators new to the field pointed out that time-consuming literature searches were needed to discover that there exists a plethora of names for essentially identical proteins (see TABLE 1 and Ni *et al.* in this volume for an overview). Experienced researchers also felt that it is awkward and error-prone to always consider all the different names for a family member. As a result, important publications appearing in specialized journals could remain virtually unknown for a

---

[a]Address for correspondence: J. Klug, Institut für Molekularbiologie und Tumorforschung (IMT), Philipps-Universität Marburg, Emil-Mannkopff-Strasse 2, D-35033 Marburg, Germany. Voice: +49 6421 28 66545; fax: +49 6421 28 65398.
Klug@IMT.Uni-Marburg.de

long time, even to competing scientists in some cases. With the advent of efficient and fast electronic search facilities, these problems could only be alleviated, but not eliminated.

Initially, many family members had been named after the tissue in which the protein was detected first, like for uteroglobin (rabbit uterus) or lacryglobin (human tears). Later, it turned out that the same protein is also expressed in other tissues in significant amounts. Soon after the discovery of uteroglobin, for example, a uteroglobin-like antigen or protein[1] was detected in the rabbit lung[2] and male genital tract.[3] By comparing numerous biochemical and physicochemical parameters, purified uteroglobin-like protein from rabbit lung and uteroglobin from the uterus were shown to be identical.[1] These early results were rediscovered ten years later when a Clara cell secretory protein (CCSP) was described in the rabbit lung,[4] which was again shown to be identical to uteroglobin.[5] From then on, the name CCSP and related versions thereof were often used by scientists investigating the mouse, rat, and human lung (TABLE 1). Lacryglobin was initially identified in a human tear proteome project,[6] but was soon after also identified as a subunit of a heterodimeric human tear protein and called lipophilin because it binds to hydrophobic adsorbents.[7] Moreover, it was cloned by homology screening with a mammaglobin probe and hence called mammaglobin B,[8] although it is also expressed in the prostate for example.[9]

Altogether, it was felt that in many instances the current nomenclature is confusing and constitutes a barrier to the dissemination of newly acquired knowledge among researchers outside as well as inside the field. Thus, there was a broad consensus that, first of all, a new generic family name should be coined. Based on the phylogenetic tree of all known family members presented by one of the conference speakers (see Ni *et al.*), it was suggested to further divide the family into subfamilies so that, for each individual member, the generic name could be complemented with a subfamily, group, and protein identifier.

There was also a consensus that the use of the nomenclature system should follow other well-established systems like the one used for the family of nuclear receptors.[10] In each new manuscript dealing with a member of the uteroglobin/Clara cell 10-kDa family of proteins, it is recommended that the protein(s) be identified by the official nomenclature name at least once in the summary and introduction. Once the assignment of nomenclature name and trivial name(s) has been made, authors are free to use the trivial name(s) of their choice for the remainder of the manuscript.

Already during the presentation of his paper on secretory lipophilins, Robert Lehrer made the first suggestion for a new family name: secretoglobins. This proposal was also discussed in the nomenclature session, but many scientists felt that such an important decision should not be rushed. Therefore, a nomenclature committee (H. M. Beier, A. Bernard, B. S. Chilton, T. P. Fleming, J. Klug, R. I. Lehrer, L. Miele, N. Pattabiraman, and G. Singh) was appointed in order to thoroughly discuss new family names and select one by the end of June 2000. Quickly, a number of suggestions were made that all referred to the common characteristics of all family members: secretion, small size, alpha-helical, and dimeric structure. It was evident that the common functional properties of the family members are not sufficiently characterized yet to allow the introduction of a function-based generic name. The name suggestions fell into two categories: single-word names and compound names. In both categories, the suffixes -globin and -globulin were often used.

TABLE 1. Compilation of trivial names assigned to members of the secretoglobin (uteroglobin/CC10) family of proteins

Subfamily	Species	Trivial Names	References	Abbreviation/Symbol
1[a]	Rabbit	Blastokinin	(11)	
		Uteroglobin	(12)	UG, UGL, UGB[b]
		Uteroglobin-like Antigen	(3)	
		Uteroglobin-like Protein	(1)	
		Clara Cell Secretory Protein	(4, 5)	CCSP
	Hamster, Hare, Horse, Macaque, Pica, Pig	Uteroglobin/Clara Cell 10-kDa Protein	(13, 14, 15, 16, 17)	UG, UGL, UGB[b] CC10
	Mouse, Rat	Clara Cell (Secretory) Protein C	(18)	
		Polychlorinated Biphenyl Binding Protein	(19)	PCB-Binding Protein
		Clara Cell 10 kDa Protein	(20)	CC10
		Clara Cell-Specific 10 kD Protein	(21)	CC10
		Clara Cell Secretory Protein	(22)	CCSP
		Clara Cell 17 kDa Protein	(23)	CC17
		Uteroglobin	(24)	UG, UGL, UGB[b]
	Human	Clara Cell-Specific 10 kD Protein	(25)	CC10
		Clara Cell 10 kDa Protein	(26, 27)	
		Polychlorinated Biphenyl Binding Protein	(28)	PCB-Binding Protein
		Protein 1/Urinary Protein 1	(29)	UP-1
		Uteroglobin	(30)	UG, UGL, UGB[b]
		Clara Cell Phospholipid-Binding Protein	(31)	HCCPBP
		Clara Cell Protein (CC16)	(32)	CC16
		Uteroglobin-like Antigen	(45)	UGL, ULA

	Species	Protein	Ref.	Abbrev.
2	Rat	Prostatic Binding Protein[c] Component 1	(33)	PBP C1
		Prostatic Binding Protein[c] Component 2	(33)	PBP C2
	Hamster	Protein encoded by female Harderian gland Clone 22	(34)	FHG22
	Human	Lipophilin A	(7)	Lpn A
		Lipophilin B	(9)	Lpn B
		Lymphoglobin[d]		YGB
3	Rat	Prostatic Binding Protein[c] Component 3	(33)	PBP C3
	Human	Mammaglobin	(35)	MGB, MGB1
		Lacryglobin	(6)	
		Lipophilin C	(7)	Lpn C
		Mammaglobin B	(8)	MGB2
	Hamster	Heteroglobin[e]		
4	Mouse	Androgen Binding Protein Alpha	(37)	ABP a
	Cat	Fel dI Chain 1	(36)	Fel dI C1
		Lacrimal Gland Protein[f]		
5	Cat	Fel dI Chain 2	(36)	Fel dI C2
6	Rat	RYD5	(38)	RYD5

NOTE: The numbering of subfamilies is arbitrary. Proteins within one species, which are shaded gray, are identical.

[a]All proteins of subfamily 1 (uteroglobins and CC10 proteins) are orthologues. [b]The official Human Gene Nomenclature Committee symbol for uteroglobin is UGB. [c]The tetrameric holoprotein prostatic binding protein (PBP, subunit structure C1-C3/C2-C3) is also called prostatein,[39] prostate alpha-protein,[40] ventral prostate protein,[41] prostate steroid-binding protein,[42] estramustine-binding protein,[43] or steroid-binding sialoglycoprotein.[44] Its subunits are called component 1, 2 and 3 (C1, C2, C3 for PBP, prostatic binding protein, and prostatic steroid-binding protein); components I, II, and III (for prostate alpha-protein); or alpha, beta, and gamma (for ventral prostate protein). The C1-C3 and C2-C3 heterodimers are called F (fast) and S (slow) for PBP, and A and B for prostate alpha-protein, respectively. Only the heterodimer C2-C3 can bind steroid hormones. [d]See the contribution of Ni et al. in this volume. [e]Unpublished (see Genbank Acc. Nos. AJ252138 and AJ252139). [f]Unpublished (see Genbank Acc. No. AF008595).

In the discussion, it was pointed out that the term globulin goes back to a one-century-old distinction between albumins and globulins. The differentiation was based on solubility in water: whereas albumins are soluble in pure water, globulins need some salt in order to dissolve. Under this criterion, uteroglobin for example must be called an albumin and not a globulin. Furthermore, the suffix -globin refers to hemoglobin (another albumin) and also has a structural meaning in the term globin-fold, which is a bundle of eight alpha-helices connected by rather short loop regions and arranged so that the helices form a pocket for binding of the heme group. This situation is very reminiscent of a uteroglobin homodimer for example. Each monomer is a four-helix bundle, and the dimer—an eight-helix bundle—forms a pocket for the binding of numerous hydrophobic molecules. Thus, the suffix -globin, in the case of the uteroglobin/Clara cell 10-kDa family of proteins, implicates dimerization behavior and correctly describes the overall structure of the common dimeric entity.

The favorite suggestions in both groups eventually were secretoglobins and haplokines among single-word names and secretory parvoglobins (or haploglobins) and secretory miniglobins among the compound names. In the final election within the nomenclature committee, the name "secretoglobin" received the majority of vote points. Thus, the first name proposed was also the final name elected. The search term "secretoglobin" does not retrieve any item in the PubMed database of the National Library of Medicine at the National Center for Biotechnology Information in Bethesda (http://www.ncbi.nlm.nih.gov). Further, the Human Gene Nomenclature Committee (http://www.gene.ucl.ac.uk/nomenclature) has officially approved the generic name "secretoglobin" and its symbol SCGB. The details of the numbering system for subfamilies, groups, and individual proteins along with recommendations for the allocation of official names for new members will be published elsewhere. It is also hoped that the name secretoglobin will be routinely used in keyword sections and listed in the Medical Subject Headings (MeSH) list of the PubMed library so that a future literature search would be fast, simple, and comprehensive.

ACKNOWLEDGMENTS

We thank the conference organizers, in particular Anil B. Mukherjee, for making the nomenclature debate an important part of the meeting.

REFERENCES

1. BEATO, M. & H.M. BEIER. 1978. Characteristics of the purified uteroglobin-like protein from rabbit lung. J. Reprod. Fertil. **53:** 305–314.
2. BEIER, H.M., C. KIRCHNER & U. MOOTZ. 1978. Uteroglobin-like antigen in the pulmonary epithelium and secretion of the lung. Cell Tissue Res. **190:** 15–25.
3. BEIER, H.M., H. BOHN & W. MÜLLER. 1975. Uteroglobin-like antigen in the male genital tract secretions. Cell Tissue Res. **165:** 1–11.
4. GUPTA, R.P., S.E. PATTON, A.M. JETTEN et al. 1987. Purification, characterization, and proteinase-inhibitory activity of a Clara-cell secretory protein from the pulmonary extracellular lining of rabbits. Biochem. J. **248:** 337–344.

5. LOPEZ DE HARO, M.S., L. ALVAREZ & A. NIETO. 1988. Evidence for the identity of antiproteinase pulmonary protein CCSP and uteroglobin. FEBS Lett. **232:** 351–353.
6. MOLLOY, M.P., S. BOLIS, B.R. HERBERT et al. 1997. Establishment of the human reflex tear two-dimensional polyacrylamide gel electrophoresis reference map: new proteins of potential diagnostic value. Electrophoresis **18:** 2811–2815.
7. LEHRER, R.I., G. XU, A. ABDURAGIMOV et al. 1998. Lipophilin, a novel heterodimeric protein of human tears. FEBS Lett. **432:** 163–167.
8. BECKER, R.M., C. DARROW, D.B. ZIMONJIC et al. 1998. Identification of mammaglobin B, a novel member of the uteroglobin gene family. Genomics **54:** 70–78.
9. ZHAO, C., T. NGUYEN, T. YUSIFOV et al. 1999. Lipophilins: human peptides homologous to rat prostatein. Biochem. Biophys. Res. Commun. **256:** 147–155.
10. NUCLEAR RECEPTORS NOMENCLATURE COMMITTEE. 1999. A unified nomenclature system for the nuclear receptor superfamily. Cell **97:** 161–163.
11. KRISHNAN, R.S. & J.C. DANIEL, JR. 1967. "Blastokinin": inducer and regulator of blastocyst development in the rabbit uterus. Science **158:** 490–492.
12. BEIER, H.M. 1968. Uteroglobin: a hormone-sensitive endometrial protein involved in blastocyst development. Biochim. Biophys. Acta **160:** 289–291.
13. SAGAL, R.G. & A. NIETO. 1998. Molecular cloning of the cDNA and the promoter of the hamster uteroglobin/Clara cell 10-kDa gene (ug/cc10): tissue-specific and hormonal regulation. Arch. Biochem. Biophys. **350:** 214–222.
14. NIETO, A. & M. LOMBARDERO. 1982. Uteroglobin-like antigens in species of Lagomorpha. Comp. Biochem. Physiol. **B71:** 511–514.
15. BEIER-HELLWIG, K., H. KREMER, B. BONN et al. 1995. Partial sequencing and identification of three proteins from equine uterine secretion regulated by progesterone. Reprod. Domest. Anim. **30:** 295–298.
16. HASHIMOTO, S., K. NAKAGAWA & K. SUEISHI. 1996. Monkey Clara cell 10 kDa protein (CC10): a characterization of the amino acid sequence with an evolutional comparison with humans, rabbits, rats, and mice. Am. J. Respir. Cell Mol. Biol. **15:** 361–366.
17. SAGAL, R.G. & A. NIETO. 1998. Cloning and sequencing of the cDNA coding for pig pre-uteroglobin/Clara cell 10 kDa protein. Biochem. Mol. Biol. Int. **45:** 205–213.
18. SINGH, G., S.L. KATYAL & S.A. GOTTRON. 1985. Antigenic, molecular, and functional heterogeneity of Clara cell secretory proteins in the rat. Biochim. Biophys. Acta **829:** 156–163.
19. NORDLUND-MÖLLER, L., O. ANDERSSON, R. AHLGREN et al. 1990. Cloning, structure, and expression of a rat binding protein for polychlorinated biphenyls: homology to the hormonally regulated progesterone-binding protein uteroglobin. J. Biol. Chem. **265:** 12690–12693.
20. HAGEN, G., M. WOLF, S.L. KATYAL et al. 1990. Tissue-specific expression, hormonal regulation, and 5'-flanking gene region of the rat Clara cell 10 kDa protein: comparison to rabbit uteroglobin. Nucleic Acids Res. **18:** 2939–2946.
21. RAY, M.K., S.W. MAGDALENO, M.J. FINEGOLD et al. 1995. Cis-acting elements involved in the regulation of mouse Clara cell–specific 10-kDa protein gene: in vitro and in vivo analysis. J. Biol. Chem. **270:** 2689–2694.
22. HACKETT, B.P., N. SHIMIZU & J.D. GITLIN. 1992. Clara cell secretory protein gene expression in bronchiolar epithelium. Am. J. Physiol. **262:** L399–L404.
23. UMLAND, T.C., S. SWAMINATHAN, W. FUREY et al. 1992. Refined structure of rat Clara cell 17 kDa protein at 3.0 Å resolution. J. Mol. Biol. **224:** 441–448.
24. SANDMÖLLER, A., A.K. VOSS, J. HAHN et al. 1991. Cell-specific, developmentally and hormonally regulated expression of the rabbit uteroglobin transgene and the endogenous mouse uteroglobin gene in transgenic mice. Mech. Dev. **34:** 57–67.
25. SINGH, G., J. SINGH, S.L. KATYAL et al. 1988. Identification, cellular localization, isolation, and characterization of human Clara cell–specific 10kD protein. J. Histochem. Cytochem. **36:** 73–80.
26. SINGH, G., S.L. KATYAL, W.E. BROWN et al. 1988. Amino-acid and cDNA nucleotide sequences of human Clara cell 10 kDa protein. Biochim. Biophys. Acta **950:** 329–337.
27. MANTILE, G., L. MIELE, E. CORDELLA-MIELE et al. 1993. Human Clara cell 10-kDa protein is the counterpart of rabbit uteroglobin. J. Biol. Chem. **268:** 20343–20351.

28. ANDERSSON, O., L. NORDLUND-MÖLLER, M. BRONNEGARD et al. 1991. Purification and level of expression in bronchoalveolar lavage of a human polychlorinated biphenyl (PCB)–binding protein: evidence for a structural and functional kinship to the multihormonally regulated protein uteroglobin. Am. J. Respir. Cell Mol. Biol. **5:** 6–12.
29. BERNARD, A., H. ROELS, R. LAUWERYS et al. 1992. Human urinary protein 1: evidence for identity with the Clara cell protein and occurrence in respiratory tract and urogenital secretions. Clin. Chim. Acta **207:** 239–249.
30. WOLF, M., J. KLUG, R. HACKENBERG et al. 1992. Human CC10, the homologue of rabbit uteroglobin: genomic cloning, chromosomal localization, and expression in endometrial cells. Hum. Mol. Genet. **1:** 371–378.
31. UMLAND, T.C., S. SWAMINATHAN, G. SINGH et al. 1994. Structure of a human Clara cell phospholipid-binding protein-ligand complex at 1.9 Å resolution. Nat. Struct. Biol. **1:** 538–545.
32. DIERYNCK, I., A. BERNARD, H. ROELS et al. 1995. Potent inhibition of both human interleukin-gamma production and biologic activity by the Clara cell protein CC16. Am. J. Respir. Cell Mol. Biol. **12:** 205–210.
33. HEYNS, W. & P. DE MOOR. 1977. Prostatic binding protein: a steroid-binding protein secreted by rat prostate. Eur. J. Biochem. **78:** 221–230.
34. DOMINGUEZ, P. 1995. Cloning of a Syrian hamster cDNA related to sexual dimorphism: establishment of a new family of proteins. FEBS Lett. **376:** 257–261.
35. WATSON, M.A. & T.P. FLEMING. 1996. Mammaglobin, a mammary-specific member of the uteroglobin gene family, is overexpressed in human breast cancer. Cancer Res. **56:** 860–865.
36. MORGENSTERN, J.P., I.J. GRIFFITH, A.W. BRAUER et al. 1991. Amino acid sequence of Fel dI, the major allergen of the domestic cat: protein sequence analysis and cDNA cloning. Proc. Natl. Acad. Sci. U.S.A. **88:** 9690–9694.
37. DLOUHY, S.R., B.A. TAYLOR & R.C. KARN. 1987. The genes for mouse salivary androgen-binding protein (ABP) subunits alpha and gamma are located on chromosome 7. Genetics **115:** 535–543.
38. DEAR, T.N., T. BOEHM, E.B. KEVERNE et al. 1991. Novel genes for potential ligand-binding proteins in subregions of the olfactory mucosa. EMBO J. **10:** 2813–2819.
39. LEA, O.A., P. PETRUSZ & F.S. FRENCH. 1979. Prostatein: a major secretory protein of the rat ventral prostate. J. Biol. Chem. **254:** 6196–6202.
40. CHEN, C., K. SCHILLING, R.A. HIIPAKKA et al. 1982. Prostate alpha-protein: isolation and characterization of the polypeptide components and cholesterol binding. J. Biol. Chem. **257:** 116–121.
41. PARKER, M.G., G.T. SCRACE & W.I. MAINWARING. 1978. Testosterone regulates the synthesis of major proteins in rat ventral prostate. Biochem. J. **170:** 115–121.
42. PARKER, M.G., R. WHITE & J.G. WILLIAMS. 1980. Cloning and characterization of androgen-dependent mRNA from rat ventral prostate. J. Biol. Chem. **255:** 6996–7001.
43. FORSGREN, B., P. BJORK, K. CARLSTROM et al. 1979. Purification and distribution of a major protein in rat prostate that binds estramustine, a nitrogen mustard derivative of estradiol-17 beta. Proc. Natl. Acad. Sci. U.S.A. **76:** 3149–3153.
44. LIMPASENI, T. & M. CHULAVATNATOL. 1986. Purification and characterization of a steroid-binding sialoglycoprotein from rat ventral prostate. Arch. Biochem. Biophys. **249:** 154–163.
45. BEIER, H.M. 1978. Control of implantation by interference with uteroglobin synthesis, release, and utilization. *In* Human Fertilization, pp. 191–200. Thieme. Stuttgart.

Closing Remarks and Future Directions

DOROTHY BERLIN GAIL

Lung Biology and Disease Program, Division of Lung Diseases,
National Heart, Lung, and Blood Institute, Bethesda, Maryland 20892-7952, USA

The organizers of this conference deserve special thanks and appreciation for bringing a broad, multidisciplinary focus and international perspective to the discussion of the uteroglobin family of proteins. Investigators from many centers throughout the world representing various disciplines and approaches brought diverse interests and points of view to the conference that will help to advance future research on the uteroglobin family of proteins.

During the presentations and discussions, the participants identified gaps in our knowledge as well as new research directions and opportunities for further progress. We bring their ideas together in this final section of the conference report. We hope that the conference and the meeting report will spark new collaborations and research to advance understanding of this fascinating protein family. Investigation of the uteroglobin protein family cuts across many areas of biology and medicine and we look forward to seeing this research move ahead.

A most important goal is clarifying the biological role(s) of uteroglobin/Clara cell secretory protein (UG/CCSP) in normal human tissues and organs and in human disease. Despite the fact that the protein is 50–70% conserved among species and binds to other molecules such as retinoids, phospholipids, and calcium, its biological function remains to be defined. The role of UG/CC10 in organ homeostasis, as an anti-inflammatory agent or a tumor suppressor, is not yet fully explained. In the lung, for example, more needs to be done to clarify the part that UG/CCSP plays in interacting with lung surfactant and protecting against injury from oxidant stress or inflammation caused by mechanical ventilation. The role of UG/CCSP in modulating cell differentiation, with potential implications for pathogenesis of lung carcinogenesis, also needs more attention. The full biologic implications of UG/CCSP interaction with fibronectin and extracellular matrix components in protection against renal disease are unknown.

There are at least 23 genes encoding 23 isoforms of UG/CCSP on human chromosome 11q12.1-13.2. In the advent of the completion of the human genome project, the sequences of these genes will be known shortly. Using a functional genomics approach to dissect out the physiologic roles of these isoforms and their role in human biology and disease is a high priority. The use of knockout animal models to address the question of UG/CCSP biologic function is already under way and is presenting an interesting challenge that should lead to important new information. Defining tissue- or cell-specific transcriptional control elements has allowed and will continue to allow development of animal models with specific defects that will yield new information about the cellular and molecular contributions of UG/CCSP *in vivo* in both health and disease.

Little is known also about the biologic function of other members of the broader protein family, including the lipophilins and mammaglobins. It is important to define functional roles for these proteins, identify new members of this family, and establish their structural and functional relatedness.

A deeper understanding of taxonomy, structural modifications, biochemistry, and tissue specificity of all uteroglobin family members is needed. Although, for example, we know a lot about the structure of a uteroglobin homodimer and how the rat prostatic binding protein (PBP) heterotetramer is configured in the prostate, we know very little about homo- or heterodimerization of many other members. The structure of uteroglobin and rat PBP are not fully understood. With the discovery of many new family members in a few species, the questions of interactive partner(s) and tissue specificity need to be addressed. Because the cysteines, which form interchain disulfide bridges, are highly conserved among the family of uteroglobin proteins, a better understanding of the biochemistry of these molecules might provide a clue to their biologic function. Integrative studies and proteomics of the uteroglobin family should provide a broad perspective of this and other needed information.

The mechanism of action by which UG/CCSP exerts its biologic effects *in vivo* needs to be determined. What are the molecular characteristics of the receptor(s) through which it functions? Identifying the physiologic ligands for UG/CCSP would likely be an important key to a full understanding of its function. Information about the transcription factors that interact to determine development and hormonal regulation of UG/CCSP gene expression is required in order to evaluate the role of uteroglobin in modulating cell differentiation.

Even before the structure and functions of the uteroglobin protein family are completely understood, investigators are addressing the future clinical application of UG/CCSP for preventing and treating human disease. Evaluation of recombinant human UG/CCSP in treating bronchopulmonary dysplasia, a chronic lung disease of premature newborn infants, is already in the early stages. UG/CCSP-derived peptides and, in the future, orally bioavailable drugs rationally designed on the basis of UG/CCSP protein structure are promising approaches. Lung diseases including neonatal respiratory distress syndrome, acute respiratory distress syndrome, asthma, complications of cystic fibrosis, and cancer are possible indications for these agents. Information on whether UG/CCSP protein plays a role in human renal disease and in neoplastic disease such as prostate or breast carcinoma is also lacking and needs to be determined. Studying the possible clinical applications of UG/CCSP as a diagnostic marker (e.g., of lung injury, lung prematurity, or the loss of UG/CCSP expression in lung cancers) is an additional area of promising future clinical application. Detection of mammaglobin protein and mRNA in primary and metastatic breast tumors suggests that this protein may be a useful marker for breast cancer. Developing more and better ways to use our knowledge of uteroglobin to advance fertility control for animals and humans has merit as an area of future investigation.

Progress was also made at the conference in developing recommendations regarding a common system of nomenclature for the uteroglobin proteins. Rapid adoption of these recommendations should ease confusion and facilitate further progress.

Index of Contributors

Abe, S., 268–279

Bally, R., 90–112
Beato, M., 25–42
Beier, H.M., 9–24, 316–320, 332–335, 348–354
Beier-Hellwig, K., 332–335
Bernard, A., 68–77, 348–354
Billing, P., 312–315
Bøe, I.M., 202–209
Braun, H., 154–165
Broeckaert, F., 68–77

Callebaut, I., 90–112
Cassel, T.N., 154–165, 300–302
Chan, C.C., 141–146
Chang, A., 181–192, 336–337
Chilton, B.S., ix–x, 166–180, 321–324, 348–354
Choi, M., 303–306
Chowdhury, B., 307–311
Classen-Linke, I., 332–335
Clippe, A., 68–77
Collee, J.M., 303–306
Colpitts, T.L., 312–315
Cordella-Miele, E., 307–311
Cowan, M.J., 193–201

Daniel, J.C., 1–8
Delettré, J., 90–112
Demaret, J.P., 90–112
DeMayo, F., ix–x, 181–192, 210–233, 249–267, 336–337, 346–347
Devine, J., 166–180
Dhanireddy, R., 325–331

Farrow, J., 338–342
Fleming, T.P., 78–89, 348–354

Gail, D.B., 355–356
Gentz, R., 25–42
Gerritsen, J., 303–306
Glasgow, B.J., 59–67
Granados, E., 312–315

Ha, C.X., 59–67
Hendrix, E., 166–180
Hermans, C., 68–77
Herrler, A., 316–320
Hewetson, A., 166–180
Hodges, S., 312–315
Hossenlopp, P., 90–112
Housset, D., 90–112
Huang, X., 193–201

Itoh, Y., 268–279

Kalff-Suske, M., 25–42
Kasparu, H., 343–345
Katyal, S.L., 43–58
Klug, J., 25–42, 348–354
Knoops, B., 68–77
Krieger, O., 343–345
Krusche, C.A., 316–320
Kundu, G.C., 210–233, 234–248, 307–311

Lee, E., 210–233
Lehrer, R.I., 59–67, 348–354
Li, Q., 141–146
Linnoila, I., 249–267
Lohnas, J., 338–342
Lund, J., 202–209
Lutz, D., 343–345

Magdaleno, S., 181–192
Malkinson, A., 249–267
Mandal, A., 234–248

Mansharamani, M., 166–180, 321–324
Mantile-Selvaggi, G., 113–127, 234–248, 307–311
Matthews, J.H., 113–127
Melby, J., 338–342
Menhart, N., 312–315
Miele, L., 113–127, 128–140, 307–311, 348–354
Momeda, K., 325–331
Moreno, J.J., 147–153
Mornon, J.P., 90–112
Mukherjee, A.B., ix–x, 113–127, 210–233, 234–248, 303–306, 307–311, 325–331
Müller-Schöttle, F., 332–335

Nguyen, T., 59–67
Ni, J., 25–42
Nord, M., 154–165, 300–302

Park, M., 181–192, 336–337, 346–347
Pattabiraman, N., 113–127, 348–354
Pilon, A.L., 280–299, 338–342
Plopper, C.G., 202–209
Poupon, A., 90–112

Ramsay, P.L., 181–192, 336–337, 346–347
Reardon, M.J., 181–192, 336–337
Reynolds, S.D., 202–209
Russell, J., 312–315

Sabourin, C., 249–267

Schageman, J., 25–42
Shelhamer, J.H., 193–201
Shen, D.F., 141–146
Shijubo, N., 268–279
Singh, G., ix–x, 43–58, 348–354
Sterzik, K., 332–335
Stripp, B.R., 202–209
Stroupe, S., 312–315
Suske, G., 154–165, 300–302
Szabo, E., 249–267

Ten Kate, L.P., 303–306
Tuaillon, N., 141–146

Wagner, H., 343–345
Ward, J., 210–233
Ward, K.B., 113–127
Watson, M.A., 78–89
Welch, R., 338–342
Welty, S.E., 181–192, 346–347
Westphal, H., 210–233
Wiese, L., 338–342
Witschi, H., 249–267

Yamaguchi, T., 268–279
Yao, X.L., 193–201
Yuan, C-J., 210–233, 234–248

Zach, O., 343–345
Zhang, Z., 210–233, 234–248, 303–306, 307–311, 325–331
Zhao, B., 181–192, 336–337, 346–347
Zhao, C., 59–67
Zheng, F., 210–233

OHIO UNIVERSITY LIBRARY

Please return this book as soon as you have finished with it. In order to avoid a fine it must be returned by the latest date stamped below. All books are subject to recall after two weeks or immediately if needed for reserve.

CF